Principles of physical chemistry

Principles
of physical chemistry

ROBERT M. ROSENBERG
Lawrence University

New York
OXFORD UNIVERSITY PRESS
1977

Copyright © 1977 by Oxford University Press, Inc.
Library of Congress Catalogue Card Number: 75-16909
Film set in Hong Kong by Asco Trade Typesetting Ltd.
Printed in the United States of America

To V.

Note to the Reader

This book uses SI units almost exclusively. The designation SI stands for *Le Système International d'Unités,* adopted by Le Bureau International des Poids et Mesures, and based on the kilogram, the meter, the second, and the ampere as the base units. Appendix A provides an introduction to the SI units and the general rules for the use of physical quantities and units. It should be perused before reading the text and consulted thereafter as particular questions arise.

Preface

This book is an exposition of the principles of physical chemistry. It is designed for an undergraduate course of two terms or two semesters at the junior or senior level, and it presupposes a two term course in general physics, preferably with calculus, a facility with differential and integral calculus, and one or two courses in chemistry beyond the introductory course. The aim of the book is to lead the reader safely through the mathematical difficulties of the subject rather than around them; if the teacher prefers, the students need not be held responsible for the detailed derivation of the hydrogenic wave functions, for example, but the solutions are there for the student who wants to know how the wave functions are obtained. Material of this kind is shaded in the text.

An attempt has been made to provide some sense of historical perspective by referring to early examples of measurements, by stressing the evolution of theories, and by clearly indicating sources. Many of the problems are based on data in the literature, and references are given in the hope that some readers will want to see the papers from which the data have been obtained. One feature of the book is the selection of exercises in the body of the text that is intended to stimulate the reader to participate in the development of the material rather than being a passive recipient of information. This feature was used by A. A. Noyes and M. Sherrill in their classic text.

The book is arranged so that quantum theory and statistical thermodynamics are presented before classical thermodynamics and the macroscopic description of chemical systems. The following alternative chapter sequences can be used by those who would prefer different placements for the quantum theory: 1, 2, 9-19, and 3-8; or 1, 2, 9-11, 3-8, and 12-19.

It is not possible to acknowledge all those who have been of assistance in the writing of this book. I am much indebted to teachers, colleagues, students, and other writers of physical chemistry texts. Peter Atkins, C. A. Arrington, Paul Gans, Lowell Hall, and Adam Hart-Davis have read and commented on substantial portions of the manuscript, and James Beatty has read and commented on the entire manuscript. James Evans used parts of the manuscript in a preliminary form in his course, and I have had the benefit of his comments on the text and of many conversations with him on the teaching of physical chemistry. Virginia Rosenberg has read the entire text in successive versions; she has not only improved the writing, but also the chemistry and the mathematics. I am grateful to each of these readers for correcting errors and for suggesting improved presentation and to John Harriman for helpful conversations. I thank the Chemistry Department of Amherst College for its gracious hospitality in the summer of 1972 and the libraries of Lawrence University, Amherst College, and the University of Wisconsin at Madison for their help in obtaining source material. Shirley Biese, Jean St. Pierre, and Margaret Rosenberg did an excellent job of typing from very difficult manuscript, and June Woods shepherded the manuscript through typing and reproduction of preliminary and final versions. The structure of Chapters 3 and 4 owes much to similar chapters in *The Principles of Physics and Chemistry,* by J. B. Brackenridge and the present author. I acknowledge the help of Carol Miller in the editing of the manuscript and the continued support and gentle prodding of Ellis Rosenberg. The index was produced in the Lawrence University Computer Center. Finally, I thank Virginia, Charles, Margaret, and Jim for their support and encouragement during several trying years.

Lawrence University   R.M.R.
Appleton, Wisconsin
September 1976

Contents

Principles of physical chemistry

1 · *Properties of gases*

The gaseous phase is a favorite subject for physicochemical investigation both because the behavior of gases is less complex than that of other phases of matter and because gases are ubiquitous—in industrial processes, in biological systems, and in the atmosphere.

The study of gases provides an excellent introduction to physical chemistry: there are examples of simple models, the interplay between model building and the gathering of empirical data, and the evolution of models. Thus, the emphasis in this chapter will be on those aspects of the properties of gases that can be discussed in terms of relatively simple models.

1-1. The experimental variables

Pressure

One has only to blow up a balloon or use a hand pump to inflate a bicycle tire to realize that a gas exerts a force on the walls of its container. It is observed that the magnitude of the force exerted by a gas on a surface is independent of the orientation of the surface, that is, of direction. A scalar quantity is therefore appropriate to deal with the phenomenon. The ratio of the magnitude of the force vector to the magnitude of the vector representing area on which the force is acting is a convenient, and easily obtained measurement called the *pressure*, P, expressed as

$$P = \frac{|\mathbf{F}|}{|\mathbf{A}|} \tag{1-1}$$

1

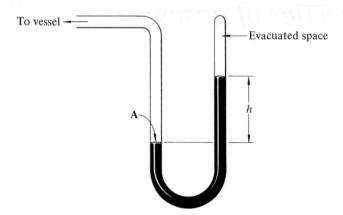

Figure 1-1. A diagram of a U-tube manometer.

A simple way to measure the pressure of a gas is to attach a U-tube *manometer* to the vessel that contains the gas, as shown in Fig. 1-1. The force exerted by the gas on the surface of the manometric fluid is equal to the magnitude of the force exerted by the column of liquid of height *h*. That is,

$$|\mathbf{F}| = P|\mathbf{A}| = h\rho g|\mathbf{A}| \qquad (1\text{-}2)$$

where ρ is the density of the fluid, g is the acceleration due to gravity, and $\mathbf{A}$ is the area of the exposed surface. Thus,

$$P = h\rho g \qquad (1\text{-}3)$$

Exercise 1-1. Verify that the units of pressure are dyn-cm^{-2} in CGS units and N-m^{-2} in SI units. (The latter is also called the Pascal, Pa.)

Practical units of pressure, based on the standard atmosphere, are taken to be that pressure that will support a column of mercury (at 0°C) 760 mm high at a g value of 9.807 m-s^{-2}. The pressure equivalent to 1 mm of such a column is 133.32 Pa.

Exercise 1-2. Calculate the pressure in dyn-cm^{-2} and N-m^{-2} of a 76.0 cm high column of mercury at 0°C. The density of mercury at 0°C is 13.6 g-cm^{-3}.

Though it is possible to discuss the pressure of a system that is changing with time, or the pressures in different regions of a nonuniform system, the pressure discussed in this section is that of a homogeneous system that is not changing with time, a system at *equilibrium*.

Volume

For an irregularly shaped vessel, the volume is determined by measuring the mass of a liquid of known density contained in the vessel. The volume of a container with a regular geometric shape can be calculated from its dimensions. Even here, however (as in a cylinder of uniform cross section), the cross-sectional area is more usually and more precisely obtained by determining the mass of a liquid of known density contained in a measured length of the container rather than by measuring the cross section directly.

Temperature

The concept of temperature arose from the physiological perception that one object is hotter than another object, that is, it is at a higher temperature. This perception is a relative and qualitative measure.

To quantify the concept, it was necessary to find a measurable property that changes with temperature. The length of a column of liquid, the volume of a uniform gas at constant pressure, the pressure of this gas at constant volume, the electrical resistance of a metal, or the resonance frequency of a quartz crystal all have been used. The first property mentioned is utilized in the design of the conventional liquid-in-glass thermometer.

In a first approximate definition of a temperature scale, a choice of two fixed points and the linear division of the interval between the fixed points must be made. The conventional fixed points for a temperature scale have been the normal freezing point and normal boiling point of water, 0 °C and 100 °C, respectively, on the Celsius scale. For example, if L_{100} is the length of the mercury column in a thermometer in steam in equilibrium with boiling water at 1.05×10^5 Pa pressure and L_0 is the length of the mercury column in the same thermometer immersed in an ice-water mixture at the same pressure, then the temperature θ of an object can be calculated from the length L_θ of the mercury column of the thermometer immersed in that object, as shown in Eq. 1-4.

$$\theta = \frac{L_\theta - L_0}{L_{100} - L_0}(100) \qquad (1\text{-}4)$$

A similar equation can be written in terms of any other thermometric property, X.

$$\theta = \frac{X_\theta - X_0}{X_{100} - X_0}(100) \qquad (1\text{-}5)$$

Experimentally, the numerical values of θ found for the temperature of a given object differ for different liquids in liquid-in-glass thermometers, and for different thermometric properties, even though they are calculated to agree at $0\,^\circ$C and $100\,^\circ$C. We shall see that the study of the properties of gases led to a temperature scale that is independent of the properties of individual substances. Like measurements of pressure, temperature measurements are valid only for systems at equilibrium.

Extensive and intensive variables

Volume is an *extensive* variable in that it is proportional to the mass of the system, other conditions being constant. Pressure and temperature, on the other hand, are *intensive* variables, in that they are characteristic of a given system and do not change where the system is simply subdivided. Other extensive variables we shall consider include heat capacity, energy, entropy, and free energy; intensive variables include viscosity, refractive index, density, as well as volume, heat capacity, energy, entropy, and free energy, per mole or per unit mass.

1-2. The simple empirical laws of gases

Boyle's Law

Robert Boyle, in the seventeenth century, carried out the first quantitative experiments on the pressure-volume behavior of gases (or the "spring of

Figure 1-2. A plot of the pressure against volume from Boyle's data in Table 1-1.

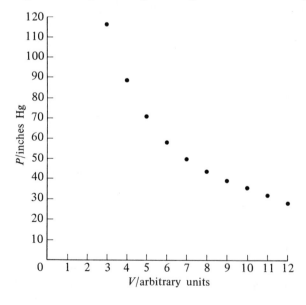

Table 1-1. Boyle's data for pressure volume behavior of gases.

Volume of gas (arbitrary units)	Pressure (inches of mercury)	Pressure-volume product
12	$29\,^2/_{16}$	$349\,^8/_{16}$
11	$31\,^{15}/_{16}$	$351\,^5/_{16}$
10	$35\,^5/_{16}$	$353\,^2/_{16}$
9	$39\,^5/_{16}$	$353\,^{13}/_{16}$
8	$44\,^3/_{16}$	$353\,^8/_{16}$
7	$50\,^5/_{16}$	$352\,^3/_{16}$
6	$58\,^{13}/_{16}$	$352\,^{14}/_{16}$
5	$70\,^{11}/_{16}$	$353\,^7/_{16}$
4	$87\,^{14}/_{16}$	$351\,^8/_{16}$
3	$117\,^9/_{16}$	$352\,^{11}/_{16}$

Table 3-1 from *The World of the Atom*, Vol. 1, edited with commentaries by Henry A. Boorse and Lloyd Motz, (C) 1966 by Basic Books, Inc., Publishers, New York, p. 49.

the air," as he described it). Table 1-1 shows some of Boyle's data. The volume of a sample of gas is given as the height of the sample in the tube.

It is clear from the data in Table 1-1 that the volume decreases, proportionally, as the pressure increases; the simplest mathematical expression that satisfies this condition is the inverse relation, as seen in Eq. 1-6.

$$V = \frac{k}{P} \tag{1-6}$$

When the pressures in Table 1-1 are plotted against their corresponding volumes (Fig. 1-2), we can fit the points on a rectangular hyperbola, the curve represented by Eq. 1-6. It is difficult, however, to decide whether the experimental points really fit such a curve, and it is better to express the equation that is being tested as a linear relation.

This is easily done, rearranging Eq. 1-6, and we obtain

$$PV = k(\theta, m') \tag{1-7}$$

in which it is clear that the "constant" k is a function of the temperature θ and the mass m' of the gas. It can be seen from Table 1-1 that the pressure-volume product is constant within $\pm 1\%$ over a fourfold range of pressures. In Fig. 1-3, the pressure-volume product is plotted against the pressure, and the values do not change with increasing pressure.

Exercise 1-3. Calculate $1/V$ for the data in Table 1-1, plot P against $1/V$, and calculate k from the slope of the best straight line through the points.

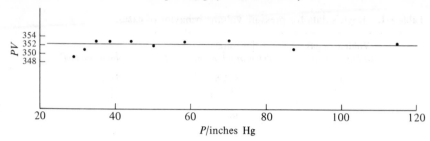

Figure 1-3. A plot of the pressure-volume product against pressure from Boyle's data in Table 1-1.

Amontons' Law

In 1702, Amontons[1] studied the pressure of a constant volume of air as a function of the temperature and developed the first gas thermometer. He observed constant decreases in sample pressure with constant decreases in temperature. He estimated that the pressure of the gas would approach and then equal zero at $-240\,°C$. Assuming that the pressure of a gas cannot be negative, he suggested that this value represents a lower limit of temperature. This was the earliest reference to absolute zero. A century later, Charles and Gay-Lussac, working independently, developed a more rigorous statement about the effect of temperature on gas pressure, obtaining $-273\,°C$ as absolute zero. The mathematical relation between pressure and temperature, now known as Charles's or Gay-Lussac's law, is

$$P = P_0(1 + \alpha\theta) \tag{1-8}$$

where P is the pressure of a sample of gas at temperature θ (°C), P_0 is the pressure of the same sample in the same volume at 0 °C, and α is very nearly a constant for all gases at sufficiently low pressure, with the value $1/273$. Because α is constant for a number of different gases, the gas thermometer is fundamentally superior to other thermometers, since the scale between fixed points is virtually independent of the gas used.

When the value $1/273$ is substituted for α in Eq. 1-8, the result is

$$P = P_0\left(1 + \frac{\theta}{273}\right)$$

$$= \frac{P_0}{273}(273 + \theta)$$

1. K. Mendelssohn, *The Quest for Absolute Zero*, McGraw-Hill Book Co., New York, 1966, pp. 10–13.

$$P = \frac{P_0}{273}(T) \qquad (1\text{-}9)$$

where $T = 273 + \theta$. Thus, when $\theta = 0\,°C$, and $P = P_0$, $T = 273$ and, when $\theta = -273\,°C$, $T = 0$, and $P = 0$. The scale of temperatures represented by T is called the *ideal gas temperature scale* or the *absolute temperature scale*. It is numerically the same as the thermodynamic temperature (Sec. 10-2) in Kelvins (K).

Exercise 1-4.   Charles's law can be expressed in terms of the volume of a sample of gas at constant pressure as

$$V = V_0(1 + \alpha\theta) \qquad (1\text{-}10)$$

Show that, if $\alpha \approx 1/273$, Eq. 1-10 also leads to an absolute zero temperature (the temperature at which $V = 0$) of $-273\,°C$.

General gas law

The direct proportionality between the pressure and the ideal gas temperature indicated in Eq. 1-9 is for a given sample of gas at constant volume. The result of Exercise 1-4 shows a similar proportionality between the volume of a sample of gas at constant pressure and the ideal gas temperature.

From Eq. 1-9.

$$\left(\frac{\partial P}{\partial T}\right)_V = \frac{P_0}{273}$$

and, from the results of Exercise 1-4.

$$\left(\frac{\partial V}{\partial T}\right)_P = \frac{V_0}{273}$$

For T as a function of P and V,

$$\delta T = \left(\frac{\partial T}{\partial P}\right)_V \delta P + \left(\frac{\partial T}{\partial V}\right)_P \delta V$$

$$= \frac{273}{P_0}\delta P + \frac{273}{V_0}\delta V$$

$$= \frac{T}{P}\delta P + \frac{T}{V}\delta V$$

Rearranging and integrating, we have

$$\ln T = \ln P + \ln V + \text{constant}$$

or

$$PV = [k'(m')]T \tag{1-11}$$

where k' is written as a function of the mass m' of the sample of gas. But, since the volume of a gas at a fixed temperature and pressure is proportional to the mass (volume is an extensive variable), $k'(m')$ can be expressed as $k''m'$, and Eq. 1-11 can be written

$$PV = k''m'T \tag{1-12}$$

Though k'' is a constant for a given gas, and independent of temperature and volume, the value of k'' varies from one gas to another.

The possibility of writing Eq. 1-12 as an equation that contains only a universal constant arises from Avogadro's principle, enunciated in 1811 by the Italian physicist, Amedeo Avogadro, though it was accepted by chemists only when Cannizzaro, in 1860, showed that the principle led to a consistent scale of atomic masses. Avogadro's principle states that a given number of molecules of any gas, at the same temperature and pressure, occupy the same volume independent of the nature of the gas. Specifically, if a mass m' of gas, equal to the molar mass M, is chosen, namely, one *mole* (a number of molecules, 6.022×10^{23}, equal to Avogadro's constant, L), V in Eq. 1-12 is replaced by $\mathscr{V} = V/n$, the volume per mole, where n is the number of moles, and $k''M$ is found to be the same for all gases and is designated R, the *universal gas constant*. Thus, the ideal gas law can be expressed

$$P\mathscr{V} = RT \tag{1-13}$$

or

$$PV = nRT \tag{1-14}$$

The ideal gas law, like the other simple gas laws and like Avogadro's principle, is only an approximation. At ordinary atmospheric pressure, however, the predicted values derived from the ideal gas law agree with the experimental values within $\pm 1\%$, so that the approximation is adequate under many circumstances. When the pressure is expressed in atmospheres, the volume in cubic decimeters, and the temperature in Kelvins, R has the approximate value 0.082 dm^3-atm-K^{-1}-mol^{-1}.

Exercise 1-5. Calculate the value of R in fundamental energy units of J-K^{-1}-mol^{-1} and erg-K^{-1}-mol^{-1}. Hint: See Exercise 1-2.

Equation 1-13 is an example of an *equation of state*; for a pure substance, when the values of *two*[2] intensive variables are specified, the values of all other intensive variables are determined (if the effects of external fields, such as gravitational, electric, or magnetic, can be neglected). The ideal gas law is thus an approximate equation of state that describes the molar volume of a gas as a function of the pressure and temperature.

Mixtures of gases

In 1801, Dalton discovered his law of partial pressures, which is applicable to mixtures of gases with approximately the same precision as that of the ideal gas law, for individual gases. In a mixture of gases, the pressure of the mixture is the sum of the pressures each gas would exert if it were the only gas present, at the same volume and the same temperature. That is,

$$P = p_1 + p_2 + \dots \tag{1-15}$$

$$= \frac{n_1 RT}{V} + n_2 \frac{RT}{V} + \dots \tag{1-16}$$

$$= (n_1 + n_2 + \dots) \frac{RT}{V} \tag{1-17}$$

where p_1 and p_2 are the partial pressures of gases 1 and 2, and n_1 and n_2 are the numbers of moles of gases 1 and 2. Equation 1-15 is particularly useful in experiments in which a gas is collected over water, since the total pressure of the system then is equal to the pressure of the gas plus the pressure of the water vapor in equilibrium with liquid water (the vapor pressure).

Exercise 1-6. Calculate the molar mass of a gas if 0.64 g of the gas occupies 0.560 dm³ at 22 °C and a total pressure of 740 mm Hg when collected over water. The vapor pressure of water is 2670 Pa at 22 °C.

1-3. Deviations from the simple empirical laws

When the pressure-volume-temperature properties of gases were studied with more precision than Boyle was able to attain, it was found that PV at constant temperature is not the same for all gases and that the dependence on pressure differs for different gases.

Some data for oxygen, neon, and carbon dioxide at 0 °C are shown in Table 1-2. The variation in the value of PV is well within the precision of

---

2. The number two is not based on any fundamental theory, but is simply a result of empirical observation.

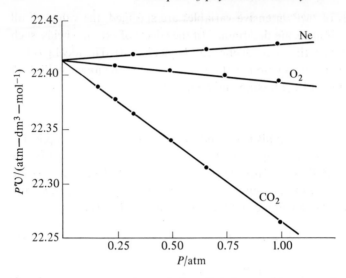

Figure 1-4. Pressure-volume product data for several gases at low pressure at 0°C from Table 1-2.

about 1% that we can expect from Boyle's law, but the data vary in one direction only. The data given in Table 1-2 are plotted in Fig. 1-4, and $P\mathcal{V}$ is seen to be an approximately linear function of P. Thus, an appropriate equation to describe the data can be derived.

$$P\mathcal{V} = RT + aP \qquad (1\text{-}18)$$

where a is a constant that is characteristic of a particular gas at a given temperature, and RT is the common intercept of all three lines at $P = 0$. Though $P\mathcal{V}$ does not have the same value for all gases, the values for different gases approach each other as the pressure decreases, and all approach the value RT, as P approaches zero. One can then write,

$$\lim_{P \to 0} P\mathcal{V} = RT \qquad (1\text{-}19)$$

and the ideal gas law becomes a "limiting law," that is, it becomes exact at the limit of zero pressure.[3]

In the form of Eq. 1-19, the ideal gas law leads to exact values of molecular mass, using gas densities as the primary experimental data, in contrast to the approximate values obtained from Eq. 1-14.

---

3. Equation 1-19 represents an interesting physical analogue of an indeterminant form in mathematics. The limit of P as P approaches zero is zero, and the limit of $1/\mathcal{V}$ as P approaches zero is zero, yet the product $P\mathcal{V} = P/(1/\mathcal{V})$ has a perfectly definite, finite limit as P approaches zero.

Table 1-2. Pressure-volume data for several gases at 0 °C[a]

	P/atm	$\mathscr{V}$/(dm^3/mol)	$P\mathscr{V}$/(atm-dm^3/mol)
Oxygen			
	1.00000	22.3939	22.3939
	0.75000	29.8649	22.3987
	0.50000	44.8090	22.4045
	0.25000	89.6384	22.4096
Neon			
	1.00000	22.4280	22.4280
	0.66667	33.6360	22.4241
	0.33333	67.2567	22.4189
Carbon dioxide			
	1.00000	22.2643	22.2643
	0.66667	33.4722	22.3148
	0.50000	44.6794	22.3397
	0.33333	67.0962	22.3654
	0.25000	89.5100	22.3775
	0.16667	134.3382	22.3897

[a] Data on oxygen and neon from Baxter and Starkweather, *Proc. Natl. Acad. Sci.*, (*U.S.*), *12*, 699 (1926); *14*, 50 (1928); Data on carbon dioxide from Millard, *Physical Chemistry for Colleges*, McGraw-Hill Book Co., New York, 1946, p. 15, by permission.

Exercise 1-7. Rearrange Eq. 1-19 with P and the density ρ as the variables at a fixed temperature. Choose the appropriate function of P and ρ to be plotted against P and to be extrapolated to $P = 0$ to obtain an exact value of M. (Hint: $\mathscr{V} = V/n = (V)(M)/m' = M/\rho$, where m' is the mass of the gas.)

The ideal gas temperature scale

The ideal gas law, as expressed in Eq. 1-19, is exact insofar as the limiting value of $P\mathscr{V}$ at zero pressure for a fixed temperature is the same for all gases. It is not, however, exact, in that the limiting value of $P\mathscr{V}$ at zero pressure is not exactly proportional to T, when T is measured utilizing a variety of thermometric properties. The way Eq. 1-19 can be made exact, and the way the temperature scale can be made independent of the properties of a particular substance, is to *define* the limiting value of $P\mathscr{V}$ at zero pressure as *the* thermometric property. To make the ideal gas scale consistent with the Celsius scale, R is given a value such that $T = 273.16$ K at the ice–water–water vapor equilibrium temperature, and $T = 273.15$ K at 0 °C.

Exercise 1-8. The $\lim\limits_{P \to 0} P\mathscr{V} = 22.4148$ dm^3-atm-mol^{-1} at 0 °C. Calculate the value of R in dm^3-atm-K^{-1}-mol^{-1} and J-K^{-1}-mol^{-1}.

It may seem that defining the temperature scale in terms of the limiting value of $P\mathscr{V}$ at zero pressure makes Eq. 1-19 not only exact, but trivial.

The triviality is only apparent, however, since the definition is not wholly arbitrary, but based on a previous relation that is already a good approximation. The process of refinement of the definition of the temperature scale, from an arbitrary scale based on the thermometric properties of particular substances, through an approximate relationship, to a scale that is independent of the properties of any particular substance is a good example of the evolution of a scientific concept.

The virial equation

The linear function of Eq. 1-18 is applicable with precision only for low pressures. At pressures approaching 10^7 Pa, the data show definite deviations from linearity. To represent such data, additional terms are needed. One representation is a power series, called a *virial equation*. Since a power series in P has been found to represent the experimental data less well than an equation in powers of $1/\mathscr{V}$, it is the latter form, first suggested by Kamerlingh-Onnes in 1901, that is commonly used. To facilitate comparisons at different temperatures, the *compressibility factor* $z = P\mathscr{V}/RT$ is usually the function that is equated to the power series, as shown in Eq. 1-20.

$$\frac{P\mathscr{V}}{RT} = 1 + B(T)\left(\frac{1}{\mathscr{V}}\right) + C(T)\left(\frac{1}{\mathscr{V}}\right)^2 + D(T)\left(\frac{1}{\mathscr{V}}\right)^4 + \ldots \quad (1\text{-}20)$$

When the virial equation in powers of $1/\mathscr{V}$ is used to fit experimental data, a better fit is obtained if odd powers beyond one are omitted. The coefficients $B(T)$, $C(T)$, and $D(T)$ are called the second, third, and fourth *virial coefficients*, respectively, and they are all functions of the temperature, as indicated by the notation used.

Second and third virial coefficients for several gases at 273 K are given in Table 1-3.

Table 1-3. Virial coefficients of gases at 273 K

Gas	$B/(10^{-3}\ dm^3\text{-}mol^{-1})$	$C/(10^{-6}\ dm^6\text{-}mol^{-2})$
He	11.9	91
Ne	10.8	220
H_2	13.8	415
N_2	−10.3	1500
O_2	−21.9	1200
CO_2	−151.2	5600
CH_4	−54.1	3000
NH_3	−345	—

J. H. Dymond and E. B. Smith, *The Virial Coefficients of Gases*, Clarendon Press, Oxford, 1969. Used by permission.

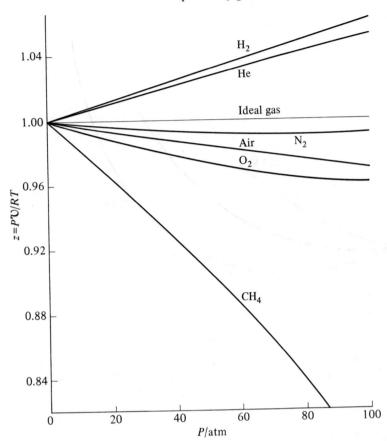

Figure 1-5. A plot of the compressibility factor as a function of pressure at 273 K for several gases up to 100 atm pressure. (By permission, from E. D. Eastman and G. R. Rollefson, *Physical Chemistry*, McGraw-Hill Book Co., New York, 1947, p. 66.)

The compressibility factor z at 273 K is plotted against P for several gases in Fig. 1-5. Though the virial equation is best expressed as a function of $1/\mathscr{V}$, it is convenient to compare different gases in a plot with P as abscissa. The negative slopes in Fig. 1-5 indicate the predominance of attractive forces, whereas positive slopes indicate the predominance of repulsive forces.

The effect of temperature on z is shown for nitrogen in Fig. 1-6. A comparison of the value of z at different temperatures at equal $\mathscr{V}$ reflects only the effect of temperature. A comparison at equal P includes the effect of varying intermolecular distance as well as the effect of varying temperature. The temperature at which the initial slope of the curve is zero is called the Boyle temperature T_B.

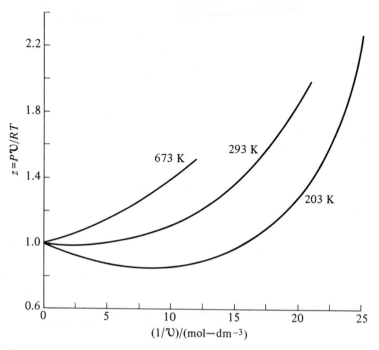

Figure 1-6. The compressibility factor of N_2 as a function of density $(1/\mathcal{V})$ at several temperatures. (By permission, from E. D. Eastman and G. K. Rollefson, *Physical Chemistry*, McGraw-Hill Book Co., New York, 1947, p. 69.)

Semiempirical equations of state

At sufficiently high pressures and sufficiently low temperatures, all gases liquefy. Many attempts have been made to devise equations of state that describe the phenomenon of liquefaction as well as to describe deviations from ideality in the gaseous state. Three of the most widely used equations of this type are those of van der Waals (Eq. 1-21), Dieterici (Eq. 1-22), and Berthelot (Eq. 1-23). They all contain two specific gas constants a and b, in addition to the universal gas constant R.

$$\left(P + \frac{n^2 a}{V^2}\right)(V - nb) = nRT \qquad (1\text{-}21)$$

$$P \exp\left(na/VRT\right)(V - nb) = nRT \qquad (1\text{-}22)$$

$$\left(P + \frac{n^2 a}{TV^2}\right)(V - nb) = nRT \qquad (1\text{-}23)$$

The form of Eqs. 1-21, 1-22, and 1-23 was undoubtedly the result of attempting to formulate, mathematically, the experimental results of Andrews, who, in 1869, described his celebrated *isotherms* of carbon dioxide, as shown in Fig. 1-7. Each curve shows a plot of pressure against specific volume for a fixed temperature, hence the name isotherm (constant temperature). For temperatures below T_c (31.1 °C) the curves show a steep portion at high pressure and low volume, representing the low compressibility characteristic of the liquid state; a roughly hyperbolic curve at low pressures and large volumes, representing the highly compressible gaseous state; and a horizontal portion of the curve, representing the conditions under which liquid and gas coexist in equilibrium. The right end of the horizontal portion of the curve corresponds to the specific volume of the liquid at the temperature of the isotherm (at the pressure at which liquid and gas are in equilibrium), the left end to the specific volume of the gas (at the same temperature and pressure). As the temperature increases, the horizontal part of the isotherm shrinks, until, at T_c, only an inflection point with a horizontal slope remains. Above T_c there are no discontinuities in the slope of the curve, consistent with the visual observation that no boundary of discontinuity appears between phases in the system. The temperature T_c is called the *critical temperature*, the pressure at the horizontal inflection point the *critical pressure* P_c, and the molar volume at that point on the isotherm the critical volume $\mathscr{V}_c$.

The relation of van der Waals' equation to the isotherms in Fig. 1-7 can be seen clearly if both sides of Eq. 1-21 are multiplied by V^2 and expanded. Thus,

$$(PV^2 + n^2 a)(V - nb) = nRTV^2$$

and

$$PV^3 - nbPV^2 + n^2 aV - n^3 ab = nRTV^2$$

or

$$PV^3 - (nbP + nRT)V^2 + n^2 aV = n^3 ab \qquad (1\text{-}24)$$

Equation 1-24 is a cubic equation in V; therefore, there should be three values of V for each value of P. As shown in Fig. 1-8, the van der Waals equation predicts three real roots below T_c, the three roots approach one another and become identical at T_c, and, above T_c, there is only one real root (and two physically insignificant complex roots). Though the critical phenomenon is clearly displayed, the van der Waals equation cannot be used to produce the discontinuity of slope that occurs in the liquid-vapor

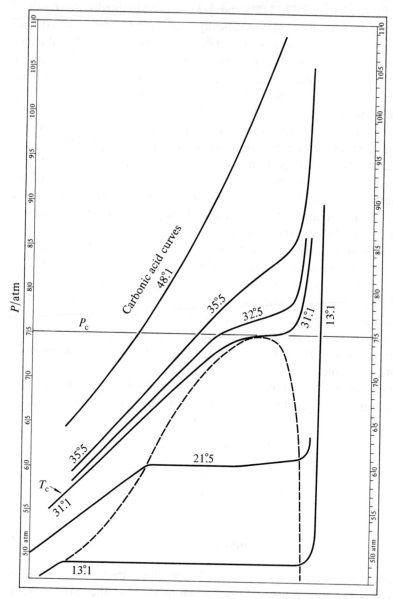

Figure 1-7. Andrews' isothermals for CO_2, with pressure in atmospheres (increasing toward the top) and volume in arbitrary units (increasing to the left). Original plot. Reported to the Royal Society on June 17, 1869, and published in *Phil. Mag.* [4], 39, 150–3 (1870). The puzzling feature of plotting gas data with volume increasing to the left seems to be an artifact of the graph's being rotated 90° to fit more conveniently on a narrow page. The text indicates that the pressure scale is the horizontal scale.

16

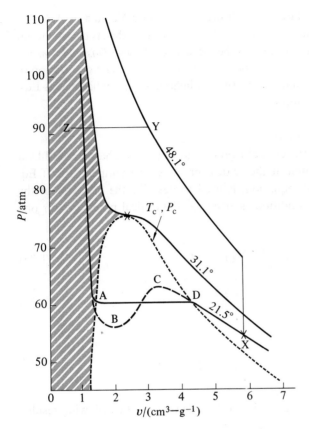

Figure 1-8. Van der Waals curve (ABCD) superimposed on Andrew's isotherms for carbon dioxide. The shaded area indicates the stability limits of a one-phase liquid system. Liquid and vapor exist together at equilibrium under the dotted curve. Above the critical isotherm there is no distinction between liquid and gas. Andrew suggested that the term vapor be used only to represent the state to the right of the dotted curve below the critical temperature. (By permission, from J. R. Partington, *An Advanced Treatise on Physical Chemistry*, Vol. 1, Longman Group, Ltd., London, 1949, p. 628.)

equilibrium region (Fig. 1-8); it predicts, instead, an oscillatory behavior. The parts of the curve between A and B and between C and D represent metastable states of liquid and vapor, respectively, but the curve between B and C is not observed, since it would indicate a state in which an increase in pressure results in an increase in volume.

The areas under the dotted curves in Figs. 1-7 and 1-8 represent regions in which two phases, liquid and gas, coexist, with a boundary of discontinuity. Nowhere else in the diagram is there any indication of such a discontinuity between states. By following the path XYZ in Fig. 1-8, it is possible to go from a state that is clearly gas to a state that is clearly liquid without ever observing a boundary between two phases. *This phenomenon is called the "continuity of state" between liquid and gas.* The description of this continuity of state, and of the critical phenomenon, is what makes Eqs. 1-21, 1-22, and 1-23 so important.

The law of corresponding states

As pointed out above, the critical isotherm in Fig. 1-7 is characterized by a horizontal inflection point at the critical pressure and temperature. If Eq. 1-21, the van der Waals equation, is used to describe the behavior of the gas, the mathematical conditions satisfied at the critical point are, for one mole of gas,

$$\left(P_c + \frac{a}{\mathscr{V}_c{}^2}\right)(\mathscr{V}_c - b) = RT_c \tag{1-25}$$

$$\left(\frac{\partial P}{\partial \mathscr{V}}\right)_{T_c} = 0 \tag{1-26}$$

and

$$\left(\frac{\partial^2 P}{\partial \mathscr{V}^2}\right)_{T_c} = 0 \tag{1-27}$$

Applying Eqs. 1-26 and 1-27 to Eq. 1-25, one obtains the following results.

$$\frac{-RT_c}{(\mathscr{V}_c - b)^2} + \frac{2a}{\mathscr{V}_c{}^3} = 0 \tag{1-28}$$

$$\frac{2RT_c}{(\mathscr{V}_c - b)^3} - \frac{6a}{\mathscr{V}_c{}^4} = 0 \tag{1-29}$$

Exercise 1-9.   Verify Eqs. 1-28 and 1-29.

If Eqs. 1-25, 1-28, and 1-29 are solved simultaneously for the three constants a, b, and R in terms of the critical constants T_c, P_c, and $\mathscr{V}_c$, the results are

$$b = \frac{\mathscr{V}_c}{3} \tag{1-30}$$

$$a = 3P_c\mathscr{V}_c{}^2 \tag{1-31}$$

$$R = \frac{8 P_c \mathscr{V}_c}{3 T_c} \tag{1-32}$$

Substitution of the values of a, b, and R in terms of the critical constants into the van der Waals equation yields the result

$$\left(\frac{P}{P_c} + \frac{3 \mathscr{V}_c^2}{\mathscr{V}^2} \right) \left(\frac{\mathscr{V}}{\mathscr{V}_c} - \frac{1}{3} \right) = \frac{8}{3} \frac{T}{T_c} \tag{1-33}$$

or, if expressed in terms of the reduced variables (relative to the critical value),

$$P_r = \frac{P}{P_c}, \ \mathscr{V}_r = \frac{\mathscr{V}}{\mathscr{V}_c}, \ T_r = \frac{T}{T_c} \tag{1-34}$$

$$\left(P_r + \frac{3}{\mathscr{V}_r^2} \right) \left(\mathscr{V}_r - \frac{1}{3} \right) = \frac{8}{3} T_r \tag{1-35}$$

Thus, if the behavior of the gas is described in terms of reduced variables, *all gases conform to the same equation, with no arbitrary constants. This is the law of corresponding states.* The same conclusion can be reached by using the Dieterici or Berthelot equations.

Exercise 1-10. Show that the Berthelot equation also leads to a universal reduced equation of state.

A consequence of the law of corresponding states (*a law that is as approximate as the semiempirical equation from which it is derived*) is that the compressibility factor, $z = P\mathscr{V}/RT$, should be the same function of P_r and T_r for all substances. Thus, estimates of the P-V-T behavior of a substance can be obtained from curves of z as a function of P_r and T_r, if the critical constants of the substance are known. If we substitute from the definitions of the reduced variables in the definition of z, the result is

$$z = \left(\frac{P_r \mathscr{V}_r}{R T_r} \right) \left(\frac{P_c \mathscr{V}_c}{T_c} \right)$$

Substituting from Eq. 1–32, we have

$$z = \frac{3 P_r \mathscr{V}_r}{8 T_r} \tag{1-36}$$

Since $\mathscr{V}_r$ is the same for all gases at fixed values of P_r and T_r, Eq. 1-36 states that z is the same for all gases at given values of P_r and T_r. If the critical constants of a gas are known, P_r and T_r can be calculated for desired

Figure 1-9. The Nelson-Obert generalized compressibility charts, with $z = P \mathscr{V} / R T$ expressed as a function of P_r, T_r, and $V'_r = \mathscr{V} P_c / R T_c$ (a) at very low reduced pressures; (b) at low reduced pressures; (c) at intermediate reduced pressures; and (d) at high reduced pressures. The values of the critical constants used in constructing these charts are given in Table 1-5. H_2, He, and NH_3 could not be correlated satisfactorily below $T_r = 2.5$. Above $T_r = 2.5$, H_2 and He fit the charts if $T_r = T / (T_c + 8)$ and $P_r = P / (P_c + 8.08 \times 10^5$ Pa$)$. (Courtesy of Professor E. F. Obert; see L. C. Nelson and E. F. Obert, *Chem. Eng.*, July, 1954, pp. 203–8.)

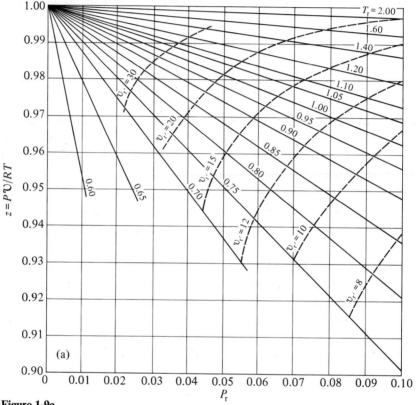

Figure 1-9a.

values of P and T, and a value of z and, thus, $\mathscr{V}$ can be obtained from the appropriate curve of Fig. 1-9.

Figure 1-10 shows data for ten gases at values of T_r from 1.00 to 2.00. It can be seen that the chart is not useful at the critical point, since the value of z varies from 0.2 to 0.4, when T_r and P_r are equal to 1.00. Su found that the average deviation of individual gases from the curves varied from 0.63%

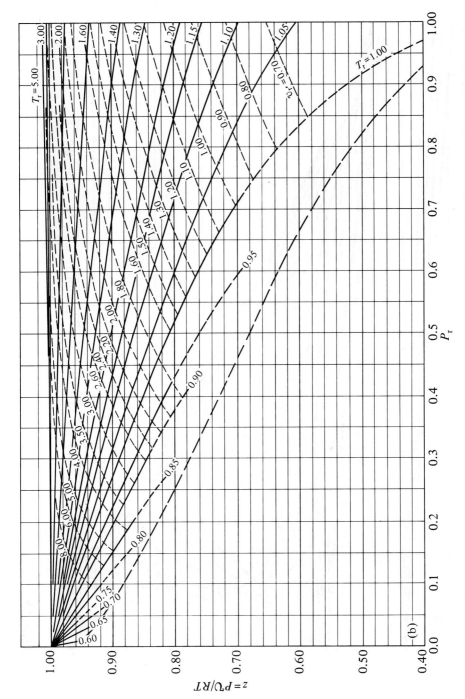

Figure 1-9b.

21

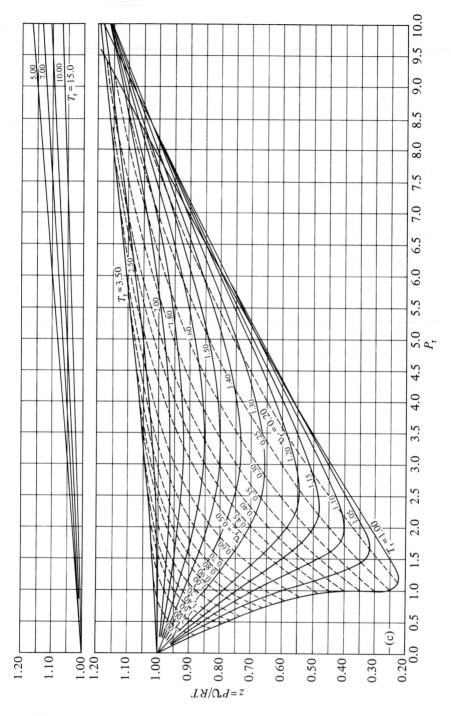

Figure 1-9c.

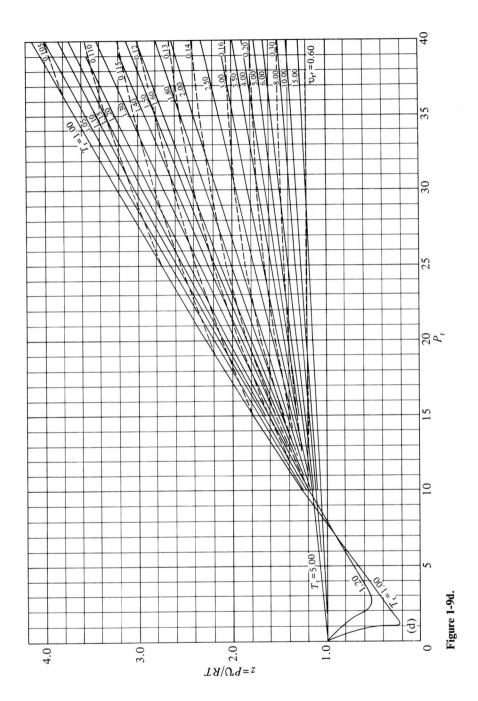

Figure 1-9d.

23

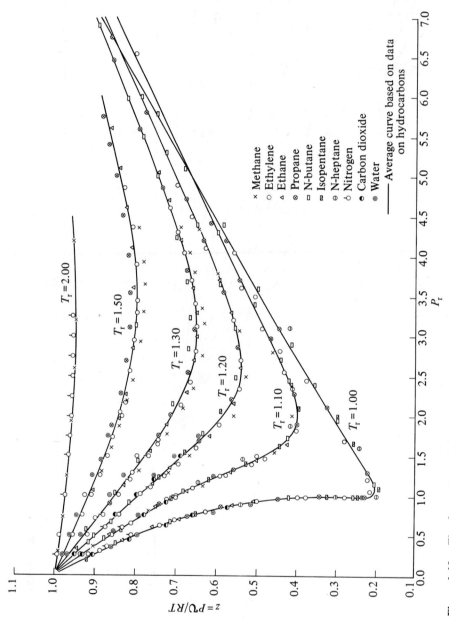

Figure 1-10. Fit of experimental data for ten gases to the law of corresponding states. [By permission, from G. J. Su, *Ind. Eng. Chem.*, **38**, 803–4 (1946).]

to 2.47%. The compressibility behavior of gases can be represented to higher precision if another parameter (in addition to P_r and T_r) is used.[4]

One of the most useful empirical equations of state is that of Beattie and Bridgeman.[5] A gas is characterized in this equation by five empirical constants A_0, a, B_0, b, and c, and the equation of state is

$$P = \frac{RT(1 - \varepsilon)}{\mathscr{V}^2} (\mathscr{V} + B) - \frac{A}{\mathscr{V}^2} \qquad (1\text{-}37)$$

where

$$A = A_0 \left(1 - \frac{a}{\mathscr{V}}\right)$$

$$B = B_0 \left(1 - \frac{b}{\mathscr{V}}\right)$$

and

$$\varepsilon = \frac{c}{\mathscr{V} T^3}$$

The constants for ten gases are given in Table 1-4; the differences between calculated and experimental values are less than 0.3% over a wide temperature range and up to pressures of the order of 10^7 Pa.

Table 1-4. Constants for the Beattie-Bridgeman equation[a]

Gas	A_0	a	B_0	b	$c/(10^4)$
He	0.0216	0.05984	0.01400	0.0	0.004
Ne	0.2125	0.02196	0.02060	0.0	0.101
A	1.2907	0.02328	0.03931	0.0	5.99
H_2	0.1975	−0.00506	0.02096	−0.04359	0.0504
N_2	1.3445	0.02617	0.05046	−0.00691	4.20
O_2	1.4911	0.02562	0.04624	0.04208	4.80
Air	1.3012	0.01931	0.04611	−0.01101	4.34
CO_2	5.0065	0.07132	0.10476	0.07235	66.00
CH_4	2.2769	0.01855	0.05587	−0.01587	12.83

By permission, from J. A. Beattie and O. C. Bridgeman, *J. Am. Chem. Soc.*, *50*, 3133–38 (1928).
[a]Units: $P/1.01 \times 10^5$ Pa, $\mathscr{V}/dm^3\text{-}mol^{-1}$, T/K.

4. K. S. Pitzer, D. Z. Lippman, R. F. Curl, Jr., C. M. Huggins, and D. E. Petersen, *J. Am. Chem. Soc.*, 77, 3433 (1955).
5. J. A. Beattie and O. C. Bridgeman, *J. Am. Chem. Soc.*, *50*, 3133-38 (1928); *49*, 1665 (1927); *Proc. Am. Acad. Arts Sci.*, *63*, 229 (1928).

Table 1-5. Critical constants of gases

Substance	T_c/K	$P_c/1.01 \times 10^5$ Pa
Air	132.53	37.17
Ammonia	405.4	111.3
Argon	150.72	47.996
Benzene	562.7	48.7
n-Butane	425.17	37.47
Isobutane	408.14	36.00
1-Butene	419.6	39.7
Carbon dioxide	304.20	72.85
Carbon monoxide	132.91	34.529
Deuterium	38.40	16.40
Ethane	305.43	48.20
Ethyl ether	467.8	35.6
Ethylene	283.06	50.50
Helium	5.19	2.26
n-Heptane	540.17	27.00
n-Hexane	507.9	29.94
Hydrogen	33.24	12.797
Hydrogen sulfide	373.7	88.8
Methane	191.05	45.79
Methyl fluoride	317.71	58.0
Neon	44.39	26.86
Nitrogen	126.26	33.54
Nitric oxide	180.3	64.6
Oxygen	154.78	50.14
n-Pentane	471.0	33.10
Isopentane	461.0	32.92
Propane	370.01	42.1
Propylene	364.92	45.61
Water	647.27	218.167
Xenon	289.81	57.89

By permission, from L. C. Nelson and E. F. Obert, *Chem. Eng.*, July, 1954, pp. 203–8.

Problems

1-1. The volume of a sample of toluene over the temperature range from 0 to 100 °C is given by the relation

$$V = V_0 \,(1 + 1.028 \times 10^{-3} \, t + 1.779 \times 10^{-6} \, t^2)$$

where V_0 is the volume at 0°C, and t is the temperature in degrees Celsius on the ideal gas scale. If the volume of toluene is used as a thermometric property, calculate the error in measuring the temperature of a substance at 67°C.

1-2. The density of gaseous propane is 1.636×10^{-2} kg-dm^{-3} at 373 K and 1.01×10^6 Pa pressure. Calculate the pressure at this temperature and density using the ideal gas law, and compare the calculated result to the experimental value. [W. W. Deschner and G. G. Brown, *Ind. Eng. Chem.*, 32, 836–40 (1940).]

1-3. The density of CO_2 at several pressures at 273 K is listed below. Calculate from these data the function derived in Exercise 1-7, plot the values as a function of P, and calculate the molar mass of CO_2 by extrapolation.

$P/1.01 \times 10^5$ Pa	$\rho/(10^{-3}$ kg-dm$^{-3})$
1.00000	1.9767
0.66667	1.3148
0.50000	0.9850
0.33333	0.6559
0.25000	0.4917
0.16667	0.3276

1-4. Deschner and Brown, *loc. cit*, determined the following values for the critical constants of propane: $\rho_c = 0.224$ kg-dm^{-3}, $T_c = 370.00$ K, and $P_c = 4.25 \times 10^6$ Pa. Use these data to obtain values for the van der Waals constants for propane. Calculate the pressure corresponding to the density and temperature given in Problem 1-2 with van der Waals equation; compare with the experimental value and with that calculated from the ideal gas law.

1-5. Derive an expression for the Boyle temperature of a van der Waals gas, the temperature at which $[\partial(P\mathscr{V})/\partial(1/\mathscr{V})]_T$ is equal to zero at $(1/\mathscr{V}) = 0$. Hint: Solve the van der Waals equation for P and multiply by $\mathscr{V}$ before differentiating. Calculate the Boyle temperature for propane.

1-6. Show that the second virial coefficient $B(T)$ is equal to $\lim\limits_{(1/\mathscr{V}) \to 0} \left[\partial z/\partial \left(\frac{1}{\mathscr{V}} \right) \right]_T$.

1-7. Obtain values of z as a function of $1/\mathscr{V}$, and calculate $B(T)$ for propane at 303 K from the following data:

$P/1.01 \times 10^5$ Pa	z	$P/1.01 \times 10^5$ Pa	z
1.000	0.9845	4.000	0.9360
2.000	0.9685	5.000	0.9190
3.000	0.9526	6.000	0.9008

1-8. Calculate $B(T)$ at 273 K for the gases for which data are given in Table 1-2, and compare with the values in Table 1-3.

1-9. At 89×10^5 Pa ($\sim P_r = 2.1$) and 448 K ($\sim T_r = 1.2$), Deschner and Brown found $z = 0.61$. Compare this result with that obtained from the curves based on the law of corresponding states in Fig. 1-9.

2 · *Kinetic theory of gases*

Having considered the empirical properties of gases in Chapter 1, we are now prepared to discuss some of the simple models, both conceptual and mathematical, that have been devised to explain those properties. For convenience, the empirical properties of gases are treated in that chapter and the theoretical models of gas behavior in the next chapter, but there was no historical separation between empirical investigation and the construction of theories. Boyle, in addition to measuring what he called "the spring of the air," speculated that air consisted of bodies "lying upon one another, as may be resembled to a fleece of wool. For this . . . consists of many slender and flexible hairs, each of which may indeed, like a little spring, be easily bent or rolled up, but will also like a spring be still endeavoring to stretch itself out again."[1] After measuring gas pressure as a function of temperature, Amontons speculated that absolute zero represented a state in which all motion ceased. But it was the mathematician Daniel Bernoulli, who, in 1738, formulated the simple kinetic theory of gases, the theory that is the conceptual basis for modern treatments of gas theory.

2-1. The simple kinetic model

In the simple kinetic model, gases are assumed to consist of particles called molecules, which have three important characteristics. *Molecules are* (a) *small compared to the distance between them, in effect, point masses,*

1. From *The World of the Atom*, Vol. I, edited with commentaries by Henry A. Boorse and Lloyd Motz, (C) 1966 by Basic Books, Inc., Publishers, New York, pp. 41–42.

(b) *moving rapidly and randomly, and* (c) *colliding elastically with each other and with the walls of their container, there being no intermolecular forces acting among them except during collisions.* These assumptions account qualitatively for the compressibility of gases and for the observation that the pressure of gases is maintained indefinitely.

Quantitative comparisons between the model and the empirical observations require that Newtonian mechanics be applied, as Bernoulli applied it, to the motion of gas molecules.

Newtonian analysis of the model

Consider a collection of N identical gas molecules, each of mass m, in a cubic container whose edges are of length l. A molecule, with velocity u, has components of velocity in the x, y, and z directions denoted by u_x, u_y, and u_z, as shown in Fig. 2-1. When a molecule rebounds from the wall of a container that has the y and z axes as edges, only the x component of the velocity changes; in the absence of friction, the wall can exert a force on the molecule only in the direction perpendicular to the plane of the wall. If the x component of the velocity of a molecule is $-u_x$ when it approaches the wall, then after the collision with the wall the x component of the velocity will be $+u_x$; since the collision is assumed to be elastic, energy is conserved, and there is only a change in the direction of the velocity. The corresponding

Figure 2-1. A coordinate system for the simple kinetic theory model showing the velocity vector **u** and the components u_x, u_y, and u_z.

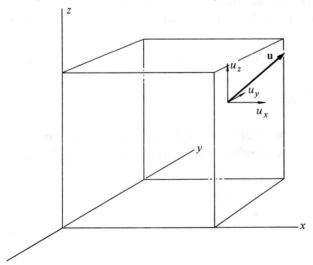

momenta before and after collision are $-mu_x$ and $+mu_x$. The time interval between successive collisions with the same wall is

$$\Delta t = \frac{2l}{u_x} \tag{2-1}$$

According to Newton's second law of motion, the magnitude F' of the force exerted by the wall on the molecule in the collision is given by the change in momentum per unit time.

$$
\begin{aligned}
F' &= \frac{mu_x - (-mu_x)}{(2l/u_x)} \\
&= \frac{2mu_x}{(2l/u_x)} \\
&= \frac{mu_x^2}{l}
\end{aligned}
\tag{2-2}
$$

The force $\mathbf{F}$ exerted by the molecule on the wall is equal in magnitude to $\mathbf{F}'$, but the two forces are opposite in direction. Thus,

$$F = \frac{mu_x^2}{l} \tag{2-3}$$

Collisions of a molecule with the walls of the container are discrete events, so the value of F from Eq. 2-3 is actually the time average value of a fluctuating force. The time average is a useful quantity when many collisions occur during the time required to measure the force.

The total force exerted on the wall is the sum of the forces exerted by the individual molecules

$$F = F_1 + F_2 + \ldots + F_i + \ldots + F_N \tag{2-4}$$

$$= \sum_{i=1}^{N} F_i \tag{2-5}$$

where Σ is the summation of all the terms F_i, with the index varying from $i = 1$ to $i = N$. Thus, the substitution of F_i from Eq. 2-3 in Eq. 2-5 yields

$$
\begin{aligned}
F &= \sum_{i=1}^{N} \frac{mu_x^2(i)}{l} \\
&= \frac{m}{l} \sum_{i=1}^{N} u_x^2(i)
\end{aligned}
\tag{2-6}
$$

Since the velocities of molecules vary considerably, it is convenient to use an average quantity to describe the behavior of all the molecules. The average of the x component of the velocity, $\overline{u_x}$, is zero, since there is no preferred direction to the motion of molecules. The average of the square of the x component of the velocity $\overline{u_x^2}$ is not zero, however, since the squares are all positive. Let us define $\overline{u_x^2}$ as

$$\overline{u_x^2} = \frac{\sum\limits_{i=1}^{N} u_x^2(i)}{N} \tag{2-7}$$

On the basis of this definition, Eq. 2-6 can be rewritten as

$$F = \frac{m}{l} N \overline{u_x^2} \tag{2-8}$$

Using Eq. 1-1, $P = F/A$, and remembering that the area of the wall is l^2, we can now derive an expression for the pressure on the wall. Thus,

$$P = \frac{|\mathbf{F}|}{|\mathbf{A}|} = \left(\frac{m}{l^3}\right) N \overline{u_x^2}$$

$$= \frac{N m \overline{u_x^2}}{V} \tag{2-9}$$

where V is the volume of the cube.

The magnitude of the velocity u of a molecule is related to the components u_x, u_y, and u_z, by the Pythagorean theorem

$$u^2 = u_x^2 + u_y^2 + u_z^2 \tag{2-10}$$

If Eq. 2-10 is summed over all the molecules in the system, and divided by N, the result is

$$\frac{\sum\limits_{i=1}^{N} u^2(i)}{N} = \frac{\sum\limits_{i=1}^{N} u_x^2(i)}{N} + \frac{\sum\limits_{i=1}^{N} u_y^2(i)}{N} + \frac{\sum\limits_{i=1}^{N} u_z^2(i)}{N}$$

or

$$\overline{u^2} = \overline{u_x^2} + \overline{u_y^2} + \overline{u_z^2} \tag{2-11}$$

Since the motion of the gas molecules is assumed to be completely random, the average squares of the velocity components in the three directions are equal, so that

$$\overline{u^2} = 3\,\overline{u_x^2} \qquad (2\text{-}12)$$

Using Eq. 2-12, we can rewrite Eq. 2-9

$$P = \frac{1}{3}\,\frac{Nm\overline{u^2}}{V} \qquad (2\text{-}13)$$

or

$$PV = \frac{1}{3}\,Nm\overline{u^2} \qquad (2\text{-}14)$$

Equation 2-14 is the *fundamental equation* of the simple kinetic theory of gases. Since the average kinetic energy $\overline{K}$ of a molecule of gas is

$$\overline{K} = \frac{1}{2}\,m\overline{u^2} \qquad (2\text{-}15)$$

Eq. 2-14 can also be written

$$PV = \frac{2}{3}\,N(\tfrac{1}{2}\,m\overline{u^2})$$

$$= \frac{2}{3}\,N\overline{K} \qquad (2\text{-}16)$$

Here, it is obvious from the fundamental equation that the pressure of a gas at constant volume is proportional to the number of gas molecules and the average kinetic energy of the molecules.

2-2. Temperature and the kinetic model

The fundamental equation (Eq. 2-14) resembles the ideal gas law, Eq. 1-14, since it contains the PV product and since the PV product is proportional to the mass of gas (Nm). The temperature does not appear as a variable in Eq. 2-14, however, because temperature is not a mechanical variable. An additional factor is required to relate the temperature and the mechanical model.

This relationship can be shown by equating the right-hand sides of Eqs. 1-14 and 2-16.

$$nRT = \frac{2}{3}\,N\overline{K} \qquad (2\text{-}17)$$

Now we can see that if the kinetic model and the empirical equations are to agree the *absolute temperature must be proportional to the average kinetic energy of the molecules.*[2]

2. The conclusion that the temperature is proportional to the average kinetic energy of the molecules was first put forward by Waterston in a paper that was submitted to the Royal Society in 1845 but was relegated to the archives and not published. Fifteen years later, the same relation was rediscovered by Clausius, Maxwell, and Boltzmann, and it was only in 1892 that Waterston's original paper was published, after it had been found by Lord Rayleigh. See H. A. Boorse and L. Motz, *World of the Atom*, Basic Books, New York, 1966, Vol. I, pp. 213–35.

If Eq. 2-17 is solved for $\overline{K}$, the result is

$$\overline{K} = \frac{3}{2} n \frac{RT}{N} \tag{2-18}$$

The number of molecules N is equal to the number of moles n multiplied by L, Avogadro's constant ($6{,}022 \times 10^{23}$ mol^{-1}). Thus,

$$\overline{K} = \frac{3}{2} \frac{n}{nL} RT$$

$$= \frac{3}{2} R \frac{T}{L}$$

$$= \frac{3}{2} kT \tag{2-19}$$

The constant k of Eq. 2-19 is the Boltzmann constant; it is the gas constant per molecule, and it is equal to R/L.

Exercise 2-1. Calculate numerical values of k in J-K^{-1} and erg-K^{-1}.

Implicit in Eq. 2-19 is the assumption that the average kinetic energy of all gases is the same at the same temperature. The derivation of Eq. 2-19 demonstrates the relationship between the universality of the Boltzmann constant and the observation that is expressed mathematically in Eq. 1-14 *that the pressure-volume product per mole is the same for all gases at the same temperature (in the limit of zero pressure)*. Deductions from the simple model obviously apply only to ideal gases (that is, gases at low pressures).

Exercise 2-2. Why is the simple kinetic theory applicable only to ideal gases?

Temperature and the velocity of molecules
If Eq. 2-15 is substituted in Eq. 2-19, the result is

$$\frac{1}{2} m \overline{u^2} = \frac{3}{2} kT \tag{2-20}$$

Solving for $\overline{u^2}$, one obtains

$$\overline{u^2} = \frac{3 kT}{m} \tag{2-21}$$

A more convenient form is obtained by multiplying the right-hand side of Eq. 2-21 by L/L,

$$\overline{u^2} = \frac{3\,L\,k\,T}{L\,m}$$

$$= \frac{3\,RT}{M} \qquad (2\text{-}22)$$

where M is the mass of one mole of the gas. The square root of $\overline{u^2}$ is called the *root mean square velocity* and is written

$$u_{RMS} = \left(\frac{3\,RT}{M}\right)^{\frac{1}{2}} \qquad (2\text{-}23)$$

Exercise 2-3. Calculate u_{RMS} for the O_2 molecule at 100 K, 200 K, 300 K, and 400 K.

Graham's law of effusion

Since the average kinetic energy is the same for all gases at a given temperature, one can write for two gases A and B

$$\frac{1}{2}m_A\overline{u_A^2} = \frac{1}{2}m_B\overline{u_B^2} \qquad (2\text{-}24)$$

or

$$\frac{\overline{u_A^2}}{\overline{u_B^2}} = \frac{m_B}{m_A} \qquad (2\text{-}25)$$

For the mass of one mole of gas, Eq. 2-25 becomes

$$\frac{\overline{u_A^2}}{\overline{u_B^2}} = \frac{M_B}{M_A} \qquad (2\text{-}26)$$

Taking the square root of both sides of Eq. 2-26, we obtain

$$\frac{u_{RMS}(A)}{u_{RMS}(B)} = \left(\frac{M_B}{M_A}\right)^{\frac{1}{2}} \qquad (2\text{-}27)$$

Equation 2-27 agrees with Graham's law of effusion, which was discovered empirically before the kinetic theory was developed. Graham's law states that *the rates of effusion of two gases at the same temperature are proportional to the concentration (or pressure) of gas and inversely proportional to the square root of the mass of one mole of the gas.* If two gases are at the same pressure, as well as the same temperature, the ratio of the effusion rates is the same as the ratio of the root mean square velocities as given by Eq. 2-27.

Exercise 2-4. Divers and astronauts sometimes use a breathing mixture of 20% O_2 and 80% He, by volume. In an experiment, this mixture is allowed to effuse through a small opening in the container. What is the composition of the first sample of gas to effuse through the opening? Assume that the sample is large enough to be representative but small enough not to change the composition of the original mixture.

2-3. Thermal energy and heat capacity

In the simple kinetic model, *gas molecules are assumed to be point masses having no intermolecular forces acting among them except during collisions.* These two assumptions imply that *there is no internal motion and that there is no intermolecular potential energy.* Thus, the total energy of a collection of gas molecules is the sum of the kinetic energies of the individual molecules. Since this energy depends only on the temperature, it is called the *thermal energy*, and designated E. Because there is no intermolecular potential energy, E is independent of the distance between molecules and, thus, of V.

The thermal energy per mole is $\mathscr{E}$. Thus,

$$E = N\bar{K} \tag{2-28}$$

and

$$\mathscr{E} = L\bar{K}$$

$$= L(^3/_2\, kT) \tag{2-29}$$

$$= {}^3/_2\, RT \tag{2-30}$$

The average thermal energy per molecule is, of course, $\frac{3}{2}\, kT$.

Heat capacity

The heat capacity[3] C of a system is defined as *the amount of heat that must be added to a system to increase the temperature* 1 K (one Kelvin). If there is a heat transfer at constant volume, the heat only increases the thermal energy of the system, and no mechanical work of expansion is done. Thus, the heat capacity at constant volume is the change in thermal energy per degree. In the notation of the calculus,

$$C_V = \left(\frac{\partial E}{\partial T}\right)_V \tag{2-31}$$

where subscript V indicates that the temperature change is carried out at

---

3. The term capacity comes from an earlier theory of heat in which heat was considered to be a fluid contained in a system. We now prefer to say that systems have a thermal energy; heat is only a form of energy that is transferred when there is a temperature difference. In many older books, heat capacity is called specific heat.

constant volume, and the derivative is a partial derivative because E is a function of two variables in general.

The corresponding molar quantity $\mathscr{C}_V$ is

$$\mathscr{C}_V = \left(\frac{\partial \mathscr{E}}{\partial T}\right)_V \tag{2-32}$$

According to Eq. 2-30,

$$\mathscr{E} = {}^3\!/_2\, RT$$

so that

$$\mathscr{C}_V = {}^3\!/_2\, R \tag{2-33}$$

Thus, the simple kinetic theory model, derived to provide a theoretical basis for the ideal gas law, *leads to the prediction that the molar heat capacity of a gas at constant volume will be independent of the temperature and will have the value of* $\frac{3}{2} R(12.5\ \text{J-K}^{-1}\text{-mol}^{-1})(2.99\ \text{cal-K}^{-1}\text{-mol}^{-1})$. The experimental values of the heat capacity of several gases as a function of temperature are shown in Fig. 2-2.

It can be seen in Fig. 2-2 that of the gases shown, only helium, a monatomic gas, has a heat capacity consistent with Eq. 2-33. The heat capacities of other monatomic gases are also consistent with Eq. 2-33, but heat capacities of polyatomic gases are not, suggesting that the presence of several atoms in a gas molecule introduces complications. Equation 2-28 suggests that the thermal energy of a set of gas molecules is simply the sum of the translational kinetic energies of the molecules. Since this assumption correctly predicts the heat capacity of monatomic gases, it appears that the internal motion of electrons do not contribute to the heat capacity. Now we will merely raise this question; we shall return to the problem when we discuss the quantum theory of heat capacities.

Heat capacity and molecular structure.   All the gases other than helium shown in Fig. 2-2 have heat capacities greater than $3/2\ R$ (except H_2 and D_2 at very low temperatures), and all the gases have at least two atoms in the molecule, which naturally leads us to speculate that *the higher heat capacities of complex molecules reflect contributions to the thermal energy from motions other than translational, with the rotational motion of the molecule as a whole, and the vibrational motion of its component atoms being likely possibilities.*

The simple kinetic theory led to the prediction that the average kinetic energy per molecule is $\frac{3}{2}\, kT$. When this result was derived, it was assumed

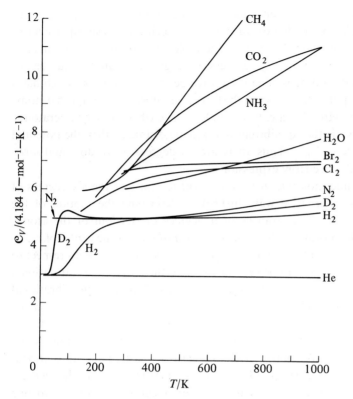

Figure 2-2. The molar heat capacity at constant volume as a function of temperature for several gases. (By permission, from E. D. Eastman and G. K. Rollefson, *Physical Chemistry*, McGraw-Hill Book Co., 1947, p. 140.)

that the mean squares of the three velocity components are equal and, therefore, that the mean kinetic energies corresponding to the three velocity components are equal. Thus, the average kinetic energy corresponding to each velocity component is $\frac{1}{2}kT$. The position of any point particle can be fixed by specifying the values of *three* coordinates; the particle is then said to have *three degrees of freedom*. These statements are combined to state the *equipartition theorem*, first suggested by Waterston for the translational kinetic energy of molecules: *To each degree of freedom of a molecule there corresponds $\frac{1}{2}kT$ of thermal energy.* An alternative statement that will be useful in treating polyatomic molecules is: *To each quadratic term in the energy* (e.g. $\frac{1}{2}mu_x{}^2$) *there corresponds $\frac{1}{2}kT$ of thermal energy.*

Since a diatomic molecule has two atoms the location of which must be specified, and since each atom's location is described by three coordinates, such a molecule has six degrees of freedom, which implies six square terms in the energy expression. For a diatomic molecule, however, the number of quadratic terms in the energy expression is greater than the number of degrees of freedom because there is a force between the atoms, which has a corresponding potential energy. If the force is assumed to be simple harmonic (that is, it obeys Hooke's law, $F = -k(X - X_0)$, where X is the interatomic distance and X_0 is the equilibrium interatomic distance), then the potential energy is $\frac{1}{2} k(X - X_0)^2$. Thus, there are *six* degrees of freedom and *seven* square terms in the energy expression.

For a diatomic molecule, it is convenient to consider the six degrees of freedom in an alternative way. Instead of describing the motion of the molecule in terms of six translational degrees of freedom of the atoms, we can view the motion of the molecule as *translation* of the center of mass, *rotation* about the center of mass, and *vibration* of the atoms with respect to the center of mass. As can be seen in Fig. 2-3, there are *three* translational degrees of freedom, *two* rotational degrees of freedom, and *one* vibrational degree of freedom, a total of *six. Only those rotations that result in a change of the coordinates of the atoms (proper motions) are considered*; rotation about the molecular axis is analogous to internal motion in a monatomic molecule and does not contribute to the thermal energy. Each translational degree of freedom corresponds to one quadratic term in the energy, $\frac{1}{2}(m_1 + m_2) u_x^2$, each rotational degree of freedom corresponds to one quadratic term in the energy, $\frac{1}{2} I\omega_x^2$, and the vibrational degree of freedom corresponds to two quadratic terms in the energy,

$$\frac{1}{2} \left(\frac{m_1 m_2}{m_1 + m_2} \right) \left(\frac{dX}{dt} \right)^2$$

and $\frac{1}{2} k(X - X_0)^2$. (See Eq. 6-20 and Eqs. 6-21 to 6-22.) Thus, the average thermal energy per diatomic molecule can be written

$$\bar{\varepsilon} = 3(\tfrac{1}{2} kT)_{\text{trans}} + 2(\tfrac{1}{2} kT)_{\text{rot}} + 2(\tfrac{1}{2} kT)_{\text{vib}} \qquad (2\text{-}34)$$

$$= \tfrac{7}{2} kT \qquad (2\text{-}35)$$

Correspondingly, the thermal energy per mole, $\mathscr{E}$, for the same kind of molecule is

$$\mathscr{E} = 3(\tfrac{1}{2} RT)_{\text{trans}} + 2(\tfrac{1}{2} RT)_{\text{rot}} + 2(\tfrac{1}{2} RT)_{\text{vib}}$$

$$= \tfrac{7}{2} RT \qquad (2\text{-}36)$$

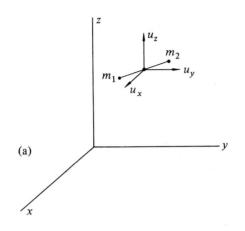

(a)

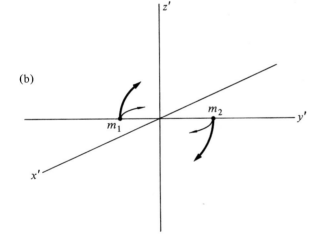

(b)

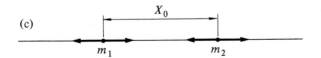

(c)

Figure 2-3. (a) The three translational degrees of freedom of a diatomic molecule; (b) the two rotational degrees of freedom of a diatomic molecule (heavy arrow, rotation about the x-axis, light arrow, rotation about the z-axis); (c) the vibrational degree of freedom of a diatomic molecule.

39

(a)

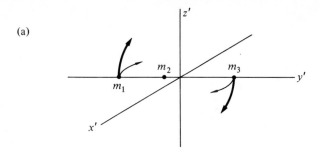

(b)

Symmetric stretch   Asymmetric stretch

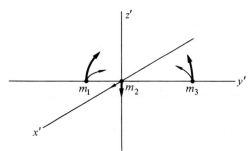

(c)

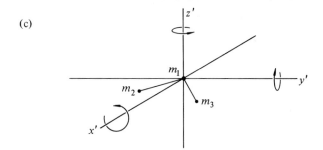

(d)

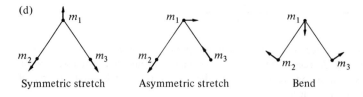

Symmetric stretch   Asymmetric stretch   Bend

40

The predicted value of $\mathscr{C}_v$ is equal to $\frac{7}{2} R$.

The value of the thermal energy predicted for a triatomic molecule by the equipartition theorem depends on whether the molecule is linear or non-linear. A triatomic molecule has nine degrees of freedom. If the molecule is linear, it has only two rotational degrees of freedom and, therefore, four vibrational degrees of freedom. If the molecule is nonlinear, there are three rotational degrees of freedom and only three vibrational degrees of freedom.

Figure 2-4(a) shows the fundamental rotations about orthogonal axes perpendicular to the molecular axis for a linear triatomic molecule. As in the case of the diatomic molecule, rotation about the molecular axis is not a rotational degree of freedom, since it does not change the coordinates of the atoms. Figure 2-4(b) shows the four fundamental vibrations of the linear triatomic molecule—the symmetric stretch, the asymmetric stretch, and the bending vibrations in two mutually perpendicular planes. In Fig. 2-4(c), it can be seen that rotation about all three coordinate axes change the coordinates of the atoms. Finally, Fig. 2-4(d) shows the three fundamental vibrations of the nonlinear triatomic molecule—the symmetric stretch, the asymmetric stretch, and the (one) bending vibration in the plane of the molecule.

The data shown in Table 2-1 indicate that only chlorine and bromine, of the gases listed, reach the predicted value of $\frac{7}{2} R (29.2$ J-K^{-1}-mol$^{-1})$ and then only at 1000 K. Hydrogen and nitrogen at 300 K have heat capacities closer to $\frac{5}{2} R (20.8$ J-K^{-1}-mol$^{-1})$, the equipartition value predicted if there is no contribution from one of vibrational or rotational motion. These results will become clear when we discuss the quantum theory of heat capacities. Hydrogen has a heat capacity much lower than $\frac{5}{2} R$ at low temperatures. The most striking departure from the equipartition prediction is that the heat capacity does vary with the temperature. That the heat capacities of gases and solids were not accurately predicted for most materials by classical kinetic theory was one of the major problems that led to the development of the quantum theory. Nevertheless, the equipartition theorem accounts for a few simple situations and for the effect of molecular complexity on the high temperature heat capacity of gases.

Figure 2-4. (a) The two rotational degrees of freedom of a linear triatomic molecule; (b) the four vibrational degrees of freedom of a linear triatomic molecule (symmetric and asymmetric stretch; the two bending vibrations are shown in the *x-y* and *y-z* planes); (c) the three rotational degrees of freedom of a nonlinear triatomic molecule; (d) the three vibrational degrees of freedom of a nonlinear triatomic molecule (symmetric and asymmetric stretch and bend).

Table 2-1. Molar heat capacities of some diatomic gases

	T/K	$\mathscr{C}_v/(\text{J-mol}^{-1}\text{-K}^{-1})$
H_2	92	13.8[a]
	300	20.5[b]
	1000	21.8[b]
N_2	300	20.5[b]
	1000	24.3[b]
Cl_2	300	25.9[b]
	1000	29.3[b]
Br_2	300	28.0[b]
	1000	29.3[b]

$$\tfrac{7}{2}R = 29.2 \text{ J-K}^{-1}; \tfrac{5}{2}R = 20.8 \text{ J-K}^{-1}$$

[a] K. Scheel and W. Heuse, *Ann. Physik, 40*, 473 (1913).
[b] Calculated from the empirical equation $\mathscr{C}_v = a + bT + cT^2 - R$, with values of a, b, and c, taken from H. M. Spencer, *Ind. Eng. Chem., 40*, 2152 (1948).

2-4. Molecular size and molecular collisions

The simple molecular model that we have used so far is based on the assumption that the molecules of a gas are point masses. We have already seen that the treatment of molecules as point masses could not explain the heat capacities of polyatomic gases. Though the assumption of finite size is not required to explain the heat capacities of monatomic gases, the modern theory of atoms describes the electrons in an atom as occupying extended regions of space about the nucleus, so that even monatomic molecules cannot be regarded as point masses.

Collision frequency

The concept of molecules with a non-zero volume makes the kinetic model more complicated, since collisions among molecules can no longer be ignored. The added complication, however, extends considerably the range of phenomena about which predictions can be made. The simplest extension of the basic model is one in which a molecule is considered to be a rigid sphere of diameter σ. *A collision is defined as an event in which the centers of two identical molecules come within a distance σ of one another.* If one considers all the molecules but one to be at rest, then the moving molecule will collide with all those molecules with centers within a cylinder of radius σ about the path of the moving molecule (Fig. 2-5).

The average number of collisions per molecule per second is given by the number of molecules contained in the cylinder that is swept out by the moving molecule in one second. That is, $n_A \pi \sigma^2 \bar{u}$, where n_A is the number of molecules

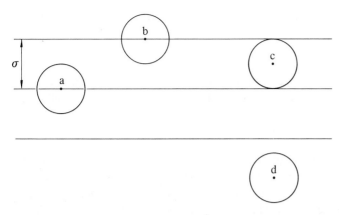

Figure 2-5. Collision path of a moving molecule a of diameter σ. It will undergo a glancing collision with b, a full collision with c, and no collision with d.

of gas A per unit volume, $\pi\sigma^2$ is the cross-sectional area of the cylinder, and $\bar{u}$ is the average speed of the moving molecule. A more complete treatment,[4] in which the relative velocities of the colliding molecules and the distribution of molecular velocities are considered, yields a factor of $\sqrt{2}$, giving $\sqrt{2}\,n_A\pi\sigma^2\bar{u}$. The total number of collisions per unit volume per unit time is $n_A/2$ times this expression, since each molecule undergoes the same number of collisions, but each collision involves two molecules. Thus, the average number of collisions per unit volume per unit time Z is

$$Z = \tfrac{1}{2}\sqrt{2}\pi\sigma^2\bar{u}\,n_A^{\,2} \tag{2-37}$$

Exercise 2-5. Calculate the number of collisions per cubic meter per second in a sample of O_2 at 273 K and a pressure of 1.01×10^5 Pa. Take the molecular diameter as 3×10^{-10} m. Use Eq. 2-64 to calculate $\bar{u}$.

Mean free path
The average distance traveled by a molecule between collisions is called the mean free path λ. The mean free path is given simply by the product of the average speed and the average time between collisions. That is,

$$\lambda = \bar{u}\,[1/(\sqrt{2}\,n_A\pi\sigma^2\bar{u})]$$
$$= 1/(\sqrt{2}\,n_A\pi\sigma^2) \tag{2-38}$$

Thus the mean free path depends only on the density of the gas. At constant density it does not depend on the temperature or pressure.

4. W. Kauzmann, *Kinetic Theory of Gases*, W. A. Benjamin, New York, 1966, pp. 175–78.

Exercise 2-6. Calculate the mean free path for the sample of O_2 described in Exercise 2-5.

The mean free path is a useful concept for the discussion of such important *transport properties* as the viscosity and the thermal conductivity of a gas.

Viscosity

To maintain a gas flow of constant velocity, a constant force must be applied. Since the force results in a steady velocity, rather than an acceleration, we can conclude from Newton's second law of motion that there is an equal opposing force in the system. The *viscosity* is a measure of this frictional force.

Consider a moving plane with a velocity **v** in the z direction in a gas. The plane is perpendicular to the x axis and parallel to a fixed plane, as indicated in Fig. 2-6.

It is found that the force required to maintain the velocity of the moving plane is proportional to the area A of the plane and the velocity gradient dv/dx along the x axis (dv/dx is assumed to be constant). The constant of proportionality depends on the nature of the gas, the temperature, and the pressure; it is called the viscosity η. These relationships are summarized in Eq. 2-39.

$$f = \eta A \frac{dv}{dx} \qquad (2\text{-}39)$$

Figure 2-6. The gradient in the x-direction of velocity in the z-direction maintained by a force $F = \eta\, A\,(dv_z/dx)$.

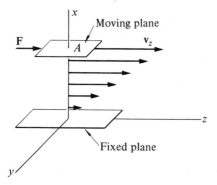

The random motion of the gas molecules along the x axis results in a net transfer of momentum (in the z direction), downward along the x axis. This transfer tends to eliminate the velocity gradient, and a force must be applied to maintain the velocity gradient.

To calculate the magnitude *of the momentum transfer* as simply as possible, let us assume that one-third of the molecules are moving in the x direction, one-third in the y direction, and one-third in the z direction, being only slightly perturbed by the directed motion in the z direction. One-sixth of the molecules are moving in the positive x direction and one-sixth in the negative x direction. Further, let us assume that all molecules are moving with the average speed $\bar{u}$. *Then the number passing through an area A at a height x in the positive x direction in unit time is the number contained in the volume A$\bar{u}$ that are moving in the positive x direction, or $\frac{1}{6}nA\bar{u}$, where n is the number of molecules per unit volume.* The same number per unit time is passing through the area A in the negative x direction. Let $v = f(x)$ be the velocity in the z direction of the molecules at the height x of area A. The molecules passing through A in the positive x direction have a velocity in the z direction characteristic of the layer in which they last encountered a collision, one mean free path below, at $x - \lambda$; that is, $v_+ = v_x - \lambda(dv/dx)$. Similarly, those molecules passing through A in the negative x direction have a velocity in the z direction, $v_- = v_x + \lambda(dv/dx)$. The corresponding momenta are

$$(mv)_+ = m[v_x - \lambda(dv/dx)]$$

and

$$(mv)_- = m[v_x + \lambda(dv/dx)]$$

The rate of momentum transfer is given by the number of molecules transferred per unit time, multiplied by the momentum transferred by each molecule. The net rate is then,

$$\frac{1}{6}nA\bar{u}\,(mv)_- - \frac{1}{6}nA\bar{u}\,(mv)_+$$

or

$$\frac{1}{6}nA\bar{u}\left[m\left(v_x + \lambda\frac{dv}{dx}\right) - m\left(v_x - \lambda\frac{dv}{dx}\right)\right]$$

or

$$\frac{1}{3}nA\bar{u}m\,\lambda\,\frac{dv}{dx}$$

But the rate of momentum transfer, according to Newton's second law, is just the frictional force. Thus,

$$f = \frac{1}{3} n A m \bar{u} \, \lambda \, \frac{dv}{dx} \qquad (2\text{-}40)$$

Equating the two expressions for f in Eq. 2-39 and Eq. 2-40 leads to an expression that relates the viscosity η of a gas to the density, the average speed, and the mean free path.

$$\eta A \, \frac{dv}{dx} = \frac{1}{3} n A m \bar{u} \, \lambda \, \frac{dv}{dx}$$

or

$$\eta = \frac{1}{3} n m \bar{u} \, \lambda \qquad (2\text{-}41)$$

A more rigorous derivation that takes into account the distribution of speeds among the molecules leads to the result[5]

$$\eta = \frac{5\pi}{32} n m \bar{u} \lambda \qquad (2\text{-}42)$$

Since $nm = \rho$, the density of the gas, Eq. 2-42 can be rewritten as

$$\eta = \frac{5\pi}{32} \rho \bar{u} \lambda \qquad (2\text{-}43)$$

At constant pressure, λ is proportional to T, ρ is inversely proportional to T, and $\bar{u}$ is proportional to $T^{\frac{1}{2}}$. Thus, from Eq. 2-43 we can predict that the viscosity of a gas is proportional to $T^{\frac{1}{2}}$. This temperatore dependence, the opposite of the commonly observed behavior of liquids, was predicted by Maxwell before experimental results had been obtained. The verification of this prediction was an early triumph of the kinetic theory model.

At constant temperature, the mean free path is inversely proportional to the pressure, and the density is directly proportional to the pressure, so that the viscosity of a gas should be independent of the pressure. This prediction too was found to be correct at low pressures.

Thermal conductivity

Consider a container of gas in which the two ends of the container are

5. W. Kauzmann, *Kinetic Theory of Gases*, W. A. Benjamin, New York, 1966, p. 202.

maintained at different temperatures, so that a linear temperature gradient dT/dz is set up. Under these conditions, it can be observed that the rate of heat flow dq/dt is given by the equation

$$\frac{dq}{dt} = \kappa A \frac{dT}{dz} \qquad (2\text{-}44)$$

where A is the cross-sectional area of the container perpendicular to the z axis, and κ is the thermal conductivity, a property of the gas.

The kinetic theory treatment of thermal conductivity is very similar to the treatment of gas viscosity, except that thermal energy, not linear momentum, is being transferred. As in the case of viscosity, the number of molecules moving in the positive z direction, through a reference plane of area A (at temperature, T), in unit time, is $\frac{1}{6} nA\bar{u}$, and the same number are moving through that plane in the negative z direction. The molecules moving in the positive z direction are characterized by a temperature $T - \lambda(dT/dz)$; those moving in the negative z direction by a temperature, $T + \lambda(dT/dz)$. The thermal energy carried by a molecule is equal to its heat capacity multiplied by the difference in temperature between the start and the finish of its mean free path. The net flow of thermal energy through the reference plane is

$$\frac{dq}{dt} = \frac{1}{6} nA\bar{u}c_v \left(T + \lambda \frac{dT}{dz} - T \right) - \frac{1}{6} nA\bar{u}\, c_v \left(T - \lambda \frac{dT}{dz} - T \right)$$

$$= \frac{1}{3} nA\bar{u}\, c_v \lambda \frac{dT}{dz} \qquad (2\text{-}45)$$

where c_v is the heat capacity per molecule. Equating the expressions for dq/dt in Eq. 2-44 and Eq. 2-45, one obtains

$$\kappa A \frac{dT}{dz} = \frac{1}{3} nA\bar{u}c_v \lambda \frac{dT}{dz}$$

or

$$\kappa = \frac{1}{3} n\bar{u}c_v \lambda \qquad (2\text{-}46)$$

In the exact expression, $\frac{1}{3}$ is replaced by $(25\pi/64)$.[6] Since n is directly proportional to the pressure and λ is inversely proportional to the pressure, Eq. 2-46 leads to the somewhat surprising conclusion that the thermal conductivity of a gas is independent of the pressure.

6. W. Kauzmann, *Kinetic Theory of Gases*, W. A. Benjamin, New York, 1966, p. 200.

Exercise 2-7. If the thermal conductivity of a gas is independent of the pressure, why is a vacuum flask evacuated to lower the thermal conductivity? Hints: What happens to the mean free path as the pressure is reduced? What happens when the mean free path becomes comparable to the distance between the walls and the pressure is reduced even more?

2-5. The Maxwell distribution of molecular speeds

The discussion thus far in the chapter has been based entirely on velocity averages for a collection of molecules, though we have mentioned that a distribution of molecular velocities exist. The nature of that distribution, named after James Clerk Maxwell, the British physicist who derived it in 1860, is given below.

Derivation

Each molecule in a sample of gas can be characterized at any instant of time by velocity components u_x, u_y, and u_z, and by a *speed* (or magnitude of velocity) u, such that (Eq. 2-10)

$$u^2 = u_x^2 + u_y^2 + u_z^2$$

Though individual molecules are continually changing velocity as a result of collisions, the *distribution of velocities* is an invariant characteristic of a given gas at a given temperature at equilibrium. The velocity components can be either positive or negative, but the speed is defined as the positive square root of u^2.

The probability of finding a molecule with an x-component of velocity between u_x and $u_x + \delta u_x$ is proportional to δu_x and depends on the value of u_x. Therefore, we express the probability as $f(u_x)\delta u_x$, with $f(u_x)$ a function to be determined. Similarly, the probability of finding a molecule with a y-component of velocity between u_y and $u_y + \delta u_y$ is $f(u_y)\delta u_y$, and the probability of finding a molecule with a z component of velocity between u_z and $u_z + \delta u_z$ is $f(u_z) \delta u_z$.[7]

Positive and negative velocity components are equally probable, since the motion is assumed to be random. The probability function must, therefore, be an even function of the velocity components. For simplicity, let us assume that

$$f(u_x) = g(u_x^2)$$

7. In this context, the expression "the probability of finding a molecule with an x component of velocity between u_x and $u_x + \delta u_x$" is equivalent to the fraction of molecules having x components of velocity between u_x and $u_x + \delta u_x$.

$$f(u_y) = g(u_y^2)$$

$$f(u_z) = g(u_z^2)$$

Since the motion of the molecules is assumed to be random, there is no preferred direction, and g must be the same function of the three velocity components. Also, since the three velocity components are independent, being mutually perpendicular, the probability of finding a molecule *with a x component of velocity between u_x and $u_x + \delta u_x$, a y component of velocity between u_y and $u_y + \delta u_y$, and a z component of velocity between u_z and $u_z + \delta u_z$ is the product of the independent probabilities*

$$g(u_x^2) \, g(u_y^2) \, g(u_z^2) \, \delta u_x \, \delta u_y \, \delta u_z$$

Exercise 2-8. Given a red die, a white die, and a green die: (a) What is the probability of obtaining a 5 on rolling the red die alone? (b) What is the probability of obtaining a 3 on rolling the white die alone? (c) What is the probability of obtaining a 4 on rolling the green die alone? (d) What is the probability of obtaining 5 on the red die, 3 on the white die, and 4 on the green die when all three dice are rolled, in a single roll? in succession?

Since the motion is random, the probability for all velocity vectors of the same magnitude, no matter what their direction, is the same. The probability is a function of u^2 only, and

$$g(u_x^2) \, g(u_y^2) \, g(u_z^2) \, \delta u_x \, \delta u_y \, \delta u_z = h(u^2) \, \delta u_x \, \delta u_y \, \delta u_z$$
$$= h(u_x^2 + u_y^2 + u_z^2) \, \delta u_x \, \delta u_y \, \delta u_z \quad (2\text{-}47)$$

Thus,

$$g(u_x^2) \, g(u_y^2) \, g(u_z^2) = h(u_x^2 + u_y^2 + u_z^2) \quad (2\text{-}48)$$

The only function that satisfies Eq. 2-48 is an exponential function; that is,

$$g(u_x^2) = A \exp(-Bu_x^2) \quad (2\text{-}49)$$

$$g(u_y^2) = A \exp(-Bu_y^2) \quad (2\text{-}50)$$

$$g(u_z^2) = A \exp(-Bu_z^2) \quad (2\text{-}51)$$

and

$$h(u^2) = A^3 \exp[-B(u_x^2 + u_y^2 + u_z^2)] \quad (2\text{-}52)$$

$$= A^3 \exp(-Bu^2) \quad (2\text{-}53)$$

where B is a positive quantity, and the negative sign is chosen so that the probability does not increase indefinitely with increasing u.

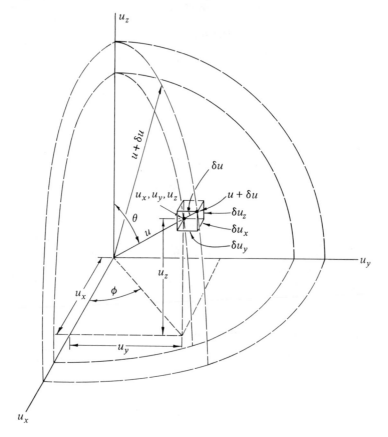

Figure 2-7. The Cartesian coordinates u_x, u_y, and u_z and the spherical polar co-ordinates u, θ, and ϕ of a velocity space, showing the volume element $\delta u_x\, \delta u_y\, \delta u_z$ and the spherical shell of volume $4\pi u^2\, \delta u$.

If we set up a Cartesian coordinate system such that the distances along the axes represent u_x, u_y, and u_z rather than x, y, and z, that coordinate system would represent a *velocity space*. Then $\delta u_x\, \delta u_y\, \delta u_z$ represents a volume element in such a space. Since $h(u^2)\, \delta u_x\, \delta u_y\, \delta u_z$ is a probability, one can say that $h(u^2)$ is a *probability density* in velocity space. (See Fig. 2-7.)

The constant A in Eq. 2-53 can be evaluated if we recall from the definition of probability that the sum of the probabilities of conceivable outcomes of an event is equal to one. The sum for a continuous function is the integral over all space. Thus,

$$1 = \int_{-\infty}^{\infty} \int_{-\infty}^{\infty} \int_{-\infty}^{\infty} h(u_x^2 + u_y^2 + u_z^2) \, du_x \, du_y \, du_z \tag{2-54}$$

Since h is an even function, the value of the integral is the same for all octants of the space. Thus,

$$1 = 8 \int_0^{\infty} \int_0^{\infty} \int_0^{\infty} h(u_x^2 + u_y^2 + u_z^2) \, du_x \, du_y \, du_z$$

$$= 8 \int_0^{\infty} \int_0^{\infty} \int_0^{\infty} A^3 \exp\left[-B(u_x^2 + u_y^2 + u_z^2)\right] du_x \, du_y \, du_z$$

$$= 8 \, A^3 \int_0^{\infty} \int_0^{\infty} \int_0^{\infty} \exp\left(-Bu_x^2\right) \exp\left(-Bu_y^2\right) \exp\left(-Bu_z^2\right) du_x \, du_y \, du_z \tag{2-55}$$

Because the integrand is a product of factors, each of which is a function of one independent variable only, Eq. 2-55 can be written as

$$1 = 8 \, A^3 \int_0^{\infty} \exp\left(-Bu_x^2\right) du_x \int_0^{\infty} \exp\left(-Bu_y^2\right) du_y \int_0^{\infty} \exp\left(-Bu_z^2\right) du_z \tag{2-56}$$

All three integrals are definite integrals of the same form, and their value is independent of the variable of integration, so that

$$1 = 8 \, A^3 \left[\int_0^{\infty} \exp\left(-Bw^2\right) dw \right]^3$$

and

$$A = \left[2 \int_0^{\infty} \exp\left(-Bw^2\right) dw \right]^{-1} \tag{2-57}$$

Tables of integrals give the value of the integral in Eq. 2-57 as $\frac{1}{2} (\pi/B)^{\frac{1}{2}}$. Thus,

$$A = \left(\frac{B}{\pi} \right)^{\frac{1}{2}} \tag{2-58}$$

Substituting in Eq. 2-53, we have

$$h(u^2) = \left(\frac{B}{\pi} \right)^{\frac{3}{2}} \exp\left(-Bu^2\right) \tag{2-59}$$

The constant B can be evaluated by calculating some average quantity that is known from simple kinetic theory, making use of the probability density $h(u^2)$. *The probability of finding a molecule with speed between u and $u + \delta u$ is the product of the probability density $h(u^2)$ and the volume of the spherical shell between u and $u + \delta u$, as illustrated in Fig. 2-7.*

Exercise 2-9. Consider the coordinate system u, θ, ϕ, a set of spherical coordinates in which

$$u_z = u \cos \theta$$

$$u_x = u \sin \theta \cos \phi$$

$$u_y = u \sin \theta \sin \phi$$

and the volume element is $u^2 \sin \theta \, \delta u \, \delta\theta \, \delta\phi$. Obtain the volume of the spherical shell between u and $u + \delta u$ by integrating the volume element between $\phi = 0$ and $\phi = 2\pi$ and between $\theta = 0$ and $\theta = \pi$.

We know that $\overline{u^2}$ is equal to $3\,kT/m$ from Eq. 2-21. To calculate $\overline{u^2}$ from the probability, the variable to be averaged u^2 is multiplied by its probability, $4\pi u^2 h(u^2)\delta u$, and the resulting expression is integrated over all possible values of the variables of integration.
Thus,

$$\overline{u^2} = 3\,\frac{kT}{m}$$

$$= \int_0^\infty u^2 h(u^2)\,(4\pi u^2)\,du$$

$$= \int_0^\infty 4\pi u^4 \left(\frac{B}{\pi}\right)^{\frac{3}{2}} \exp\left(-Bu^2\right) du$$

$$= \frac{4B^{\frac{3}{2}}}{\pi^{\frac{1}{2}}} \int_0^\infty u^4 \exp\left(-Bu^2\right) du$$

The latter integral can be evaluated from a table of integrals and has the value $(3/8B^2)\,(\pi/B)^{\frac{1}{2}}$.
Thus,

$$\overline{u^2} = \left(\frac{4B^{\frac{3}{2}}}{\pi^{\frac{1}{2}}}\right)\left(\frac{3\pi^{\frac{1}{2}}}{8B^{\frac{5}{2}}}\right)$$

$$= \frac{3}{2B} = \frac{3kT}{m} \qquad\qquad (2\text{-}60)$$

Therefore,

$$B = \frac{m}{2kT} \qquad\qquad (2\text{-}61)$$

By substitution in Eq. 2-59, we obtain

$$h(u^2) = \left(\frac{m}{2\pi k T}\right)^{\frac{3}{2}} \exp\left(-\frac{mu^2}{2kT}\right) \qquad\qquad (2\text{-}62)$$

Once B is evaluated, it is also possible to calculate other average values, such as the average speed $\bar{u}$.

$$\begin{aligned}
\bar{u} &= \int_0^\infty uh(u^2)(4\pi u^2)\,du \\
&= \int_0^\infty 4\pi u^3 \left(\frac{m}{2\pi kT}\right)^{\frac{3}{2}} \exp\left(-\frac{mu^2}{2kT}\right) du \\
&= 4\pi \left(\frac{m}{2\pi kT}\right)^{\frac{3}{2}} \int_0^\infty u^3 \exp\left(-\frac{mu^2}{2kT}\right) du
\end{aligned} \tag{2-63}$$

From tables of integrals, we find that the integral in Eq. 2-63 is $\frac{1}{2}(2kT/m)^2$, so that

$$\begin{aligned}
\bar{u} &= \frac{4\pi}{2}\left(\frac{m}{2\pi kT}\right)^{\frac{3}{2}}\left(\frac{2kT}{m}\right)^2 \\
&= \left(\frac{8kT}{\pi m}\right)^{\frac{1}{2}}
\end{aligned} \tag{2-64}$$

It can be seen that $\bar{u}^2$ is less than $\overline{u^2}$, as it should be.

From Eq. 2-62, then, the probability of finding a molecule with a speed between u and $u + \delta u$ is

$$4\pi u^2 \left(\frac{m}{2\pi kT}\right)^{\frac{3}{2}} \exp\left(-\frac{mu^2}{2kT}\right)\delta u$$

The probability of finding a molecule with speed between u and $u + \delta u$ is equal to the fraction of molecules with speed between u and $u + \delta u$. Thus,

$$\frac{\delta N_u}{N} = \frac{4}{\pi^{\frac{1}{2}}}\left(\frac{m}{2kT}\right)^{\frac{3}{2}} u^2 \exp\left(-\frac{mu^2}{2kT}\right)\delta u \tag{2-65}$$

where N_u is the number of molecules with speeds between zero and u, δN_u is the number of molecules with speeds between u and $u + \delta u$, and N is the total number of molecules. Eq. 2-65 is the *Maxwell distribution of molecular speeds*.

Graphical representations

A common graphical representation of the Maxwell distribution is shown in Fig. 2-8. *The ordinate is the function* $(4/\pi^{\frac{1}{2}})(m/2kT)^{\frac{3}{2}} u^2 \exp(-mu^2/2kT)$, *which, when multiplied by the velocity interval* δu, *gives the fraction of molecules with speeds between* u *and* $u + \delta u$. As can be seen from Eq. 2-65, this function is equal to $(1/N)(dN_u/du)$.

Fig. 2-8 also shows the effect of temperature and molecular mass on the

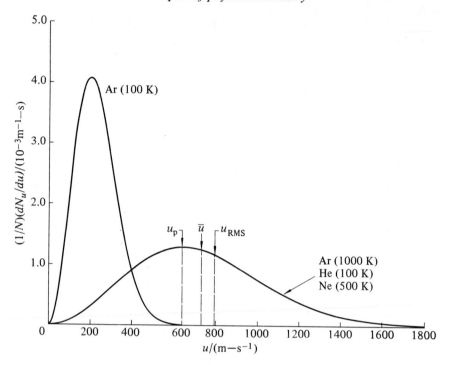

Figure 2-8. The Maxwell speed distribution for argon at 100 and 1000 K, helium at 100 K, and neon at 500 K.

shape of the distribution curve. *As would be expected from the dependence of the probability function on the ratio m/T the distribution is concentrated at low speeds both for low temperatures and for high molecular mass.*

The most probable speed can be obtained by finding the maximum value of the probability function. Thus,

$$\frac{d}{du}\left(\frac{1}{N}\frac{dNu}{du}\right) = \left(\frac{4}{\pi^{\frac{1}{2}}}\right)\left(\frac{m}{2kT}\right)^{\frac{3}{2}}\left[(2u)\exp\left(\frac{-mu^2}{2kT}\right)\right.$$

$$\left. + u^2\exp\left(\frac{-mu^2}{2kT}\right)\left(-\frac{mu}{kT}\right)\right]$$

$$= 0$$

$$2u_\mathrm{p} = \frac{mu^3{}_\mathrm{p}}{kT}$$

and

$$u_\mathrm{p} = \left(\frac{2kT}{m}\right)^{\frac{1}{2}} \tag{2-66}$$

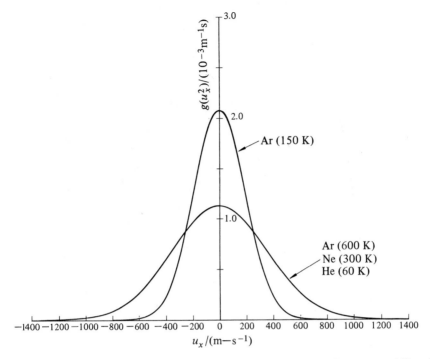

Figure 2-9. The Maxwell velocity distribution in one dimension for argon at 150 and 600 K, helium at 60 K, and neon at 300 K.

where u_p is the most probable speed.

The derivative is also equal to zero in the limit as u equals zero or ∞, but these are asymptotic minima and not maxima.

Exercise 2-10. Verify whether the extrema of the probability function are maxima or minima.

The most probable speed, the average speed, and the root-mean-square speed are indicated in Fig. 2-8.

The most probable value of the velocity components can be obtained by applying the same procedure to the probability function $g(u_x^2)$, which is, according to Eqs. 2-49, 2-58, and 2-61,

$$g(u_x^2) = \left(\frac{m}{2\pi kT}\right)^{\frac{1}{2}} \exp\left(-\frac{mu_x^2}{2kT}\right) \tag{2-67}$$

Thus,

$$\frac{dg}{du_x} = \left(\frac{m}{2\pi kT}\right)^{\frac{1}{2}} \exp\left(-\frac{mu_x^2}{2kT}\right)\left(\frac{mu_x}{kT}\right) = 0$$

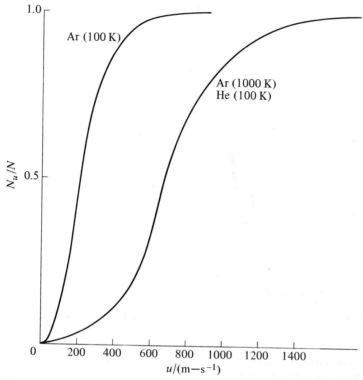

Figure 2-10. The cumulative Maxwell speed distribution for argon at 100 K and 1000 K and helium at 100 K.

so that

$$(u_x)_p = 0 \qquad (2\text{-}68)$$

The value of the derivative is also equal to zero in the limit of u_x equal to ∞ or $-\infty$, but these are asymptotic minima and not maxima. A graph of the probability function for u_x is shown in Fig. 2-9.

Though the most probable value for each velocity component is zero, there is a zero probability that all three components will be zero simultaneously for the same molecule. Thus, there is no inconsistency between a most probable *velocity component* that is zero and a most probable *speed* that is non-zero.

The probability of finding a molecule with speed less than u, or, what is equivalent, the fraction of molecules with speed less than u, is equal to the area under the curve in Fig. 2-8 between $u = 0$ and $u = u$. This can be demonstrated analytically by integrating the function $(1/N)(dN_u/du)$ from zero to u. That is,

$$\int_0^u \left(\frac{1}{N}\right)\left(\frac{dN_u}{du}\right)du = \frac{1}{N}\int_0^u \left(\frac{dN_u}{du}\right)du$$

$$= \frac{N_u}{N}$$

Figure 2-10 shows the curves for N_u/N for the same elements and temperatures as Fig. 2-8. The most probable speed and the general shape of the distribution curve are read more easily from Fig. 2-8, but both forms provide the same information.

It is only recently that it has been possible to verify directly the Maxwell

Figure 2-11. The intensity of a beam of thallium atoms as a function of speed. The points are experimental results, and the curve is calculated from the Maxwell speed distribution: 870 K and 0.43 Pa (o), u_p is equal to 376 m-sec^{-1}; 944 K and 2.8 Pa (▲), u_p is equal to 395 m-sec^{-1}. The corresponding calculated speeds are 376 m-sec^{-1} and 392 m-sec^{-1}, respectively. [By permission, from R. C. Miller and P. Kusch, *Phys. Rev.*, 99, 1314–21 (1955).]

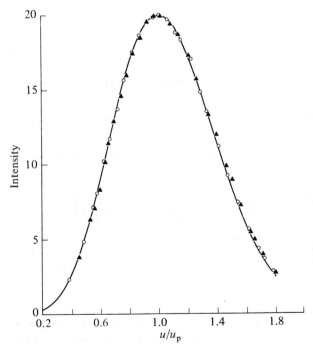

distribution of speeds with accuracy, though many deductions from the distribution have been verified. Figure 2-11 illustrates the results of experiments by Miller and Kusch for beams of thallium vapor at 870 K and 0.43 Pa and at 944 K and 2.8 Pa. They used as a velocity selector a rotating cylinder in which were cut 750 grooves at such an angle that only molecules within a narrow speed interval would reach the detector at the end of the cylinder.

The Distribution of kinetic energy

If the kinetic energy ε is substituted for the speed in Eq. 2-65, the distribution of kinetic energy among molecules is obtained. Since

$$u = \left(\frac{2\varepsilon}{m}\right)^{\frac{1}{2}}$$

and

$$\delta u = (2m\varepsilon)^{-\frac{1}{2}}\,\delta\varepsilon$$

the resulting equation is

$$\frac{\delta N_\varepsilon}{N} = \frac{4}{\pi^{\frac{1}{2}}}\left(\frac{m}{2kT}\right)^{\frac{3}{2}}\left(\frac{2\varepsilon}{m}\right)\exp\left(-\frac{\varepsilon}{kT}\right)(2m\varepsilon)^{-\frac{1}{2}}\,\delta\varepsilon$$

$$= 2\left(\frac{\varepsilon}{\pi}\right)^{\frac{1}{2}}\left(\frac{1}{kT}\right)^{\frac{3}{2}}\exp\left(-\frac{\varepsilon}{kT}\right)\delta\varepsilon \qquad (2\text{-}69)$$

The probability function for the energy distribution, $(1/N)\,(dN_\varepsilon/d\varepsilon)$ is plotted in Fig. 2-12.

As with the average kinetic energy of molecules, the distribution of kinetic energy depends only on the temperature and is independent of the molecular mass. The most probable kinetic energy is the energy at the maximum of the probability function. Thus,

$$\frac{d}{d\varepsilon}\left(\frac{1}{N}\frac{dN_\varepsilon}{d\varepsilon}\right) = \frac{d}{d\varepsilon}\left[2\left(\frac{\varepsilon}{\pi}\right)^{\frac{1}{2}}\left(\frac{1}{kT}\right)^{\frac{3}{2}}\exp\left(-\frac{\varepsilon}{kT}\right)\right]$$

$$= \frac{2}{\pi^{\frac{1}{2}}}\left(\frac{1}{kT}\right)^{\frac{3}{2}}\left[\varepsilon^{\frac{1}{2}}\exp\left(-\frac{\varepsilon}{kT}\right)\left(-\frac{1}{kT}\right)+\frac{1}{2}\varepsilon^{-\frac{1}{2}}\exp\left(-\frac{\varepsilon}{kT}\right)\right]$$

$$= 0$$

The derivative is equal to zero in the limit as $\varepsilon \to \infty$, but this is an asymptotic minimum. In the limit as $\varepsilon \to 0$, the slope approaches ∞, because of the $\varepsilon^{-\frac{1}{2}}$ term.

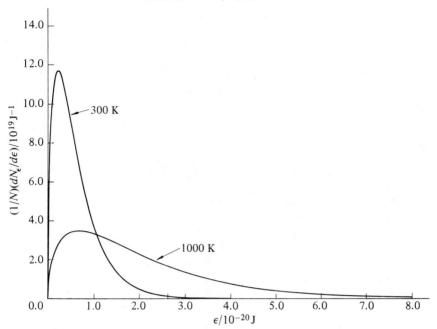

Figure 2-12. The Maxwell kinetic energy distribution at 1000 and 300 K.

The maximum occurs when

$$\varepsilon_p^{\frac{1}{2}} \exp\left(-\frac{\varepsilon_p}{kT}\right)\left(\frac{1}{kT}\right) = \frac{1}{2} \varepsilon_p^{-\frac{1}{2}} \exp\left(-\frac{\varepsilon_p}{kT}\right)$$

$$2\varepsilon_p^{\frac{1}{2}} = (kT)\,\varepsilon_p^{-\frac{1}{2}}$$

and

$$\varepsilon_p = \frac{kT}{2} \tag{2-70}$$

The average kinetic energy can be calculated from the distribution function, just as the average speed and the average value of u^2 were calculated. Thus,

$$\bar{\varepsilon} = \int_0^\infty \varepsilon \frac{2}{\pi^{\frac{1}{2}}}\left(\frac{1}{kT}\right)^{\frac{3}{2}} \varepsilon^{\frac{1}{2}} \exp\left(-\frac{\varepsilon}{kT}\right) d\varepsilon$$

$$= \frac{2}{\pi^{\frac{1}{2}}}\left(\frac{1}{kT}\right)^{\frac{3}{2}} \int_0^\infty \varepsilon^{\frac{3}{2}} \exp\left(-\frac{\varepsilon}{kT}\right) d\varepsilon \tag{2-71}$$

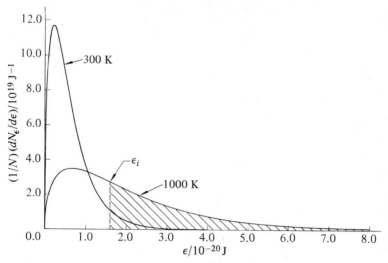

Figure 2-13.   The Maxwell distribution of kinetic energy, showing the effect of temperature on the fraction of molecules with $\varepsilon > \varepsilon_i$, where $\varepsilon_i = 1.6 \times 10^{-20}$ J.

From tables of integrals we find that the integral in Eq. 2-71 has the value $(3\pi^{\frac{1}{2}}/4)(kT)^{\frac{5}{2}}$ so that

$$\bar{\varepsilon} = \frac{2}{\pi^{\frac{1}{2}}}\left(\frac{1}{kT}\right)^{\frac{3}{2}}\left(\frac{3\pi^{\frac{1}{2}}}{4}\right)(kT)^{\frac{5}{2}}$$

$$= \frac{3}{2}kT \tag{2-72}$$

in agreement with the deduction from simple kinetic theory.

The fraction of molecules having a kinetic energy greater than some critical value is an important factor in the consideration of the temperature dependence of the rates of chemical and biological processes, the vapor pressure of liquids and solids, the viscosity of liquids, and the equilibrium constants for chemical reactions. This fraction can be calculated from the energy distribution formula.[8]

In fig. 2-13, the area under the curve to the right of the energy ε_i represents the fraction of molecules that have energy greater than ε_i. Though energy at the maxima at 300 K and 1000 K differ only by a factor of 3.3, it can be seen that the area in question is much greater at 1000 K than at 300 K.

8. W. Kauzmann, *Kinetic Theory of Gases*, W. A. Benjamin, New York, 1966, pp. 156–61.

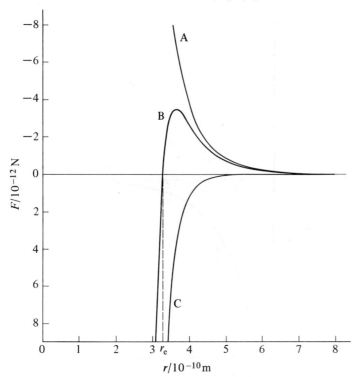

Figure 2-14. The intermolecular force-distance curves for hydrogen molecules. Curve A represents the attractive force, curve C represents the repulsive force, and curve B represents the net force. The vertical dashed line represents the equilibrium intermolecular distance, r_e, at which $F = 0$. (By permission, from E. D. Eastman and G. K. Rollefson, *Physical Chemistry*, McGraw-Hill Book Co., New York, 1947, p. 93.)

Exercise 2-11. The fraction of molecules with energy greater than ε_i is given approximately by the expression $\exp(-\varepsilon_i/kT)$, for $\varepsilon_i > \varepsilon_p$. Calculate the fraction of molecules with energy greater than 10^{-19} J at 300 K and 1000 K, and compare with the average energy at each of these temperatures.

2-6. Intermolecular forces.

In Sec. 1-3, it was pointed out that the empirical properties of gases, when investigated over a wide enough range of temperatures and pressures, indicate the existence of forces between gas molecules, both attractive forces and repulsive forces. Though the detailed discussion of these forces requires a basis in quantum mechanics, which we shall discuss in the next few chapters, we can present a common description of these forces in terms of intermolecular potential energy functions.

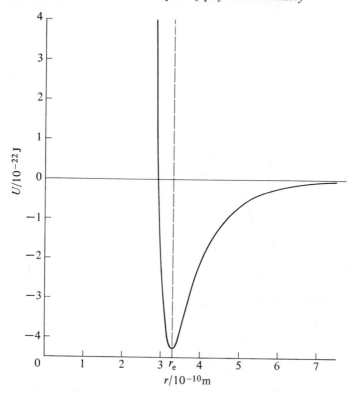

Figure 2-15. The potential energy-distance curve for hydrogen molecules. The vertical dashed line represents the equilibrium internuclear distance, r_e, at which the intermolecular potential energy is a minimum. (By permission, from E. D. Eastman and G. K. Rollefson, *Physical Chemistry*, McGraw-Hill Book Co., New York, 1947. p. 94.)

The basic relation between a force and the corresponding potential energy is given in Eq. 2-73 for a one-dimensional potential field and in Eq. 2-74 for a three-dimensional potential field.

$$|\mathbf{F}| = -\frac{du(r)}{d(r)} \tag{2-73}$$

$$F_x = -\frac{\partial u(x, y, z)}{\partial x}$$

$$F_y = -\frac{\partial u(x, y, z)}{\partial y} \tag{2-74}$$

$$F_z = -\frac{\partial u(x, y, z)}{\partial z}$$

where r is the distance between molecular centers, and x, y, and z are the components of the displacement between molecular centers. Thus, an increase of potential energy with increasing intermolecular distance represents a negative attractive force. A decrease in potential energy with increasing intermolecular distance represents a positive repulsive force. The relationship between the intermolecular force and the intermolecular potential energy for hydrogen molecules is shown in Figs. 2-14 and 2-15. The vertical dashed lines in each case represent the equilibrium intermolecular distance, the point at which the force is equal to zero and the potential energy is at a minimum. A common analytical representation of the curves in Figs. 2-14 and 2-15 is that proposed by Lennard-Jones and shown in Eqs. 2-75 and 2-76.

$$u(r) = - Ar^{-6} + Br^{-12} \tag{2-75}$$

$$|\mathbf{F}|(r) = - 6\,Ar^{-7} + 12\,Br^{-13} \tag{2-76}$$

It can be seen from the exponents of r in the equations that both the attractive and the repulsive forces are short range and that the repulsive forces are extremely short range. From the steepness of the repulsive portions of the force curve and the potential energy curve, it can be seen that a "molecular diameter" could be defined with reasonable precision, but such a quantity should not be interpreted as indicating a rigid boundary for the molecule.

Problems

2-1. Derive the fundamental equation of the simple kinetic theory for an ideal monatomic gas constrained to move in two dimensions. The appropriate experimental variables are the temperature T, the area A, and the surface pressure π, with dimensions of force per unit length. This model is applicable to some gases absorbed on solid surfaces. The relation of the temperature and the average kinetic energy should be used to relate the mechanical model to thermal variables.

2-2. What is u_{RMS} for argon in two dimensions at 100 K? at 1000 K? How do these values compare with those of argon at the same temperatures in three dimensions?

2-3. Calculate the equipartition value of the molar heat capacity at constant area for argon in two dimensions. What is the corresponding value for a diatomic gas?

2-4. Calculate the number of collisions per argon molecule per unit time at 300 K and a pressure of 1.00×10^5 Pa. Assume a value of 3.7×10^{-10} m for the collision diameter. Calculate the total number of collisions per unit time per unit volume. Calculate the mean free path.

2-5. Calculate the collision frequency per particle for a group of 1000 particles of mass 1×10^{-3} kg and diameter 1×10^{-2} m in a volume of 1 m^3 at 300 K. Calculate the mean free path. Is the latter large enough compared to the molecular diameter that the model is somewhat reasonable?

2-6.   The viscosity of argon at 273 K and 1.01×10^5 Pa is 20.99×10^{-6} kg-m^{-1}-s^{-1}. Estimate the hard-sphere collision diameter from these data.

2-7.   The thermal conductivity of argon at 273 K and 1.01×10^5 Pa is 9.42×10^{-4} J-m^{-1}-s^{-1}-K^{-1}. Estimate the hard-sphere collision diameter from these data.

2-8.   Derive an expression for the Maxwell speed distribution for a gas constrained to move in two dimensions.

2-9.   The following values represent the viscosity of argon at 1.01×10^5 Pa over a range of temperature. (NBS Circular 564, "Tables of Thermal Properties of Gases", 1955, p. 128.)

$\eta/(10^{-5}$ kg-s^{-1}-m^{-1})	0.830	1.243	1.625	1.974	2.294	2.590	2.867
T/K	100	150	200	250	300	350	400

Use these data to test the temperature dependence deduced from Eq. 2-42. Hint: Take logarithms of η and T and find the slope of the resulting straight line. For further discussion, see W. Kauzmann, *Kinetic Theory of Gases*, W. A. Benjamin, New York, 1966, pp. 213–16.

2-10. What is the prediction of Eq. 2-43 for the temperature dependence of the viscosity of a gas at constant density?

3 · Transition to quantum theory

During the latter half of the nineteenth century, a number of experimental observations were made that could not be explained by classical theory, either qualitatively or quantitatively. Among these were the temperature dependence of the heat capacity of molecules and solids (Sec. 2-3), the spectral distribution of energy in black-body radiation, the photoelectric effect, and the discrete spectra emitted by excited atoms. We shall consider each of these phenomena and the new interpretations provided by Planck, Einstein, and Bohr.

3-1. Black-body radiation[1]

It is common knowledge that the color of light emitted from a heated solid depends on the temperature and that the frequency of the light emitted shifts toward the blue region of the spectrum as the temperature increases. When such radiation is generated inside a black body (that absorbs radiation perfectly and emits it internally without any loss), it is found that the matter of the black body comes to equilibrium with the radiation, so that the distribution of energy with frequency is independent of the nature of the matter of the black body. This discovery, made by Kirchhoff in 1859, indicates that the phenomenon of black-body radiation illustrates some fundamental aspect of the interaction between matter and radiation.

We expect the energy density in a frequency interval of the spectrum from

1. H. A. Boorse and L. Motz, *The World of the Atom*, Vol. I, Basic Books, New York, 1966, pp. 462–501. Max Born's Obituary of Planck and Planck's Nobel Prize address.

v to $v + \delta v$ to depend on the temperature, on the frequency, and to be proportional to the frequency interval δv. The function $\rho(v, T)$ was defined so that $\rho(v, T)\delta v$ represents the energy density (SI units of J-m^{-3}) in the frequency interval from v to $v + \delta v$. The total energy density of radiation in the black body is given by $\int_0^\infty \rho\, dv$. Stefan discovered, experimentally, that this integral is proportional to T^4, a fact subsequently proved theoretically by Boltzmann. The proportionality constant σ is called the Stefan-Boltzmann constant.

Wien, basing his conclusions on an analogy with the Maxwell distribution of energy among gas molecules, suggested that ρ has the form

$$\rho = \alpha v^3 \exp\left(-\frac{\beta v}{T}\right) \tag{3-1}$$

As illustrated in Fig. 3-1, Wien's formula fits the experimental data at high frequencies but deviates considerably at low frequencies.

By treating the radiation in the cavity as a superposition of standing waves, Rayleigh and Jeans derived from classical electromagnetic theory an expression for the density of standing waves with frequency between v and $v + \delta v$ equal to $8\pi v^2 \delta v/c^3$. They concluded from classical kinetic theory that the average energy per standing wave must be equal to kT, the equipartition value for a vibrational degree of freedom. The combination led to the expression

$$\rho = \frac{8\pi v^2}{c^3} kT \tag{3-2}$$

Equation 3-2 fits the experimental data at low frequencies. At high frequencies, however, it leads to the "ultraviolet catastrophe." It predicts an energy density that increases without limit as frequency increases, thus contradicting the law of conservation of energy as well as the experimental data on black-body radiation.

Planck knew that the Wien and Rayleigh-Jeans expressions were adequate in different limited ranges of values of v. He sought a single expression that he could use for interpolation, one that would reduce to these equations in the proper range of values of v. Through a consideration of the relation between the energy and entropy of the radiation, Planck derived the expression

$$\rho = \alpha v^3 \left[\exp\left(\frac{\beta v}{T}\right) - 1\right]^{-1} \tag{3-3}$$

At high values of v, $\exp(\beta v/T)$ is large with respect to 1, and Eq. 3-3 reduces to

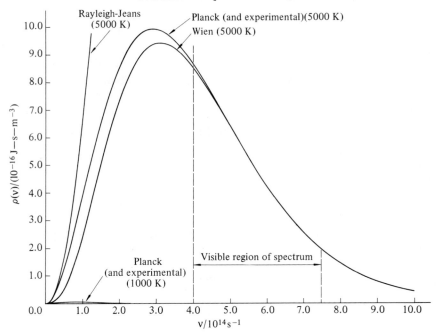

Figure 3-1. (a) The spectral distribution of energy density in black-body radiation, experimental and Planck's equation, compared to the Rayleigh-Jeans and Wien expressions at 5000 K, and showing the experimental curve at 1000 K. (b) A logarithmic plot of $\rho(v)$ against v at 5000 and 1000 K.

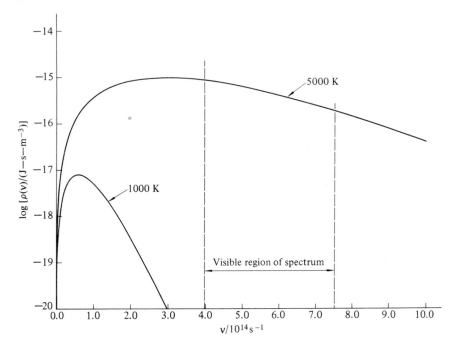

$$\rho = \alpha v^3 \exp\left(-\frac{\beta v}{T}\right)$$

the Wien formula, Eq. 3-1. The limiting form at small values of v can be obtained by expanding $\exp(\beta v/T)$ in a McLaurin series in $\beta v/T$ and neglecting higher powers.

$$\exp(\beta v/T) = 1 + (\beta v/T) + \frac{(\beta v/T)^2}{2!} + \ldots$$

so that

$$\rho = \frac{\alpha v^3}{1 + (\beta v/T) - 1} = \frac{\alpha}{\beta} v^2 T \qquad (3\text{-}4)$$

Comparing Eq. 3-4 with the Rayleigh-Jeans expression, Eq. 3-2, we see that they are the same if

$$\frac{\alpha}{\beta} = \frac{8\pi k}{c^3}$$

and

$$\alpha = \frac{8\pi k \beta}{c^3}$$

The Planck expression can then be written

$$\rho = \frac{8\pi k \beta}{c^3} \frac{v^3}{[\exp(\beta v/T)] - 1} \qquad (3\text{-}5)$$

Numerical values of k and β were obtained, using the precise data of Lummer and Pringsheim and Rubens and Kurlbaum. This was the first experimental determination of Boltzmann's constant.

Having obtained an expression for the energy distribution in black-body radiation, Planck was still faced with the problem of finding a theoretical model to explain the results. He accepted the Rayleigh-Jeans expression for the density of standing waves of $8\pi v^2 \, \delta v/c^3$, but he chose to calculate the average energy of the oscillators in the wall of the cavity that interact with the radiation rather than the average energy of the standing waves, having shown that the two must be equal at equilibrium. He did not accept the equipartition value of kT for the average energy, since this *frequency-independent* value led to the Rayleigh-Jeans expression.

Instead, he used a method suggested by Boltzmann to calculate the average energy of the oscillators, in which Boltzmann had assumed, as a temporary expedient in the calculation, that the energy of an oscillator takes on discrete values, integral multiples of a basic unit ε_0. (Once the computation of the average is completed, ε_0 can be allowed to become infinitesimally small, in agreement with the classical assumption that the energy of a mechanical system can vary continuously.)

Then Planck assumed that the probability of an oscillator having an energy ε_j is proportional to $\exp(-\varepsilon_j/kT)$, a form reminiscent of the exponential in the Maxwell distribution, Eq. 2-69. The probability is then equal to

$$\frac{\exp\left(-\varepsilon_j/kT\right)}{\sum\limits_j \exp\left(-\varepsilon_j/kT\right)}$$

and the average energy of the collection of oscillators is the sum of products in which each energy is multiplied by its probability. That is,

$$\bar{\varepsilon} = \frac{\sum \varepsilon_j \exp\left(-\varepsilon_j/kT\right)}{\sum \exp\left(-\varepsilon_j/kT\right)} \tag{3-6}$$

The resulting value for $\bar{\varepsilon}$ is (see Eq. 8-25)

$$\bar{\varepsilon} = \frac{\varepsilon_0}{\exp\left(\varepsilon_0/kT\right) - 1} \tag{3-7}$$

When Planck combined his expression for the average energy of the oscillators with the Rayleigh-Jeans expression for the density of standing waves, he obtained the following expression for ρ.

$$\rho = \frac{8\pi v^2}{c^3} \frac{\varepsilon_0}{[\exp\left(\varepsilon_0/kT\right)] - 1} \tag{3-8}$$

Equating ρ in Eq. 3-8 with ρ in Eq. 3-5, we can see that

$$\varepsilon_0 = k\beta v = hv \tag{3-9}$$

in which h is Planck's constant, the "quantum of action," with the dimensions (energy)(time). Equation 3-5 then becomes

$$\rho = \frac{8\pi h v^3}{c^3} \frac{1}{[\exp\left(hv/kT\right)] - 1} \tag{3-10}$$

If Planck had continued with Boltzmann's procedure, allowing h to approach zero, and the energy of an oscillator to vary continuously, Eq. 3-10 would have reduced to the Rayleigh-Jeans expression, which is valid only for small values of v. To avoid this, he made the revolutionary assumption that *the energy of an oscillator in a black body could change only by discrete amounts proportional to the frequency, and not continuously.*

3-2. The photoelectric effect

Though Planck was aware of the revolutionary implications of his treatment of the black-body radiation problem, he carefully restricted himself to the conclusion that the energy of the oscillators is quantized. When Einstein read Planck's papers, however, he leaped to the next step, that the radiation field itself is quantized, consisting of light quanta called photons.

In 1905 Einstein first applied the concept of quantized radiation to the photoelectric effect, in which a beam of light falling on a metal surface causes a beam of electrons to be ejected. Classical wave theory predicts that the energy of the electrons ejected from the metal depends on the intensity of the light beam. For a very low intensity of light, an electron would require considerable time to acquire enough energy from the wave to escape the metal surface. Lenard's experiments, however, clearly showed that the energy of the electrons ejected depends on the frequency but not the intensity of the incident light, that there is no detectable delay between the illumination of the surface by a low intensity light beam and the ejection of an electron and that there is a threshold frequency below which no electrons are ejected.

Figure 3-2. A scheme of the photoelectric effect experiment. The voltage at which the current through the galvanometer G is equal to zero is recorded.

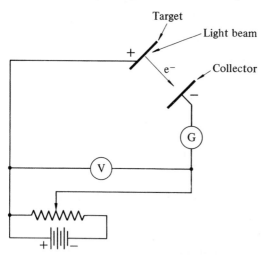

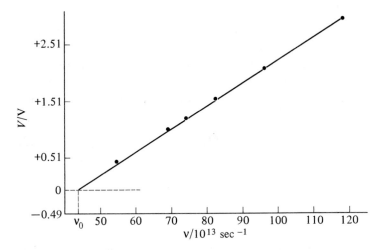

Figure 3-3. A sample of Millikan's data for the retarding potential required to prevent electron flow in the photoelectric experiment as a function of frequency of incident light. [*Phys. Rev.*, 7, 362 (1916).]

Einstein assumed that an electron at the surface of the metal can be removed with the expenditure of energy equal to W, the ionization energy of the solid. A photon with energy hv transfers that energy to an electron, and $hv - W$ is the kinetic energy with which the electron leaves the metal surface. If the frequency of the light is less than $v_0 = W/h$, no electrons will leave the surface.

Figure 3-2 shows schematically a device for determining the kinetic energy of electrons leaving a metal plate. The quantity measured is the retarding voltage V required to prevent an electron from reaching the collector plate. The electrical work done, equal in magnitude to the kinetic energy of the electrons, is Ve, where e is the magnitude of the charge of the electron. Thus,

$$Ve = \tfrac{1}{2} mv^2 = hv - W = hv - hv_0$$

Thus,
$$V = \frac{h}{e}v - \frac{h}{e}v_0 \tag{3-11}$$

In 1916, Millikan published the results of several years' experiments that had been designed to test Einstein's prediction. Some of his data are shown in Fig. 3-3. From the slope and the previously determined value of e, he obtained a value for h of 6.56×10^{-34} J-s. This compares well with the value of 6.60×10^{-34} J-s obtained by Planck from data on black-body radiation.

3-3. Heat capacities

Einstein saw that Planck's ideas could also be applied to the heat capacity problem. He visualized a monatomic solid as a collection of oscillators with three vibrational degrees of freedom. For such a system, as we have seen in Sec. 2-3, classical equipartition theory suggests an average energy of kT per degree of freedom, or $3kT$ per atom. Thus, for one mole of oscillators, the molar energy is

$$\mathscr{E} = 3LkT \tag{3-12}$$

and

$$\begin{aligned}\mathscr{C}_v &= 3Lk \\ &= 3R\end{aligned} \tag{3-13}$$

where R is the molar gas constant. This result is consistent with the empirical generalization of Dulong and Petit, which is valid only at high temperatures.

Einstein assumed that each atom in the solid vibrated with the same fundamental frequency, v_0, and that the energies of the atoms could only be integral multiples of the value $\varepsilon_0 = hv_0$. He was then able to use Planck's results for $\bar{\varepsilon}$, the average energy of an oscillator, to calculate the molar energy and the heat capacity. That is,

$$\begin{aligned}\mathscr{E} &= 3L\bar{\varepsilon} \\ &= 3L\frac{hv_0}{[\exp(hv_0/kT)] - 1}\end{aligned} \tag{3-14}$$

and

$$\begin{aligned}\mathscr{C}_v &= \left(\frac{\partial\mathscr{E}}{\partial T}\right)_v \\ &= 3Lk\left(\frac{hv_0}{kT}\right)^2 \frac{\exp(hv_0/kT)}{([\exp(hv_0/kT)] - 1)^2}\end{aligned} \tag{3-15}$$

Einstein's theory explained the examples then known in which the heat capacity of a solid is less than the classical value of $3R$ given by the law of Dulong and Petit. It also was used to predict the temperature dependence of the heat capacity down to 0 K, and these predictions were quickly confirmed in the low-temperature heat capacity measurements of Nernst and his students. Slight discrepancies at temperatures close to 0 K were resolved by refinements due to Debye and Born and von Karman. (See Sec. 8-3.)

The classical equipartition model assumes that each oscillator can absorb the energy kT at all temperatures, so that heat capacity should be inde-

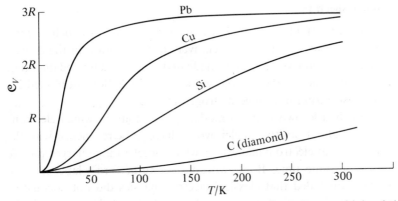

Figure 3-4. The molar heat capacity at constant volume (as a multiple of the gas constant R) of several solid elements (lead, copper, silicon, and carbon) as a function of temperature. (By permission, from E. D. Eastman and G. K. Rollefson, *Physical Chemistry*, McGraw-Hill Book Co., New York, 1947, p. 151.)

pendent of temperature. If, as Einstein suggested, the oscillators can only absorb energy in units of $h v_0$, then they absorb very little energy when kT is small compared to $h v_0$, and the heat capacity decreases with decreasing temperature. The form of the variation of heat capacity with temperature is the same for all substances, but the temperature at which the decrease in heat capacity becomes important depends on the value of v_0 for each substance.

The fundamental frequency of a harmonic oscillator is given by $v_0 = (1/2\pi)(k_f/m)^{\frac{1}{2}}$, where k_f is the Hooke's law constant of the oscillator, and m is the mass. Thus, as shown in Fig. 3-4, the more strongly bound the atoms in the solid, and the lighter the atoms, the higher the temperature at which the heat capacity decreases from the classical equipartition value.

Einstein's formula reduces to the classical result at temperatures that are high enough so that $h v_0$ is much less than kT. This can be seen by expanding the exponentials in Eq. 3-15 in the McLaurin series. Then,

$$\mathscr{C}_v = 3Lk \frac{[(h v_0/kT)^2 + (h v_0/kT)^3 + \ldots]}{[(h v_0/kT) + \frac{1}{2!}(h v_0/kT)^2 + \ldots]^2}$$

At sufficiently high values of T, $h v_0/kT$ is small enough that powers higher than two can be neglected, and

$$\mathscr{C}_v = 3Lk = 3R \tag{3-16}$$

in agreement with experiment and classical theory.

3-4. Atomic spectra [2]

Though the work of Planck and Einstein proved to be of epoch-making importance in physics, it attracted relatively little attention in the decade after publication. Little progress was made in developing a unified quantum theory along the lines of their initial work, until Niels Bohr published his paper on the spectrum of atomic hydrogen in 1913.

Rutherford had known that his model of the atom, in which electrons move in orbits about positive nuclei, would have to be reconciled with the models of classical electrodynamics, in which accelerating charges radiate energy. At the same time, Bohr, working in Rutherford's laboratory in Manchester, concluded that classical electrodynamics did not adequately describe systems of atomic dimensions. His conclusions led him to apply the quantum concepts of Planck and Einstein to atomic particles.

To follow Bohr's argument, consider an electron of mass m and charge $-e$ moving in a circular orbit about a proton of charge $+e$. By using Coulomb's law to express the magnitude of the force between the electron and the proton, and Newton's second law to equate that force to the product of the mass of the electron and its centripetal acceleration, we obtain the equation

$$\frac{e^2}{4\pi\varepsilon_0 r^2} = \frac{mv^2}{r} \tag{3-17}$$

in which ε_0 is the permittivity (dielectric constant) of free space, and r is the radius of the orbit of the electron. If both sides of Eq. 3-17 are multiplied by $r/2$, the expression

$$K = \frac{1}{2}mv^2 = \frac{e^2}{8\pi\varepsilon_0 r} \tag{3-18}$$

is obtained for the kinetic energy of the electron. Since the potential energy of the electron in the field of the nucleus is

$$U = \frac{-e^2}{4\pi\varepsilon_0 r} \tag{3-19}$$

its total energy, $E = K + U$, is given by

$$E = -\frac{e^2}{8\pi\varepsilon_0 r} = -K \tag{3-20}$$

2. H. A. Boorse and L. Motz, *The World of the Atom*, Vol. I, Basic Books, New York, 1966, pp. 734–65. Bohr's classic paper on atomic structure *Phil. Mag.*, 26, 1–19 (1926), with biographical notes and commentary.

We also know that the classical consideration of a particle of mass m rotating in a circle of radius r with a frequency f gives a kinetic energy of

$$K = \tfrac{1}{2} I (2\pi f)^2 \tag{3-21}$$

where I, the moment of inertia, equals mr^2. Thus,

$$E = -K = -(\tfrac{1}{2} mr^2)(2\pi f)^2$$
$$= -2m(\pi r f)^2 \tag{3-22}$$

(In more precise treatment, the motion of the proton and electron about the center of mass of the system would have to be considered, but the difference between these treatments is negligible for our purposes.)

Equations 3-20 and 3-22 describe classical orbits of the electron about the hydrogen nucleus. According to classical theory, the electron in its orbit should radiate large amounts of energy and spiral into the nucleus. Such radiation is not observed, and atoms are stable. Because the quantum constant h has the same dimensions as angular momentum, Bohr selected as the stable states of the electron those in which the angular momentum is an integral multiple of $h/2\pi$. This quantum condition yields the relation for the angular momentum M

$$M = mr^2 (2\pi f) = \frac{Nh}{2\pi} \quad N = 1, 2, 3, \ldots \tag{3-23}$$

This gives

$$f = \frac{Nh}{m(2\pi r)^2}$$

and

$$f^2 = \frac{N^2 h^2}{m^2 (2\pi r)^4} \tag{3-24}$$

Substituting for f^2 from Eq. 3-24 and for r from Eq. 3-20 into Eq. 3-22, we obtain

$$E = -\frac{me^4}{8\varepsilon_0^2 h^2 N^2} \tag{3-25}$$

Thus, Bohr had described the stationary states of an electron in a hydrogen atom entirely in terms of known constants. He also assumed that the monochromatic radiation emitted from excited hydrogen atoms results from the transition between stationary states and that frequency of the emitted radiation is related to the energy difference by the Bohr frequency relation

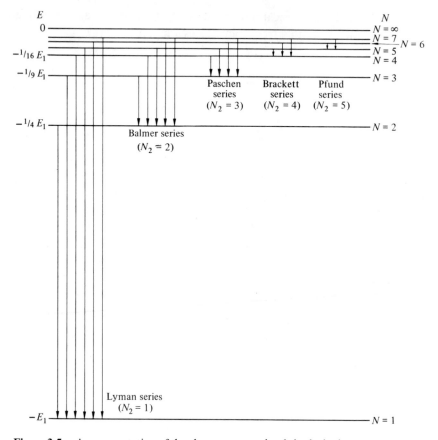

Figure 3-5.   A representation of the electron energy levels in the hydrogen atom, showing the transitions associated with the spectral series.

$$hv = \Delta E \qquad (3\text{-}26)$$

Combining Eqs. 3-25 and 3-26, we obtain

$$v = \frac{\Delta E}{h}$$

$$= -\frac{me^4}{8\varepsilon_0^2 h^3} \left(\frac{1}{N_1^{\,2}} - \frac{1}{N_2^{\,2}} \right)$$

$$= \frac{me^4}{8\varepsilon_0^2 h^3} \left(\frac{1}{N_2^{\,2}} - \frac{1}{N_1^{\,2}} \right), N_1 > N_2 \qquad (3\text{-}27)$$

As Bohr knew, Balmer had described the frequencies of the lines in the visible spectrum of hydrogen in 1885, using the empirical expression

$$v = 3.291 \times 10^{15} \left(\frac{1}{4} - \frac{1}{N^2} \right) \quad N = 3, 4, \ldots \quad (3\text{-}28)$$

The constant multiplier of

$$\left(\frac{1}{4} - \frac{1}{N^2} \right)$$

calculated from Bohr's formula with values of the fundamental constants available in 1913 was 3.1×10^{15}. Later refinement of the values of these constants led to agreement within the very small experimental error of spectroscopic measurements.

With the expression given in Eq. 3-27, Bohr was able to account for the Balmer series in the visible region with $N_2 = 2$, for the Paschen series in the infrared with $N_2 = 3$, and he predicted additional series in the ultraviolet with $N_2 = 1$ and in the infrared with $N_2 > 3$, all of which were later observed. Figure 3-5 shows the energy levels of the hydrogen atom and the transitions that give rise to the various spectroscopic series.

Bohr's model could be used to predict the frequencies of the spectral lines of hydrogen, to obtain a reasonable value for the size of the hydrogen atom, and to calculate the ionization energy of hydrogen. Yet, the quantization condition is an arbitrary one, and the lowest energy state of hydrogen is one with zero angular momentum rather than $h/2\pi$. It was ten years after Bohr's paper that these difficulties were resolved by the wave mechanics of de Broglie, Schrödinger, Heisenberg, and Born.

3-5. Wave-particle duality[3]

Planck was able to derive an exact description of the spectral energy distribution of black-body radiation by assuming that the oscillators in the black body could exchange energy with radiation only in integral multiples of hv_0. Einstein was able to describe the photoelectric effect exactly by assuming that light consists of particles called photons of energy hv. He was also able to describe the temperature dependence of the heat capacity of crystals by assuming that the atomic oscillators could exchange thermal energy only in multiples of hv_0. Bohr was able to calculate the frequencies

3. H. A. Boorse and L. Motz, *The World of the Atom*, Vol. II, Basic Books, New York, 1966, pp. 1041–59. De Broglie's Nobel Prize address and biographical notes and commentary.

of the spectral lines of hydrogen by assuming that the angular momentum of the electron in the atom must be an integral multiple of $h/2\pi$. In all these cases, explanations were provided for problems that had eluded classical solutions, with their continuous variation of energy, of momentum, and of the wave motion of light. Yet, the explanations were incomplete, and they provided no coherent framework for physical theory to take the place of the classical theory, which they contradicted. Louis de Broglie provided the basis for this framework in 1924 when he suggested that material particles, like light, should exhibit both particle and wave properties.

For every material particle, there should be an "associated wave," with an appropriate frequency and wavelength. There should be a relation between the particle and wave characteristics of all matter similar to Einstein's relation, $E = h\nu$, which relates the wave and particle properties of radiation.

De Broglie assumed that there is a wave motion associated with all free particles and that it could be described by such a function as

$$\Psi_0 = A_0 \sin 2\pi\nu_0 \, t_0 \tag{3-29}$$

where A_0 is the amplitude of the wave, ν_0 is its frequency, and t_0 is the time, all within a frame of reference *fixed* with respect to the particle.[4] He associated with the wave motion an energy, $E = h\nu_0$, which he equated to $m_0 c^2$ as in the Einstein mass-energy equivalence, where m_0 is the rest mass of the particle and c is the speed of light. Thus, he obtained

$$\nu_0 = \frac{m_0 c^2}{h} \tag{3-30}$$

In order to describe the wave motion as seen by an observer in a frame of reference moving at a constant velocity v, with respect to the particle, De Broglie applied the Lorentz-Einstein transformation of special relativity theory to the terms of Eq. 3-29. The time t in the observer's frame of reference is related to t_0 by

---

4. Alternative possible representations include

$$\Psi_0 = A_0 \cos 2\pi\nu_0 \, t_0$$

and the linear combination

$$\Psi_0 = A_0 [\cos 2\pi\nu_0 \, t_0 - i \sin 2\pi\nu_0 \, t_0]$$
$$= A_0 \exp [-i(2\pi\nu_0 \, t_0)]$$

where $i = \sqrt{-1}$.

$$t_0 = \frac{t - (vx/c^2)}{(1 - v^2/c^2)^{\frac{1}{2}}} \tag{3-31}$$

where x is the position of the particle as measured in the observer's frame of reference. Substituting Eq. 3-31 in Eq. 3-29, we obtain

$$\Psi = A_0 \sin 2\pi v_0 \left(t - \frac{vx}{c^2} \right) \left(1 - \frac{v^2}{c^2} \right)^{-\frac{1}{2}} \tag{3-32}$$

A wave moving through space can be described by the expression

$$\Psi = A_0 \sin 2\pi v \left(t - \frac{x}{u} \right) \tag{3-33}$$

where v is the frequency of the wave, and u is the *phase velocity* of the wave, the speed at which an individual point on the wave moves through space (usually taken as a peak or a trough for convenience).[5] If we compare Eq. 3-33 with Eq. 3-32, we see that the frequency seen by the observer is related to the frequency in the frame of reference of the moving particle by

$$v = v_0 \left(1 - \frac{v^2}{c^2} \right)^{-\frac{1}{2}} \tag{3-34}$$

and the phase velocity is

$$u = \frac{c^2}{v} \tag{3-35}$$

Exercise 3-1. Show that Eq. 3-34 and Eq. 3-35 follow from Eq. 3-32 and Eq. 3-33.

Equations 3-30 and 3-34 combine to give

$$v = \left(\frac{m_0 c^2}{h} \right) \left(1 - \frac{v^2}{c^2} \right)^{-\frac{1}{2}} \tag{3-36}$$

5. The complex exponential alternative to the representation in Eq. 3-32 is

$$\Psi = A_0 \exp \left[-2\pi i v_0 \left(t - \frac{vx}{c^2} \right) \left(1 - \frac{v^2}{c^2} \right)^{\frac{1}{2}} \right]$$

and for Eq. 3-33

$$\Psi = A_0 \exp \left[-2\pi i v \left(t - \frac{x}{u} \right) \right]$$

Special relativity theory relates the mass m of a moving particle to the rest mass m_0 by

$$m = m_0 \left(1 - \frac{v^2}{c^2} \right)^{-\frac{1}{2}}$$

(3-37)

so that

$$v = \frac{mc^2}{h}$$

(3-38)

describes the frequency of the wave associated with a moving particle. Since

$$u = v\lambda$$

(3-39)

relates the phase velocity u to the frequency v and the wavelength λ for any wave, substitution of Eq. 3-38 in Eq. 3-39 yields

$$u = \frac{mc^2 \lambda}{h}$$

$$= \frac{c^2}{v}$$

Thus,

$$\lambda = \frac{h}{mv} = \frac{h}{p}$$

(3-40)

Equation 3-40 is the de Broglie relation between the wavelength associated with a material particle and the momentum of the particle.

Exercise 3-2. Calculate the velocity at which an electron would have a wavelength of 1×10^{-10} m. What is the kinetic energy in joules? How large is the factor $(1 - v^2/c^2)$ in this case?

Exercise 3-3. Calculate the wavelength of a 1-kg particle moving with a velocity of 100 m-s^{-1}. Why are no diffraction effects observed with such a particle?

Since v must be less than c, which is considered an upper limit to the speed of material particles, it can be seen from Eq. 3-35 that u, the phase velocity, is greater than c. Another velocity associated with wave motion is the *group velocity*, V, the velocity with which a beat or modulation moves in a group of waves that differ slightly in frequency.

The group velocity is defined as

$$V = \frac{dv}{d(1/\lambda)}$$

(3-41)

which can be expressed as

$$V = \frac{dv/d\upsilon}{d(1/\lambda)/d\upsilon} \tag{3-42}$$

Differentiation of Eq. 3-36 gives

$$\frac{dv}{d\upsilon} = \frac{m_0 \upsilon}{h}\left(1 - \frac{\upsilon^2}{c^2}\right)^{-\frac{3}{2}} \tag{3-43}$$

From Eqs. 3-40 and 3-37, we see that since

$$\frac{1}{\lambda} = \frac{m_0 \upsilon}{h}\left(1 - \frac{\upsilon^2}{c^2}\right)^{-\frac{1}{2}}$$

$$\frac{d(1/\lambda)}{d\upsilon} = \frac{m_0}{h}\left(1 - \frac{\upsilon^2}{c^2}\right)^{-\frac{3}{2}} \tag{3-44}$$

Exercise 3-4. Verify Eq. 3-43 and Eq. 3-44.

Thus,

$$V = \upsilon \tag{3-45}$$

so the group velocity of the waves also equals the velocity of the particles.

Exercise 3-5. Calculate the phase velocity of the electron discussed in Exercise 3-2.

Shortly after de Broglie's work appeared, it was confirmed by experiments published by Davisson and Germer in the United States and by G. P. Thomson in Scotland. Davisson and Germer directed beams of electrons at thin nickel crystals and detected diffraction patterns from which they calculated wavelengths for the electron in agreement with de Broglie's equation. Thomson passed electron beams through powdered solids and obtained diffraction patterns similar to those found by Debye and Scherrer with X-rays.

If de Broglie's concept is valid, we should be able to apply it to the photon as well as to other particles and calculate the momentum of a photon. The momentum p is defined classically as the product $m\upsilon$, so from Eq. 3-37,

$$p = m_0 \upsilon \left(1 - \frac{\upsilon^2}{c^2}\right)^{-\frac{1}{2}} \tag{3-46}$$

For a photon, $\upsilon = c$, and Eq. 3-46 implies that the limit of p as υ approaches c is infinite; the limit can be finite only if m_0, the rest mass of the photon, is zero. Since Eq. 3-46 is an indeterminate form, an alternative way to calculate the momentum must be found. If the value of m from Eq. 3-38 and the value of $v = c$ are substituted in the definition of momentum,

$$p = mv = \frac{hv}{c} = \frac{h}{\lambda} \qquad (3\text{-}47)$$

Equation 3-47 was verified in the year de Broglie published his work, when Arthur Compton reported the results of his experiments on the scattering of X-rays by electrons. He found that the frequency of the scattered X-rays depended on the angle of scatter. He was also able to account for the angular dependence of the frequency by assuming conservation of the energy of the incident photon,

$$hv = hv' + \tfrac{1}{2}\,mv^2$$

and conservation of the momentum of the incident photon,

$$\left(\frac{hv'}{c}\right) \sin \theta = mv \sin \phi$$

and

$$\left(\frac{hv'}{c}\right) \cos \theta + mv \cos \phi = \frac{hv}{c}$$

where v' is the frequency of the X-ray scattered at angle θ, and v is the velocity of the electron scattered at angle ϕ (Fig. 3-6).

De Broglie's concept can also be applied to Bohr's quantization conditions for the electron in the hydrogen atom. If it is assumed that the wave associated with an electron moving in a stable orbit about the nucleus is a standing wave, in which successive cycles reinforce one another, we have the relation

$$2\pi r = n\lambda$$

Figure 3-6. A scheme of the Compton effect. The X-ray beam scattered in the collision of an X-ray photon with a stationary electron has a frequency less than that of the incident beam, which is a function of the scattering angle, θ.

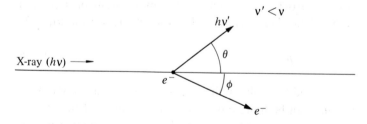

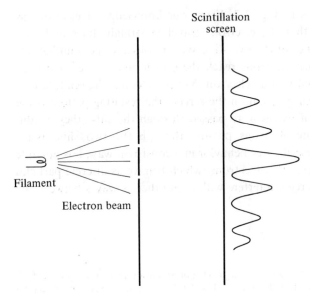

Figure 3-7. The interference produced when a beam of electrons passes through two slits, as evidenced by the variation in beam intensity with position on a plane behind the slits.

since the condition for reinforcement is that the circumference of the circular orbit be an integral multiple of the wavelength.

Then

$$2\pi r = n\lambda$$

$$= \frac{nh}{mv} \tag{3-48}$$

and

$$mvr = n\left(\frac{h}{2\pi}\right) \tag{3-49}$$

where mvr is the angular momentum of the electron.

The "mystery" of wave-particle duality, as Feynman described it,[6] can be seen clearly in a hypothetical analogue of the electron diffraction experiment. Consider a system in which a beam of electrons is passed through two closely

6. R. P. Feynman, R. B. Leighton, and M. Sands, *The Feynman Lectures on Physics*, Addison-Wesley, Vol. III, Reading, 1965, p. 1-1.

spaced narrow slits, as in Fig. 3-7. From our knowledge of electrons as particles, we assume that the electrons travel in straight lines and pass through one or the other of the two slits. We can observe the scintillations that are produced when electrons reach the screen, and we see that each scintillation is produced by one electron. Yet if we count the scintillations as a function of vertical position on the screen, the resulting pattern is the one we would expect if waves were passed through the slits; they exhibit interference among the electrons passing through the two slits. Wave mechanics is able to describe this behavior in a consistent way; it is not able to explain the "mystery," that electrons, which behave as discrete particles when they strike the screen, interfere with each other as waves between the slits and the screen.

Problems

3-1. The Wien expression for $\rho(v, T)$ is a good representation at high frequency. Calculate the value of ρ at 1000 K and $v = 1 \times 10^{13} s^{-1}$ (in the ultraviolet), using the Wien expression and the Rayleigh-Jeans expression. Are the units correct for an energy density per unit frequency interval? How do the two values compare at a frequency of $3 \times 10^{14} s^{-1}$ at 1000 K?

3-2. Millikan found the threshold frequency for the emission of photoelectrons from metallic sodium to be $43.9 \times 10^{13} s^{-1}$. Calculate the ionization energy of the metal. He found that the slope of the voltage-frequency line is 4.11×10^{-15} V-s. Compute the value of h.

3-3. The characteristic frequency v_0 for copper in the Einstein model is $6.25 \times 10^{12} s^{-1}$ and that for carbon (diamond) is $3.61 \times 10^{13} s^{-1}$. Calculate the value of T at which $hv_0 = kT$ for copper and diamond, and compare with the results in Fig. 3-4.

3-4. Calculate a value of the constant in Balmer's equation (Eq. 3-28), using Bohr's relationship (Eq. 3-27).

4 · Waves mechanics and atomic structure

De Broglie's suggestion was the clue to the development of a comprehensive theory of wave mechanics, but it did not provide a general method for the solution of physical problems comparable to that of classical mechanics, nor was the physical significance of the wave associated with particles clear. Erwin Schrödinger dealt with the first of these difficulties when in 1926 he began publishing a series of papers on the Schrödinger wave equation and some solutions to it.[1] Max Born, in the same year, clarified the second by his probability interpretation of the waves.

4-1. Schrödinger's wave mechanics

We shall use Schrödinger's formulation as a set of postulates from which the properties of matter on the atomic scale can be derived. The equations that follow, however, may make the form chosen for the equation more plausible. For a particle moving in the x direction consider a wave function of the form

$$\Psi = \exp\left[-2\pi i\left(vt - \frac{x}{\lambda}\right)\right] \tag{4-1}$$

Exercise 4-1.   Verify that Eq. 4-1 is equivalent to the exponential form in footnote 5, Chapter 3 (p. 79).

---

1. Werner Heisenberg developed an alternative mathematical formulation of quantum mechanics, matrix mechanics, at the same time Schrödinger carried out his work on wave mechanics. Schrödinger later showed that the two methods were mathematically equivalent. We shall limit our discussion to the Schrödinger formulation.

Differentiation of Eq. 4-1 with respect to x at constant t yields

$$\left(\frac{\partial \Psi}{\partial x}\right)_t = \frac{2\pi i}{\lambda} \exp\left[-2\pi i\left(vt - \frac{x}{\lambda}\right)\right] \tag{4-2}$$

$$= \left(\frac{2\pi i}{\lambda}\right)\Psi \tag{4-3}$$

If the de Broglie expression for λ is substituted in Eq. 4-3, we obtain

$$\left(\frac{\partial \Psi}{\partial x}\right)_t = \frac{2\pi i p_x}{h}\Psi$$

or

$$\frac{h}{2\pi i}\left(\frac{\partial \Psi}{\partial x}\right)_t = p_x \Psi \tag{4-4}$$

where p_x is the x component of the momentum. Differentiation of Eq. 4-1 with respect to t at constant x yields

$$\left(\frac{\partial \Psi}{\partial t}\right)_x = -2\pi i v \exp\left[-2\pi i\left(vt - \frac{x}{\lambda}\right)\right]$$

$$= -2\pi i v \Psi \tag{4-5}$$

If the Einstein expression for v is substituted in Eq. 4-5, we obtain

$$\left(\frac{\partial \Psi}{\partial t}\right)_x = \frac{-2\pi i E}{h}\Psi$$

or

$$-\frac{h}{2\pi i}\left(\frac{\partial \Psi}{\partial t}\right)_x = E\Psi \tag{4-6}$$

Equations 4-4 and 4-6 can be interpreted in terms of the *operators* Schrödinger used to transform classical expressions about particles into the equations of wave mechanics. From Eq. 4-4 we see that the operator $(h/2\pi i)$ $(\partial/\partial x)_t$ acting on Ψ yields the momentum p_x multiplied by Ψ, and $(h/2\pi i)$ $(\partial/\partial x)_t$ is called the momentum operator $\hat{p}$. Similarly, in Eq. 4-6, the operator $(-h/2\pi i)(\partial/\partial t)_x$ acting on Ψ yields the energy E multiplied by Ψ, and $(-h/2\pi i)(\partial/\partial t)_x$ is called the energy operator $\hat{E}$.

Exercise 4-2. Calculate the derivatives $(\partial^2 \Psi/\partial t^2)_x$ and $(\partial^2\Psi/\partial x^2)_t$ for the function defined in Eq. 4-1, and show that Ψ is a solution of the classical wave equation

$$\left(\frac{\partial^2 \Psi}{\partial x^2}\right)_t = \frac{1}{u^2}\left(\frac{\partial^2 \Psi}{\partial t^2}\right)_x$$

where u is the phase velocity of the wave.

The Schrödinger postulates

The postulates of wave mechanics can be derived heuristically by using de Broglie's relation in classical wave equations. They cannot be proved but must be judged by the agreement between experimental data and the predictions derived from the Schrödinger postulates. Wave mechanics has, in fact, been successfully used as a basis for predicting a wide range of atomic and nuclear phenomena. Even classical mechanics appears from these equations for the conditions under which classical mechanics is valid.

POSTULATE I   For every state of a physical system, a function Ψ of the coordinates and the time can be written. Knowledge of this function permits a complete description of the state of the system. This function must be a 'well-behaved'' function, in that it must be finite, continuous, single-valued, and have a continuous first derivative.[2] The function can be real or complex.

POSTULATE II   The product of Ψ and Ψ^*, where Ψ^* is the complex conjugate of Ψ, represents a *probability density* for a given state of the system such that $\Psi\Psi^*\, dx\, dy\, dz$ represents the probability of finding a particle in the volume element $dx\, dy\, dz$. The product $\Psi\Psi^*$ is taken so that the probability is a real number. Since $\Psi\Psi^*$ represents a probability density, it is necessary that

$$\int_{-\infty}^{\infty}\int_{-\infty}^{\infty}\int_{-\infty}^{\infty} \Psi\Psi^*\, dx\, dy\, dz = 1 \qquad (4\text{-}7)$$

Equation 4-7 is called the *normalization* condition. It can be seen from Eq. 4-7 that Ψ has the dimensions of $(V)^{-\frac{1}{2}}$.

The interpretation of $\Psi\Psi^*$ as a probability is one of Born's major contributions to quantum mechanics.

POSTULATE III   Corresponding to every observable $F(x, y, z, p_x, p_y, p_z)$, there is a linear operator $\hat{F}$ such that the average of F over a large number of measurements made when a particle is in the state described by Ψ is given by

$$\langle F \rangle = \int \Psi^* \hat{F} \Psi\, dx\, dy\, dz \qquad (4\text{-}8)$$

where $\langle F \rangle$ is the *expectation value*, and the integration is taken over the whole range of each variable.

To obtain the operator $\hat{F}$ for the variable F, the position variables of F are replaced by the corresponding position operators that represent multipli-

---

2. The requirement of a continuous first derivative does not apply to certain problems in which the boundary conditions impose a discontinuity. There are also cases in which Ψ is not finite, but we shall not consider them. See L. Pauling and E. B. Wilson, *Introduction to Quantum Mechanics*, McGraw-Hill Book Co., New York, 1935, p. 58.

cation by the position variable. The momentum variables are replaced by $(h/2\pi i)\,(\partial/\partial x)$, $(h/2\pi i)\,(\partial/\partial y)$, or $(h/2\pi i)\,(\partial/\partial z)$. For the energy, E, the corresponding operator is $(-h/2\pi i)\,(\partial/\partial t)$. The operations of multiplication and differentiation so defined are applied to a function Ψ.

If, when an operator $\hat{F}$ is applied to a function Ψ, the result is a constant multiple of Ψ, the constant is called an *eigenvalue* or *characteristic value* of F for the state described by the function Ψ, which is called an *eigenfunction* or *characteristic function* of F. When

$$\hat{F}\Psi_n = F_n\Psi_n$$

and F_n is a constant, then F_n is an eigenvalue, so Eq. 4-8 can be written as

$$
\begin{aligned}
\langle F \rangle &= \int \Psi_n{}^* \hat{F}\Psi_n \, dx \, dy \, dz \\
&= \int \Psi_n{}^* F_n\Psi_n \, dx \, dy \, dz \\
&= F_n \int \Psi_n{}^* \Psi_n \, dx \, dy \, dz \\
&= F_n
\end{aligned}
\tag{4-9}
$$

By the same reasoning,

$$
\begin{aligned}
\langle F^2 \rangle &= \int \Psi_n{}^* \hat{F}\hat{F}\Psi_n \, dx \, dy \, dz \\
&= F_n{}^2
\end{aligned}
\tag{4-10}
$$

Any set of numbers such that the average of the squares is equal to the square of the average has zero variance. Thus, eigenvalues are sharply defined, quantized values and can be precisely determined by experiment. Other variables of the system in the same state can be described only by probability statements; experimental measurements result in a wide range of values.

Exercise 4-3. Verify Eq. 4-10.

POSTULATE IV   The sum of the potential energy and the kinetic energy of a system can be expressed as a function of the coordinates and the momenta, which is called the Hamiltonian H. The Hamiltonian operator $\hat{H}$ operating on Ψ must give the same results as the energy operator $\hat{E}$ operating on Ψ. The resulting equation, the *time-dependent Schrödinger equation*,

$$\hat{H}\Psi = \hat{E}\Psi \tag{4-11}$$

describes how a system changes with time.

For a single particle moving in one dimension

$$H = \frac{1}{2} m v_x^2 + U(x)$$

$$= \left(\frac{p_x^2}{2m}\right) + U(x) \qquad (4\text{-}12)$$

so that the Hamiltonian operator is

$$\hat{H} = \frac{1}{2m} \left(\frac{h}{2\pi i}\right)\left(\frac{\partial}{\partial x}\right)\left(\frac{h}{2\pi i}\right)\left(\frac{\partial}{\partial x}\right) + U(x)$$

$$= \frac{1}{2m} \left(-\frac{h^2}{4\pi^2}\right)\left(\frac{\partial^2}{\partial x^2}\right) + U(x)$$

$$= -\frac{h^2}{8\pi^2 m} \left(\frac{\partial^2}{\partial x^2}\right) + U(x) \qquad (4\text{-}13)$$

Equation 4-11 becomes

$$-\frac{h^2}{8\pi^2 m} \left(\frac{\partial^2 \Psi}{\partial x^2}\right) + U(x)\Psi = -\frac{h}{2\pi i}\left(\frac{\partial \Psi}{\partial t}\right) \qquad (4\text{-}14)$$

Solutions to Eq. 4-14 are functions that permit the calculation of expectation values

Separation of the Schrödinger equation
If we express Ψ, which is a function of x and t, as a product of two functions $\psi(x)$ and $T(t)$, then

$$\frac{\partial \Psi}{\partial x} = T(t) \frac{d\psi}{dx}$$

and

$$\frac{\partial^2 \Psi}{\partial x^2} = T(t) \frac{d^2 \psi}{dx^2} \qquad (4\text{-}15)$$

Similarly,

$$\frac{\partial \Psi}{\partial t} = \psi(x) \frac{dT}{dt} \qquad (4\text{-}16)$$

Substituting from Eqs. 4-15 and 4-16 into Eq. 4-14, we obtain

$$-\frac{h^2}{8\pi^2 m} T \frac{d^2 \psi}{dx^2} + u(x) T\psi = -\frac{h\psi}{2\pi i} \frac{dT}{dt}$$

Dividing each term by $T\psi$, we have

$$-\frac{h^2}{8\pi^2 m}\frac{1}{\psi}\frac{d^2\psi}{dx^2} + U(x) = -\frac{h}{2\pi i}\frac{1}{T}\frac{dT}{dt} \qquad (4\text{-}17)$$

The left side of Eq. 4-17 is a function only of x, and the right side is a function only of t. Since these variables are independent, the equality can hold for all values of x and t only if both sides are equal to a constant, which we call E. The separation is not so easily carried out if U is a function of t as well as x.
Then,

$$-\frac{h}{2\pi i}\frac{1}{T}\frac{dT}{dt} = E \qquad (4\text{-}18)$$

and

$$-\frac{h^2}{8\pi^2 m}\frac{1}{\psi}\frac{d^2\psi}{dx^2} + U(x) = E \qquad (4\text{-}19)$$

The solution of Eq. 4-18 is

$$T = (\text{constant}) \exp\left(-2\pi i E\,\frac{t}{h}\right) \qquad (4\text{-}20)$$

Exercise 4-4.   Verify that Eq. 4-20 represents a solution of Eq. 4-18.

Equation 4-19 can be rearranged to give

$$-\frac{h^2}{8\pi^2 m}\frac{d^2\psi}{dx^2} + U(x)\psi = E\psi \qquad (4\text{-}21)$$

which is Schrödinger's equation for stationary states of the system. Thus,

$$\Psi = \psi T$$
$$= (\text{constant})\,\psi\,\exp\left(-2\pi i E\,\frac{t}{h}\right)$$

or, if we absorb the constant in the function ψ

$$\Psi = \psi\,\exp\left(-2\pi i E\,\frac{t}{h}\right) \qquad (4\text{-}22)$$

The probability density $\Psi^*\Psi$ is then

$$\Psi^*\Psi = \psi^*\,\exp\left(2\pi i E\,\frac{t}{h}\right)\,\psi\,\exp\left(-2\pi i E\,\frac{t}{h}\right)$$
$$= \psi^*\psi \qquad (4\text{-}23)$$

Thus, the probability density depends only on the time-independent part of the wave functions (when U is not a function of t, as it would be, for example, for an electron in an oscillating electromagnetic field).

For a particle in field-free space, $U(x) = 0$, and Eq. 4-21 becomes

$$\frac{d^2\psi}{dx^2} = -\frac{8\pi^2 mE}{h^2}\psi \tag{4-24}$$

A solution to Eq. 4-24 is

$$\psi = A \exp\left[2\pi i(2mE)^{\frac{1}{2}}\frac{x}{h}\right] \tag{4-25}$$

where A is a constant to be determined by the normalization requirement.

Exercise 4-5. Verify that Eq. 4-25 is a solution of Eq. 4-24.

A corresponding time-dependent solution is

$$\Psi = \psi \exp\left(-2\pi iE\frac{t}{h}\right)$$

$$= A \exp\left[2\pi i(2mE)^{\frac{1}{2}}\frac{x}{h}\right]\exp\left(-2\pi iE\frac{t}{h}\right) \tag{4-26}$$

If we apply the momentum operator $(h/2\pi i)\,(\partial/\partial x)$ to Ψ, the result is

$$\hat{p}_x\Psi = \frac{h}{2\pi i}\left[\frac{2\pi i(2mE)^{\frac{1}{2}}}{h}\right]\Psi$$

$$= (2mE)^{\frac{1}{2}}\Psi \tag{4-27}$$

Since m and E are constants of the motion of the particle, Eq. 4-27 shows that Ψ is an eigenfunction of the momentum and that $(2mE)^{\frac{1}{2}}$ is an eigenvalue of the momentum. That the classical value of the momentum is a constant in the absence of an external force is consistent with this description. If the position operator $\hat{x}$ is applied to Ψ from Eq. 4-26, the result is

$$\hat{x}\,\Psi = x\,\Psi \tag{4-28}$$

The expectation value of x is

$$\langle x \rangle = \int_{-\infty}^{\infty}\Psi^*\,\hat{x}\,\Psi\,dx$$

$$= \int_{-\infty}^{\infty}\Psi^*\,x\,\Psi\,dx$$

$$= \int_{-\infty}^{\infty} A^* \exp \left[-2\pi i (2mE)^{\frac{1}{2}} \frac{x}{h} \right] \exp \left(2\pi i E \frac{t}{h} \right)$$

$$\times \, x \, A \exp \left[2\pi i (2mE)^{\frac{1}{2}} \frac{x}{h} \right] \exp \left(-2\pi i E \frac{t}{h} \right) dx$$

$$= \int_{-\infty}^{\infty} A A^* x \, dx$$

$$= A A^* \left[\frac{x^2}{2} \right]_{-\infty}^{\infty} = 0$$

indicating that the particle is symmetrically distributed with respect to the origin. More precisely, since the probability density $\Psi^*\Psi$ is equal to $A A^*$, there is a constant probability of finding the particle at any point in space. Thus x is not an eigenvalue, in contrast to p_x.

The normalization of the wave function for a free particle is a difficult problem, beyond the scope of this text.[3]

4-2. One-electron atoms

The triumph of Bohr's theory of atomic structure was its correct prediction of the frequencies of the spectral lines of hydrogen. But more comprehensive theory was required to combine this achievement with the ability to predict the experimental data for a wider range of systems in which Bohr's theory led to incorrect results or could not be applied. In this section, we discuss the application of Schrödinger's wave mechanics to the one-electron atom. We will see how these principles can be applied, in a more approximate way, to many-electron atoms.

In Schrödinger's treatment, the one-electron atom is viewed as a spherical system, with the electron and nucleus revolving about the same center of mass. This model is equivalent to an electron of reduced mass μ revolving about a stationary nucleus,

$$\mu = \frac{(m_n)(m)}{m_n + m} \tag{4-29}$$

where m_n is the mass of the nucleus, and m is the mass of the electron. Since m_n for the lightest element, hydrogen, is 1840 times as great as m, we shall approximate μ by m.

The Schrödinger equation

To solve the Schrödinger equation (Eq. 4-11) for the one-electron atom,

3. See L. Pauling and E. B. Wilson, Jr., *Introduction to Quantum Mechanics*, McGraw-Hill Book Co., New York, 1935, p. 92.

we must express the energy in Hamiltonian form. In Cartesian coordinates, the kinetic energy of the electron is

$$K = \frac{p_x^2}{2m} + \frac{p_y^2}{2m} + \frac{p_z^2}{2m} \tag{4-30}$$

The potential energy of an electron of charge $-e$ in the field of a nucleus of charge $+Ze$, where Z is the atomic number, is

$$U(x, y, z,) = \frac{-Ze^2}{4\pi\varepsilon_0 r} \tag{4-31}$$

$$= \frac{-Ze^2}{4\pi\varepsilon_0 (x^2 + y^2 + z^2)^{\frac{1}{2}}} \tag{4-32}$$

The quantity r is the distance of the electron from the nucleus, and x, y, and z are the Cartesian coordinates of the electron if the nucleus is taken to be at the origin. Thus,

$$H(x, y, z, p_x, p_y, p_z) = \frac{p_x^2}{2m} + \frac{p_y^2}{2m} + \frac{p_z^2}{2m} - \frac{Ze^2}{4\pi\varepsilon_0 r} \tag{4-33}$$

The operator $\hat{H}$ is, then

$$\hat{H} = -\frac{h^2}{8\pi^2 m}\left(\frac{\partial^2}{\partial x^2} + \frac{\partial^2}{\partial y^2} + \frac{\partial^2}{\partial z^2}\right) - \frac{Ze^2}{4\pi\varepsilon_0 r} \tag{4-34}$$

$$= -\frac{h^2}{8\pi^2 m}\nabla^2 - \frac{Ze^2}{4\pi\varepsilon_0 r} \tag{4-35}$$

where ∇^2 (del-squared) is a symbol for the differential operators in parentheses in Eq. 4-34. Substituting the appropriate expressions for $\hat{H}$ and $\hat{E}$ in Eq. 4-11 we obtain

$$-\frac{h^2}{8\pi^2 m}\nabla^2\Psi - \frac{Ze^2}{4\pi\varepsilon_0 r}\Psi = -\frac{h}{2\pi i}\frac{\partial\Psi}{\partial t} \tag{4-36}$$

The Schrödinger equation for the one-electron atom (Eq. 4-36), like the Schrödinger equation for a single dimension (Eq. 4-14), can be separated into two parts, one dependent only on the coordinates, the other dependent only on the time. The resultant time-independent equation, from which the stationary states of the atom are obtained, is

$$-\frac{h^2}{8\pi^2 m}\nabla^2\psi - \frac{Ze^2}{4\pi\varepsilon_0 r}\psi = E\psi \tag{4-37}$$

Exercise 4-6. Set Ψ in Eq. 4-36 equal to $[T(t)][\Psi(x, y, z)]$ and, by taking the appropriate derivatives, and substitution in Eq. 4-36, show that Eq. 4-37 can be obtained.

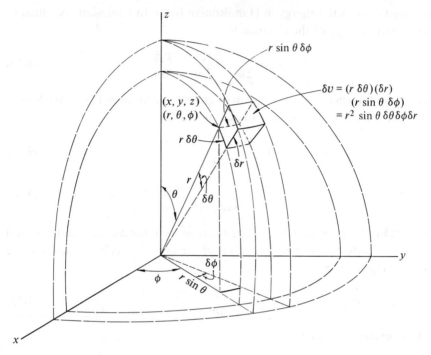

Figure 4-1. The spherical and Cartesian coordinate systems for the one-electron atom problem, showing the element of volume in spherical coordinates.

Writing Eq. 4-37 in detail, we have

$$-\frac{h^2}{8\pi^2 m}\left(\frac{\partial^2 \psi}{\partial x^2}+\frac{\partial^2 \psi}{\partial y^2}+\frac{\partial^2 \psi}{\partial z^2}\right)-\frac{Ze^2\psi}{4\pi\varepsilon_0(x^2+y^2+z^2)^{\frac{1}{2}}}=E\psi$$

$$(4-38)$$

The solution of Eq. 4-38, a partial differential equation, is a function of three variables. It would be convenient to separate the equation into three *ordinary* differential equations of one independent variable each. This cannot be done by assuming that $\psi(x, y, z) = X(x)\, Y(y)\, Z(z)$, because of the $(x^2 + y^2 + z^2)^{\frac{1}{2}}$ term in the potential energy.

Like many problems about physical systems, this problem can be simplified if we use a coordinate system that reflects the symmetry of the system. Since the atom is spherically symmetric, the appropriate coordinate system to use is spherical polar coordinates, as indicated in Fig. 4-1, for which

$$z = r \cos \theta$$

$$x = r \sin \theta \cos \phi$$

and

$$y = r \sin \theta \sin \phi$$

where θ is the angle between the z axis and the radius vector, and ϕ is the angle between the x axis and the projection of the radius vector on the x-y plane. The range of θ is from zero to π, and the range of ϕ is from zero to 2π.

In spherical coordinates, Eq. 4-38 becomes

$$\frac{\partial^2 \psi}{\partial r^2} + \frac{2}{r} \frac{\partial \psi}{\partial r} + \frac{1}{r^2} \frac{\partial^2 \psi}{\partial \theta^2} + \frac{\cos \theta}{r^2 \sin \theta} \frac{\partial \psi}{\partial \theta} + \frac{1}{r^2 \sin^2 \theta} \frac{\partial^2 \psi}{\partial \phi^2}$$

$$+ \frac{8\pi^2 m}{h^2} \left(E + \frac{Ze^2}{4\pi\varepsilon_0 r} \right) \psi = 0 \tag{4-39}$$

Separation of variables

If we assume that ψ can be expressed as a product of three functions, each a function of only one variable,

$$\psi(r, \theta, \phi) = R(r) \, \ominus (\theta) \, \Phi(\phi) \tag{4-40}$$

then the derivatives

$$\frac{\partial \psi}{\partial r} = \ominus \, \Phi \, \frac{dR}{dr}$$

$$\frac{\partial^2 \psi}{\partial r^2} = \ominus \, \Phi \, \frac{d^2 R}{dr^2}$$

$$\frac{\partial \psi}{\partial \theta} = R \, \Phi \, \frac{d \ominus}{d\theta}$$

$$\frac{\partial^2 \psi}{\partial \theta^2} = R \, \Phi \, \frac{d^2 \ominus}{d\theta^2}$$

and

$$\frac{\partial^2 \psi}{\partial \phi^2} = R \, \ominus \, \frac{d^2 \Phi}{d\phi^2}$$

can be substituted in Eq. 4-39. The result,

$$\ominus \, \Phi \, \frac{d^2 R}{dr^2} + \frac{2}{r} \, \ominus \, \Phi \, \frac{dR}{dr} + \frac{1}{r^2} \, R \, \Phi \, \frac{d^2 \ominus}{d\theta^2}$$

$$+ \frac{\cos \theta}{r^2 \sin \theta} \, R \, \Phi \, \frac{d \ominus}{d\theta}$$

$$+ \frac{1}{r^2 \sin^2 \theta} R \ominus \frac{d^2 \Phi}{d\phi^2}$$

$$+ \frac{8\pi^2 m}{h^2} \left(E + \frac{Ze^2}{4\pi\varepsilon_0 r} \right) R \ominus \Phi = 0 \qquad (4\text{-}41)$$

can be multiplied by $r^2/(R \ominus \Phi)$, to give, upon separation of angular and radial terms,

$$\frac{r^2}{R} \frac{d^2 R}{dr^2} + \frac{2r}{R} \frac{dR}{dr} + \frac{8\pi^2 m r^2}{h^2} \left(E + \frac{Ze^2}{4\pi\varepsilon_0 r} \right) =$$

$$- \frac{1}{\ominus} \frac{d^2\ominus}{d\theta^2} - \frac{\cos\theta}{\sin\theta} \frac{1}{\ominus} \frac{d\ominus}{d\theta} - \frac{1}{\sin^2\theta} \frac{1}{\Phi} \frac{d^2\Phi}{d\phi^2} \qquad (4\text{-}42)$$

Again, both sides of Eq. 4-42 must be equal to a constant, which we shall write $l(l + 1)$, for reasons that will be clear later. The radial equation, which depends only on r, is then

$$\frac{r^2}{R} \frac{d^2 R}{dr^2} + \frac{2r}{R} \frac{dR}{dr} + \frac{8\pi^2 m r^2}{h^2} \left(E + \frac{Ze^2}{4\pi\varepsilon_0 r} \right) = l(l + 1)$$

If this is multiplied by R/r^2 and rearranged, we obtain

$$\frac{d^2 R}{dr^2} + \frac{2}{r} \frac{dR}{dr} + \frac{8\pi^2 m}{h^2} \left(E + \frac{Ze^2}{4\pi\varepsilon_0 r} \right) R - \frac{l(l + 1)}{r^2} R = 0 \qquad (4\text{-}43)$$

The angular equation, which depends only on θ and ϕ, is

$$\frac{1}{\ominus} \frac{d^2\ominus}{d\theta^2} + \frac{\cos\theta}{\sin\theta} \frac{1}{\ominus} \frac{d\ominus}{d\theta} + \frac{1}{\sin^2\theta} \frac{1}{\Phi} \frac{d^2\Phi}{d\phi^2} = -l(l + 1)$$

If this equation is multiplied by $\sin^2\theta$ and rearranged, we obtain

$$- \frac{1}{\Phi} \frac{d^2\Phi}{d\phi^2} = \frac{\sin^2\theta}{\ominus} \frac{d^2\ominus}{d\theta^2} + \frac{\sin\theta\cos\theta}{\ominus} \frac{d\ominus}{d\theta} + l(l + 1)\sin^2\theta \qquad (4\text{-}44)$$

Again both sides of Eq. 4-44 must be equal to a constant, which we call m_l^2. Thus

$$\frac{d^2\Phi}{d\phi^2} = -m_l^2 \Phi \qquad (4\text{-}45)$$

and

$$\frac{\sin^2\theta}{\ominus} \frac{d^2\ominus}{d\theta^2} + \frac{\sin\theta\cos\theta}{\ominus} \frac{d\ominus}{d\theta} + l(l + 1)\sin^2\theta = m_l^2$$

which, when multiplied by $\ominus/\sin^2\theta$ and rearranged, becomes

$$\frac{d^2\ominus}{d\theta^2} + \frac{\cos\theta}{\sin\theta}\frac{d\ominus}{d\theta} + \left[l(l+1) - \frac{m_l^2}{\sin^2\theta} \right]\ominus = 0 \qquad (4\text{-}46)$$

The angular functions

Some solutions to Eq. 4-45 are

$$\Phi = \exp(\pm im_l\phi) \qquad (4\text{-}47)$$

$$\Phi = \sin m_l\phi \qquad (4\text{-}48)$$

or

$$\Phi = \cos m_l\phi \qquad (4\text{-}49)$$

Exercise 4-7.   Verify that Eqs. 4-47, 4-48, and 4-49 are solutions of Eq. 4-45.

For all these functions, if Φ is to be single valued and continuous, $\Phi(\phi + 2\pi)$ must be equal to $\Phi(\phi)$. For example,

$$\exp[im_l(\phi + 2\pi)] = \exp[im_l\phi]$$

so

$$\exp(2\pi im_l) = 1 \qquad (4\text{-}50)$$

Since

$$\exp(2\pi im_l) = \cos(2\pi m_l) + i\sin(2\pi m_l) = 1$$

m_l can have only integral values.

$$m_l = 0, \pm 1, \pm 2, \pm 3, \ldots \qquad (4\text{-}51)$$

Exercise 4-8.   Show that m_l must have integral values for Eqs. 4-48 and 4-49 as well as Eq. 4-47.

We see that m_l, the *magnetic quantum number*, has quantized values as a result of the requirements that the wave function be well-behaved.

Just as the operator for linear momentum in the x direction is $(h/2\pi i)$ $(\partial/\partial x)$, the operator for angular momentum of rotation about the z axis is $(h/2\pi i)$ $(\partial/\partial\phi)$. The application of this operator to $\exp(im_l\phi)$ in Eq. 4-47 gives

$$\frac{h}{2\pi i}\frac{d\Phi}{d\phi} = \left(\frac{h}{2\pi i}\right)(im_l)\exp(im_l\phi)$$

$$= m_l \left(\frac{h}{2\pi} \right) \Phi \qquad (4\text{-}52)$$

As we have seen, this means that Φ in Eq. 4-47 is an *eigenfunction* of the z component of the angular momentum, and the variable p_ϕ has the *eigenvalues*

$$p_\phi = m_l \left(\frac{h}{2\pi} \right) \qquad m_l = 0, \pm 1, \pm 2, \pm 3, \ldots \qquad (4\text{-}53)$$

This result is one of the striking advances of the Schrödinger formulation over the Bohr model. The ground state for all one-electron atoms has zero angular momentum, a prediction of the Schrödinger equation but inconsistent with the Bohr model.

Exercise 4-9. Show that the Φ in Eq. 4-48 and Eq. 4-49 are not eigenfunctions of the z component of the angular momentum.

Equation 4-46 is a form of Legendre's equation, the solutions of which are the associated Legendre functions, polynomials in $\cos \theta$. These functions appear in many physical problems that have spherical symmetry.[4] Consider the series

$$\Theta = (\sin \theta)^a [A_0 + A_1 \cos \theta + \ldots + A_k (\cos \theta)^k + \ldots]$$

$$= (\sin \theta)^a \sum_{k=0}^{\infty} A_k (\cos \theta)^k \qquad (4\text{-}54)$$

where $a = |m_l|$ and the coefficients A_k are to be determined. If Eq. 4-54 is used to make the appropriate substitutions in Eq. 4-46, the following result is obtained.

$$[-a^2 \sum_{k=0}^{\infty} A_k (\cos \theta)^k - a \sum_{k=0}^{\infty} A_k (k+1)(\cos \theta)^k - (a+1) \sum_{k=0}^{\infty} kA_k (\cos \theta)^k$$

$$- \sum_{k=0}^{\infty} A_k k(k-1)(\cos \theta)^k - \sum_{k=0}^{\infty} kA_k (\cos \theta)^k + l(l+1) \sum_{k=0}^{\infty} A_k (\cos \theta)^k]$$

$$+ [\sum_{k=0}^{\infty} A_k k(k-1)(\cos \theta)^{k-2}] = 0 \qquad (4\text{-}55)$$

If the summations in Eq. 4-55 are to equal zero for all values of $\cos \theta$, the coefficient of each power of $\cos \theta$ must equal zero. From this requirement, we can obtain the recursion relation between the coefficients A_k and A_{k+2}

$$A_{k+2} = A_k \frac{(k+a)(k+a+1) - l(l+1)}{(k+1)(k+2)} \qquad (4\text{-}56)$$

4. R. V. Churchill, *Fourier Series and Boundary Value Problems*, McGraw-Hill Book Co., New York, 1941, Chap. IX.

Thus, suitable choices of A_0 and A_1 will provide values for all other coefficients.

The infinite series in Eq. 4-54 converges to a finite sum for $-1 < \cos\theta < 1$. For this to be a well-behaved solution to the Schrödinger equation, Θ must be finite at $\theta = 0$ and π, when $\cos\theta = 1$ and -1, respectively. This requires that the series have a finite number of terms; that is, A^{k+2} becomes zero for some value of k. This condition is fulfilled for

$$(a + k)(a + k + 1) = l(l + 1)$$

or

$$l = a + k = |m_l| + k \qquad (4\text{-}57)$$

The definition of k as an integer and the quantum requirement for m_l result in l being an integer greater than or equal to $|m_l|$. The possible values of l, the *angular momentum* quantum number, and m_l, the *magnetic* quantum number, are

$$l = 0, 1, 2, \ldots \qquad (4\text{-}58)$$

$$m_l = 0, \pm 1, \pm 2, \ldots, \pm l \qquad (4\text{-}59)$$

Thus, there are $2l + 1$ values of m_l for each value of l, and, for each Θ function, the index of the last term of the polynomial is

$$k = l - |m_l|$$

Table 4-1 gives the normalized solutions $\Theta \Phi$, Φ from Eq. 4-47 and Θ from Eq. 4-54.

Table 4-1. Normalized angular functions $\Theta \Phi$ for a one-electron atom

	l^a	m_l	$2\pi^{\frac{1}{2}}\,\Theta\,\Phi$
s	0	0	1
p_0	1	0	$(3)^{\frac{1}{2}}\cos\theta$
$p_{\pm 1}$	1	± 1	$(3/2)^{\frac{1}{2}}\sin\theta\exp(\pm i\phi)$
d_0	2	0	$(5^{\frac{1}{2}}/2)(3\cos^2\theta - 1)$
$d_{\pm 1}$	2	± 1	$(15/2)^{\frac{1}{2}}\sin\theta\cos\theta\exp(\pm i\phi)$
$d_{\pm 2}$	2	± 2	$(15/8)^{\frac{1}{2}}\sin^2\theta\exp(\pm 2i\phi)$

a The values $l = 0, 1, 2,$ and 3 conventionally are designated s, p, d, and f, respectively.

An alternative set of angular functions, using Eqs. 4-48 and 4-49 for Φ, is given in Table 4-2. These are real functions and may be more easily visualized. They are plotted in Fig. 4-2 as polar diagrams in the plane in which the function has its maximum value.

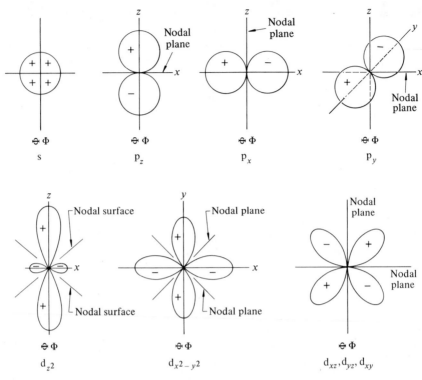

Figure 4-2. The angular functions $\Theta\,\Phi$ as polar coordinate plots in the plane in which the function has a maximum value. For s, p_x, p_y, p_z, d_{z^2}, $d_{x^2-y^2}$, d_{yz}, d_{xz}, and d_{xy} functions.

Table 4-2. Alternative angular functions $\Theta\Phi$ for a one-electron atom

	l	$|m_l|$	$4\pi^{\frac{1}{2}}\,\Theta\Phi$
s	0	0	2
p_z	1	0	$2(3)^{\frac{1}{2}}\cos\theta$
p_x	1	1	$2(3)^{\frac{1}{2}}\sin\theta\cos\phi$
p_y	1	1	$2(3)^{\frac{1}{2}}\sin\theta\sin\phi$
d_{z^2}	2	0	$(5)^{\frac{1}{2}}(3\cos^2\theta-1)$
d_{xz}	2	1	$(15)^{\frac{1}{2}}\sin 2\theta\cos\phi$
d_{yz}	2	1	$(15)^{\frac{1}{2}}\sin 2\theta\sin\phi$
$d_{x^2-y^2}$	2	2	$(15)^{\frac{1}{2}}\sin^2\theta\cos 2\phi$
d_{xy}	2	2	$(15)^{\frac{1}{2}}\sin^2\theta\sin 2\phi$

Exercise 4-10. Show that the functions p_x and p_y are linear combinations of p_1 and p_{-1}.

Exercise 4-11. Show that the functions $d_{x^2-y^2}$ and d_{xy} are linear combinations of d_2 and d_{-2} and that d_{xz} and d_{yz} are linear combinations of d_1 and d_{-1}.

Angular functions and angular momentum

The angular momentum vector **M** of the electron is defined classically by the vector product

$$\mathbf{M} = \mathbf{r} \times \mathbf{p} \tag{4-60}$$

The components of **M** are

$$M_x = yp_z - zp_y \tag{4-61}$$

$$M_y = zp_x - xp_z \tag{4-62}$$

$$M_z = xp_y - yp_x \tag{4-63}$$

and

$$M^2 = M_x{}^2 + M_y{}^2 + M_z{}^2 \tag{4-64}$$

The corresponding operators are

$$\hat{M}_x = \frac{h}{2\pi i}\left(y\frac{\partial}{\partial z} - z\frac{\partial}{\partial y}\right)$$

$$\hat{M}_y = \frac{h}{2\pi i}\left(z\frac{\partial}{\partial x} - x\frac{\partial}{\partial z}\right)$$

$$\hat{M}_z = \frac{h}{2\pi i}\left(x\frac{\partial}{\partial y} - y\frac{\partial}{\partial x}\right)$$

When converted to spherical coordinates, the results are

$$\hat{M}_x = \frac{h}{2\pi i}\left(\sin\theta\frac{\partial}{\partial\theta} + \frac{\cos\theta\cos\phi}{\sin\theta}\frac{\partial}{\partial\phi}\right)$$

$$\hat{M}_y = \frac{h}{2\pi i}\left(-\cos\phi\frac{\partial}{\partial\theta} + \frac{\cos\theta\sin\phi}{\sin\theta}\frac{\partial}{\partial\phi}\right)$$

$$\hat{M}_z = \frac{h}{2\pi i}\frac{\partial}{\partial\phi}$$

$$(4\text{-}65)$$

and

$$\hat{M}^2 = -\frac{h^2}{4\pi^2}\left(\frac{\partial^2}{\partial\theta^2} + \frac{\cos\theta}{\sin\theta}\frac{\partial}{\partial\theta} + \frac{1}{\sin^2\theta}\frac{\partial^2}{\partial\phi^2}\right)$$

We have already seen, from Eq. 4-52, that, when $\hat{M}_z$ operates on Φ in the

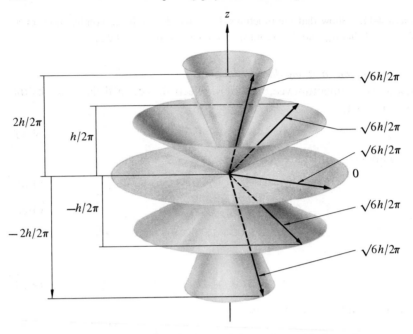

Figure 4-3. The loci of the quantized orientations of an orbital angular momentum vector with $l = 2$ in an external magnetic field whose direction defines the z-axis. (By permission, from P. W. Atkins, *Molecular Quantum. Mechanics*, Clarendon Press, Oxford, 1970.)

complex exponential form, we obtain the eigenvalues of $M_z = m_l(h/2\pi)$. Neither $\hat{M}_x$, nor $\hat{M}_y$, acting on $\ominus \Phi$ leads to eigenvalues. We can see from Eq. 4-44 that $\hat{M}^2$ acting on $\ominus \Phi$ (in either the real or the complex form) leads to the eigenvalue equation

$$\hat{M}^2 \ominus \Phi = \left(\frac{-h^2}{4\pi^2}\right)\left[-l(l+1)\right] \ominus \Phi \qquad (4\text{-}66)$$

The eigenvalues of M^2 are then

$$M^2 = l(l+1)\left(\frac{h}{2\pi}\right)^2 \qquad (4\text{-}67)$$

and the magnitude of M is given by

$$|M| = [l(l+1)]^{\frac{1}{2}}\frac{h}{2\pi} \qquad (4\text{-}68)$$

Hence the designation of l as the angular momentum quantum number.

Atoms are spherically symmetric, so that the designation of the z axis is entirely arbitrary without an external reference. Thus, states that differ only in m_l, which describes the z component of the angular momentum, cannot be distinguished. An electron with angular momentum has a magnetic field and can interact with an external magnetic field. In the presence of an external magnetic field we define the positive z axis as the direction of the magnetic field. Figure 4-3 shows the possible orientations of the angular momentum vector of an electron with $l = 2$, with respect to an external magnetic field. The loci of the vector are shown as cones or as a circle because the x and y components are not eigenvalues; only the total angular momentum and the z component have sharp values.

Exercise 4-12. Verify that $\ominus \Phi$ is not an eigenfunction of $\hat{M}_x$ or $\hat{M}_y$.

Exercise 4-13. Verify that $\ominus \Phi$ is an eigenfunction of $\hat{M}^2$.

Exercise 4-14. Calculate the magnitude of the total orbital angular momentum of an s electron, a p electron, and a d electron.

Exercise 4-15. Calculate the angle that the angular momentum vector makes with respect to the z axis if $l = 2$ and $m_l = 1$.

The radial functions
The solution to Eq. 4-43, like the solution to Eq. 4-46, is a series solution. Before a series is chosen, it is helpful if we consider the behavior of the wave function for large values of r. For $\int_0^\infty RR^* dr$ to be finite, the wave function must approach zero asymptotically for very large r. For very large values of r, Eq. 4-43 reduces to

$$\frac{d^2 R}{dr^2} + \frac{8\pi^2 mE}{h^2} R = 0 \tag{4-69}$$

If we let

$$-\alpha^2 = \frac{8\pi^2 mE}{h^2} \tag{4-70}$$

a solution of Eq. 4-69 is, as we have seen earlier,

$$R = \exp(\pm \alpha r)$$

The exponential with the positive exponent increases without limit as r increases and thus is an unsatisfactory solution, so we shall use only the negative exponent.

Returning then to the general equation, let us write

$$R = W(r) \exp(-\alpha r) = \frac{W(r)}{\exp(\alpha r)} \tag{4-71}$$

where

$$W(r) = r^l (A_0 + A_1 r + \ldots + A_k r^k + \ldots)$$

$$= r^l \sum_{k=0}^{\infty} A_k r^k$$

is the form obtained by Schrödinger.

When Eq. 4-71 and the appropriate derivatives are substituted in Eq. 4-43, and the coefficients of like powers of r are collected and equated to zero, the following recursion formula is obtained:

$$A_k = -2 A_{k-1} \frac{\alpha(l+k) - (\pi m e^2 Z / \varepsilon_0 h^2)}{(l+k)(l+k+1) - l(l+1)} \tag{4-72}$$

The series $W(r)$ not only increases without limit as r increases, it increases with r more rapidly than $\exp(\alpha r)$ at large r, and $R(r)$ becomes infinite. If, however, $W(r)$ is truncated at some finite value of k, $\exp(\alpha r)$ will increase with r more rapidly than $W(r)$ at large r, and the ratio $W(r)/\exp(\alpha r)$ will go to zero.

$W(r)$ will terminate at k if the numerator in Eq. 4-72 goes to zero. Therefore

$$\alpha(l+k) = \frac{\pi m e^2 Z}{\varepsilon_0 h^2}$$

and

$$\alpha^2 = \frac{\pi^2 m^2 e^4 Z^2}{\varepsilon_0^2 h^4 (l+k)^2} \tag{4-73}$$

Since l and k are both integers, with $l \geqslant 0$ and $k > 0$, the sum $l + k = n$ must be an integer greater than zero. It is called the *principal quantum number*. If we substitute for α^2 from Eq. 4-70 in Eq. 4-73, we obtain

$$-\frac{8\pi^2 m E}{h^2} = \frac{\pi^2 m^2 e^4 Z^2}{\varepsilon_0^2 h^4 n^2}$$

and

$$E = -\frac{m e^4 Z^2}{8 \varepsilon_0^2 h^2 n^2} \tag{4-74}$$

It is in this manner that the Bohr expression for the energy states of the electron in a one-electron atom has been derived from the postulates of wave mechanics and that quantization has resulted from the requirements that the wave function be well behaved. Also, it can be seen from Eq. 4-74 that the energy depends only on the principal quantum number n and is independent of l and m_l. The independence of l is limited to the case of a one-electron atom, but the independence of m_l is characteristic of the spherical symmetry of the model. Wave functions with the same energy are called *degenerate*, and the *degeneracy* of an energy level is the number of wave functions corresponding to that level.

Since the minimum value of k is 1, n must be equal to or greater than $l + 1$. Thus,

$$n = 1, 2, 3, \ldots \tag{4-75}$$

and

$$l = 0, 1, 2, \ldots, n - 1 \tag{4-76}$$

The possible values of the quantum numbers of a one-electron atom, together with the corresponding normalized radial functions, are given in Table 4-3 for n up to 3. The symbol a_0, the radius of the first Bohr orbit, is equal to $\varepsilon_0 h^2/(\pi m e^2)$ and $\rho = Z r/a_0$.

Table 4-3. Radial functions for a one-electron atom

n	l	m_l		R_{nl}	
1	0	0	1s	2	
2	0	0	2s	$(2 - \rho)$	
2	1	$0, \pm 1$	2p	$(\frac{1}{3})^{\frac{1}{2}} \rho$	
3	0	0	3s	$2[1 - (2\rho/3) + (2\rho^2/27)]$	$\left(\dfrac{Z}{na_0}\right)^{\frac{3}{2}} \exp(-\rho/n)$
3	1	$0, \pm 1$	3p	$[4(2)^{\frac{1}{2}}/9](1 - \rho/6)\rho$	
3	2	$0, \pm 1, \pm 2$	3d	$(2/27)(2/5)^{\frac{1}{2}} \rho^2$	

Graphical representations

The wave function ψ is a function of three variables, so that a complete graphical representation would require four dimensions. A variety of partial representations have been devised to provide an impression of the nature of the function. Atoms are spherical systems, and it is useful to see how ψ varies with r, the distance from the nucleus. For the s functions, which are dependent only on r, such a description is complete. For the functions with nonzero angular momentum, p, d, and f, the description as a function of r is useful, but it must be complemented with a description of the angular dependence.

Figure 4-4 shows the radial dependence of ψ for the hydrogen atom ($Z = 1$)

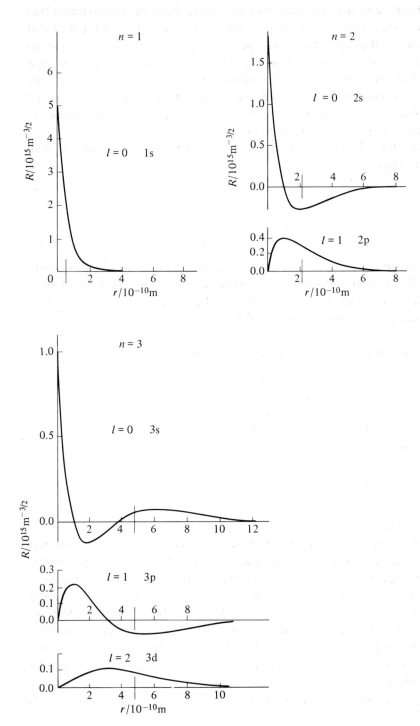

Figure 4-4. Radial dependence of ψ for the hydrogen atom, for $n = 1$, 2, and 3. (By permission, from G. Herzberg, *Atomic Spectra and Atomic Structure*, Dover Publications, New York, 1944, p. 40.)

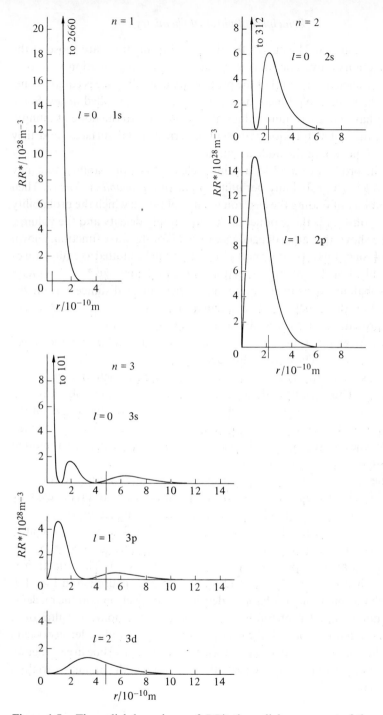

Figure 4-5. The radial dependence of RR^*, the radial component of the probability density, for the hydrogen atom, for $n = 1$, 2, and 3. (By permission, from G. Herzberg, *Atomic Spectra and Atomic Structure*, Dover Publications, New York, 1944, p. 43.)

for $n = 1$, 2, and 3. The curves for the p and d functions are those of the radial functions only. The limiting value of each wave function is zero as r increases, since the integral of the probability over all space is equal to one. Loci at which the wave functions are equal to zero are called *nodes*. Some functions have nodes at finite values of r in addition to the node at infinity. Nodes in the radial function represent spherical nodal surfaces, and they number $n-l$, including the node at infinity.

Since the product of $\psi\psi^*$ by a volume element is a probability, one may say that $\psi\psi^*$ is a probability per unit volume or a *probability density*. Then the probability of finding the electron in any volume in which the probability density is constant is the product of the probability density and the volume. Figure 4-5 shows the radial dependence of RR^* for the wave functions shown in Fig. 4-4, since this is the only part of $\psi\psi^*$ that can be plotted as a function of one variable. For the spherical functions, the s functions, $\psi\psi^* = (^1/_{4\pi}) RR^*$ is the probability density for any volume element at distance r from the nucleus. For the functions with angular dependence, $(^1/_{4\pi}) RR^*$ is the average probability density at distance r from the nucleus.

For spherical functions, the s functions, it is useful to find the *radial probability density*. The probability density is a function only of r for an s function, so that $\psi\psi^*$ is a constant over any spherical shell of infinitesimal thickness δr. Thus, the probability of finding an electron in a spherical shell between r and $r + \delta r$ is the product of $\psi\psi^*$ and $4\pi r^2 \delta r$, the volume of the spherical shell. Then the radial probability density is the probability divided by the thickness of the shell δr, $4\pi r^2 \psi\psi^*$ or $r^2 RR^*$. This quantity is plotted in Fig. 4-6 for the 1s, 2s, and 3s functions.

For the p and d functions, the probability of finding an electron in a spherical shell between r and $r + \delta r$ can be found by multiplying $\psi\psi^*$ by $r^2 \sin\theta \, \delta r \, \delta\theta \, \delta\phi$, the volume element in spherical coordinates, and integrating the product over the angular variables at a constant value of r. The quantity $r^2 RR^*$, which is equal to the result of the integration divided by δr, represents an average probability density over the shell. This function is plotted against r for the 2p, 3p, and 3d functions of hydrogen in Fig. 4-7.

Though the maximum probability density for an electron is at the nucleus, the maximum radial probability density occurs at a nonzero value of r. For each n and for $l = n - 1$, the maximum radial probability density occurs at the Bohr orbit for the value of n, shown as a short vertical line in each figure.

The total probability density $\psi\psi^*$ also depends on the angular variables θ and ϕ, since

$$\psi\psi^* = RR^* \Theta\Theta^* \Phi\Phi^*$$

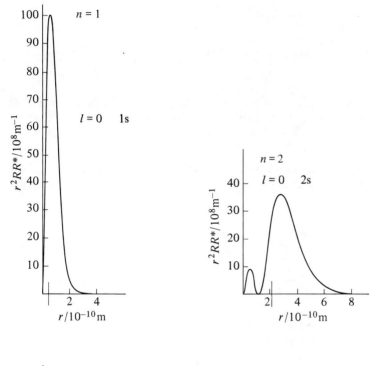

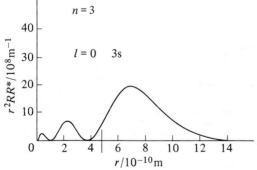

Figure 4-6. The radial probability density, $r^2\psi\psi^*$, for the 1s, 2s, and 3s functions of hydrogen. (By permission, from G. Herzberg, *Atomic Spectra and Atomic Structure*, Dover Publications, New York, 1944, p. 43.)

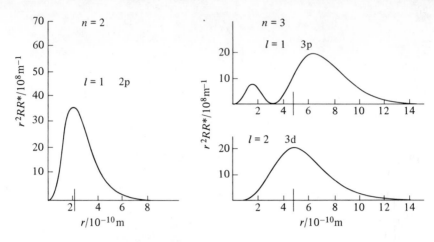

Figure 4-7. The averaged radial probability density, $r^2 RR^*$, for the 2p, 3p, and 3d functions. (By permission, from G. Herzberg, *Atomic Spectra and Atomic Structure*, Dover Publications, New York, 1944, p. 43.)

Figure 4-8. The angular probability density $\Theta^2 \Phi^2$ for s, p_x, p_y, p_z, d_{z^2}, $d_{x^2-y^2}$, d_{yz}, d_{xz} and d_{xy} functions, as polar coordinate plots in the plane in which each function has its maximum value.

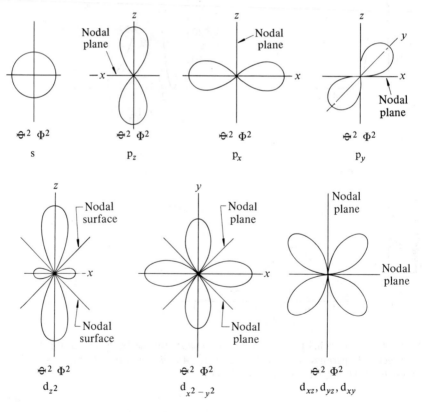

110

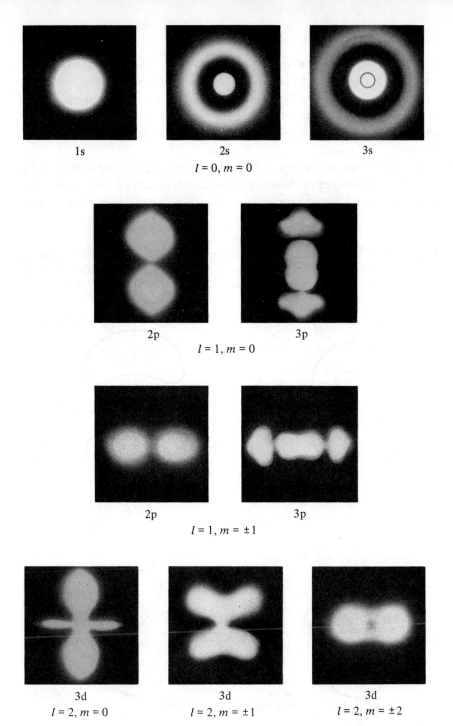

1s   **2s**   **3s**

$l = 0, m = 0$

2p   **3p**

$l = 1, m = 0$

2p   **3p**

$l = 1, m = \pm 1$

3d   **3d**   **3d**

$l = 2, m = 0$ $l = 2, m = \pm 1$ $l = 2, m = \pm 2$

Figure 4-9. A representation of the over-all probability density $\psi\psi^*$ in which the intensity of the light is proportional to $\psi\psi^*$. [Photographed from a spinning mechanical model by H. E. White, *Phys. Rev.*, *37*, 1416 (1931). Used by permission.]

Figure 4-2 shows polar diagrams of the real functions $\Theta\Phi$ from Table 4-2 for $l = 0$, 1, and 2 drawn in the plane in which the function has a maximum value. The distance from the origin to a point on a curve at a given angle is proportional to the value of the angular function at that angle. Figure 4-8 shows similar diagrams for $\Theta^2\Phi^2$, which represents the angular probability density.

The nodal lines in Fig. 4-8 and Fig. 4-2 represent planar or conical nodal surfaces. The number of these nodal surfaces is equal to l, so that the total number of nodal surfaces for any wave function is equal to n. As we saw in

Figure 4-10. Contours of constant ψ^2 at several values of f, the probability of finding an electron within the contour, for $2p_z$, $3p_z$, $3d_{z^2}$, and $3d_{x^2-y^2}$ functions. [By permission, adapted from E. A. Ogryzlo and G. B. Porter, *J. Chem. Educ.*, **40**, 256 (1963).]

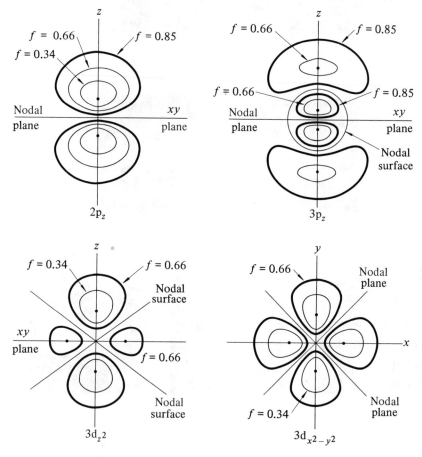

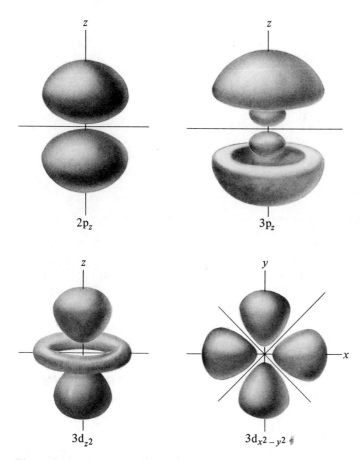

Figure 4-11. Contour surfaces of constant ψ^2 for $2p_z$, $3p_z$, $3d_{z^2}$, and $3d_{x^2-y^2}$ functions. [By permission, adapted from E. A. Ogryzlo and G. B. Porter, *J. Chem. Educ.*, *40*, 256 (1963).]

Eq. 4-74, the energy of the electron increases with increasing value of n, so that the number of nodes of a wave function is a rough indication of the energy. Since the number of angular nodes is equal to l, and since the angular momentum of the electron increases with increasing l, the relative number of radial and angular nodes indicates the relative importance of radial and angular momentum for a given wave function.

The over-all probability density $\psi\psi^*$ is a function of three variables, and, thus, it is impossible to represent graphically. Figure 4-9 shows photographs in which the light intensity represents the magnitude of $\psi\psi^*$ for

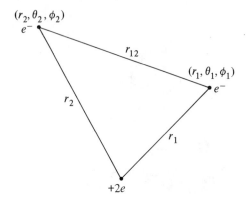

Figure 4-12. Spherical coordinates of the electrons in the helium atom.

$n = 1, 2$, and 3, using the complex functions for Φ. Another representation that is useful in relating atomic structure to chemical problems is that of contour surfaces of constant $\psi\psi^*$ within which $\int\psi\psi^*dv$ has a value close enough to 1 to be an approximate surface for the region in which the electron can be found. Projection of such surfaces on a plane is shown in Fig. 4-10, and sketches of three-dimensional models of these contour surfaces are shown in Fig. 4-11. It is clear from these figures that the polar diagrams of the angular functions do not adequately represent them for $n > l + 1$.

4-3. Many-electron atoms. The Pauli principle

Even for the helium atom, solutions to the Schrödinger equation cannot be expressed exactly in terms of analytical functions, such as those in Tables 4-2 and 4-3. This problem reflects here, as does its analogue in classical mechanics, the inability to solve in closed form the differential equation that describes the interaction of three bodies.

If we let r_1, θ_1, ϕ_1 represent the coordinates of electron 1 about the helium nucleus and let r_2, θ_2, ϕ_2 be the coordinates of electron 2, as indicated in Fig. 4-12, the Hamiltonian operator for the electrons in the helium atom is

$$\hat{H} = -\frac{h^2}{8\pi^2 m}(\nabla_1{}^2 + \nabla_2{}^2) - \frac{Z\,e^2}{4\pi\varepsilon_0}\left(\frac{1}{r_1} + \frac{1}{r_2}\right) + \frac{e^2}{4\pi\varepsilon_0 r_{12}} \quad (4\text{-}77)$$

where $-(h^2/8\pi^2 m)\,\nabla_1{}^2$ is the kinetic energy operator for electron 1 in spherical coordinates (See Eq. 4-37 and Eq. 4-39), a similar term appears for electron 2, the next two terms represent the potential energy of each electron in the field of the nucleus, and the last term is the interelectronic potential energy, where r_{12} is the distance between the electrons.

The resulting Schrödinger equation for the stationary states is

$$-\frac{h^2}{8\pi^2 m}(\nabla_1^2\psi + \nabla_2^2\psi) - \frac{Z e^2}{4\pi\varepsilon_0}\left(\frac{1}{r_1} + \frac{1}{r_2}\right)\psi + \frac{e^2}{4\pi\varepsilon_0 r_{12}}\psi$$

$$= E\psi \qquad (4\text{-}78)$$

In the absence of the r_{12} term, this equation could be separated into two equations of the form of Eq. 4-39 with solutions that are products $\psi(1)\,\psi(2)$ in which $\psi(1)$ is a *one-electron function* of the coordinates of electron 1 and $\psi(2)$ is a one-electron function of the coordinates of electron 2. These one-electron functions, each a product of a radial function from Table 4-3 and an appropriate angular function from Table 4-1 or Table 4-2, are called *orbitals*, characterized by three quantum numbers n, l, and m_l, and given the conventional designation 1s, $2p_x$, $3d_{z^2}$, etc. The product of one-electron functions is called the *electron configuration* of the atom.

Exercise 4-16. Show that, for Eq. 4-78, without the term involving r_{12}, $\psi = \psi_1$ $(r_1, \theta_1, \phi_1)\,\psi_2\,(r_2, \theta_2, \phi_2)$ would be a solution, that the equation would separate into two, and that the separation constant $E = E_1 + E_2$, where E_1 and E_2 are the eigenvalues from the two equations.

The self-consistent field

A commonly used approach to the problem of many-electron atoms is the self-consistent field method devised by Hartree in 1928. This method assumes that the wave function for the atom can be expressed as a product of one-electron wave functions and that each electron moves in a spherically averaged field of all the other electrons in the atom. To compute the field of the electrons, their wave functions must be known. Hartree assumed trial wave functions for all the electrons. From these functions, he calculated an average potential and the resulting wave function for each electron. The process was repeated until the resulting wave function and trial wave function agreed.

The one-electron functions used were similar to those obtained for the one-electron atom. Since the potential is assumed to be spherically symmetrical, the angular part of the wave function is the same as that for the one-electron atom. Thus the designations s, p, and d are still valid; only the radial part of the wave functions differ. The radial part of each orbital is obtained by numerical integration of the radial equation; the result is therefore a table of values of $R(r)$ rather than an analytical function. The wave functions can still be designated by the three quantum numbers n, l, and m_l.

For given values of n and l, the value of m_l does not affect the energy

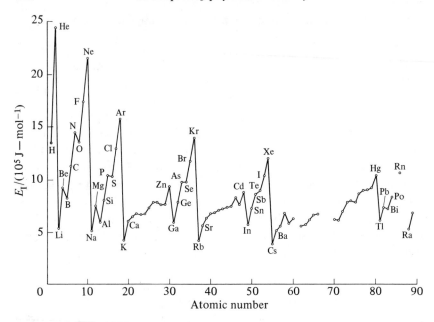

Figure 4-13. Periodic variation of the first ionization energy (E_1) of the elements with atomic number. [Drawn from data of C. E. Moore, "Atomic Energy Levels," NBS Circular, 467, Vol. III, 1958, and unpublished revisions by C. E. Moore; see also R. W. Kiser, "Tables of Ionization Potentials", U.S. AEC, TID, 6142 (1960).]

because m_l denotes only the orientation of an orbital in space. For a given value of n, the value of l does affect the energy. As Fig. 4-6 and Fig. 4-7 show, electrons with lower values of l are more likely to be found close to the nucleus. Thus, they are shielded less from the nucleus by other electrons and have a lower energy. In the absence of interelectron interaction, all functions of a given n correspond to the same energy.

The Pauli exclusion principle

The choice of wave functions for the electrons of a many-electron atom requires two additional concepts, electron spin and the Pauli principle. From what we have said thus far, one could well suggest that the ground state of each element can be represented by $\psi = (1s)^Z$ where Z is the number of electrons. If all three electrons in lithium were in the 1s state, one might expect the ionization energy to be greater than that of helium because of the greater nuclear charge in lithium despite the increased interelectron repulsion. An examination of the ionization energies of hydrogen, helium, and lithium in Table 4-4 shows, however, that lithium has a much smaller ionization energy than helium, indicating that lithium does not have the configuration $(1s)^3$. Thus, three quantum numbers are not adequate to explain the periodicity in ionization energies shown in Table 4-4 and Fig. 4-13 or the periodic variation of the chemical properties of the elements.

Pauli, in 1924, proposed that a complete description of the state of an electron required, in addition to the three quantum numbers discussed thus far, a fourth quantum number, not related to a classical motion, with two possible values. He further proposed, as a postulate, the Pauli exclusion principle, that two electrons in an atom could not have the same values of all four quantum numbers. With this principle, the qualitative wave-mechanical basis of the electron configuration of the elements and of the periodic table was established.

Electron spin

In 1926, Uhlenbeck and Goudsmit proposed a physical model for Pauli's two-valued quantum number. They suggested that the electron has, in addition to the angular momentum due to its orbital motion, an *intrinsic angular momentum*, which they called *spin*, and a corresponding magnetic moment. The z component of this angular momentum has two possible values $\frac{1}{2}h/2\pi$ and $(-\frac{1}{2})h/2\pi$. The spin quantum number for this z component, m_s, can have the values $\frac{1}{2}$ or $-\frac{1}{2}$. The value of the z component of the spin angular momentum is $m_s h/2\pi$, and the magnitude of the spin angular momentum is $[s(s + 1)]^{\frac{1}{2}}(h/2\pi)$, where s has the value $\frac{1}{2}$ and m_s takes on the $(2s + 1)$ values $s, s - 1, \ldots, -s$. Without some external reference, it is not possible to distinguish between a spin of $\frac{1}{2}$ and a spin of $-\frac{1}{2}$; in the presence of a magnetic field the two should be distinguishable, since electrons of one spin would be aligned with the magnetic field, while those of the opposite spin would be aligned against the magnetic field.

Exercise 4-17. Calculate the angle between the spin angular momentum vector of an electron with $m_s = \frac{1}{2}$ and the external magnetic field.

The concept of spin was originally introduced into quantum mechanics as an additional assumption to explain experimental results. Dirac later showed that spin is a necessary consequence of a relativistically correct theory and derived the value $\frac{1}{2}$ for s.

The experiment that demonstrated the effect of a magnetic field on electrons of opposite spin had been done by Stern and Gerlach in 1922, without any reference to spin. They passed a beam of silver atoms through

Table 4-4. Ionization energies of the elements (10^6 J-mol^{-1})[a]

Atomic number	Element	Subshell	I_1	I_2	I_3	I_4
1	H		1.312			
2	He	1s	2.371	5.248		
3	Li		0.520	7.295	11.81	
4	Be	2s	0.899	1.756	14.842	21.00
5	B		0.807	2.425	3.658	25.02
6	C		1.086	2.352	4.618	6.222
7	N		1.402	2.855	4.576	7.48
8	O	2p	1.313	3.387	5.295	7.47
9	F		1.681	3.372	6.044	8.40
10	Ne		2.080	3.953	6.0	9.4
11	Na		0.496	4.562	6.910	9.54
12	Mg	3s	0.737	1.450	7.729	10.54
13	Al		0.577	1.816	2.744	11.58
14	Si		0.786	1.576	3.230	4.35
15	P		1.011	1.902	2.911	4.96
16	S	3p	0.999	2.25	3.38	4.56
17	Cl		1.250	2.296	3.820	5.16
18	Ar		1.520	2.665	3.88	5.77
19	K		0.419	3.05	4.4	5.88
20	Ca	4s	0.589	1.145	4.91	6.5
21	Sc		0.631	1.235	2.388	7.13
22	Ti		0.658	1.309	2.650	4.17
23	V	3d	0.650	1.413	2.828	4.6
24	Cr		0.652	1.591	2.986	4.8
25	Mn		0.717	1.509	3.250	—
26	Fe		0.759	1.561	2.956	—
27	Co		0.758	1.645	3.230	—
28	Ni		0.736	1.752	3.430	—
29	Cu		0.745	1.957	3.553	—
30	Zn		0.906	1.733	3.830	—
31	Ga		0.579	1.979	2.962	6.19
32	Ge		0.762	1.537	3.300	4.41
33	As		0.946	1.794	2.734	4.83
34	Se	4p	0.941	2.05	3.1	4.1
35	Br		1.139	2.08	3.56	4.56
36	Kr		1.351	2.350	3.56	—

[a] Values from C. E. Moore, "Atomic Energy Levels", NBS Circular 467, Vol. III, 1958; see also R. W. Kiser, "Tables of Ionization Potentials", U.S. AEC, TID 6142, 1960, and unpublished revisions by C. E. Moore.

a strong inhomogeneous magnetic field. Classical theory predicted that the beam of silver atoms would spread out in the field because it was thought the magnetic moments of the atoms could take on any orientation. Stern and Gerlach actually observed a distinct splitting into two beams. Since the silver atom has a single s electron outside a closed shell, it has zero orbital angular momentum, and the splitting was clear evidence for the "space quantization" that results from the presence of spin angular momentum.

If one takes into account the spin of the electron, the one-electron wave function is a *spin orbital*, and it is a product of an orbital function and a spin function, where the latter can be either α or β, corresponding to the two possible values of the spin quantum number. Thus, the spin orbital $2p\alpha$ refers to an electron in a 2p state with a spin of $+\frac{1}{2}$.

Antisymmetrization of the wave function

Since the electrons in a many-electron system are indistinguishable, the wave function in a many-electron system should be such that $\psi\psi^*$ remains unchanged when the four quantum numbers that represent the spin orbitals of any two electrons are interchanged. That is,

$$\psi^2(1, 2) = \psi^2(2, 1)$$

and

$$\psi(1, 2) = \pm\psi(2, 1) \tag{4-79}$$

Equation 4-79 states that the wave function must be either symmetric (no change in sign) or antisymmetric (a change in sign) when the interchange is made.

The symmetric function leads to the *Bose-Einstein* statistics (Sec. 8-5), and particles with properties described by these statistics are called *bosons*. Photons, and atoms, ions, nuclei, and molecules with an even number of electrons, protons, and neutrons are bosons and have integral spin. The antisymmetric function leads to the *Fermi-Dirac* statistics (Sec. 13-2), and particles with properties described by these statistics are called *fermions*. Electrons, protons, and neutrons, as well as nuclei, atoms, ions, and molecules with an odd number of electrons, protons, and neutrons are fermions and have a half-integral spin. The striking differences in properties between ^{4}He and ^{3}He have been explained on this basis: that ^{4}He is a boson and ^{3}He a fermion.[5] The relationship between spin and statistics has been derived from relativistic quantum electrodynamics by Pauli, but no elementary explanation of the relationship has been developed.[6]

For an atom with one electron in each of two spin orbitals ψ_A and ψ_B, neither the product $\psi_A(1)\,\psi_B(2)$ nor $\psi_A(2)\,\psi_B(1)$ satisfies the antisymmetry requirement or the indistinguishability requirement. The linear combination

$$\psi_A(1)\psi_B(2) - \psi_A(2)\psi_B(1) \tag{4-80}$$

5. K. Mendelssohn, *The Quest for Absolute Zero*, McGraw-Hill Book Co., New York, 1966, p. 244.

6. R. P. Feynman, R. B. Leighton, and M. Sands, *The Feynman Lectures on Physics*, Addison-Wesley, Vol. III, Reading, 1965, p. 4-3.

does satisfy both requirements. A simple way to represent the function in Eq. 4-80 is as the determinant

$$
\begin{vmatrix}
\psi_A(1) & \psi_A(2) \\
\psi_B(1) & \psi_B(2)
\end{vmatrix}
\tag{4-81}
$$

For any product of one-electron functions, the determinant form analogous to Eq. 4-81 ensures satisfaction of the antisymmetry requirement, because an interchange of any two rows or two columns of a determinant results in a change of sign. The interchange of two rows or two columns of the determinant is equivalent to the interchange of two electrons between orbitals.

If any two electrons are represented by the same spin orbitals, so that they have four identical quantum numbers, the rows of the determinant are dependent. In this case the value of the determinant is zero. Thus, the Pauli exclusion principle in its elementary form is a consequence of the anti-symmetry requirement. Since a spin orbital is used to express the probability of finding an electron in a region of space, for every electron there is a region in which the probability of finding another electron of like spin is essentially zero. This region is sometimes called a *Fermi hole*.

When the self-consistent field calculation is applied to antisymmetrized wave functions rather than to simple products of spin orbitals, it is called the Hartree-Fock method.

The vector-model of angular momentum

The electron configuration, a product of spin orbitals, is an incomplete description of the state of a many-electron atom because it includes several angular momentum states of different energy. Consideration of the way in which angular momentum vectors of electrons couple permits us to distinguish among these states. The simplest model, called Russell-Saunders coupling, assumes that the orbital angular momentum vectors of the electrons couple to form a resultant orbital angular momentum vector **L**, that the spin angular momentum vectors couple to form a resultant spin angular momentum vector **S**, and that the two resultants couple to form a total angular momentum vector **J** for the atom. In each case, coupling arises from the interaction of the magnetic moments associated with the angular momentum vectors.

For two electrons, the model assumes that only those relative orientations of the orbital angular momentum vectors are possible that give the resultant quantum number L the quantized values

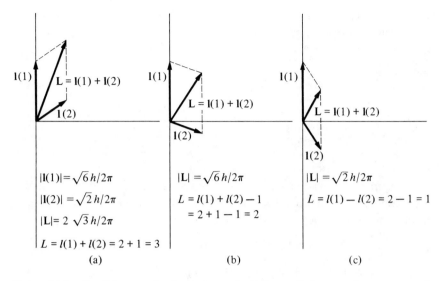

Figure 4-14. Possible quantized relative orientations of the angular momentum vectors with $l(1) = 2$ and $l(2) = 1$, according to the Russell-Saunders coupling. (a) $L = 3$, (b)$L = 2$, (c)$L = 1$.

$$[l(1) + l(2)], [l(1) + l(2) - 1], [l(1) + l(2) - 2],\ldots, |l(1) - l(2)|$$

Figure 4-14 shows the possible relative orientations of two angular momentum vectors with $l(1) = 2$ and $l(2) = 1$.

Exercise 4-18. Calculate the angle between $l(1)$ and $l(2)$ for $L = 2$.

Consider a two-electron atom with the configuration (2p)(3p). The angular momentum quantum numbers are $l(1) = 1$ and $l(2) = 1$. The possible values of m_l, the magnetic quantum number that describes the z component of the angular momentum, are $m_l(1) = 1, 0, -1$, and $m_l(2) = 1, 0, -1$. Table 4-5 shows the possible value of M_l, the z component of the resultant orbital angular momentum, as the nine

$$\{[2l(1) + 1][2l(2) + 1]\}$$

combinations $m_l(1) + m_l(2)$.

There is an appropriate subset $(2L + 1)$ of this set of nine values of M_l, corresponding to each possible value of L.

Thus, we have $M_l = 2, 1, 0, -1, -2$, corresponding to the resultant $L = 2$; $M_l = 1, 0, -1$, corresponding to $L = 1$; and $M_l = 0$, corresponding to $L = 0$. The symbols for $L = 0, 1, 2, \ldots$ are S, P, D, F, G $\ldots$.

Table 4-5. Addition of the z-component of orbital angular momentum according to Russell-Saunders coupling for the configuration (2p) (3p)

$m_l(1)$	1	1	0	-1	0	1	-1	0	-1
$m_l(2)$	1	0	1	1	0	-1	0	-1	-1
M_l	2	1	1	0	0	0	-1	-1	-2

Exercise 4-19. Find the possible values of L from the possible values of M_l when $l(1) = 3$ and $l(2) = 1$.

The spin angular momenta are added as vectors just as the orbital angular momenta are. Table 4-6 shows the possible values of M_s, the z component of the resultant spin angular momentum, as the four

$$\{[2s(1) + 1][2s(2) + 1]\}$$

combinations $m_s(1) + m_s(2)$. There is an appropriate subset $(2S + 1)$ of this set of four values of M_s corresponding to each possible value of S, the resultant spin angular momentum quantum number.

Table 4-6. Addition of the z-component of spin angular momentum according to Russell-Saunders coupling for the configuration (2p)(3p)

$m_s(1)$	$\frac{1}{2}$	$\frac{1}{2}$	$-\frac{1}{2}$	$-\frac{1}{2}$
$m_s(2)$	$\frac{1}{2}$	$-\frac{1}{2}$	$\frac{1}{2}$	$-\frac{1}{2}$
M_s	1	0	0	-1

Thus, we have $M_s = 1, 0, -1$ for $S = 1$, and $M_s = 0$ for $S = 0$. In general, the resultant S can have only the values $[s(1) + s(2)]$ down to $|s(1) - s(2)|$ in integral steps. The quantity $2S + 1$ defines the *multiplicity* of a group of states with a given value of L. The state for which $S = 0$ is called a *singlet* state, and the states for $S = 1$ are called a *triplet*.

The possible values of J, the total angular momentum quantum number, are obtained from the possible combinations of M_l and M_s to give M_J, in a similar fashion to that used in obtaining L and S. There are $2S + 1$ possible values of J for each value of L; $(L + S), (L + S - 1), \ldots |L - S|$. The set of $(2L + 1)(2S + 1)$ states corresponding to a given value of L and S is called a *term*. A term is written with a letter designation for the value of L and a left superscript representing the value of $2S + 1$. Thus, for a (2p) (3p) configuration, Table 4-7 shows the possible terms. The ^3S state is actually a single state, since for $L = 0$, $J = S$ is the only possible value. The superscript nevertheless indicates the value of S.

Table 4-7. Possible terms for a (2p)(3p) configuration

L	0	0	1	1	2	2
S	0	1	0	1	0	1
Term	^1S	^3S	^1P	^3P	^1D	^3D
Number of states	1	3	3	9	5	15

If the two p electrons have the same principal quantum number, only 15 combinations of M_l and M_s are possible, instead of 36, due to the Pauli exclusion principle. For $L = 2$, which includes states for which $m_l(1) = m_l(2)$, S must be equal to zero, so that only ^1D is possible, and not ^3D. The ^3S and ^1P terms are not allowed by the Pauli principle, but this cannot be demonstrated simply.

In a closed shell, such as $(ns)^2$, $(np)^6$, or $(nd)^{10}$, the Pauli exclusion principle requires that there be an equal number of electrons with $m_s = \frac{1}{2}$ and $m_s = -\frac{1}{2}$, so that $M_s = 0$. Similarly, for every electron with a positive value of m_l, there is an electron with a negative but equal value of m_l, so that $M_l = 0$. Thus, all closed shell atoms have a single term ^1S, and the angular momentum states of the atom are determined only by the open shell, or valence, electrons. A closed shell has spherical symmetry.[7]

The Energy of Angular Momentum States

If the state of electrons in atoms were accurately described by their interaction with the nucleus and with the spherically averaged field of the other electrons, as in the simplest form of the self-consistent field method, then all the terms of a configuration that we have described in the previous section would have the same energy. Since atomic spectra show more complex structure than is predicted by this simple model, we must assume that there are energy differences among terms.

Consider an atom with the configuration (1s) (2s). From our previous considerations, $L = 0$, and S can have the values 1 or 0, so that the possible terms are ^1S and ^3S. For a two-electron problem, the wave function can be factored into a spatial function and a spin function. Since the total wave function must be antisymmetric with respect to interchange of electrons, a symmetric spatial function must be combined with an antisymmetric spin function and vice versa.

Exercise 4-20. Verify that the configuration $(1s)(2s)$ has the terms ^1S and ^3S.

---

7. J. C. Slater, *Quantum Theory of Atomic Structure*, McGraw-Hill Book Co., Vol. I, New York, 1960, p. 182, p. 234.

The possible spin functions are

$$\alpha (1) \, \alpha (2)$$

$$\beta (1) \, \beta (2) \tag{4-82}$$

$$\alpha (1) \, \beta (2) + \beta (1) \, \alpha (2)$$

and

$$\alpha (1) \, \beta (2) - \beta (1) \, \alpha (2) \tag{4-83}$$

where the functions in Eq. 4-82 are symmetric, corresponding to $M_s = 1$, -1, and 0, respectively, $S = 1$, and the function in Eq. 4-83 is antisymmetric, corresponding to $M_s = 0$, $S = 0$.

The possible spatial functions are

$$\psi_{1s} (1) \, \psi_{2s} (2) + \psi_{1s} (2) \, \psi_{2s} (1) \tag{4-84}$$

and

$$\psi_{1s} (1) \, \psi_{2s} (2) - \psi_{1s} (2) \, \psi_{2s} (1) \tag{4-85}$$

The four possible spin orbitals formed as products of the orbital functions and the spin functions are

$$[\psi_{1s} (1) \, \psi_{2s} (2) - \psi_{1s} (2) \, \psi_{2s} (1)] \begin{Bmatrix} \alpha (1) \, \alpha (2) \\ \beta (1) \, \beta (2) \\ \alpha (1) \, \beta (2) + \alpha (2) \, \beta (1) \end{Bmatrix} \tag{4-86}$$

and

$$[\psi_{1s} (1) \, \psi_{2s} (2) + \psi_{1s} (2) \, \psi_{2s} (1)] \, [\alpha (1) \, \beta (2) - \alpha (2) \, \beta (1)] \tag{4-87}$$

Equation 4-86 represents the wave functions for a triplet of states in which $S = 1$ and the resultant spin vector has three possible orientations with respect to the z axis, whereas Eq. 4-87 represents a singlet state in which $S = 0$.

Since the 1s and 2s functions are both spherically symmetric, the probability function is dependent only on r_1 and r_2, the distances of the two electrons from the nucleus. In this case, the function to be calculated is the probability of finding electron 1 at a distance r_1 and electron 2 at a distance r_2. A convenient way to represent this function graphically, as Dickens and Linnett have done,[8] is to plot contour lines of constant ψ^2 as a function of r_1/a_0 and r_2/a_0, where a_0 is the Bohr radius. Such a plot is shown for the singlet in Fig. 4-15 and for the triplet in Fig. 4-16. It can be seen that there

8. J. W. Linnett and P. G. Dickens, *Quart. Rev. Chem. Soc.*, 11, 291 (1957).

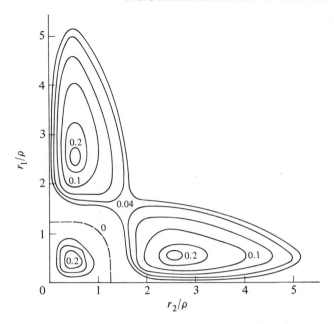

Figure 4-15. Contour diagrams for the probability of finding electron 1 at r_1 and electron 2 at r_2 for the helium atom in the singlet state of the (1s)(2s) configuration, as calculated from the wave function in Eq. 4-87. [By permission, from P. G. Dickens and J. W. Linnett, *Quart. Rev. Chem. Soc., 11*, 291 (1957).]

is a zero probability that $r_1 = r_2$ for the triplet function, indicating that there is a tendency for electrons of like spin not to occupy the same space. In contrast, there is a rather high probability that $r_1 = r_2$ for the singlet function, indicating that there is a tendency for electrons of opposite spin to be close together.

The calculations of the probability were made with one-electron wave functions that did not take into account the electrostatic repulsion between electrons or the magnetic properties related to electron spin. Thus, the difference between the spatial distributions is a consequence *only* of the antisymmetry requirement and the resultant difference between the orbital function of the singlet and triplet states. Since electrons do repel one another, consideration of the interelectronic repulsion leads to a lower energy for the triplet states and a higher energy for the singlet state. The kind of behavior described here forms the basis for *Hund's rule: For different terms of a configuration, the term with the greatest multiplicity has the lowest energy.* It is also true that, for terms of the same multiplicity, terms with higher values of L have lower energy.

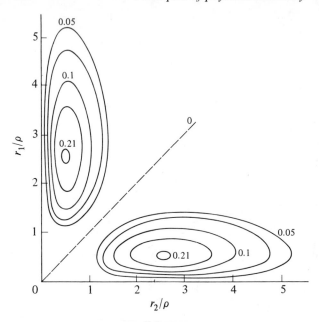

Figure 4-16. Contour diagrams for the probability of finding electron 1 at r_1 and electron 2 at r_2 for the helium atom in the triplet state of the (1s)(2s) configuration, as calculated from the wave function in Eq. 4-86. [By permission, from P. G. Dickens and J. W. Linnett, *Quart. Rev. Chem. Soc.*, *11*, 291 (1957).]

The energy difference between singlet and triplet states has sometimes been called "exchange energy" on the assumption that the wave functions in which the orbitals of electron 1 and electron 2 are interchanged represent different states. Linnett's suggested description, *spin-correlation energy*, seems to be a more useful way to indicate the effect of spin on the relationship between the positions of electrons.

Although the conclusions given in the preceding paragraphs are consistent with calculations using approximate wave functions, such as those in Eq. 4-86 and 4-87, it has been found that exact calculations for the excited states of helium result in a smaller average interelectronic distance for the triplet state than for the singlet. The lower energy of the triplet is due to greater nuclear attraction rather than to decreased interelectron repulsion.[9] From this result, we can see that caution must be exercised in drawing conclusions from approximate treatments.

9. R. L. Snow and J. L. Bills, *J. Chem. Educ.*, 51, 585–86 (1974).

Magnetic properties of electrons in atoms

In classical electromagnetic theory, any charged particle that has angular momentum also has a magnetic moment. The magnetic moment vector μ and the angular momentum vector **M** are related by the equation

$$\mu = \frac{q}{2m}\,\mathbf{M} \tag{4-88}$$

where q is the charge on the particle and m is its mass. The magnetic effects of the orbital motion of electrons in atoms are described by the classical relationship in the form

$$\mu_l = -\frac{e}{2m}\,\mathbf{L} \tag{4-89}$$

where μ_l is the orbital magnetic moment of the electrons, and L is the resultant orbital angular momentum vector. But, the spin angular momentum leads to a magnetic moment

$$\mu_s = -2\frac{e}{2m}\,\mathbf{S} \tag{4-90}$$

where **S** is the resultant spin angular momentum. The factor of 2 in Eq. 4-90 was at first an empirical result, but was later shown by Dirac to be a result of his relativistic theory of electron spin.[10] The magnitudes of the orbital and spin magnetic moments are

$$\mu_l = -\frac{e}{2m}\,\sqrt{L(L+1)}\,\frac{h}{2\pi}$$

$$= -\sqrt{L(L+1)}\,\mu_B \tag{4-91}$$

and

$$\mu_s = -2\sqrt{S(S+1)}\,\mu_B \tag{4-92}$$

where μ_B, the Bohr magneton, a natural unit of atomic magnetic moment, is equal to $eh/4\pi m$.

10. Precise measurements by Kusch and others have led to a value of 2.002319 instead of 2, and this result has been confirmed by Feynman and others on the basis of quantum electrodynamics. For simplicity, we shall continue to use 2. Further details can be found in P. Kusch, *Phys. Today*, 19(2), 23(1966). Variation in the value of this quantity for electrons in molecules accounts partly for the usefulness of electron spin resonance. (Sec. 6-5).

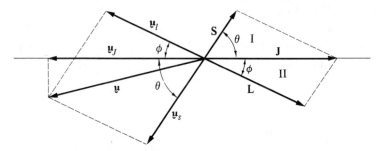

Figure 4-17. Geometric relationship among the resultant orbital and spin angular momentum vectors and magnetic moment vectors in Russell-Saunders coupling.

The resultant magnetic moment is

$$\mu = \mu_l + \mu_s$$

$$= -\frac{e}{2m}(L + 2S) \tag{4-93}$$

$$= -\frac{e}{2m}(J + S) \tag{4-94}$$

and is not colinear with the resultant angular momentum J. The geometric relationships are indicated in Fig. 4-17. When an applied magnetic field is small enough not to appreciably perturb the Russell-Saunders coupling, it is the projection μ_J of μ along the line of J that can be related to the effect of the magnetic field. The magnitude of the projection of μ along the line of J is given by

$$\mu_J = \mu_l \cos\phi + \mu_s \cos\theta \tag{4-95}$$

The values of $\cos\phi$ and $\cos\theta$ can be obtained from the law of cosines and the magnitudes of J, S, and L. In triangle I of Fig. 4-17

$$L(L + 1) = S(S + 1) + J(J + 1) - 2[S(S + 1) J(J + 1)]^{\frac{1}{2}} \cos\theta \tag{4-96}$$

Similarly, from triangle II,

$$S(S + 1) = L(L + 1) + J(J + 1) - 2[L(L + 1) J(J + 1)]^{\frac{1}{2}} \cos\phi \tag{4-97}$$

If we substitute for $\cos\theta$ and $\cos\phi$ from Eq. 4-96 and 4-97, and for the magnitudes of μ_l and μ_s from Eq. 4.91 and Eq. 4-92, then Eq. 4-95 becomes

$$\mu_J = \mu_B [J(J + 1)]^{\frac{1}{2}} \left\{ 1 + \frac{J(J + 1) + S(S + 1) - L(L + 1)}{2J(J + 1)} \right\} \tag{4-98}$$

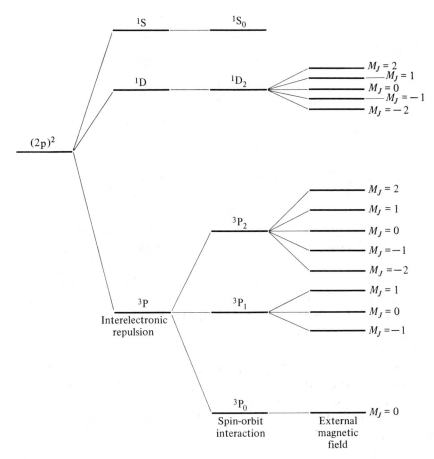

Figure 4-18. Scheme of energy separation of the terms of a $(2p)^2$ configuration in Russell-Saunders coupling due to interelectronic repulsion, with further splitting of the triplet term by spin-orbit magnetic interactions and Zeeman splitting in an external magnetic field.

Exercise 4-21. Show that Eq. 4-98 reduces to Eq. 4-91 when S, the resultant spin angular momentum is equal to zero, and to Eq. 4-92 when L, the resultant orbital angular momentum is equal to zero.

The factor in braces in Eq. 4-98 is usually symbolized by g and is called the *Landé g factor* or the *gyromagnetic ratio.*

The energy of interaction between the magnetic moment μ_J and the magnetic field **B** is given by

Principles of physical chemistry

$$\Delta E = \mu_J B \cos \delta \tag{4-99}$$

where B is the magnitude of the magnetic field and δ is the angle between J and B. Since $M_J (h/2\pi)$ is the z component of the total angular momentum,

$$\cos \delta = \frac{M_J (h/2\pi)}{[J(J+1)]^{\frac{1}{2}}(h/2\pi)}$$

$$= \frac{M_J}{[J(J+1)]^{\frac{1}{2}}} \tag{4-100}$$

Substituting from Eqs. 4-100 and 4-98, in Eq. 4-99, we obtain

$$\Delta E = g M_J \mu_B B \tag{4-101}$$

Using Eq. 4-101, one can see that, in a magnetic field, the degeneracy of the $2J + 1$ states corresponding to a given value of J in an atomic term can be resolved.

Figure 4-18 shows how the independent states of a $(2p)^2$ configuration are separated under the influence of internal interactions and an external magnetic field. As was indicated above, the energy of the three terms of this configuration differ as a result of differences in interelectronic repulsion due to spin correlation.

A smaller splitting of each term occurs as a result of magnetic interaction between the orbital and spin angular momentum. The subscript to the term symbol for each of these levels represents the value of J. Finally, in the presence of an external magnetic field, each level is split into $2J + 1$ levels corresponding to the possible values of M_J.

Problems

4-1. Consider a system in which an electron of mass m and charge $-e$ is constrained to move in a circle of constant radius R about a stationary nucleus of charge $+Z e$. Set up the time-independent Schrödinger equation in two dimensions, and convert it to polar coordinates, with $x = R \cos \phi$ and $y = R \sin \phi$. Solve the equation for the condition that

$$E + \left(\frac{Z e^2}{4\pi\varepsilon_0 R}\right) > 0.$$

Choose a solution that is related directly to the angular momentum. Apply the Schrödinger postulates and obtain a quantum number. Show that the resulting solution is an eigenfunction of the angular momentum operator $\hat{M}_z$. Derive an expression for the quantized energy levels. What is the significance of the two terms?

4-2. Plot polar diagrams of (a) p_z in the x-z plane, (b) s in the x-y plane, (c) d_{z^2} in the x-z plane, (d) $d_{x^2-y^2}$ in the x-y plane, and (e) d_{xz} in the x-z plane.

4-3. Calculate the value of r at which $r^2 RR^*$ is a maximum for the 1s, 2p, and 3d functions. How do they relate to the corresponding Bohr orbits?

4-4. Calculate the difference in energy between adjacent Zeeman levels in Fig. 4-18 when the atoms are in a magnetic field of 4 Tesla. If a transition between these levels were to be observed by absorption of electromagnetic radiation, what would be the frequency? In what part of the electromagnetic spectrum would it be found?

5 · Wave mechanics and molecular structure

Just as with the many-electron atom, the complete Schrödinger equation for molecules cannot be solved in closed form. We shall be concerned, therefore, with some of the approximate methods that have been developed to deal with the wave mechanics of molecules, with the strengths and weaknesses of each, and with their application to a variety of elements and simple compounds.

5-1. The Born-Oppenheimer approximation

If we neglect the motion of the molecule as a whole in space, then the total stationary state wave function ψ_t for a molecule is a function of the coordinates of the various nuclei and electrons relative to a convenient center in the molecule. The Schrödinger equation for the molecule, analogous to Eq. 4-102 for the helium atom, contains several terms for the kinetic energy of the nuclei and electrons as well as terms to represent the potential energy between pairs of nuclei, between each electron and the field of the nuclei, and between pairs of electrons. The energy E_t is an eigenvalue of the wave function ψ_t.

The motion of the nuclei is much slower than that of the electrons because of their greater mass, and Born and Oppenheimer, in 1927, suggested that the electrons adjust rapidly enough to the motion of the nuclei that electronic wave functions and energies could be calculated at successive fixed positions of the nuclei, as if the nuclei were at rest. The eigenvalues E_e obtained from the calculation represent the kinetic and potential energy of the electrons,

132

but the potential energy only of the nuclei. The eigenfunctions obtained, ψ_e, are functions of the electron coordinates for a fixed value of the nuclear coordinates, and differ from ψ_t.

The values of E_e are functions of the nuclear coordinates rather than constants of the system. They are substituted as potential energy terms into the Schrödinger equation for the motion of the nuclei. The solutions ψ_n of this equation then give a description of the rotational and vibrational motion of the molecule. These solutions will be discussed in Chapter 6.

The Born-Oppenheimer approximation utilizes the values E_e obtained by combining the potential energies of the nuclei with respect to each other, and to the electrons, and the kinetic energy of the electrons to describe the motion of the nuclei. From the Hellmann-Feynman theorem, the force acting on the nuclei is just that calculated by classical electrostatics utilizing the charges and positions of the nuclei and the electron distribution determined by the *exact* wave function. Thus, we assume that once a satisfactory electron distribution has been found using wave mechanics, the forces can be approximated by applying classical considerations.

Though the complete Schrödinger equation for the molecule is not exactly separable into nuclear and electronic motion, for most systems ψ_t can be approximated by

$$\psi_t = \psi_e \psi_n \tag{5-1}$$

5-2. Molecular orbitals

The hydrogen molecule ion
The simplest molecular system is the hydrogen molecule ion, consisting of two protons and one electron. The Schrödinger equation for the hydrogen molecule ion can be separated and solved exactly (within the Born-Oppenheimer approximation) in elliptical coordinates.[1] These solutions, however, do not provide a convenient basis for approximate solutions of more complex molecules. We shall, therefore, refer to values obtained from the exact solutions only to compare them to simpler but approximate methods.

In the *molecular orbital* method, an approximate wave function for a molecule containing many electrons is expressed as a product of one-electron functions or *orbitals*. In this respect, it is analogous to the self-consistent field method for many-electron atoms. In contrast to the atomic problem, however, there is no exact one-electron function available as a starting point,

1. D. R. Bates, K. Ledsham, and A. L. Stewart, *Phil. Trans. Roy. Soc. London, A 246*, 215 (1953).

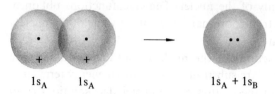

$1s_A \quad 1s_A \qquad\qquad 1s_A + 1s_B$

Figure 5-1. Schematic contour surface of constant ψ^2 for $1s_A + 1s_B$. (Adapted by permission, from A. C. Wahl, Argonne Natl. Lab. Report 7076, 1965.)

except for $H_2{}^+$. The most common procedure is to use *linear combinations* of *atomic orbitals*, usually with adjustable parameters, as the starting point for a molecular orbital; hence the common acronym LCAO-MO. We can see some of the strengths and weaknesses of the LACO-MO method by applying it to $H_2{}^+$, where it can be compared to the results of exact calculations. Also, in the case of $H_2{}^+$, the method can be tested without the complication of interelectronic repulsion.

A trial wave function frequently used to represent the $H_2{}^+$ molecule is a linear combination of the hydrogen $1s$ functions,

$$\psi = \psi_A + \psi_B$$
$$= 1s_A + 1s_B \tag{5-2}$$

where $1s_A$ and $1s_B$ are functions of the electron's distance from nucleus A and nucleus B, respectively. As R, the internuclear distance, increases without limit at one extreme, a satisfactory representation should lead to a hydrogen atom in the $1s$ state and a proton; at the other extreme, as R approaches zero, to a helium ion in the $1s$ state.

We can write Eq. 5-2 as

$$\psi = N_1 \exp\left(\frac{-Z\,r_A}{a_0}\right) + N_2 \exp\left(\frac{-Z\,r_B}{a_0}\right) \tag{5-3}$$

Figure 5-2. Schematic contour curve of constant ψ^2 for $1s_A + 1s_B$. (Adapted by permission, from A. C. Wahl, Argonne Natl. Lab. Report 7076, 1965.)

$1s_A \qquad\qquad 1s_B \qquad\qquad 1s_A + 1s_B$

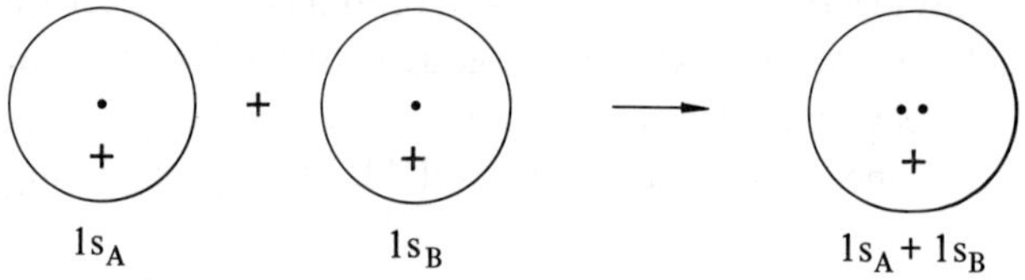

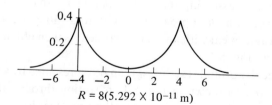

$R = 8(5.292 \times 10^{-11} \, \text{m})$

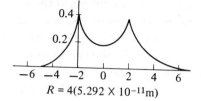

$R = 4(5.292 \times 10^{-11} \text{m})$

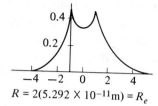

$R = 2(5.292 \times 10^{-11} \text{m}) = R_e$

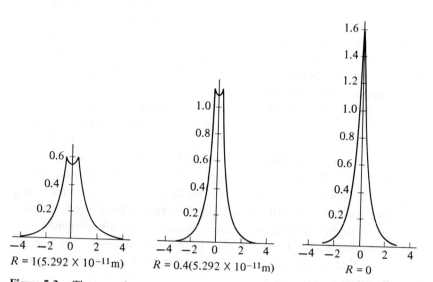

$R = 1(5.292 \times 10^{-11} \text{m})$ $R = 0.4(5.292 \times 10^{-11} \text{m})$ $R = 0$

Figure 5-3. The wave function $1s_A + 1s_B$ plotted as a function of position along the internuclear (z) axis at several values of R. Vertical scale, relative electron density; horizontal, $z/5.292 \times 10^{-11}$ m. (By permission, from J. C. Slater, *Quantum Theory of Matter*, McGraw-Hill Book Co., New York, 1951.)

using expressions from Table 4-3, where N_1 and N_2 are normalization constants. As R becomes very large, that is as r_A approaches infinity and r_B remains small, Eq. 5-3 does in fact reduce to the hydrogen $1s$ function. As R approaches zero, that is as $r_A = r_B$, Eq. 5-3 would become a $1s$ function for the He^+ ion if Z were to become 2. This change in the value of Z can be accomplished by assuming that Z is a parameter that varies continuously from a value of 1 as R approaches infinity to a value of 2 as R approaches zero. Good agreement with an exact treatment is then obtained, as shown in Table 5-1 and Fig. 5-6.

The molecular orbital $(1s_A + 1s_B)$ can be represented by a contour surface of constant ψ^2, as shown in Fig. 5-1. A section through this surface in a plane containing the internuclear axis (the z axis) is shown in Fig. 5-2. The solid surface in Fig. 5-1 can be obtained by rotating the curve in Fig. 5-2 about the z axis. Since ψ is a function of three variables (a convenient choice for a linear diatomic molecule is cylindrical polar coordinates, where z represents position along the internuclear axis, ρ distance from the axis, and ϕ the angle about the axis), a plot of ψ would require four dimensions.

Let us consider the nuclei A and B at some fixed distance R apart. The point midway between them is a convenient origin 0. In order to plot ψ for an electron as a function of z, the distance along the internuclear axis measured from 0, we recall that the function has cylindrical symmetry, so ψ is independent of ϕ. If we also choose a fixed value of ρ, the distance of the electron from the axis, the value of ψ as a function of z can be plotted. Examples of such graphs are shown in Fig. 5-3 for several values of R. For other values of ρ, the curve will have the same shape, and, at any fixed value for z, ψ decreases exponentially with increasing ρ.

The graph for the equilibrium internuclear distance R_e shows that the value for ψ^2 at a point along the z axis midway between the nuclei is almost as large as ψ^2 at the nuclei. Also, the value of ψ^2 at the nuclei is larger than that in separated atoms. Recent detailed calculations by Ruedenberg and his students have shown that the energy of H_2^+ is less than the sum of the energies of a proton and a hydrogen atom largely because the probability of finding the electron near the nuclei in H_2^+ is greater than in the free atoms. These results do not support the commonly made qualitative statement that energy lowering results from the high probability of finding an electron between the nuclei, where its mutual attraction for both nuclei overcomes the internuclear repulsion.[2]

2. M. J. Feinberg, K. Ruedenberg, and E. L. Mehler, *Adv. Quant. Chem.*, 5, 28–98 (1970). M. J. Feinberg and K. Ruedenberg, *J. Chem. Phys.*, 54, 1495–1511 (1971).

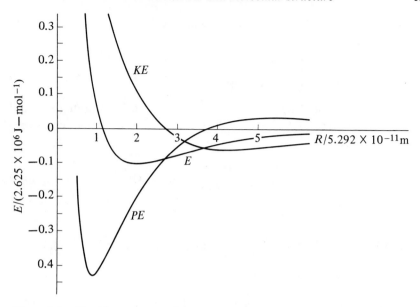

Figure 5-4. The kinetic, potential, and total energies (*KE*, *PE*, and *E*) of the ground state of H_2^+ as functions of the internuclear distance. The zero of energy is that of a hydrogen atom and a proton at infinite distance. Nuclear repulsion is included. (By permission, from J. C. Slater, *Quantum Theory of Matter*, McGraw-Hill Book Co., New York, 1951.)

The results of Slater's calculations[3] from the exact wave function of the kinetic energy of the electron, the potential energy of the protons and the electron, and the total energy, as functions of R, are shown in Fig. 5-4 for H_2^+. The state of zero energy is taken to be a hydrogen atom and a proton at infinite R.

An explanation of the potential energy changes comes from electrostatics, according to the Hellmann-Feynman theorem, but the changes in kinetic energy of the electron must come from a quantum mechanical model. Consider a particle in a cubic container with edge length a and the origin of the coordinate system at one corner. We assume that the potential energy is constant inside the container and can be set equal to zero. We also assume that the potential energy becomes infinite at the walls of the container, so that ψ equals zero at the walls. This model is described as *a particle in a three-dimensional infinite potential well*. The Schrödinger equation for

3. J. C. Slater, *Quantum Theory of Matter*, McGraw-Hill Book Co., New York, 1951, p. 406.

stationary states of such a particle is obtained by expanding Eq. 4-24 to describe a function of three variables.

$$-\frac{h^2}{8\pi^2 m}\left[\frac{\partial^2 \psi}{\partial x^2} + \frac{\partial^2 \psi}{\partial y^2} + \frac{\partial^2 \psi}{\partial z^2}\right] = E\,\psi(x, y, z) \qquad (5\text{-}4)$$

If we assume that ψ is a product $X(x)\ Y(y)\ Z(z)$, and substitute for ψ and $\partial^2 \psi/\partial x^2$, $\partial^2 \psi/\partial y^2$, and $\partial^2 \psi/\partial z^2$, Eq. 5-4 can be separated into three equations of the form

$$-\frac{h^2}{8\pi^2 m}\left(\frac{d^2 X}{dx^2}\right) = E_x\, X(x) \qquad (5\text{-}5)$$

where $E = E_x + E_y + E_z$.

Exercise 5-1.   Show that Eq. 5-4 separates into three equations of the form of Eq. 5-5.

A general solution of Eq. 5-5 is

$$X = A \sin k\,x + B \cos kx \qquad (5\text{-}6)$$

where

$$k = (8\pi^2 m E_x/h^2)^{\frac{1}{2}} \qquad (5\text{-}7)$$

Exercise 5-2.   Show by substitution that Eq. 5-6 is a solution of Eq. 5-5.

Since ψ and therefore X must equal zero at $x = 0$, B must equal zero. Since ψ and therefore X must equal zero at $x = a$, ka must be an integral multiple of π. Thus,

$$k = n_x\frac{\pi}{a} \qquad n_x = 1, 2, 3, \ldots \qquad (5\text{-}8)$$

and the solutions to Eq. 5-5 consistent with the Schrödinger postulates are

$$X = A \sin n_x\,\pi\,\frac{x}{a} \qquad (5\text{-}9)$$

The corresponding solutions for the y and z equations are

$$Y = A \sin n_y\,\pi\,\frac{y}{a} \qquad (5\text{-}10)$$

and

$$Z = A \sin n_z\,\pi\,\frac{z}{a} \qquad (5\text{-}11)$$

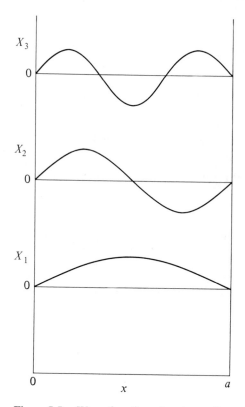

Figure 5-5. Wave functions for the x dimension of a particle in an infinite three-dimensional potential well. The three lowest levels are shown. The length of one side of the well is a.

These solutions are the same as the equations for the amplitude of standing waves on a vibrating string, the ends of which are fixed at a distance a. The significance of the solution is the same in both cases; a standing wave, or a stationary state in a bound quantum system, is possible only if there is an integral number of half wavelengths between the boundaries of the system. Figure 5-5 shows the dependence of X on x for $n_x = 1, 2$, and 3.

From Eq. 5-7 and Eq. 5-8 we find that

$$E_x = \frac{n_x^2 h^2}{8ma^2} \tag{5-12}$$

Thus the energy of the particle is quantized, and there is a lowest energy, the *zero-point energy*, which is not zero.

Exercise 5-3. Show that X is identically equal to zero for $n_x = 0$.

The values of E_y and E_z are given by equations of the same form as Eq. 5-12, so that

$$E = E_x + E_y + E_z = \frac{(n_x{}^2 + n_y{}^2 + n_z{}^2)\,h^2}{8\,ma^2} \qquad (5\text{-}13)$$

It can be seen from Eq. 5-13 that the energy, which is equal to the kinetic energy, increases as the size of the container decreases. This relation explains the changes in kinetic energy as two protons come together under the influence of an electron to form $H_2{}^+$.

Figure 5-6. Kinetic plus potential energy of the electron in $1s_A + 1s_B$. Comparison of the simple LCAO function, the variation function in which the effective nuclear charge is a function of R, and the exact function. Nuclear repulsion is not included. The zero of energy is that of a free electron, the horizontal axis is the energy of an electron in the ground state of the H atom, and $-2.0\,(2.625 \times 10^6)$ J-mol^{-1} is the energy of an electron in the ground state of He$^+$. (By permission, from J. C. Slater, *Quantum Theory of Matter*, McGraw-Hill Book Co., New York, 1951.)

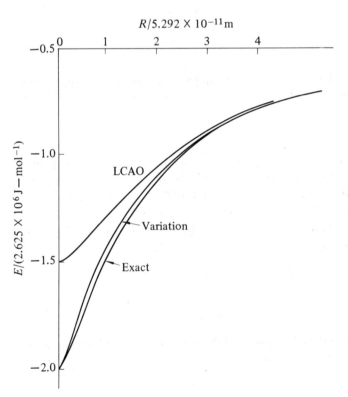

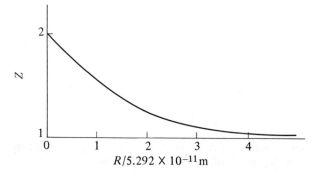

Figure 5-7. The variation of effective nuclear charge with R required to achieve the energy curve in Fig. 5-6. (By permission, from J. C. Slater, *Quantum Theory of Matter*, McGraw-Hill Book Co., New York, 1951.)

As the proton and the hydrogen atom come closer together, the initial *increase* in potential energy and *decrease* in kinetic energy, is seen as a decrease in total energy, due to an increase in the size of the region in which the electron moves—from a region entirely about one nucleus to a region about two well-separated nuclei. As R is decreased further, the electron is effectively constrained to move in a smaller region of space, both because electron density is shifted from the region outside the nuclei to the region between the nuclei and because the electron density about each nucleus contracts sharply. This effect is seen in going from a to b to c in Fig. 5-3. The consequent *decrease* in potential energy and *increase* in kinetic energy leads to a decrease in the total energy. The contraction of electron density about the two nuclei is found to be quantitatively more important than the increase in electron density between the nuclei. As the distance between the nuclei continues to decrease, the energy begins to increase, due to an increase in kinetic energy, while the potential energy is still decreasing. It is only when the internuclear distance becomes less than one-half the equilibrium internuclear distance that internuclear repulsion produces an increase in potential energy.

Though the simple molecular orbital function in Eq. 5-3 provides a reasonable approximation to the total energy when the nuclear charge Z is assumed constant and equal to 1, it deviates sharply from the exact energy at small R, and it predicts a minimum in the kinetic energy and an increase in the potential energy at the equilibrium distance. If we assume that Z is a variable *effective* nuclear charge, instead of a constant, a better approximation is obtained. For example, Fig. 5-6 shows the energy approximation obtained when Z is chosen to minimize $\overline{E}$ at each value of R. Internuclear repulsion is omitted. The values of Z required are shown in Fig. 5-7. The increase in

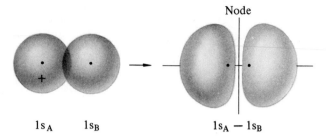

Figure 5-8. Schematic contour surface of constant ψ^2 for $1s_A - 1s_B$. (Adapted by permission, from A. C. Wahl, Argonne Natl. Lab. Report 7076, 1965.)

the effective nuclear charge is consistent with the contraction of electron density about the nuclei indicated in Fig. 5-3.

The method described agrees fairly well with the exact calculation for the total energy, but, since the energy involved in chemical reactions is a small fraction of the total energy, a small error in the total energy may result in a large error in the dissociation energy, as seen in Table 5-1.

Table 5-1. Calculated values of the equilibrium internuclear distance, the total energy, and the dissociation energy of H_2^+.

	Exact calculation	Simple LCAO[a]	Variation LCAO[b]	Experimental data
$R_e/10^{-10}$ m	1.06	1.32	1.06	1.06
$D_e/9.65 \times 10^4$ J-mol^{-1}	2.79	1.76	2.25	2.79
Total energy/9.65×10^4 J-mol^{-1}	16.39	15.36	15.85	16.39

[a] J. W. Linnett, *Wave Mechanics and Valency*, Methuen, London, 1960, p. 91.
[b] B. Finkelstein and G. Horowitz, *Z. Physik*, *48*, 118 (1928).

An alternative to the trial wave function in Eq. 5-2 is

$$\psi = 1s_A - 1s_B \tag{5-14}$$

Figure 5-9. Contour curve of constant ψ^2 for $1s_A - 1s_B$. (Adapted by permission, from A. C. Wahl, Argonne Natl. Lab. Report 7076, 1965.)

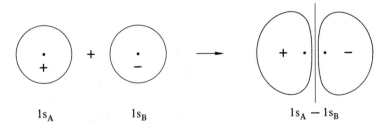

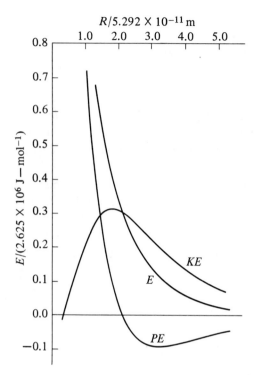

Figure 5-10. Kinetic, potential, and total anti-binding energy (*KE*, *PE*, and *E*) for $1s_A - 1s_B$ as a function of *R*. Nuclear repulsion is included. Zero as in Fig. 5-4. [By permission, from M. J. Feinberg, K. Ruedenberg, and E. L. Mehler, *Adv. Quant. Chem.*, *5*, 28–98 (1970).]

The wave functions add with opposite sign, and they interfere destructively, as would waves of opposite phase. The resulting value of ψ^2 in the region between the nuclei is small, and there is a node midway between the nuclei, as shown in Fig. 5-8 and Fig. 5-9. Ruedenberg's curves for the kinetic energy, the potential energy, and the total energy as a function of internuclear distance are shown in Fig. 5-10. Initially, as the nuclei approach, the kinetic energy increases and the potential energy decreases as the electron density contracts about the two nuclei. As the internuclear distance decreases further, the electron density between the nuclei decreases and the electron density outside the nuclei increases, resulting in an increase in potential energy and a decrease in kinetic energy. The total energy increases steadily as the internuclear distance decreases, indicating that the wave function in Eq. 5-14 represents an unstable state. The molecular orbital is therefore called an *antibonding orbital*.

Allen has pointed out that hydrogen is almost unique among the elements with respect to orbital contraction when it forms bonds.[4] It remains for detailed calculations, such as Ruedenberg's, to be carried out for bonds involving other atoms to show whether Ruedenberg's conclusions are generally applicable in all details.

Excited states of H_2^+. *Many-electron homonuclear diatomic molecules*
Linear combinations of higher energy atomic orbitals are used to construct molecular orbitals of higher energy. A number of systems have been devised to describe these molecular orbitals.

One system is based on the atomic orbitals of the separated atoms. The small Greek letters σ, π, δ, ϕ are used to denote the angular momentum of the electron about the internuclear axis, representing $m_l = 0$, ± 1, ± 2, and ± 3 units of magnitude $h/2\pi$. A subscript g (gerade) or u (ungerade) to the angular momentum designation refers to the symmetry of the wave function on inversion through the center of the molecule. If $\psi(x, y, z) = \psi(-x, -y, -z)$, the subscript is g. If $\psi(x, y, z) = -\psi(-x, -y, -z)$, the subscript is u. The antibonding orbitals, such as $(1s_A - 1s_B)$, are indicated by an asterisk. An asterisk is also used to represent antisymmetry with respect to reflection in a plane perpendicular to the internuclear axis at the point midway between the nuclei. The angular momentum quantum number and the symmetry characteristics are conserved in the transformation from atomic orbitals to molecular orbitals. The symmetries depend on the way in which the atomic orbitals are combined and cannot be inferred from the atomic orbitals alone. For H_2^+, the lowest state is $\sigma_g(1s)$, as is illustrated (near the right) in Fig 5-11, whereas $(1s_A - 1s_B)$ is $\sigma_u^*(1s)$.

A second system involves designations appropriate to the united atom that would be formed at $R = 0$. Angular momentum and symmetry are conserved in this system also. For H_2^+, the designation of the lowest energy molecular orbital is $1s\sigma_g$, whereas the lowest antibonding orbital is $2p\sigma_u^*$. These designations are shown (near the left) in Fig. 5-11.

A third system, which appears in the center of Fig. 5-11, simply designates orbitals of a given angular momentum and symmetry in order of increasing energy. Since σ_u and π_g always involve antibonding orbitals, the asterisk is not used for homonuclear molecules. This system takes into account the fact that a more complete molecular orbital treatment for diatomic molecules uses linear combinations of all the atomic orbitals of a given symmetry and angular momentum to form the molecular orbitals; therefore, neither the separated atoms nor the united atom are indicated.

4. L. C. Allen, in *Quantum Theory of Atoms, Molecules, and the Solid State* (P.-O. Lowdin, ed.), Academic Press, New York, 1966.

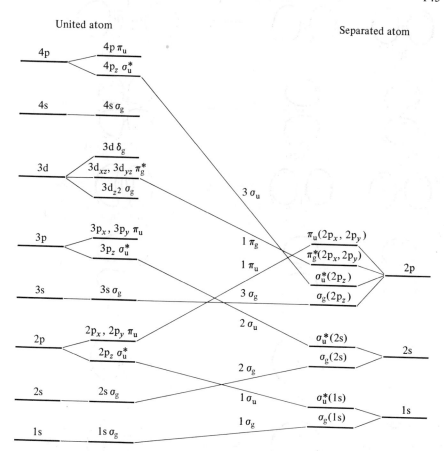

Figure 5-11. Correlation diagram relating molecular orbitals of diatomic molecules to separated atoms and united atoms. [Adapted by permission, from R. S. Mulliken, *Rev. Mod. Phys., 4, 1* (1932).]

The three systems are illustrated in Fig. 5-11; the vertical placement of the orbitals represents relative energy schematically. The hydrogen 2s and 2p orbitals have the same energy, but, for other homonuclear diatomic molecules, for which Fig. 5-11 also applies, they are separated as shown. In Figure 5-12 are shown the approximate contour lines of constant ψ^2 for the separated atoms and the molecular orbitals of Fig. 5-11.

In the orbital approximation, it is assumed that the wave function of a many-electron system, a function of the coordinates of all the electrons, can be expressed as a product of one-electron functions. For molecules, as for atoms, the *electron configuration* is a product of these one-electron functions.

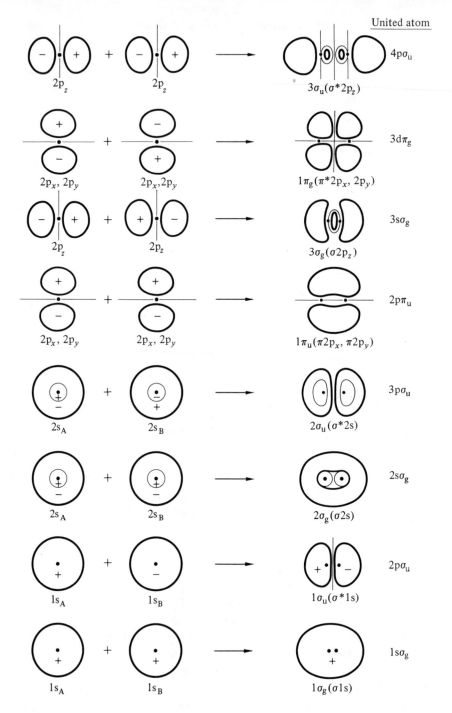

Figure 5-12. Schematic contour curves of constant ψ^2 for molecular orbitals of diatomic molecules. (Adapted by permission, from A. C. Wahl, Argonne Natl. Lab. Report 7076, 1965.)

The molecular orbitals are arranged in order of increasing energy, and the electrons are assigned to orbitals starting with the lowest energy orbital, in accord with the Pauli principle and Hund's rules. The orbitals of Fig. 5-12 are arranged roughly in order of increasing energy, but there is some variation in the relative energy of the $3\sigma_g$ and the $1\pi_u$ in different molecules.

Thus, H_2^+, with one electron, has a ground state configuration $1\sigma_g$. The molecule H_2, with two electrons, has the configuration $(1\sigma_g)^2$. The helium molecule ion He_2^+ has the configuration $(1\sigma_g)^2 (1\sigma_u)$ and would be expected to have some stability (it does), since it has two bonding electrons and one antibonding electron. But He_2, with the configuration $(1\sigma_g)^2 (1\sigma_u)^2$, is unstable because the destabilizing effect of two antibonding electrons is greater than the stabilizing effect of two bonding electrons.

A new question arises with B_2. Its configuration is $(1\sigma_g)^2 (1\sigma_u)^2 (2\sigma_g)^2 (2\sigma_u)^2 (1\pi_u)^2$. The $1\pi_u$ orbital has an angular momentum of magnitude $h/2\pi$, and m_l can be either $+1$ or -1, so that there are two $1\pi_u$ orbitals of equal energy. According to Hund's rule, the stable arrangement occurs with the two electrons in separate orbitals with parallel spin. Thus, the configuration is $(1\sigma_g)^2 (1\sigma_u)^2 (2\sigma_g)^2 (2\sigma_u)^2 (1\pi_{u,\,x}) (1\pi_{u,\,y})$.

Exercise 5-4. Write the electron configurations for the diatomic molecules Li_2, C_2, N_2, O_2, F_2, and Ne_2. Which molecules would be expected to be stable? Relate your answer to the conventional Lewis structures for N_2, O_2, F_2, and Ne_2. Predict configurations and structures for O_2^+, O_2^-, and O_2^{2-}, and find values of bond length and bond energy to check your predictions.

A term, for a molecule as for an atom, describes the electronic state more precisely than a configuration. For a diatomic molecule, we have the notation given in Table 5-2, where Λ is the resultant orbital angular momentum quantum number and M_l is the resultant angular momentum about the z axis.

Table 5-2. Symbols for the orbital angular momentum states of diatomic molecules

Λ	M_l	Term symbol
0	0	Σ
1	$\pm h/2\pi$	Π
2	$\pm 2h/2\pi$	Δ

A left superscript indicates the multiplicity $(2S + 1)$, where S is the resultant spin angular momentum in units of $h/2\pi$. A right subscript g or u is added for homonuclear diatomic molecules, depending on whether there are an even or odd number, respectively, of electrons with u orbitals. A right superscript is negative if a plane in which the z axis lies can be found such that the total wave function is antisymmetric with respect to this plane; if not, it is positive. For Δ and Π states, which are doubly degenerate, one state is positive and the other is negative, so the notation is not used. A $^1\Sigma$ state has a spin function that is antisymmetric with respect to the interchange of electrons, so its spatial function must be symmetric. The $^3\Sigma$ states have symmetric spin functions, so their spatial function must be antisymmetric. (The spatial function and the spin function can be separated only in the case of two electrons.) That this requirement of the Pauli principle results in $^1\Sigma^+$ and $^3\Sigma^-$ states is shown below.

Consider a molecule with a $(\pi)^2$ configuration, in which one electron has $m_l = 1$ and the other has $m_l = -1$. The ϕ-dependent parts of the molecular orbital are then $\exp i\phi$ and $\exp(-i\phi)$. Two possible products are

$$\exp i\phi\,(1) \exp(-i\phi)\,(2)$$

and

$$\exp(-i\phi)\,(1) \exp i\phi\,(2)$$

To obtain a wave function that is consistent with the indistinguishability of electrons, we use the following linear combinations of these products.

$$\Phi_+ = \exp(i\phi)\,(1)\exp(-i\phi)\,(2) + \exp(-i\phi)\,(1)\exp(i\phi)\,(2) \quad (5\text{-}15)$$

$$\Phi_- = \exp(i\phi)\,(1)\exp(-i\phi)\,(2) - \exp(-i\phi)\,(1)\exp(i\phi)\,(2) \quad (5\text{-}16)$$

Reflection in a plane in which the z axis lies is equivalent to the replacement of ϕ by $-\phi$. Thus, Φ_+, which is symmetric with respect to an interchange of electrons, is also symmetric with respect to reflection, and Φ_-, which is antisymmetric with respect to an interchange of electrons, is also antisymmetric with respect to reflection.[5]

Exercise 5-5. Show that the possible terms for the lowest energy configuration of O_2 are $^3\Sigma_g^-$, $^1\Sigma_g^+$, and $^1\Delta_g$.

Heteronuclear diatomic molecules

If the two atoms that unite to form a heteronuclear diatomic molecule are sufficiently alike, the molecular orbital designations in Fig. 5-11 can be used to describe the electron configuration of the molecule. For example, NO has the same number of electrons as O_2^+, and the configuration would be $(1\sigma)^2$

---

5. J. N. Murrell, S. F. A. Kettle, and J. M. Tedder, *Valence Theory*, John Wiley & Sons, Ltd., London, 1965, pp. 143–44.

$(1\sigma^*)^2 (2\sigma)^2 (2\sigma^*)^2 (1\pi)^4 (3\sigma)^2 (1\pi^*)$. Similarly, CO is isoelectronic with N_2, so its configuration would be $(1\sigma)^2 (1\sigma^*)^2 (2\sigma)^2 (2\sigma^*)^2 (1\pi)^4 (3\sigma)^2$. Since the atoms are not identical, the linear combination of atomic orbitals that is used to approximate the molecular orbitals does not have identical coefficients for the atomic orbitals of the two atoms. For example, for CO, one might use

$$3\sigma = c_1 (2s)_C + c_2 (2s)_O \tag{5-17}$$

with 1σ and 2σ the 1s orbitals of C and O.

The values of c_1 and c_2 are chosen to minimize the energy corresponding to the wave function in Eq. 5-17, since the *variation theorem*[6] states that the wave function of a given form with the lowest energy is the best approximation of that form to the correct wave function.

The energy corresponding to the trial wave function 3σ in Eq. 5-17 is given by the expectation value of the Hamiltonian function, which is, for an unnormalized real wave function

$$\langle H \rangle = \frac{\int (3\sigma) \, \hat{H} \, (3\sigma) \, dv}{\int (3\sigma) \, (3\sigma) \, dv} \tag{5-18}$$

The Hamiltonian function for this system includes terms for the potential energy of nuclear repulsion, the potential energy of attraction between each electron-nucleus pair, the potential energy of interelectron repulsion, and the kinetic energy of the electrons. The nuclear repulsion is a constant for a fixed internuclear distance, and it is frequently omitted from the initial calculation.

If we substitute from Eq. 5-17 in Eq. 5-18, we obtain

$$E = \langle H \rangle = \frac{\int [c_1 \, (2s)_C + c_2 \, (2s)_O] \, \hat{H} \, [c_1 \, (2s)_C + c_2 \, (2s)_O] \, dv}{\int [c_1 \, (2s)_C + c_2 \, (2s)_O] \, [c_1 \, (2s)_C + c_2 \, (2s)_O] \, dv}$$

$$= \frac{c_1^2 \, H_{11} + 2c_1 c_2 \, H_{12} + c_2^2 \, H_{22}}{c_1^2 \, S_{11} + 2c_1 c_2 \, S_{12} + c_2^2 \, S_{22}} \tag{5-19}$$

where
$$H_{11} = \int (2s)_C \, \hat{H} \, (2s)_C \, dv \tag{5-20}$$

$$H_{22} = \int (2s)_O \, \hat{H} \, (2s)_O \, dv \tag{5-21}$$

$$H_{12} = \int (2s)_C \, \hat{H} \, (2s)_O \, dv = \int (2s)_O \, \hat{H} \, (2s)_C \, dv \tag{5-22}$$

$$S_{12} = \int (2s)_C \, (2s)_O \, dv = \int (2s)_O \, (2s)_C \, dv \tag{5-23}$$

6. J. C. Slater, *Quantum Theory of Atomic Structure*, McGraw-Hill Book Co., Vol. I, New York, 1960, p. 110.

and

$$S_{11} = S_{22} = \int (2s)_{\mathrm{C}}^2 \, dv = \int (2s)_{\mathrm{O}}^2 \, dv = 1 \qquad (5\text{-}24)$$

The quantity H_{11} is just the energy of a 2s orbital in an isolated C atom, denoted E_1, and H_{22} is the energy E_2 of a 2s orbital in an isolated oxygen atom. Thus,

$$E = \frac{c_1^2 \, E_1 + c_2^2 \, E_2 + 2 \, c_1 c_2 \, H_{12}}{c_1^2 + c_2^2 + 2 \, c_1 c_2 \, S_{12}} \qquad (5\text{-}25)$$

To determine the minimum energy, Eq. 5-25 is differentiated with respect to c_1 and c_2, and the two partial derivatives are set equal to zero. The resulting equations are

$$c_1 \, (E_1 - E) + c_2 \, (H_{12} - ES_{12}) = 0$$

$$(5\text{-}26)$$

$$c_1 \, (H_{12} - ES_{12}) + c_2 \, (E_2 - E) = 0$$

Exercise 5-6. Verify Eq. 5-26. Hint: When setting the partial derivatives equal to zero to find the energy minimum, retain the denominator so that E will be in the final equations.

Equations 5-26 have a non-zero solution only if the determinant of the coefficients of c_1 and c_2 equals zero. Thus,

$$\begin{vmatrix} (E_1 - E)(H_{12} - ES_{12}) \\ (H_{12} - ES_{12})(E_2 - E) \end{vmatrix} = 0 \qquad (5\text{-}27)$$

This is a quadratic equation in E, with two real roots. If $E_1 < E_2$, one value of E will be less than E_1, and the other will be greater than E_2, as shown in Fig. 5-13, where the value of the determinant is plotted against E. Since we are interested in the values of E for which the determinant equals zero, the two equations in Eq. 5-26 are dependent, and we can obtain only a ratio c_1/c_2 when we substitute each of the values of E into Eq. 5-26 and solve the resulting equations.

The lower value of E, for which c_1 and c_2 have the same signs, represents a bonding orbital. The higher value of E, for which c_1 and c_2 have opposite signs, represents an antibonding orbital.

When c_1 is zero for one solution E, and c_2 is zero for the other, the orbitals represent an ionic bond, with both electrons on one atom. When c_1 equals c_2, the orbitals represent a nonpolar covalent bond, with the electron pair

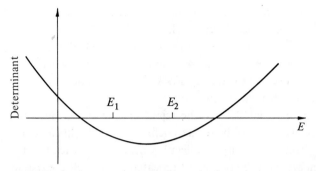

Figure 5-13. A plot of the value of the determinant in Eq. 5–27 as a function of E. (By permission, from C. A. Coulson, *Valence*, Clarendon Press, Oxford, 2nd ed., 1961.)

shared equally between the two atoms. Variation between these extremes permits description of the entire range of polarity, usually as percent ionic character of the bond.

Self-consistent field calculations

Self-consistent field calculations provide the best simple molecular orbital wave functions. The calculations are much more difficult for molecules than for atoms, because the molecule lacks spherical symmetry. The trial molecular wave function is an *antisymmetrized* product of one-electron spin orbitals, usually written in determinant form, in which the spatial part of each spin orbital is a linear combination of atomic orbitals. Rather than the tables of numerical values that represent the best atomic wave functions, approximate analytic functions, for which integrals can be evaluated, are used.

The orbitals most commonly used to construct the linear combination are Slater orbitals and Gaussian orbitals. The former are a product of an angular function, such as s, p_z, or d_{z^2}, for example, and a quantity of the form $r^k \exp(-\alpha r)$ to approximate the numerical tabulation of the Hartree-Fock radial function for the orbital.[7] The Gaussian orbitals have a radial part of the form $\exp(-ar^2)$. Though many more of them are required to construct a satisfactory linear combination than is the case for Slater orbitals, the calculation of the required integrals is simpler.[8]

For example, the product of one-electron spin orbitals for Li_2 might be $(1\sigma_g\alpha)(1\sigma_g\beta)(1\sigma_u\alpha)(1\sigma_u\beta)(2\sigma_g\alpha)(2\sigma_g\beta)$. Each σ orbital would be expressed

7. J. C. Slater, *Quantum Theory of Molecules*, Vol. I, McGraw-Hill Book Co., 1963, p. 274.
8. *Ibid.*, p. 399.

as a linear combination of many lithium atomic orbitals that are cylindrically symmetric about the z (internuclear) axis, such as $1s$, $2s$, and $2p_z$. The more orbitals that are used in the linear combination, the more closely the resulting wave function approaches the best function that can be obtained by the self-consistent field method, the Hartree-Fock function, and the more computer time is required for the calculation.

The potential energy calculated from the trial functions is put into the Hamiltonian operator used to optimize the coefficients in the linear combination of atomic orbitals for each molecular orbital. This leads to a modified potential energy, and the process is repeated until a self-consistent wave function is obtained.

The calculation must be repeated for a number of values of R to obtain curves, such as the curves shown in Fig. 5-6. These calculations predict the equilibrium internuclear distance accurately. The results for the dissociation energy, however, are poor because the molecular wave function for every electron is delocalized over the molecule; thus, there is as much chance that two electrons in an orbital are on the same atom as that one electron is on each atom. At very large values of R, therefore, dissociation into ions is as probable as dissociation into atoms. The latter is seen experimentally. We can express this result another way by saying that the self-consistent field calculation is in error because it obtains the interelectron repulsion from a potential energy based on the average positions of all the other electrons in the molecule rather than their instantaneous positions, thereby neglecting the tendency of electrons to stay apart. The difference between the exact wave function and the Hartree-Fock function is called the *correlation error.*

The standard method to improve the self-consistent field results to take instantaneous electron repulsion into account is *configuration interaction.* Here we use a wave function that is a linear combination of the self-consistent field wave function and others formed from excited electron configurations. Such calculations are now feasible, with the use of high speed computers, but they require large amounts of computer time. Das and Wahl have developed a method called *optimal valence configuration,* which uses only those excited configurations that reflect changes in correlation as the molecule is formed from the atoms and that ensure that the dissociation results in atoms in their ground state. The most precise and accurate results are obtained by explicit inclusion of the interelectronic distance r_{12} as an independent variable in the wave function, but this is difficult for molecules other than H_2. The results shown in Table 5-3 are from calculations of several levels of complexity for the equilibrium internuclear distance and dissociation energy for H_2 and F_2. Figure 5-14 shows potential energy curves

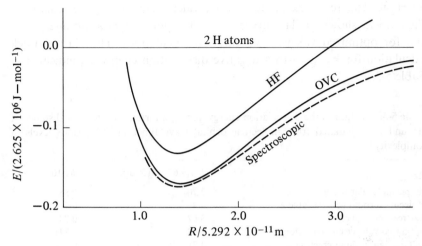

Figure 5-14. Potential energy curves for H_2 calculated from a Hartree-Fock wave function and an optimized valence configuration wave function compared to a curve calculated from spectroscopic data. [By permission, from G. Das and A. C. Wahl, *J. Chem. Phys.*, *44*, 87 (1966).]

Figure 5-15. Potential energy curves for F_2 calculated from a full configuration interaction wave function (x) and an optimized valence configuration wave function (o) compared to a curve calculated from spectroscopic data (–). [By permission, from G. Das and A. C. Wahl, *J. Chem. Phys.*, *56*, 3532 (1972).]

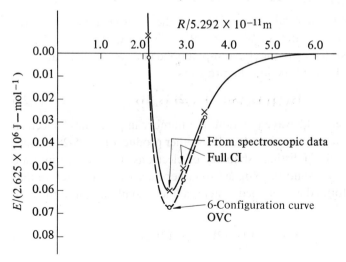

for H_2 for Hartree-Fock and for optimal valence configuration calculations. Figure 5-15 shows similar curves for F_2 for full configuration interaction and for optimized valence configuration calculations. The Hartree-Fock calculation for F_2 leads to a negative dissociation energy, as indicated in Table 5-3.

Table 5-3. Values of the dissociation energy and equilibrium internuclear distance of H_2 and F_2 calculated from molecular orbital wave functions of various levels of complexity[a]

H_2	$D_e/(9.65 \times 10^4$ J-mol$^{-1})$	$R_e/10^{-10}$ m
Simple molecular orbital	2.70	0.85
Molecular orbital with variable Z	3.49	0.732
Hartree-Fock function	3.62	0.74
Optimized valence configuration	4.63	0.741
Best configuration interaction	4.71	0.741
James-Coolidge (r_{12}, with 13 terms)	4.72	0.740
Kolos-Roothaan (r_{12}, with 50 terms)	4.7467	0.74127
Experimental	4.7466 ± 0.0007	0.741
F_2		
Hartree-Fock function	-1.37	1.32
Optimized valence configuration	1.67	1.41
Experimental	1.68	1.42

[a] Data for F_2 and optimized valence configuration for H_2 from A. C. Wahl and G. Das, *Adv. Quant. Chem.*, **5**, 261–96 (1970); A. C. Wahl, in *Theoretical Chemistry* (W. B. Brown, ed.), Butterworths, London, 1972. Other data for H_2 from A. D. McLean, A. W. Weiss, and M. Yoshimine, *Revs. Mod. Phys.*, **32**, 211 (1960).

5-3.   The valence bond method

The hydrogen molecule

The first wave-mechanical calculation for the hydrogen molecule was described by Heitler and London in 1927. They expressed the Lewis-Langmuir-Kossel theory of the electron-pair bond in wave-mechanical terms, choosing as trial wave function the products

$$1s_A(1)\ 1s_B(2); \quad 1s_A(2)\ 1s_B(1) \tag{5-28}$$

where 1s represents the wave function, the number in parentheses electron 1 or 2, and the subscript nucleus A or B. Neither product of Eq. 5-28 satisfies the requirement for indistinguishability of electrons. In order to indicate that either electron could be assigned to either nucleus, and therefore that the two atoms formed a combined system, a linear combination of the two, resulting in

$$\psi_1 = 1s_A(1)\ 1s_B(2) + 1s_A(2)\ 1s_B(1) \tag{5-29}$$

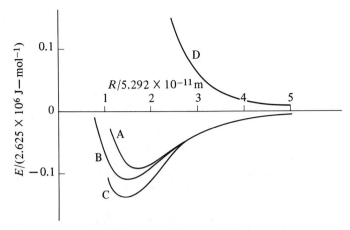

Figure 5-16. Total energy as a function of internuclear distance for the Heitler-London wave function of H_2. Curve A, the simple function in Eq. 5-29 with fixed nuclear charge; B, the function with varying nuclear charge; C, curve calculated from spectroscopic data; D, curve for Eq. 5-30. (By permission, from C. A. Coulson, *Valence*, Clarendon Press, Oxford, 2nd ed., 1961.)

and

$$\psi_2 = 1s_A(1)\,1s_B(2) - 1s_A(2)\,1s_B(1) \tag{5-30}$$

is required. As before, the symmetric function ψ_1 must be associated with an antisymmetric spin function, a singlet, and the antisymmetric function ψ_2 must be associated with a symmetric spin function, a triplet.

Like the molecular orbital method, the valence bond method expresses a wave function of the coordinates of two electrons as a product of one-electron functions. But, the valence bond function is a product of the atomic orbitals and reflects the localized nature of the orbital in contrast to the delocalized molecular orbital. The energy corresponding to wave function ψ_1 (Eq. 5-29) appears as curve A in Fig. 5-16. The equilibrium internuclear distance of 86.9 pm and the dissociation energy of 3.00×10^5 J-mol^{-1} are a considerable improvement over the values obtained with the simple molecular orbital treatment. The main advantage of the Heitler-London wave function is that it emphasizes the high probability of the electrons being centered about different nuclei, thus minimizing interelectronic repulsion.

Like the simple molecular orbital treatment, the simple valence bond treatment leads to a decrease in kinetic energy of the electrons and an increase in potential energy as the molecule is formed, in contrast to the results of exact calculations. If the nuclear charge Z in ψ_1 is varied to

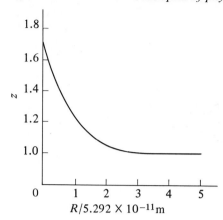

Figure 5-17. The variation in effective nuclear charge required to obtain curve B in Fig. 5-16. (By permission, from C. A. Coulson, *Valence*, Clarendon Press, Oxford, 2nd ed., 1961.)

minimize the value of $\bar{E}$ at each value of R, it varies from 1.0 at infinite R to 1.625 at $R = 0$, as shown in Fig. 5-17. With this variation in Z taken into account, curve B in Fig. 5-16 is obtained. Not only is this curve in closer agreement with observation (curve C) than curve A, but the variation in Z leads to the correct relative contributions of changes in potential energy and kinetic energy to the formation of the molecule.

Further refinements improve the agreement with the observed energy curve. A linear combination of the 1s and $2p_z$ functions "polarizes" the orbital toward the second nucleus. "Ionic" terms in which both electrons are near one nucleus may be incorporated. Angular correlation of the two electrons can be introduced by adding $2p_x$ and $2p_y$ terms to the wave function. The results are shown in Table 5-4.

Table 5-4. Values of the dissociation energy and equilibrium internuclear distance of H_2 calculated from valence bond wave functions of various levels of complexity[a]

	$D_e/9.65 \times 10^4$ J-mol^{-1}	$R_e/10^{-10}$ m
Heitler-London	3.14	0.869
Heitler-London with variable Z	3.76	0.743
Variable Z and polarization	4.02	0.74
Variable Z and ionic terms	4.00	0.749
Variable Z, ionic and polarization	4.10	0.749
Variable Z, ionic, polarization, and angular[b]	4.25	0.74
Experimental	4.7466 ± 0.0007	0.74127

[a] C. A. Coulson, *Valence*, Clarendon Press, Oxford, 1952, p. 121.
[b] J. O. Hirschfelder and J. W. Linnett, *J. Chem. Phys., 18,* 130 (1950).

As with the molecular orbital wave function, the linear combination with the negative sign, ψ_2, represents the first excited orbital. For this orbital, curve D in Fig. 5-16 shows the variation of E with R.

5-4. Comparison of valence bond and molecular orbital methods

The simple molecular orbital wave function for H_2 is

$$\psi_{1,MO} = [1s_A(1) + 1s_B(1)] [1s_A(2) + 1s_B(2)]$$

$$= 1s_A(1)1s_B(2) + 1s_B(1)1s_A(2) + 1s_A(1)1s_A(2)$$

$$+ 1s_B(1)1s_B(2) \tag{5-31}$$

The first two terms of Eq. 5-31 are just the Heitler-London function ψ_1; the last two terms are "ionic" terms, in which both electrons are near one nucleus. An essential defect of the simple molecular orbital wave function is that the latter terms are of equal weight with the covalent terms. A simple modification of Eq. 5-31 is to apply a coefficient λ (less than 1) to the last two terms, so that

$$\psi'_{1,MO} = \psi_{HL} + \lambda[1s_A(1)\,1s_A(2) + 1s_B(1)\,1s_B(2)] \tag{5-32}$$

The resulting function is identical to the valence bond function in which "resonance" with ionic terms is included. In the valence bond method, resonance refers to a procedure by which two partial descriptions of the state of a molecular system are combined to improve the description. This process leads to an improvement in the calculated energy.

The modification of a simple molecular wave function to include a variable effective nuclear charge, polarization, and angular correlation can be applied to either the molecular orbital or valence bond functions. The simplest configuration interaction treatment of H_2 utilizes the configurations $(1\sigma_g)^2$ and $(1\sigma_u{}^*)^2$, so that

$$\psi_{CI} = [1s_A(1) + 1s_B(1)] [1s_A(2) + 1s_B(2)]$$

$$+ C[1s_A(1) - 1s_B(1)] [1s_A(2) - 1s_B(2)] \tag{5-33}$$

It can be shown that ψ_{CI} is also equivalent to $\psi'_{1,MO}$.

Exercise 5-7. Find the mathematical relation between λ in Eq. 5-32 and C in Eq. 5-33.

When the simplest forms of each method are used, the valence bond method leads to better agreement with experiment than the molecular orbital method. Both the molecular orbital and valence bond approaches can be modified to correct the bias resulting from, respectively, too great and too little a probability that the electrons are close to each other. When the corrections are sufficiently elaborated, the treatments yield identical results. Less computational effort is required to use the modified molecular orbital treatment. It is therefore used for most rigorous quantitative calculations. Recent developments have made rigorous valence bond calculations comparable to molecular orbital calculations in computational requirements. [See J. M. Norbeck and G. A. Gallup., *J. Am. Chem. Soc.*, **96**, 3386–93, (1974).]

Many-electron molecules
A simple molecular orbital wave function for a many-electron molecule can be written as a single determinant. It satisfies the requirement for indistinguishability of electrons and the Pauli principle, since it changes sign upon interchange of electrons. This determinant includes products of one-electron functions for all the permutations of spin and spatial functions among the electrons.

For example, the p electrons of the N_2 molecule can be described by the molecular wave function

$$\begin{vmatrix} 1\pi_x\alpha(1) & 1\pi_x\beta(1) & 1\pi_y\alpha(1) & 1\pi_y\beta(1) & 3\sigma\alpha(1) & 3\sigma\beta(1) \\ \cdot & \cdot & \cdot & \cdot & \cdot & \cdot \\ \cdot & \cdot & \cdot & \cdot & \cdot & \cdot \\ \cdot & \cdot & \cdot & \cdot & \cdot & \cdot \\ 1\pi_x\alpha(6) & 1\pi_x\beta(6) & 1\pi_y\alpha(6) & 1\pi_y\beta(6) & 3\sigma\alpha(6) & 3\sigma\beta(6) \end{vmatrix} \quad (5\text{-}34)$$

in which each column of the determinant corresponds to a spin orbital and each row corresponds to an electron. Expansion of the determinant yields a linear combination of the 6! products that include all the possible permutations of electrons among the spin orbitals.

The valence bond function for N_2, using the p_z, p_x, and p_y orbitals of each N atom to form one σ bond and two π bonds, would be a determinant of the form

$$\begin{vmatrix} p_{zA}\,\alpha(1) & p_{zB}\,\beta(1) & p_{xA}\,\alpha(1) & p_{xB}\,\beta(1) & p_{yA}\alpha(1) & p_{yB}\,\beta(1) \\ \cdot & \cdot & \cdot & \cdot & \cdot & \cdot \\ \cdot & \cdot & \cdot & \cdot & \cdot & \cdot \\ \cdot & \cdot & \cdot & \cdot & \cdot & \cdot \\ p_{zA}\,\alpha(6) & p_{zB}\,\beta(6) & p_{xA}\,\alpha(6) & p_{xB}\,\beta(6) & p_{yA}\alpha(6) & p_{yB}\,\beta(6) \end{vmatrix} \quad (5\text{-}35)$$

One of these determinants, however, does not include products with all the possible permutations of spin and electron for each orbital. For a system of n electron-pair bonds, 2^n such determinants are needed, 2^3 for N_2, for example.

As a result of the development of large, fast computers, and of new computational techniques, self-consistent field calculations have been carried out for diatomic molecules as large as KrF.[9] These calculations provide satisfactory electron distributions but poor potential energy curves and dissociation energies. Such methods as optimized valence configuration, in which computer time is minimized by a careful choice of excited configurations, provide chemically valid potential energy curves and dissociation energies even though the absolute values of the energies may not be correct. This is accomplished by the inclusion of only those orbitals for which the error is equal for the atom and the molecule so that a valid energy difference is obtained.

Equivalent Orbitals

Molecular orbitals for polyatomic molecules as normally presented are completely delocalized; in the LCAO approximation, they consist of linear combinations of all the atomic orbitals of the appropriate symmetry. Chemists, however, are accustomed to thinking of localized bonds, and chemical bonds between particular atoms have been found to have significantly similar characteristics, such as bond lengths and bond energies, in a variety of molecules. The valence bond method was devised to describe localized chemical bonds, but the molecular orbital method requires transformation into localized orbitals to make explicit the chemical bonds that are implicit in the delocalized orbitals. *Equivalent orbitals*, introduced by Lennard-Jones and his coworkers, are localized bond orbitals or lone-pair orbitals that are obtained as linear combinations of delocalized molecular orbitals.[10]

The simplest example of the application of equivalent orbitals is to the He_2 molecule. The simple molecular orbital wave function for this four-electron molecule is the determinant

$$\left| (1\,\sigma_g\alpha)\,(1\,\sigma_g\beta)\,(1\,\sigma_u{}^*\alpha)\,(1\,\sigma_u{}^*\beta) \right| \tag{5-36}$$

9. H. F. Schaefer, *The Electronic Structure of Atoms and Molecules*, Addison-Wesley, Reading, 1972, p. 146.
10. J. A. Pople, *Quart. Revs.*, *11*, 273 (1957).

In the LCAO approximation,

$$1\sigma_g = 1s_A + 1s_B$$

and

$$1\sigma_u^* = 1s_A - 1s_B$$

Without changing the value of the determinant (except for a multiplicative constant that is determined by normalization), we may substitute for the molecular orbitals $1\sigma_g$ and $1\sigma_u^*$ the equivalent orbitals χ_+ and χ_-, where

$$\chi_+ = 1\sigma_g + 1\sigma_u^* = 1s_A$$

and

$$\chi_- = 1\sigma_g - 1\sigma_u^* = 1s_B$$

Thus, the determinant can now be represented, in a compact notation, as

$$\left|(1s_A\,\alpha)(1s_A\,\beta)(1s_B\,\alpha)(1s_B\,\beta)\right| \tag{5-37}$$

or the wave function of two isolated He atoms. The delocalized molecular orbitals are transformed into the localized equivalent orbitals. Similarly, any filled pair of bonding and antibonding molecular orbitals of the same principal quantum number and angular momentum can be transformed into equivalent nonbonding lone pair orbitals localized on the atoms.

A similar treatment of the N_2 molecule results in the transformation of the 3σ, $1\pi_x$, and $1\pi_y$ molecular orbitals of Eq. 5-34 into the equivalent orbitals

$$\chi_1 = (3\sigma) + 2^{\frac{1}{2}}(1\pi_x)$$

$$\chi_2 = (3\sigma) - \frac{2^{\frac{1}{2}}}{2}(1\pi_x) + \frac{6^{\frac{1}{2}}}{2}(1\pi_y) \tag{5-38}$$

$$\chi_3 = (3\sigma) - \frac{2^{\frac{1}{2}}}{2}(1\pi_x) - \frac{6^{\frac{1}{2}}}{2}(1\pi_y)$$

Since the σ orbital is cylindrically symmetric, the directional properties of the equivalent orbitals are derived entirely from the relative weights of the π orbitals. Figure 5-18 illustrates the geometric relations that determine the coefficients in the linear combination. For χ_1, the orbital has its maximum angular probability in the x direction, there is no contribution from the π_y orbital, and the coefficient of the π_x orbital is chosen so that its square will be twice the square of the coefficient of the 3σ orbital. For χ_2 and χ_3 the magnitude of the ratio of $\pi_x : \pi_y = \frac{1}{2} : \frac{\sqrt{3}}{2}$. The coefficients of π_x and π_y are

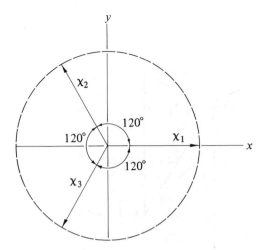

Figure 5-18. The geometric relationship among the trigonal equivalent orbitals and the π_x and π_y orbitals.

chosen so that the sum of their squares will be twice the square of the coefficient of the 3σ orbital. For χ_2, they are $-\dfrac{\sqrt{2}}{2}$ and $\dfrac{\sqrt{6}}{2}$, respectively. For χ_3 the magnitudes are the same, but both coefficients have a negative sign. The sum of the squares of the coefficients of the π orbitals is twice the square of the coefficient of the σ orbital because each equivalent orbital has $\frac{2}{3}\pi$ characteristic and $\frac{1}{3}\sigma$ characteristic. Each of these orbitals can describe a pair of electrons of opposite spin and thereby represent one of three equally spaced "banana" bonds between the nitrogen atoms. This mode of description makes explicit the consequence of the Pauli principle that electrons of like spin do not occupy the same region of space, whereas electrons of opposite spin do.

The Ne molecule, with a configuration $2s^2 2p^6$, can equally well be described in terms of four equivalent tetrahedral orbitals, as shown in Eq. 5-39.

$$t_1 = (2s) + (2p_x) + (2p_y) + (2p_z)$$

$$t_2 = (2s) + (2p_x) - (2p_y) - (2p_z) \tag{5-39}$$

$$t_3 = (2s) - (2p_x) - (2p_y) + (2p_z)$$

and

$$t_4 = (2s) - (2p_x) + (2p_y) - (2p_z)$$

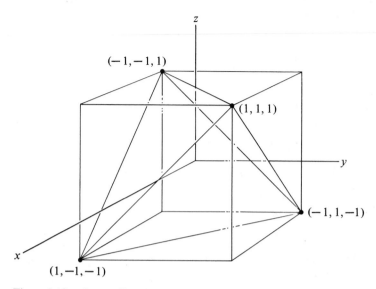

Figure 5-19. A coordinate system for tetrahedral symmetry obtained by placing the vertices of a tetrahedron at the corners of a cube with the center at the origin and the faces parallel to the coordinate planes. Length of cube edge is two units.

The geometric basis for Eq. 5-39 can be seen from a consideration of a cube centered on the origin, with sides parallel to the coordinate planes, and with edges two units in length, as shown in Fig. 5-19. A regular tetrahedron is inscribed in the cube, joining the vertices marked $(1,1,1)$, $(1,-1,-1)$, $(-1,-1,1)$, and $(-1,1,-1)$. A vector from the origin to any of these points has x, y, and z components that are equal in magnitude. The square root of the magnitude, with the sign, of the component, gives the coefficient of the appropriate p orbital. The sum of the squares of the coefficients of the p orbitals is 3, giving the proper weight relative to the s orbital. Since in an atom (e.g., Ne) there is no internal reference on which a coordinate system can be based, this description is consistent with a spherically symmetric expression for the probability density.

The molecules HF, H_2O, NH_3, and CH_4, which are isoelectronic with Ne, can also be described in terms of the tetrahedral orbitals in Eq. 5-39. If the bond in HF is taken along the direction of t_1, a linear combination of the orbital t_1 and the hydrogen $1s$ can represent the bonding orbital, whereas t_2, t_3, and t_4 represent lone pairs on the F atom.

Exercise 5-8. Describe the bonds and lone pairs in H_2O, NH_3, and CH_4 in terms of the orbitals t_1, t_2, t_3, and t_4.

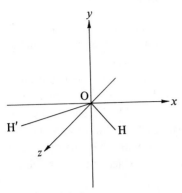

Figure 5-20. Coordinate system used to describe the molecular orbitals of the H_2O molecule.

Figure 5-21. Schematic contour surfaces for the bonding and lone-pair molecular orbitals of the H_2O molecule.

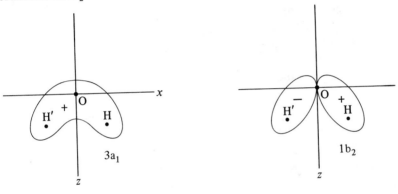

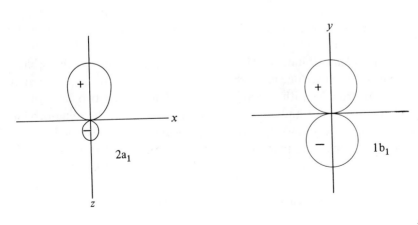

The molecular orbitals of H_2O and their corresponding equivalent orbitals provide a good example of the use of simple symmetry arguments in the choice of appropriate linear combinations of atomic orbitals for the molecular orbitals. Only those atomic orbitals can be combined that have the same symmetry with respect to the x-z and y-z planes, the planes of symmetry of the molecule as shown in Fig. 5-20. Table 5-5 shows the five occupied molecular orbitals of H_2O (constructed from the oxygen 1s, 2s, $2p_x$, $2p_y$, and $2p_z$ and H1s and H'1s) together with the corresponding symmetry; Figure 5-21 shows schematic contour surfaces for the bonding and lone pair orbitals.

Table 5-5.　LCAO molecular orbitals of H_2O

Orbital	Symmetry	Linear combinations	Description
$1a_1$	Totally symmetric	$(1s)_O$	Oxygen inner shell
$2a_1$	Totally symmetric	$(2s)_O - (2p_z)_O$	Symmetric lone pair
$1b_2$	Antisymmetric with respect to reflection in y-z plane	$(2p_x)_O + [(1s)_H - (1s)_{H'}]$	Antisymmetric bonding
$3a_1$	Totally symmetric	$[(2s)_O + (2p_z)_O] +$ $[(1s)_H + (1s)_{H'}]$	Symmetric bonding
$1b_1$	Antisymmetric with respect to reflection in x-z plane	$2(p_y)_O$	Antisymmetric lone pair

F. O. Ellison and H. Shull, *J. Chem. Phys., 23*, 2348 (1955)

The symbols for the orbitals refer to the symmetry properties with respect to reflection, analogous to the σ and π designations for diatomic molecules. Orbitals of a given symmetry are numbered in order of increasing energy, when the energies are known. The energies used here were obtained by Ellison and Shull.[11]

The equivalent orbitals for H_2O are linear combinations of the orbitals in Table 5-5. The two lone pair orbitals combine to form two approximately tetrahedral orbitals, with a maximum probability along two lines in the y-z plane, one above and below the x-z plane. The two bonding orbitals form two equivalent orbitals in the x-z plane along the O-H bonds; these, together with the lone pair orbitals, constitute four tetrahedral orbitals. The relation to the tetrahedral orbitals, shown in Fig. 5-19, can be seen by

11. F. O. Ellison and H. Shull, *J. Chem. Phys., 23*, 2348 (1955).

rotating the cube in Fig. 5-19 by $45°$ about the z axis. The components of the vectors corresponding to the bonding orbitals would be $(\sqrt{2}, 0, 1)$ and $(-\sqrt{2}, 0, 1)$ whereas the components of the vectors corresponding to the lone pair orbitals would be $(0, \sqrt{2}, -1)$ and $(0, -\sqrt{2}, -1)$.

Exercise 5-9. Show that the coefficients derived from the vector components given above satisfy the conditions given earlier for sp^3 equivalent tetrahedral orbitals.

Detailed calculations for the water molecule have shown that the bonding and nonbonding orbitals are not completely independent, so this description is an approximation.[12] The equivalent orbital description of methane is more applicable and convenient because it corresponds directly with the chemists conventional model of the covalent bond. The equivalent orbitals are linear combinations of each tetrahedral C orbital described by Eq. 5-39 with a 1s orbital of a H atom. In contrast, the delocalized molecular orbital description for methane requires that each bonding molecular orbital have a contribution from all four H 1s orbitals.

There are two alternative localized orbital models for both ethylene and acetylene. The ethylene $>CH_2$ is isoelectronic with O, and a C-H bond can be viewed as comparable to an O lone pair orbital. The other C orbitals can form one σ and one π localized molecular orbital. Alternatively, the linear combinations of the σ and π orbitals can yield two equivalent banana orbitals. The acetylene $>C$-H is isoelectronic with N, and the C-H bond can be viewed as comparable to an N lone pair orbital. The other three orbitals of each C atom can form one σ and two π localized molecular orbitals. Alternatively, linear combinations of the σ and π orbitals can yield three equivalent banana orbitals. The σ-π model has recently been shown to give better localization for ethylene than the banana bond model, which requires more interaction between bonding and nonbonding orbitals.[13]

Exercise 5-10. Describe the linear combinations of atomic orbitals that lead to the σ-π and banana bond models of ethylene and acetylene. Take the z axis as the molecular axis, and take the x-z plane as the plane of the ethylene molecule.

Equivalent orbitals, as developed by Lennard-Jones and his coworkers, are limited to molecules in which each molecular orbital describes two

12. D. Eisenberg and W. Kauzmann, *The Structure and Properties of Water*, Oxford University Press, New York, 1969, pp. 27–34.
13. C. Trindle and O. Sinanoglu, "Semiempirical orbital localization and its chemical applications," in O. Sinanoglu and K. B. Wiberg, eds. *Sigma Molecular Orbitals*, Yale University Press, New Haven, 1970, p. 211.

electrons of opposite spin. Linnett has developed a method in which electrons of opposite spin are described by separate spatial orbitals, a method called *nonpairing*.[14] This procedure has the effect of building electron correlation into a simple wave function and thereby improving the calculated energy. Also, it is possible to construct equivalent orbital descriptions for molecules with unpaired electrons, such as O_2 or NO.

For example, the molecular orbitals of the N_2 molecule were transformed into three banana bond orbitals, each describing a pair of electrons, in Eq. 5-38. In a nonpairing description, the molecular orbital configuration is expressed as

$$[(1\pi_{uy}\alpha)(1\pi_{ux}\alpha)(3\sigma_g\alpha)][(1\pi_{uy}\beta)(1\pi_{ux}\beta)(3\sigma_g\beta)]$$

and the orbitals of opposite spin are transformed separately to equivalent banana bond orbitals, each describing one electron. The orbitals of different spin are spatially independent, thus including an element of electron correlation into the wave function.

The paramagnetic nature of oxygen, with its two unpaired electrons, was explained simply by a molecular orbital model; it could not be explained by a simple valence bond model. The nonpairing method provides a simple localized description in which there are four bonding electrons between the nuclei, three of one spin and one of the other. The decrease in bonding electrons from six in N_2 to four in O_2 results from the addition of two electrons to the antibonding orbitals $1\pi_{gx}{}^*$ and $1\pi_{gy}{}^*$. In the equivalent orbital description, these orbitals combine with the corresponding bonding orbitals of the same spin to produce unshared orbitals. Thus, electrons of one spin form three bonding banana orbitals, whereas for electrons of the other spin only one bonding sigma orbital remains.

Exercise 5-11. Describe the nonpairing equivalent orbitals of O_2, with five electrons of one spin and seven electrons of the other spin with principal quantum number 2.

Exercise 5-12. Describe the nonpairing equivalent orbitals of NO, with one less electron than O_2. How would you describe the number of bonds between N and O in this molecule?

The molecule diborane B_2H_6 has a geometry

14. J. W. Linnett, *Electronic Structure of Molecules*, Methuen, London, 1964.

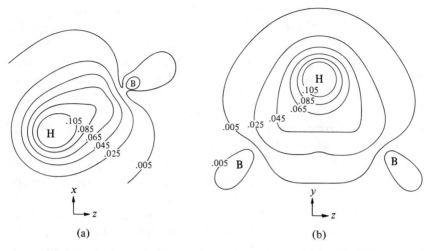

Figure 5-22. Contour curves in the plane of the nuclei for the localized (a) H–B and, (b) B—H—B orbitals of diborane obtained by localization of self-consistent field orbitals. [By permission, from E. Switkes, R. M. Stevens, W. N. Lipscomb, and M. D. Newton, *J. Chem. Phys.*, *51*, 2085 (1969).]

with the outer hydrogens in a plane perpendicular to the plane formed by the boron atoms and the central hydrogens. This structure cannot be described in Lewis-Langmuir electron-pair terms or in simple valence bond terms because there are not enough valence electrons in the molecule to provide electron pairs between each pair of bonded atoms. The molecular orbital model can provide a description of this molecule, since it does not depend on localized bonds. An equivalent orbital treatment, due to Longuet-Higgins, suggests that there are two-electron bonds between each boron atom and the outer hydrogens, whereas the bonding pairs holding the central hydrogens to two boron atoms extend over three atoms. This view has been verified by the detailed calculations of Switkes and his coworkers, who carried out a self-consistent field calculation and then transformed the results into localized equivalent orbital form. They found evidence for a localized outer B-H bond and a three-center bridge bond joining each central hydrogen atom to two boron atoms, as seen in Fig. 5-22.[15]

Hybridization
The valence bond approach to the geometry of many-electron molecules

15. E. Switkes, R. M. Stevens, W. N. Lipscomb, and M. D. Newton, *J. Chem. Phys.*, *51*, 2085 (1969).

was developed by Slater and Pauling in 1931. They provided the first quantum-mechanical explanation of the angular structure of H_2O and the pyramidal structure of NH_3, pointing out that a maximum overlap between the H 1s orbitals and the p orbitals of O or N would occur if the bonds were at 90° to one another. The discrepancy between this prediction and the experimental bond angles of 104.5° for H_2O and 107° for NH_3 was explained by assuming repulsion between the H atoms.

Their most striking contribution was to the classic problem of the tetravalence and tetrahedral geometry of carbon. The carbon atom can be described as ^3P, with an electron configuration $(1s)^2 (2s)^2 (2p)^2$. This description leads to the prediction of a divalent atom with two bonds about 90° apart. Slater and Pauling suggested the possibility of promotion of a 2s electron to the vacant p orbital, leading to a ^5S valence state having an electron configuration $(1s)^2 (2s) (2p)^3$ if all four valence electrons are of the same spin. The energy required for this promotion, about 400 kJ-mol^{-1}, is more than regained by the formation of two additional bonds. Though the geometry of the resulting bonds is inherent in the sp^3 configuration, it is not obvious from this description.

Four electrons of parallel spin in four orbitals are described by a determinant, which for sp^3 is symbolized by

$$\left|(2s\alpha) (2p_x\alpha) (2p_y\alpha) (2p_z\alpha)\right| \tag{5-40}$$

The wave function is unchanged when each column is replaced by a linear combination of the original orbitals. Pauling and Slater showed that the following linear combinations described orbitals with maximum probability densities arranged at tetrahedral angles and with identical radial functions.

$$t_1 = (2s) + (2p_x) + (2p_y) + (2p_z)$$
$$t_2 = (2s) + (2p_x) - (2p_y) - (2p_z) \tag{5-41}$$
$$t_3 = (2s) - (2p_x) - (2p_y) + (2p_z)$$
$$t_4 = (2s) - (2p_x) + (2p_y) - (2p_z)$$

These tetrahedral sp^3 hybrid orbitals are identical in form to the tetrahedral equivalent orbitals of Ne given in Eq. 5-39. A contour surface of constant ψ^2 for a C tetrahedral orbital is shown in Fig. 5-23. The valence bond wave function for CH_4 requires 2^4 determinants of eight rows and eight columns each. The rows represent the four tetrahedral hybrid orbitals and the four H 1s orbitals. Each pairing of a tetrahedral orbital with a 1s orbital describes two electrons of opposite spin. The hydrides of N, O, and F can be similarly described, except that for each molecule two groups of tetrahedral orbitals

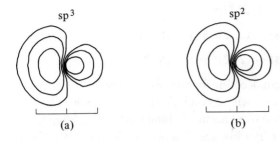

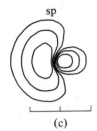
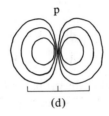

Figure 5-23. Contour curves of constant ψ^2 for (a) sp³, (b) sp², and (c) sp hybrid orbitals of carbon companed to (d) a p orbital. [By permission, from S. L. Holmgren and J. S. Evans, *J. Chem. Educ.*, *51*, 189–91 (1974).]

are required, one bonding and one lone pair. The relative proportions of s and p orbitals in the two groups also differ slightly. The angles between lone pair orbitals are a little larger than those between bonding orbitals in such a situation.

Since the ground state configuration of B is $(1s)^2 (2s)^2 (2p)$, one might expect that B would have a valence of one. Yet it is well known that boron forms three bonds in a plane at 120° from one another. Slater and Pauling suggested the promotion of a 2s electron to a vacant 2p orbital, giving a valence configuration of $(2s) (2p)^2$, where all three electrons have parallel spin. The determinant of the three orbitals can equally well be expressed in terms of three sp² hybrid orbitals, in which the maximum probability densities are in a plane at 120° from each other. If the orbitals are in the *x-y* plane, and one orbital is directed along the *x* axis, the linear combinations for the trigonal hybrids are of the same form as those given in Eq. 5-38 for the banana bonds of nitrogen; they are shown in Eq. 5-42. Linear combinations of each of these orbitals with a σ orbital of an atom like F leads to bond orbitals that account for the geometry of compounds such as BF_3.

$$tr_1 = (2s) + 2^{\frac{1}{2}} (2p_x)$$

$$tr_2 = (2s) - (2^{\frac{1}{2}}/2)(2p_x) + (6^{\frac{1}{2}}/2)(2p_y) \qquad (5\text{-}42)$$

$$tr_3 = (2s) - (2^{\frac{1}{2}}/2)(2p_x) - (6^{\frac{1}{2}}/2)(2p_y)$$

A similar hybridization can account for the linear structure of covalent compounds formed by elements with a configuration of s^2, such as Be and Hg. Promotion of an s electron to a vacant p orbital leads to a configuration of (s) (p). As with the other hybrids, the description in terms of s and p orbitals, in which both electrons are on both sides of the nucleus, is replaced by a description in terms of sp hybrids, each of which is concentrated on one side of the nucleus. The expressions for these orbitals, separated by $180°$, are

$$d_1 = s + p$$
$$d_2 = s - p \qquad (5\text{-}43)$$

The hybrid description makes the probability that electrons of like spin will not occupy the same region of space explicit without changing the wave function of the central atom.

Pauling was also able to show that a central atom configuration of d^2sp^3 can be described in terms of six orbitals pointing to the vertices of a regular octahedron. The extension of the concept of hybridization to d orbitals provided the initial theoretical basis for understanding the structure of transition metal complexes.

Conjugated and aromatic molecules

A localized description fails for molecules with highly polar bonds, because the electrostatic interaction of the dipoles makes the bonds dependent on each other. Another important class of molecules for which a localized description is inadequate is that in which the molecules contain conjugated double bonds. In the valence bond method, this problem is met by invoking *resonance* among a number of valence bond *structures* to describe the molecule. These structures, such as the Kekulé structures for benzene, are not to be considered real molecular structures, but only bases for the construction of approximate wave functions for the molecule, which has *one* structure that is described by the total wave function. Since each "structure" in the valence bond method is represented by a number of determinants, the description of the molecule is cumbersome, even at the simplest level.

Simple molecular orbitals, which start from the viewpoint that electrons are delocalized, describe conjugated molecules more easily. The Hückel treatment of benzene provides an example of the simple molecular orbital method. The framework of the molecule is taken to be a regular hexagon of carbon atoms; σ C-C and C-H bonds are linear combination of C sp^2 hybrids, with two neighboring C atoms and a H 1s. Linear combinations of the remaining p orbitals of the C atoms, with maximum probability density above and below the plane of the ring, are of the form

$$\psi = c_1 p_1 + c_2 p_2 + \ldots + c_6 p_6 \qquad (5\text{-}44)$$

and constitute the π orbitals of the molecule. The six electrons described by these orbitals move in the potential field of the nuclei and the σ electrons and the average field of the other π electrons. The values of the coefficients in the linear combinations and the energy of the molecular orbitals is calculated by a variation procedure like that used in Eq. 5-18 for CO. The form of the Hamiltonian operator is not specified, because the resulting integrals are evaluated by comparison with experiment rather than by calculation. At the simplest level, the integrals S_{AB} are set equal to zero, the integrals H_{AB} are set equal to zero for all but adjacent atoms, and all remaining integrals H_{AB} are set equal to the same constant β. The integrals H_{AA} are taken equal to the energy E_0 of a p orbital localized on a single C atom in the ring.

The resulting set of simultaneous linear equations is

$$
\begin{array}{llll}
C_1 (E_0 - E) + C_2 \beta & & + C_6 \beta = 0 \\
C_1 \beta \quad + C_2 (E_0 - E) + C_3 \beta & & = 0 \\
\quad C_2 \beta \quad + C_3 (E_0 - E) + C_4 \beta & & = 0 \\
\quad\quad C_3 \beta \quad + C_4 (E_0 - E) + C_5 \beta & & = 0 \\
\quad\quad\quad C_4 \beta \; + C_5 (E_0 - E) + C_6 \beta = 0 \\
C_1 \beta & + C_5 \beta + C_6 (E_0 - E) = 0
\end{array}
$$

The six by six determinant in $E_0 - E$ leads to six roots with the values

$$E = E_0 + 2\beta; \quad E = E_0 + \beta; \quad E = E_0 - \beta; \quad E = E_0 - 2\beta \quad (5\text{-}45)$$

The second and third levels are doubly degenerate. The six π electrons occupy the three orbitals of lowest energy, so that the total π energy is

$$2(E_0 + 2\beta) + 4(E_0 + \beta) = 6E_0 + 8\beta$$

If each electron pair had been localized in the region of one bond, with the same simplifying assumptions, the total energy would be $6E_0 + 6\beta$. The difference, 2β, is the *delocalization energy*. A numerical value for β is obtained by comparing the experimental heat of formation of benzene with a value calculated for a single Kekulé structure from tabulated bond energies. The test of the Hückel method is to see whether the value of β obtained for one molecule leads to consistent results for other conjugated molecules.

All the coefficients for the lowest bonding molecular orbital of benzene are equal and positive, consistent with the symmetry of the molecule. The

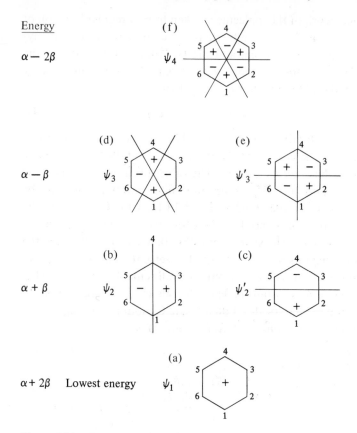

Figure 5-24. Representation of the symmetry of the Hückel molecular orbitals of benzene and their relative energy.

wave function has the same sign at all C atoms in the ring, as shown schematically in Fig. 5-24a. Since it is a π function, like all the benzene molecular orbitals, it is antisymmetrical with respect to the plane of the benzene ring. The degenerate bonding orbitals have the symmetry shown in Fig. 5-24b and 5-24c, and the functions are

$$\psi_2 = p_2 + p_3 - p_5 - p_6 \qquad (5\text{-}46)$$

and

$$\psi_2' = 2p_1 + p_2 - p_3 - 2p_4 - p_5 + p_6 \qquad (5\text{-}47)$$

The degenerate antibonding orbitals have the symmetry shown in Fig. 5-24d and 5-24e, and the functions are

$$\psi_3 = 2p_1 - p_2 - p_3 + 2p_4 - p_5 - p_6 \qquad (5\text{-}48)$$

and

$$\psi_3' = p_2 - p_3 + p_5 - p_6 \qquad (5\text{-}49)$$

(A number of alternative linear combinations of the two functions given for each degenerate pair are also used.) The highest energy antibonding orbital has the symmetry shown in Fig. 5-24f, and the function is

$$\psi_4 = p_1 - p_2 + p_3 - p_4 + p_5 - p_6 \qquad (5\text{-}50)$$

As in previous examples, the energy increases with increasing number of nodes in the wave function or decreasing wavelength of the wave. Since this system has rotational symmetry, one can also say that the angular momentum increases with increasing energy.

The Hückel model has been used extensively as a basis for theoretical organic chemistry, and it has replaced the valence bond resonance method that was dominant for many years. Because the approximations used in the model are so drastic, much effort has been expended to produce a more rigorous but still empirical model. In such *semiempirical* models, the self-consistent field solution is approximated by neglecting some of the integrals that are difficult to evaluate or by assigning them values on the basis of experimental data. Though these calculations are not as reliable as the *ab initio*, self-consistent field calculations we discussed earlier, they can yield useful results for a given class of compounds, if the errors are consistent over the group. Also, calculations are feasible for larger molecules for the semiempirical models. Further details of Hückel calculations[16] and semiempirical calculations[17] can be found in the references cited in the footnotes. As mentioned earlier, recent rigorous valence bond calculations are comparable to corresponding molecular orbital calculations.[18]

5-5. Graphical representation

As molecular quantum mechanical calculations became more sophisticated, they also became more abstract and further removed from the chemist's simple pictures of chemical bonds in molecules. Fortunately, the improve-

16. J. D. Roberts, *Molecular Orbital Calculations*, W. A. Benjamin, New York, 1962; A. Streitweiser, Jr., *Molecular Orbital Theory for Organic Chemists*, John Wiley & Sons, New York, 1961.
17. J. A. Pople and D. L. Beveridge, *Approximate Molecular Orbital Theory*, McGraw-Hill Book Co., New York, 1970; J. N. Murrell and A. J. Harget, *Semiempirical Self-consistent-field Molecular Orbital Theory of Molecules*, Wiley-Interscience, London, 1972.
18. J. M. Norbeck and G. Gallup, *J. Am. Chem. Soc.*, 96, 3386–93 (1974).

ment in computing capability also brought with it the capacity to convert masses of quantitative data into graphical form, which might be grasped by those who did not understand the calculations.

We have already seen examples of contour curves of constant ψ^2 as descriptions of atomic orbitals and simple molecular orbitals. In 1966, A. C. Wahl[19] published a set of contour diagrams of the molecular orbitals and the total electron density for Hartree-Fock calculations of the diatomic molecules of the first ten elements. They are shown in Fig. 5-25. More recently, Wahl has provided graphical representations of the results of his optimized valence configuration calculations. Figure 5-26 shows the contours of total electron density for H_2 as a function of internuclear distance. Figure 5-27 shows the contours for the orbitals used in the optimized valence configuration calculation for H_2 at large R and at R_e, showing the relative contribution of each at each distance.

In an attempt to go beyond the two-dimensional representation, Bordass and Linnett[20] and Streitweiser and Owen[21] have provided computer-generated graphs of ψ for orbitals and electron density for molecules as a function of position in the plane of some or all the nuclei in the molecule. The most striking of the results of Streitweiser and Owen are the plots that show the difference in electron density between the molecule and the separated atoms, as shown for H_2 and He_2 in Fig. 5-28. The major difference between the stable H_2 and the unstable He_2 is the electron density between the nuclei; both molecules exhibit increased electron density about the nuclei compared to the free atoms.

Problems

5-1. Calculate the energy difference between the $n_x = n_y = n_z = 1$ level and the $n_x = n_y = 1, n_z = 2$ level (a) for an electron confined to a cube 1×10^{-10} m on an edge, and (b) for an electron confined to a cube 1 m on an edge. What is the relationship between quantization of energy and the constraint on a particle?

5-2. From Eq. 5-27, find the two values of E for which the determinant equals zero.

19. A. C. Wahl, *Science*, *151*, 961 (1966); A. C. Wahl and M. T. Wall, *Four Wall Charts of Atomic and Molecular Structure*, McGraw-Hill Book Co., New York, 1969.
20. W. T. Bordass and J. W. Linnett, *J. Chem. Educ.*, *47*, 672–75 (1970).
21. A. Streitweiser, Jr., and P. H. Owens, *Orbital and Electron Density Diagrams*, Macmillan, New York, 1973.

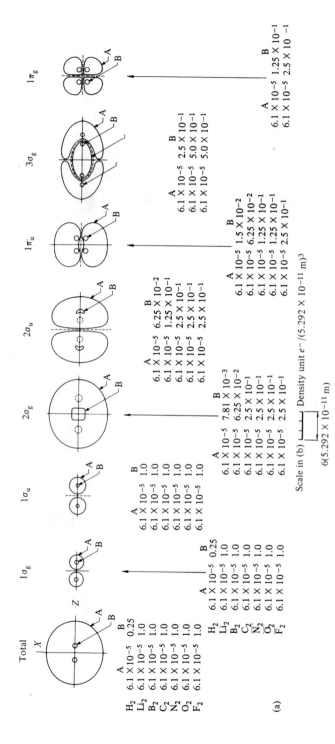

Figure 5-25 (a).

(b)

Molecule	Total	$1\sigma_g$	$1\sigma_u$	$2\sigma_g$

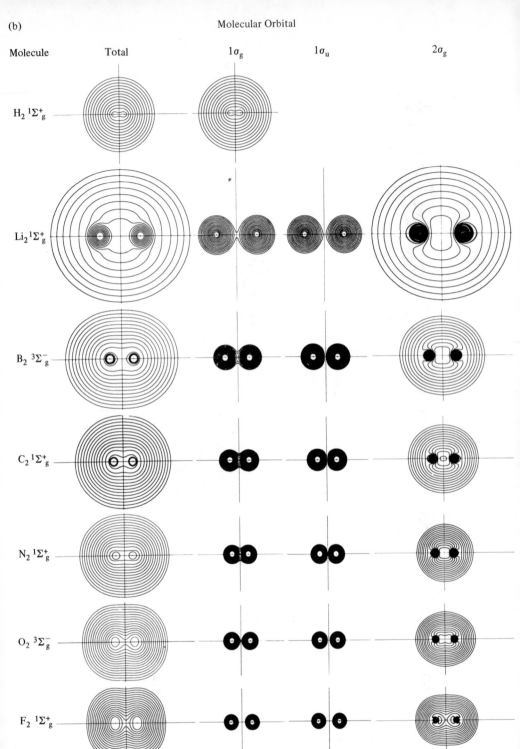

$H_2\ ^1\Sigma_g^+$

$Li_2\ ^1\Sigma_g^+$

$B_2\ ^3\Sigma_g^-$

$C_2\ ^1\Sigma_g^+$

$N_2\ ^1\Sigma_g^+$

$O_2\ ^3\Sigma_g^-$

$F_2\ ^1\Sigma_g^+$

176

Figure 5-25. Contour curves of constant ψ^2 for Hartree-Fock molecular orbitals and total electron density of the bound homonuclear diatomic molecules of the first nine elements. (a) Key to density contours. The diagrams indicate the general structure of each plot. A, lowest contour value plotted; B, highest contour value plotted in each molecule (except for the contours that rise to $e^-/(bohr)^3$ inside the $2\sigma_g$ and $2\sigma_u$ node). Adjacent contour lines differ by a factor of two. Thus, all contours are members of the geometric progression $2^{-N} e^-/(bohr)^3$, where N runs from 0 to 14. All plots are in a plane passing through the two nuclei. Dotted lines, nodal surface. (b) Comparison of molecular orbital density. Contour diagrams of the electron densities characteristic of the shell model of the molecules as listed. Both the total molecular density and the constituent shell densities are displayed at the experimental internuclear distance of each molecule. (He_2, Be_2, and Ne_2, members of this homonuclear series, are not bound in their ground state and therefore are not displayed.) (By permission, from A. C. Wahl, Argonne Natl. Lab. Report 7076, 1965.) [See also, A. C. Wahl, *Science*, *151*, 961 (1966). One bohn equals 5.292×10^{-11} m.]

Molecular Orbital

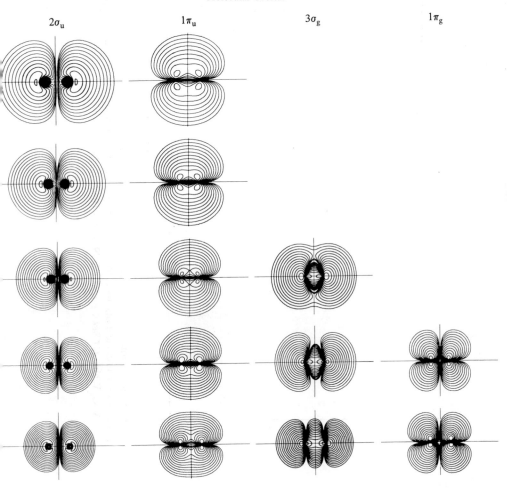

177

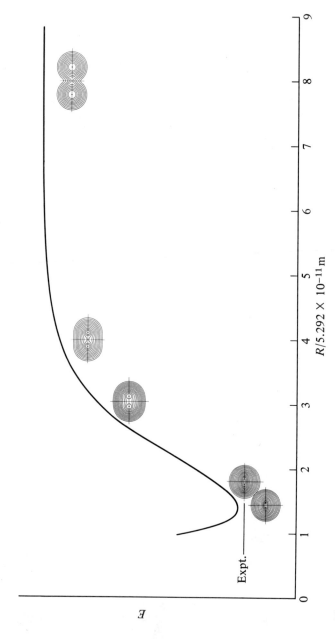

Figure 5-26. Contours of total electron density from an optimized valence configuration wave function for H_2 as a function of internuclear distance, together with the calculated potential energy function. [By permission, from G. Das and A. C. Wahl, *J. Chem. Phys.*, **47**, 2934–42 (1967).]

E

Expt.

$R/5.292 \times 10^{-11}\,\mathrm{m}$

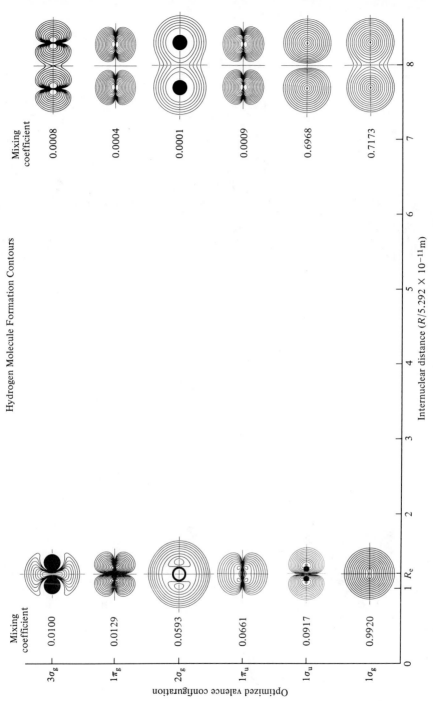

Figure 5-27. Contours of the orbitals making up the optimized valence configuration wave function of H_2 at large R and at R_e. [By permission, from G. Das and A. C. Wahl, *J. Chem. Phys.*, *47*, 2934–42 (1967).]

Hydrogen Molecule Formation Contours

Mixing coefficient

Optimized valence configuration	Mixing coefficient
$3\sigma_g$	0.0100
$1\pi_g$	0.0129
$2\sigma_g$	0.0593
$1\pi_u$	0.0661
$1\sigma_u$	0.0917
$1\sigma_g$	0.9920

Mixing coefficient (large R)

$3\sigma_g$	0.0008
$1\pi_g$	0.0004
$2\sigma_g$	0.0001
$1\pi_u$	0.0009
$1\sigma_u$	0.6968
$1\sigma_g$	0.7173

Internuclear distance $(R/5.292 \times 10^{-11}\mathrm{m})$

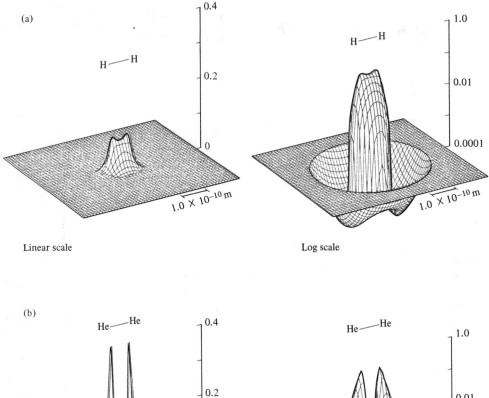

(a)

H——H

Linear scale

Log scale

(b)

He——He

Linear scale

Log scale

Figure 5-28. Plots of the difference in electron density between a molecule and separated atoms as a function of position in the plane of the internuclear axis for (a) H_2 and (b) He_2. (Reprinted by permission of Macmillan Publishing Co. Inc., from *Orbital and Electron Density Diagrams*, by A. Streitweiser and P. H. Owens, Copyright © 1973 by Macmillan Publishing Co., Inc.)

180

6 · Molecular structure and molecular spectra

Quantum-mechanical methods can be used to determine molecular structure from the data of molecular spectroscopy. In turn, the postulates of wave mechanics are supported by successful prediction of experimentally observed molecular spectra. Here, we begin our discussion of molecular spectra with the simplest ones, those spectra related to the rotational motion of molecules.

Figure 6-1. Spherical coordinate system for a reduced mass μ at a fixed distance r from the center of rotation: $x = r \sin \theta \cos \phi$; $y = r \sin \theta \sin \phi$; $z = r \cos \theta$.

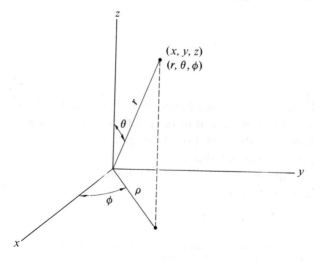

6-1. Rotational spectra

The absorption spectrum of a diatomic gas in the far infrared and microwave regions is a series of lines, spaced nearly evenly on a frequency scale. It is produced by transitions between quantized rotational energy levels. Precise values for the moments of inertia, and, therefore, for interatomic distances, can be calculated from the spacing of the rotational absorption lines.

The rigid rotor model

Consider a diatomic molecule with fixed internuclear distance r. Its center of mass may be taken as the center of a spherical coordinate system similar to that shown in Fig. 6-1. Since by definition of the center of mass, $m_1 r_1 = m_2 r_2$, the moment of inertia of the molecule is

$$I = m_1 r_1{}^2 + m_2 r_2{}^2$$
$$= (m_1 + m_2) r_1 r_2 \qquad (6\text{-}1)$$

From the same definition, it can be seen that

$$r_1 = \frac{m_2}{m_1 + m_2} (r_1 + r_2)$$

$$= \frac{m_2 r}{m_1 + m_2}$$

and

$$r_2 = \frac{m_1 r}{m_1 + m_2}$$

so that

$$I = \frac{m_1 m_2}{m_1 + m_2} r^2$$
$$= \mu r^2 \qquad (6\text{-}2)$$

where μ is the reduced mass. Thus, the rotation of a diatomic molecule about its center of mass is seen to be equivalent to the rotation of a mass μ at a distance r from its center of rotation; that is, to the motion of a particle of mass μ on the surface of a sphere of radius r.

The stable states of a diatomic molecule can be found by solving the time-independent Schrödinger equation[1] (Eq. 4-19)

$$\hat{H}\psi = \varepsilon\psi$$

---

1. Henceforth, we shall distinguish between the energy ε of a molecule, the energy E of a macroscopic system, and the molar energy $\mathscr{E}$ of a system.

Since the potential energy of a diatomic molecule is zero in the absence of an external field,

$$\hat{H} = -\frac{h^2}{8\pi^2\mu}\left[\frac{\partial^2}{\partial x^2} + \frac{\partial^2}{\partial y^2} + \frac{\partial^2}{\partial z^2}\right]$$

and

$$-\frac{h^2}{8\pi^2\mu}\left[\frac{\partial^2\psi}{\partial x^2} + \frac{\partial^2\psi}{\partial y^2} + \frac{\partial^2\psi}{\partial z^2}\right] = \varepsilon\psi \tag{6-3}$$

The form of Eq. 6-3 for a particle of mass μ moving on the surface of a sphere of radius r,[2]

$$\frac{\partial^2\psi}{\partial\theta^2} + \frac{\cos\theta}{\sin\theta}\frac{\partial\psi}{\partial\theta} + \frac{1}{\sin^2\theta}\frac{\partial^2\psi}{d\phi^2} = -\frac{8\pi^2\mu r^2}{h^2}\varepsilon\psi \tag{6-4}$$

is reminiscent of the angular equation for a one-electron atom (Eq. 4-44). It can be separated if we assume that $\psi(\theta, \phi)$ can be expressed as a product $\ominus(\theta)\,\Phi(\phi)$. On making the appropriate substitutions for ψ and its derivatives and multiplying by $\sin^2\theta$, we obtain

$$\frac{\sin^2\theta}{\ominus}\frac{d^2\ominus}{d\theta^2} + \frac{\sin\theta\cos\theta}{\ominus}\frac{d\ominus}{d\theta} + \frac{8\pi^2\mu r^2\,\varepsilon\sin^2\theta}{h^2} = -\frac{1}{\Phi}\frac{d^2\Phi}{d\phi^2} \tag{6-5}$$

Both sides of Eq. 6-5 must be equal to a constant, which we call $m_j{}^2$. The equations for $\ominus$ and Φ then are

$$\frac{d^2\ominus}{d\theta^2} + \frac{\cos\theta}{\sin\theta}\frac{d\ominus}{d\theta} + \left[\frac{8\pi^2\mu r^2\,\varepsilon}{h^2} - \frac{m_j{}^2}{\sin^2\theta}\right]\ominus = 0 \tag{6-6}$$

and

$$\frac{d^2\Phi}{d\phi^2} = -m_j{}^2\,\Phi \tag{6-7}$$

Equation 6-7 is identical to Eq. 4-45 for the one-electron atom, and the solutions are those given by Eqs. 4-47, 4-48, and 4-49. As before, m_j must have integral values to ensure a single-valued and continuous wave function. Also, m_j is a quantum number for the z component of angular momentum for the complex exponential solutions (Eq. 4-47), and $m_j{}^2$ is a quantum number for the square of the z component of the angular momentum for the real solutions (Eqs. 4-48 and 4-49) as well as for the complex solutions.

Equation 6-6 has the same form as Eq. 4-46, if

$$\frac{8\pi^2\mu r^2\,\varepsilon}{h^2} = J(J+1) \tag{6-8}$$

2. J. W. Linnett, *Wave Mechanics and Valency*, Methuen, London, 1960, p. 27.

where we have used J for the rotational quantum number to distinguish it from the angular momentum quantum number for the electron. Again, the solutions to Eq. 6-6 are well behaved only if J is a positive integer or zero, and only if $J \geqslant |m_J|$. Thus, there are $2J + 1$ values of m_J for each value of J, and each solution of Eq. 6-6 is $(2J + 1)$-fold degenerate. From Eq. 6-8, we can see that the possible quantized values of the rotational energy of a diatomic molecule are

$$\varepsilon_{\text{rot}} = J(J + 1)\frac{h^2}{8\pi^2 \mu r^2}$$

$$= J(J + 1)\frac{h^2}{8\pi^2 I} \tag{6-9}$$

Exercise 6-1. Set up and solve the Schrödinger equation for the rotation of a rigid diatomic molecule constrained to the x-y plane, and obtain the energy levels.

Since ε_{rot} is related to the angular momentum M by

$$\varepsilon_{\text{rot}} = \frac{M^2}{2I}$$

Eq. 6-9 leads to the relation

$$M^2 = J(J + 1)\frac{h^2}{4\pi^2}$$

$$= J(J + 1)\left(\frac{h}{2\pi}\right)^2 \tag{6-10}$$

and J is a quantum number for the square of the total angular momentum. The magnitude of the angular momentum is

$$M = [J(J + 1)]^{\frac{1}{2}}\left(\frac{h}{2\pi}\right) \tag{6-11}$$

which has the same form as Eq. 4-68.

Exercise 6-2. Show that $\psi = \Phi\Theta$, a solution of Eq. 6-4, is an eigenfunction of the square of the angular momentum by applying the operator given in Eq. 4-65.

The energy absorbed in the transition from $J = J''$ to $J = J'$ is obtained from Eq. 6-9 as

$$\varepsilon' - \varepsilon'' = \frac{h^2}{8\pi^2 I}[J'(J' + 1) - J''(J'' + 1)] \tag{6-12}$$

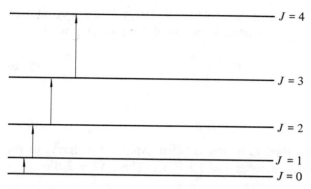

Figure 6-2. Scheme of the rotational energy levels of a diatomic molecule. The transitions shown are those occurring on absorption of radiation.

Electromagnetic radiation affects the rotational motion of a molecule only if the molecule has a permanent dipole moment. Also, since photons of electromagnetic radiation are bosons, with integral spin (Sec. 4-3), they have a spin angular momentum equal to $h/2\pi$. When a transition between rotational levels occurs by absorption of electromagnetic radiation, angular momentum is conserved, and the rotational quantum number must change by $\Delta J = \pm 1$. There is an exception, for diatomic molecules, when a molecule has nonzero angular momentum about the internuclear axis, as in the case of NO. For such molecules, ΔJ can also equal zero.

For $J' = J'' + 1 = J + 1$, Eq. 6-12 reduces to

$$\Delta\varepsilon = \frac{h^2}{4\pi^2 I}(J + 1) \tag{6-13}$$

Exercise 6-3. The molecule HI has a bond length of 1.63×10^{-10} m. Calculate the value of $\Delta\varepsilon$ in Eq. 6-13 for HI, for J values of 0, 1, and 2, in joules.

The spectral lines corresponding to these transitions will have the frequency ν given by

$$\nu = \frac{\Delta\varepsilon}{h} = \frac{h}{4\pi^2 I}(J + 1), \quad J = 0, 1, 2, \ldots \tag{6-14}$$

The spacing between adjacent spectral lines is

$$\Delta\nu = \frac{h}{4\pi^2 I} \tag{6-15}$$

a constant determined by the moment of inertia of the molecule. The scheme in Fig. 6-2 represents the rotational energy levels and allowed transitions. It is customary to define a rotational constant, B, given by

$$B = \frac{h}{8\pi^2 I} \qquad (6\text{-}16)$$

so that

$$\Delta v = 2B \qquad (6\text{-}17)$$

The frequency of absorbed radiation is often expressed in terms of the related quantity, the wavenumber $\tilde{v} = 1/\lambda = v/c$. Then, $\Delta\tilde{v} = 2B/c$.

Exercise 6-4. Carbon monoxide has absorption lines in the microwave region that are spaced 1.15×10^{11} Hz apart. Calculate the internuclear distance in CO. What is $\Delta\tilde{v}$ for this spectrum?

At higher values of J, the experimentally observed rotational spectra of diatomic molecules deviate from the usually observed even spacing of lines; this deviation is due to centrifugal stretching. The effect is small, and we shall not attempt to treat the model that takes this effect into account.

Linear polyatomic molecules show the same spectral behavior in the far infrared and microwave regions as do diatomic molecules. Since these molecules have only a single moment of inertia, the two internuclear distances or more in a single molecule cannot be calculated. Additional information can be obtained from observations on molecules in which heavy isotopes have been substituted, if we assume that the bond lengths have not changed appreciably. A similar problem exists for nonlinear polyatomic molecules, since these molecules have, at most, three independent moments of inertia, and usually more than three bond distances and bond angles must be determined. The use of isotopic substitution here, as well, sometimes provides the additional information required.

Exercise 6-5. The center of mass x_0 of a linear triatomic molecule is given by the relation $x_0 = (m_1 x_1 + m_2 x_2 + m_3 x_3)/m$, where x represents the position on an arbitrary coordinate scale and $m = m_1 + m_2 + m_3$. The moment of inertia is $I = m_1(x_1 - x_0)^2 + m_2(x_2 - x_0)^2 + m_3(x_3 - x_0)^2$. The rotational constants B' for HCN and DCN are 1.47791 cm^{-1} and 1.20770 cm^{-1}, respectively [H. C. Allen, E. D. Tidwell, and E. K. Plyler, *J. Chem. Phys.*, **25**, 302 (1956)]. Verify that these are consistent with the internuclear distances $r_{CH} = 1.066 \times 10^{-10}$ m and $r_{CN} = 1.153 \times 10^{-10}$ m.

Experimental methods

In the spectral regions in which absorption due to rotational transitions

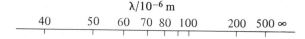

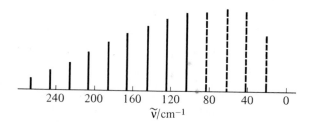

Figure 6-3. Absorption lines in the pure rotation spectrum of HCl (g). [By permission, from M. Czerny, *Z. Physik, 34,* 227 (1925).]

occurs, the far infrared and microwave regions, observations are made only with some difficulty. In a conventional, far infrared spectrophotometer, a diffraction grating to disperse radiation from a heated object and a narrow slit to isolate narrow spectral regions are used. The intensity of black-body radiation at long wavelengths is proportional to $1/\lambda^4$, so the far infrared radiation is of low intensity and difficult to measure accurately in such a spectrophotometer. Figure 6-3 shows one of the earliest examples of a pure rotation spectrum, in this case, for HCl (g).

Exercise 6-6. Calculate the frequency and wavelength corresponding to a wavenumber $\tilde{\nu} = 100 \text{ cm}^{-1}$.

Recently a Fourier transform spectrophotometer has been developed. "White" radiation is passed through the sample, using two light paths of slightly different length. Interference between the two beams produces a Fourier transform of the absorption spectrum, as a variable difference in path length is scanned. A computer then generates the absorption spectrum from the experimental Fourier transform. The intensity of radiation at the detector is greater than in a conventional spectrophotometer, and noise is averaged over the entire spectral range, rather than over a narrow region, to make this method much more sensitive than the conventional one.[3]

The widespread use of microwave techniques was a result of the develop-

3. C.C. Costain, "Molecular rotation", in *Physical Chemistry, An Advanced Treatise* Vol. IV (H. Eyring, D. Henderson, and W. Jost, ed.), Academic Press, New York, 1970, pp. 63-65; E. D. Becker and T. C. Farrar, *Science, 178,* 361–68 (1972).

ment of radar during and after World War II. A tuneable Klystron oscillator has been used as a source of radiation, and the frequencies of the absorption lines can be measured quite precisely. Harmonic generation extends the range of frequencies available, so that measurements from 8 GHz (1GHz = 1×10^9 Hz) to 300 GHz are available routinely. Some measurements have been made at frequencies up to 600 GHz. Typical measured frequencies and calculated bond lengths from microwave measurements are given in Table 6-1.

Table 6-1. Observed microwave absorption frequencies and calculated equilibrium internuclear distances for some diatomic molecules.

Molecule	$v(J = 0 \rightarrow 1)/$ GHz	$r_e/10^{-10}$m
$H^{35}Cl$	625.91924 ± 0.00052	1.27455
$D^{35}Cl$	323.29577 ± 0.00013	1.27457
$H^{79}Br$	500.67524 ± 0.00026	1.41460
$H^{81}Br$	500.51941 ± 0.00026	1.41460
HI	385.29327 ± 0.00070	1.60914

W. Gordy and R. L. Cook, *Microwave Molecular Spectra*, Wiley-Interscience, New York, 1970, p. 87.

6-2. Vibration-rotation spectra

Vibrational energy levels are much more widely spaced than rotational energy levels, and the spectra corresponding to transitions between vibrational levels appear in the infrared region of the spectrum. It was pointed out in Sec. 5-1 that the value of the electronic energy, E_{elec}, calculated as a function of the nuclear positions, is used for the potential energy in the Schrödinger equation for the nuclear motion. It is more common to obtain a potential energy curve from observed vibrational spectra and to use the resulting curve to test values of E_{elec} calculated as a function of internuclear distance, as in Figs. 5-14, 5-15, and 5-16.

The Simple Harmonic Oscillator Model

Consider the following model of a diatomic molecule: if the vibration is assumed to be harmonic, the force on the nuclei can be written

$$F = -kx \qquad (6\text{-}18)$$

where $x = r - r_e$ (r is the internuclear distance and r_e is the equilibrium internuclear distance), and k is the *Hooke's law constant* or *force constant*. The force can also be expressed in terms of the potential energy U; that is,

$$F = -\frac{dU}{dx} = -kx \qquad (6\text{-}19)$$

Upon integration of Eq. 6-19, we obtain

$$\int_{U=0}^{U} \frac{dU}{dx} dx = \int_{x=0}^{x} kx\,dx$$

or

$$U = \tfrac{1}{2} kx^2 \tag{6-20}$$

The kinetic energy K of the oscillator is

$$K = \frac{1}{2} m_1 \left(\frac{dr_1}{dt}\right)^2 + \frac{1}{2} m_2 \left(\frac{dr_2}{dt}\right)^2 \tag{6-21}$$

where r_1 and r_2 are the coordinates of m_1 and m_2 with respect to the center of mass of the molecule, and $r_1 + r_2 = r$. It follows from Eq. 6-21 that

$$K = \frac{1}{2} \mu \left(\frac{dx}{dt}\right)^2$$

$$= \frac{p_\mu^2}{2\mu} \tag{6-22}$$

where μ is the reduced mass.

Exercise 6-7. Verify Eq. 6-22. Hint: See Eqs. 6-1 and 6-2.

The total energy of the oscillator is then

$$\varepsilon = \frac{p_\mu^2}{2\mu} + \frac{1}{2} kx^2 \tag{6-23}$$

and the Hamiltonian operator is

$$\hat{H} = -\frac{h^2}{8\pi^2 \mu} \frac{d^2}{dx^2} + \frac{1}{2} kx^2 \tag{6-24}$$

Thus, the Schrödinger equation for the harmonic oscillator model is

$$-\frac{h^2}{8\pi^2 \mu} \frac{d^2\psi}{dx^2} + \frac{1}{2} kx^2 \, \psi = \varepsilon\psi \tag{6-25}$$

Like the equations for Θ and R of the one-electron atom (see Sec. 4-2), Eq. 6-25 has a polynomial solution, but it is helpful to examine an asymptotic solution before finding the polynomial solutions.

Rearranging Eq. 6-25, we obtain

$$\frac{d^2\psi}{dx^2} = \left(-\frac{8\pi^2\,\mu\varepsilon}{h^2} + \frac{4\pi^2\,\mu k}{h^2}\,x^2\right)\psi \tag{6-26}$$

$$= (-\beta + \alpha^2\,x^2)\,\psi \tag{6-27}$$

where

$$\beta = 8\pi^2\mu\,\frac{\varepsilon}{h^2}$$

and

$$\alpha^2 = 4\pi^2\mu\,\frac{k}{h^2}$$

At large enough values of x, the first term in the parentheses can be neglected, and

$$\frac{d^2\psi}{dx^2} = \alpha^2 x^2 \psi \tag{6-28}$$

An approximate solution to Eq. 6-28 is

$$\psi = \exp\left(-\frac{\alpha x^2}{2}\right) \tag{6-29}$$

for values of x large enough that $\alpha^2 x^2 \gg \alpha$.

The corresponding solution with a positive exponential is rejected because ψ must not increase without limit as x increases.

Exercise 6-8. Show that Eq. 6-29 is a solution of Eq. 6-28 for high enough values of x.

For the solution to Eq. 6-25, let us write

$$\psi = P(x)\exp\left(-\alpha\,\frac{x^2}{2}\right) = P(x)/\exp\left(\alpha\,\frac{x^2}{2}\right) \tag{6-30}$$

where

$$P(x) = \sum_{v=0}^{\infty}\ A_v x^v$$

When Eq. 6-30 and the appropriate derivative are substituted in Eq. 6-27, and the coefficients of like powers of x are collected and equated to zero, the following recursion formula is obtained

$$A_{v+2} = -A_v\,\frac{\beta - \alpha(2v + 1)}{(v + 1)(v + 2)} \tag{6-31}$$

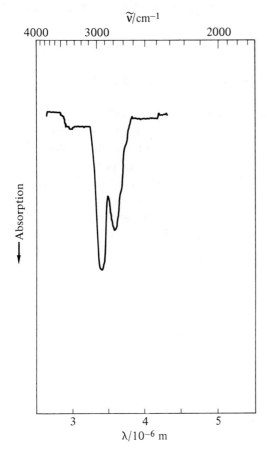

Figure 6-4. The vibrational absorption band of HCl (g) at low resolution.

$P(x)$ not only increases without limit, it increases more rapidly than exp $(\alpha x^2/2)$ as x increases. But, if $P(x)$ is truncated at some finite value of v, the ratio $P(x)/\exp(\alpha x^2/2)$ will approach zero as x becomes very large. $P(x)$ will terminate at v if the numerator in Eq. 6-31 is equal to zero, giving

$$\beta = \alpha(2v + 1)$$

or

$$\frac{8\pi^2 \mu \varepsilon}{h^2} = \left(\frac{4\pi^2 \mu k}{h^2}\right)^{\frac{1}{2}} (2v + 1)$$

Thus, the stationary state values of ε are

$$\varepsilon_v = (2v + 1)\frac{h}{4\pi}\left(\frac{k}{\mu}\right)^{\frac{1}{2}} \tag{6-32}$$

Since the fundamental frequency v_0 of a classical harmonic oscillator is equal to $[1/2\pi] (k/\mu)^{\frac{1}{2}}$, Eq. 6-32 can be written as

$$\varepsilon_v = (v + \tfrac{1}{2}) h v_0 \tag{6-33}$$

where v is vibrational quantum number corresponding to the index of the final term of the polynomial $P(x)$.

Thus we see that the energy of a quantum harmonic oscillator is quantized; only certain eigenvalues of the energy are consistent with the Schrödinger postulates. As was true of the particle in the potential well (Eq. 5-12), the oscillator has a nonzero zero-point energy, in this case equal to $\frac{1}{2}h v_0$.

Equation 6-33 describes a set of equally spaced energy levels. Absorption of a quantum of electromagnetic radiation and transition to a higher level can occur only for $\Delta v = 1$ for a harmonic oscillator and only if the oscillation of the molecule produces a changing dipole moment. If $\Delta v = 1$,

$$\Delta \varepsilon = \varepsilon_{v+1} - \varepsilon_v = (v + \tfrac{3}{2}) h v_0 - (v + \tfrac{1}{2}) h v_0$$
$$= h v_0 \tag{6-34}$$

and a diatomic molecule should exhibit a single absorption peak in the infrared region independent of the value of v for the level from which the transition occurs. Figure 6-4 shows such a peak for HCl gas at low resolution. Since $v_0 = [1/(2\pi)] (k/\mu)^{\frac{1}{2}}$, the frequency of the absorption peak can be used to calculate the force constant, a rough measure of the strength of the chemical bond. Values of k for several diatomic molecules are given in Table 6-2.

Table 6-2. Force constants of some diatomic molecules[a]

Molecule	$k/(10^2\,\text{N-m}^{-1})$
HF	9.659
HCl	5.163
HBr	4.114
HI	3.142
DF	9.662
DCl	5.163
DBr	4.116
DI	3.142
CO	19.017
NO	15.947

T. Shimanouchi, "The molecular force field", in *Physical Chemistry, an Advanced Treatise*, Vol. IV (D. Henderson, ed.), Academic Press, New York, 1970, p. 245
[a] These values were calculated by taking anharmonicity into account (see below).

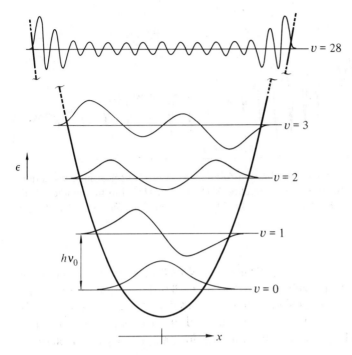

$v = 28$

$v = 3$

$v = 2$

$v = 1$

ϵ

hv_0

$v = 0$

x

Figure 6-5. The harmonic oscillator wave functions for several values of v, superimposed on the quantized energy levels and the potential energy curve. (By permission, adapted from P. W. Atkins, *Molecular Quantum Mechanics*, Clarendon Press, Oxford, 1970, p. 63.)

Exercise 6-9. Calculate the value of $\Delta\varepsilon$ in Eq. 6-34 for HI in joules, and compare with the rotational energy level spacing calculated in Exercise 6-3. Calculate the frequency, wavenumber, and wavelength of the vibrational absorption band.

Figure 6-5 shows the wave functions for several values of the quantum number v for the harmonic oscillator, superimposed on the corresponding energy levels and the potential energy curve. The probability of finding a classical oscillator is maximal at the ends of its oscillation, when the kinetic energy is zero. It can be seen from Fig. 6-5 that the quantum oscillator approaches this behavior only at high quantum numbers. The quantum oscillator has a nonzero probability of being outside the potential well, a probability that is zero for the classical oscillator. This property arises from the continuity of the wave function at the boundary of the potential well.

Rotational fine structure
At higher resolution, as shown in Fig. 6-6, a fine structure becomes evident

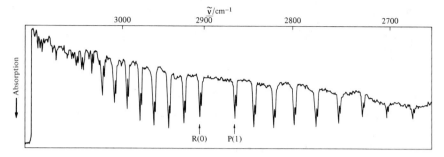

Figure 6-6. The fundamental vibration-rotation band of HCl (g) at high resolution. Each peak is split because of the presence of $H^{35}Cl$ and $H^{37}Cl$ in the sample. [E. K. Plyler, A. Danti, L. R. Blaine, and E. D. Tidwell, *J. Res. Nat. Bur. Standards* (*U.S.*), *A64*, 29 (1960).]

in the spectrum. These close, evenly spaced lines correspond to rotational transitions of the molecule. One can assume that the motion of a molecule is described by superimposing the harmonic oscillator and rigid rotor models, with no interaction between the vibrational and rotational motions. This assumption leads to a set of vibration-rotation energy levels whose energy is

$$\varepsilon = \left(v + \frac{1}{2}\right)hv_0 + J(J + 1)\frac{h^2}{8\pi^2 I} \tag{6-35}$$

Absorption transitions are limited to $\Delta v = 1$ and $\Delta J = \pm 1$ for linear molecules. Except for molecules with a nonzero electronic angular momentum, such as NO, ΔJ cannot equal zero, so absorption is observed at higher and lower frequencies, but not at v_0, the center of the peak.

If we take J to be the rotational quantum number of the lower vibrational level, we can write for $\Delta J = 1$,

$$\Delta\varepsilon = hv_0 + (J + 1)\frac{h^2}{4\pi^2 I} \tag{6-36}$$

and

$$v = v_0 + (J + 1)\frac{h}{4\pi^2 I} \tag{6-37}$$

where $J = 0, 1, 2, \ldots$. The absorption lines described by Eq. 6-37 are called the R branch of the vibrational band. The line at the center is called the Q branch. It appears, as noted above, only for the exceptional diatomic molecule. Similarly, for $\Delta J = -1$

$$\Delta\varepsilon = h\nu_0 - J\frac{h^2}{4\pi^2 I} \tag{6-38}$$

and

$$\nu = \nu_0 - J\frac{h}{4\pi^2 I} \tag{6-39}$$

where $J = 1, 2, 3, \ldots$. The absorption lines described by Eq. 6-39 are called the P branch of the vibrational band.

Exercise 6-10.   Verify Eq. 6-36 and Eq. 6-38.

The experimentally determined spacing between rotational lines in a vibration-rotation peak is not uniform. Most of the variation can be accounted for by a consideration of the interaction between rotational and vibrational motion, that is, the dependence of the moment of inertia of the molecule on the vibrational quantum number (see below). A lesser contribution results from deviations from the rigid-rotor and harmonic-oscillator models.

The Morse potential function
The potential energy increases without limit as x increases for a harmonic oscillator, whereas the potential energy of a diatomic molecule approaches a finite limit that is the sum of the energies of the isolated atoms. Thus, it is clear that we cannot use the harmonic-oscillator model to describe the vibrational behavior of molecules at high vibrational energies.

In 1929, Morse suggested a potential function

$$U(x) = D[1 - \exp(-ax)]^2 \tag{6-40}$$

for the vibrational motion of a diatomic molecule. It is equal to zero at $x = 0$, $(r = r_e)$, and it approaches D as a limit as r increases without limit, so that D represents the experimental dissociation energy D_0 plus the zero-point energy $\frac{1}{2}h\nu_0$. As r approaches zero, the function approaches a finite limit rather than increasing without limit. Small values of r are not important experimentally, and this empirical function is quite useful, as can be seen from Fig. 6-7, where the experimental curve and the Morse function are shown for HCl. The harmonic-oscillator model is a good approximation for most molecules at room temperature, at which they are in their lowest vibrational levels.

The significance of the parameter a in the Morse function can be seen if we differentiate Eq. 6-40 to obtain d^2U/dx^2 at $x = 0$, namely

$$\left[\frac{d^2 U}{dx^2}\right]_{x=0} = 2a^2 D \qquad (6\text{-}41)$$

At $x = 0$, the harmonic approximation is valid, so

$$\frac{d^2 U}{dx^2} = k \qquad (6\text{-}42)$$

and

$$a = \left(\frac{k}{2D}\right)^{\frac{1}{2}} \qquad (6\text{-}43)$$

where k is the force constant of the molecule.

Exercise 6-11.   Verify Eqs. 6-41, 6-42, and 6-43.

The anharmonic oscillator
It can be seen from Fig. 6-8 that the harmonic approximation fits the experi-

Figure 6-7.   The Morse curve (– – –) compared to the potential energy curve calculated from spectroscopic data (——), for HCl. (From *Molecular Spectra and Molecular Structure*, Vol. 1, by G. Herzberg. © 1950 by Litton Educational Publishing, Inc. Reprinted by permission of Van Nostrand Reinhold Company.)

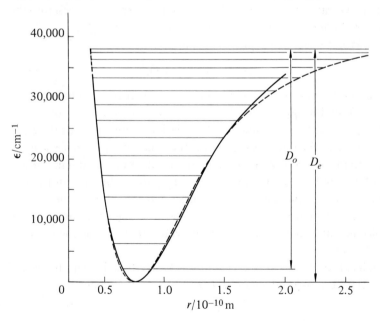

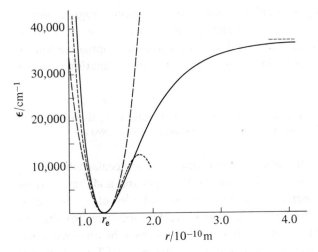

Figure 6-8. The cubic approximation to the potential energy curve, for HCl (…), compared to the experimental curve (——), and the harmonic approximation (– – –). (From *Molecular Spectra and Molecular Structure*, Vol. 1, by G. Herzberg. © 1950 by Litton Educational Publishing, Inc. Reprinted by permission of Van Nostrand Reinhold Company.)

mental potential energy curve only over a short range near the minimum. A better approximation is obtained if a cubic term is added to the potential energy function, as shown in Eq. 6-44.

$$U = \tfrac{1}{2} k_e x^2 - g x^3 \tag{6-44}$$

The solution of the Schrödinger equation, with this potential energy function, is more difficult than for the harmonic oscillator, and we shall merely state the resulting expression for the energy eigenvalues.[4]

$$\varepsilon_v = (v + \tfrac{1}{2})h\nu_e - (v + \tfrac{1}{2})^2 h\nu_e x_e \tag{6-45}$$

Here x_e is the anharmonicity constant, and ν_e is the fundamental frequency of a classical oscillator that best fits the experimental potential curve at the equilibrium internuclear distance; it can be written

$$\nu_e = \left(\frac{1}{2\pi}\right)\left(\frac{k_e}{\mu}\right)^{\frac{1}{2}} \tag{6-46}$$

4. See G. Herzberg, *Molecular spectra and Molecular structure*, Vol. I, *Spectra of Diatomic Molecules*, D. Van Nostrand, Princeton, 2nd ed., 1950, pp. 92–93. Henceforth referred to as Vol. I.

The force constant k_e is calculated from the anharmonic approximation. For HCl, the value of k_e for the anharmonic approximation[5] is 5.163×10^2 N-m^{-1}, whereas the value of k for the harmonic approximation[6] is 4.806×10^2 N-m^{-1}. The relative fit of the cubic approximation is shown graphically in Fig. 6-8.

Exercise 6-12.   Calculate a value of v_e for HCl from the value of k_e given above, and compare to the value of v_0 obtained from the absorption at $\tilde{v} = 2890$ cm^{-1}.

One consequence of the anharmonic character of molecular vibrations is that the selection rule, $\Delta v = \pm 1$, is not strictly applicable and that overtone spectra appear at higher frequencies close to multiples of v_0, but at an intensity lower than the fundamental one. Such a spectrum for the first harmonic of HCl is shown in Fig. 6-9. Table 6-3 shows the observed wavenumber of the harmonics in the vibration spectrum of HCl, together with values calculated from the harmonic and cubic approximations.

Table 6-3.   Vibrational bands of HCl

Transition	$\tilde{v}$/cm^{-1} (obs.)	$\tilde{v}$/cm^{-1} (calc.) (harmonic)	$\tilde{v}$/cm^{-1} (calc.) (cubic)
$v = 0 \rightarrow v = 1$ (fundamental)	2,885.9	2,885.9	2,885.7
$v = 0 \rightarrow v = 2$ (2nd harmonic)	5,668.1	5,771.8	5,668.2
$v = 0 \rightarrow v = 3$ (3rd harmonic)	8,347.0	8,657.7	8,347.5
$v = 0 \rightarrow v = 4$ (4th harmonic)	10,923.1	11,543.6	10,923.6
$v = 0 \rightarrow v = 5$ (5th harmonic)	13,396.6	14,429.5	13,396.5

From *Molecular Spectra and Molecular Structure*, Vol. 1, by G. Herzberg. © 1950 by Litton Educational Publishing, Inc. Reprinted by permission of Van Nostrand Reinhold Co.

Experimental values for the frequencies of the fundamental and the harmonics in the vibrational spectrum, e.g., the values in Table 6-3, are required to calculate the constants in Eq. 6-45.

Exercise 6-13.   Derive from Eq. 6-45 an equation for the wavenumber of the vibrational bands in HCl as a quadratic function of v. Plot $\tilde{v}$ against v and $\tilde{v}/v$ against v, and calculate v_e and x_e from the slope and intercept of the *linear* graph. Use the method of least squares to obtain values of v_e and x_e analytically.

5. T. Shimanouchi, "The molecular force field", in *Physical Chemistry, an Advanced Treatise*, Vol. IV, (D. Henderson, ed.), Academic Press, New York, 1970, p. 245.
6. G. Herzberg, Vol. I.

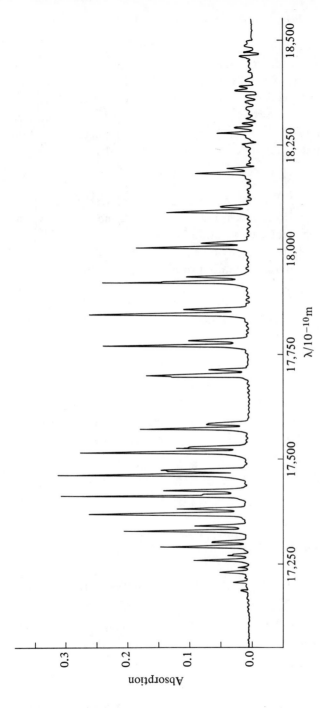

Figure 6-9. The first harmonic of the vibration-rotation band of HCl (g) $v = 0 \rightarrow v = 2$. The double peaks are due to the presence of $H^{35}Cl$ and $H^{37}Cl$ in the sample. (Courtesy of J. S. Evans, Lawrence University.)

Vibration-rotation interaction

As was pointed out above, the rotational fine structure of a vibrational spectrum does not consist of equally spaced lines. The data can be accounted for by considering the dependence of the moment of inertia on the vibrational quantum number. Thus, the constant $B = h/(8\pi^2 I)$ of Eq. 6-16 must be replaced by the expression $B_v = h/(8\pi^2 I_v)$, where I_v is explicitly a function of the vibrational quantum number v. It can be shown[7] that

$$B_v = B_e - (v + \tfrac{1}{2})c\alpha_e \qquad (6\text{-}47)$$

in which $B_e = h/(8\pi^2 I_e)$, I_e is the moment of inertia corresponding to the equilibrium internuclear distance, c is the speed of light, and α_e is a constant that is a measure of the vibration-rotation interaction.

When the interaction of vibration and rotation is taken into account, the energy levels of an anharmonic vibrating rotor are given by the expression

$$\varepsilon = (v + \tfrac{1}{2})h\nu_e - (v + \tfrac{1}{2})^2 h\nu_e x_e + J(J + 1) B_v h \qquad (6\text{-}48)$$

The frequencies of the lines of a vibration-rotation band for the transition from $v = v''$ to $v = v'$ are

$$\nu = \nu_0 + J'(J' + 1) B_v' - J''(J'' + 1) B_v'' \qquad (6\text{-}49)$$

The frequency for a transition in which there is no change in rotational quantum number is ν_0. Since, by the selection rules for diatomic molecules, $\Delta J = \pm 1$, there is no line at ν_0. There is an R branch for $\Delta J = 1$ and a P branch for $\Delta J = -1$. If J represents the rotational quantum number in the lower vibrational state,

$$\nu_R = \nu_0 + (B_v' - B_v'')J^2 + (3B_v' - B_v'')J + 2B_v' \quad J = 0, 1, 2\ldots \quad (6\text{-}50)$$

$$\nu_P = \nu_0 + (B_v' - B_v'')J^2 - (B_v' + B_v'')J \quad J = 1, 2\ldots \qquad (6\text{-}51)$$

Exercise 6-14. Verify Eqs. 6-50 and 6-51.

Equations 6-50 and 6-51 can be combined into a single equation by letting $J = m - 1$ in Eq. 6-50 and $J = -m$ in Eq. 6-51. The result is

$$\nu = \nu_0 + (B_v' - B_v'')m^2 + (B_v' + B_v'')m \qquad (6\text{-}52)$$

Exercise 6-15. Verify Eq. 6-52.

7. L. Pauling and E. B. Wilson, *Introduction to Quantum Mechanics*, McGraw-Hill Book Co., New York, 1935, p. 274.

Experimental frequencies can be fitted to a quadratic expression in m, thus permitting the calculation of v_0, B_v', and B_v''. The values of v_e and x_e in Eq. 6-48, and, thus, k_e, can be calculated from the values of v_0 for the fundamental band and several harmonics.

The constants B_e and α_e can be calculated by fitting the values of B_v to Eq. 6-47. From B_v and B_e, the corresponding values of I_v and r_v and I_e and r_e can be calculated. Since I_e is calculated for the equilibrium internuclear distance at the minimum in the potential energy curve, the value of r_e calculated from I_e has a simple physical significance. In contrast, I_v is calculated from a rotational constant that is an average value over the vibration of the molecule. Thus,

$$B_v = \frac{h^2}{8\pi^2} \frac{1}{I_v}$$

$$= \frac{h^2}{8\pi^2 \mu} \left\langle \frac{1}{r^2} \right\rangle \tag{6-53}$$

and

$$r_v = \left\langle \frac{1}{r^2} \right\rangle^{-\frac{1}{2}} \neq r_e \tag{6-54}$$

where $\langle \ \rangle$ indicates an average quantity. In particular, the interatomic distance calculated from rotational spectra in the far infrared and microwave regions is r_0, which is not equal to r_e. Values of r_0 and r_e for several diatomic molecules are shown in Table 6-4. It can be seen that only for light atoms, such as H, are there significant differences between r_e and r_0.

Table 6-4. Values of the equilibrium internuclear distance, r_e, and the average internuclear distance of the lowest vibrational level, r_0, for some diatomic molecules

Molecule	$r_e/10^{-10}$m	$r_0/10^{-10}$m
HF	0.9170	0.9257
DF	0.9171	0.9234
HCl	1.2745	1.2837
DCl	1.2744	1.2813
I^{35}Cl	2.3209	2.3236
I^{37}Cl	2.3209	2.3235

V. W. Laurie and D. R. Herschbach, *J. Chem. Phys.*, 37, 1687 (1962)

Raman spectra

When a molecule is exposed to electromagnetic radiation, the action of the oscillating electric field **E** causes the electrons of the molecule to oscillate, and an oscillating, induced dipole moment **P** is obtained. This moment is related to the field through the polarizability α, as shown in Eq. 6-55.[8]

$$\mathbf{P} = \alpha \, \mathbf{E} \qquad\qquad (6\text{-}55)$$

The induced oscillating dipole, in turn, emits electromagnetic radiation of the same frequency as the incident radiation. This phenomenon, known as Rayleigh scattering, is an elastic interaction between the radiation and the molecule, with no net gain or loss of energy by the molecule.

The scattered radiation is emitted in all directions, and its intensity is inversely proportional to the fourth power of the wavelength. As a result, radiation of short wavelengths is more prominent in scattered light than radiation of longer wavelengths. This radiation makes visible the suspended particles in a gas or liquid in the path of a light beam. Thus, the sky appears blue because the light seen is the scattered light, in which the shorter wavelengths are prominent, whereas the sunset is red because the light seen is that light which is not scattered. Since the polarizability of a molecule increases with increasing number of electrons, the intensity of scattered radiation increases with increasing mass, and light scattering can be used to determine the molecular mass of dissolved macromolecules, such as proteins.

In 1928, Raman and his students observed that, in addition to the Rayleigh scattering, there was scattered light of a different frequency from the incident light. The frequency v of the Raman scattering was found to be

$$v = v' + \frac{\varepsilon_m - \varepsilon_n}{h} \qquad\qquad (6\text{-}56)$$

where v' is the incident frequency and ε_m and ε_n are two vibration-rotation levels of the molecule. The Raman scattering is an inelastic interaction, since the molecule may gain or lose energy in the process. The difference $(v - v')$ is seen to be independent of the frequency of the incident light. The light is not absorbed in the sense of exciting the molecule to a higher electronic state; incident light of a frequency high enough to excite the molecule to a higher electronic state does not lead to Raman scattering.

Rotational Raman scattering occurs only if the polarizability of the molecule changes during a molecular rotation; that is, only if the polarizability is anisotropic and thus dependent on the orientation of the molecule with respect to the direction of the incident beam. The diatomic molecules, therefore, show rotational Raman spectra, whereas symmetric molecules,

---

8. **P** and **E** are vectors in three-dimensional space, and α is a tensor.

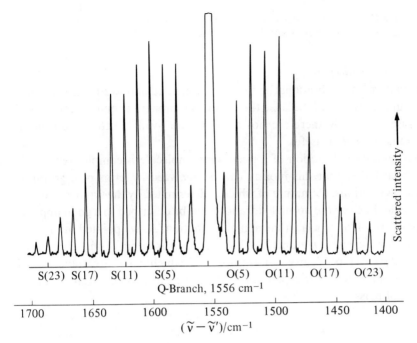

Figure 6-10. A laser-Raman spectrum showing a rotation-vibration band of O_2. The peaks labeled O correspond to $\Delta J = 2$; the peaks labeled S correspond to $\Delta J = -2$. [By permission, from J. J. Barrett and N. I. Adams, III, *J. Opt. Soc. Am.*, *58*, 311 (1968).]

like CH_4, do not. The selection rule for rotational Raman scattering is $\Delta J = 0, \pm 2$. Since Raman scattering is a two-photon process, the change in angular momentum due to the photons can be in the same direction or in opposite directions, leading to the given selection rule.

Vibrational Raman scattering occurs only if polarizability changes during a vibration. Thus, the vibration of homonuclear diatomic molecules, which is not active in the infrared, is Raman active. The selection rules are $\Delta v = +1$ and $\Delta J = 0, +2$. If a molecule has a center of symmetry, the *exclusion rule* states that a vibration that is active in the infrared is inactive in Raman scattering, and vice versa. Thus, the two are complementary. Figure 6-10 shows a Raman vibration-rotation band of O_2 obtained with a laser exciting source. The high intensity of a laser source and its narrow bandwidth permit detection of weak Raman lines close to the intense Rayleigh scattering at the wavelength of the source. The spectrum in Fig. 6-10 was obtained from a sample in a volume of $10^{-14} m^3$.

Exercise 6-16. Obtain an expression for the Raman difference $(v - v')$ for $\Delta v = 1$ and $\Delta J = 2$, assuming the rigid-rotor, harmonic-oscillator model.

Polyatomic Molecules

We shall treat the vibrational and rotational spectra of polyatomic molecules at a simple level, both because the rigorous treatment is quite complex and because frequently there are not enough data available to make a more rigorous treatment useful.

The vibrational motion of a polyatomic molecule is complicated, but it has been found that this motion can be analyzed in terms of a group of normal vibrations, the nature of which is determined by the symmetry of the molecule. Each of the normal vibrations can be described to a good approximation as a Hooke's law (harmonic) oscillation. As we pointed out in Sec. 2-3, a molecule with n atoms has $3n$ degrees of freedom. Thus, a triatomic molecule has nine degrees of freedom, of which three are translational degrees of freedom, corresponding to the three Cartesian coordinates of the center of mass.

If the molecule is linear, it has two rotational degrees of freedom, as illustrated in Fig. 2-4a, and four vibrational degrees of freedom, as illustrated in Fig. 2-4b, Since a symmetric molecule, such as CO_2, has no dipole moment, its rotational motion does not result in absorption in the far infrared and microwave region. Similarly, since the symmetric stretching vibration does not change the dipole moment, there is no infrared absorption corresponding to that vibration. There is a Raman line attributable to this vibration, however. The asymmetric stretching vibration of CO_2 and the two degenerate bending vibrations do change the dipole moment, and their infrared absorption bands can be observed. The asymmetric stretching vibration, which results in an oscillating dipole moment along the molecular axis, has the same selection rules as the diatomic molecule, $\Delta v = \pm 1$ and $\Delta J = \pm 1$, so that the absorption band shows a P branch and an R branch but no Q branch. The bending vibrations, which produce an oscillating dipole moment perpendicular to the molecular axis, have the selection rules $\Delta v = \pm 1$, $\Delta J = 0, \pm 1$, so that the absorption band has a Q branch as well as a P and an R branch. The moment of inertia can be calculated from the rotational fine structure of these bands, just as it is for diatomic molecules.

Exercise 6-17. Show that determination of a single moment of inertia is sufficient to calculate the bond lengths in CO_2.

The center of a vibration-rotation band, whether in infrared or Raman spectra, represents the frequency of the corresponding normal mode of vibration to the harmonic-oscillator approximation.

A simple model that permits the calculation of force constants for the normal vibrations of a triatomic molecule is the *valence-bond force field* model. In this model, the vibrational potential energy depends on the deviation of bond lengths and bond angles from their equilibrium values. In the notation used by Herzberg,[9] the potential energy U of CO_2 is

$$U = \tfrac{1}{2}[k_1\,(Q_1^2 + Q_2^2) + k_3\,(\phi_a^2 + \phi_b^2)] \tag{6-57}$$

in which Q_1 is the change in r (CO), Q_2 is the change in r' (CO), ϕ_a and ϕ_b are the angles of bending in two perpendicular planes through the molecular axis, k_1 is the bond-stretching force constant, and k_3 is the force constant for a bending vibration. The bending force constant is calculated from the bending frequency, and the stretching force constant is calculated from the two stretching frequencies. The comparison of the two values of k_1 is a test of the validity of the model. Table 6-5 gives the observed stretching frequencies and the values of k_1 calculated from them. The bending force constant reflects the strongly directional nature of covalent bonds.

Table 6-5. Values of the stretching force constant of CO_2 calculated from observed vibrational bands

		$k_1(10^2\,\text{N-m}^{-1})$
v_1 (symmetric stretch)	$1337\ \text{cm}^{-1}$	16.8
v_2 (asymmetric stretch)	$667\ \text{cm}^{-1}$	14.2

From *Molecular Spectra and Molecular Structure*, Vol. 2, by G. Herzberg. © 1945 by Litton Educational Publishing, Inc. Reprinted by permission of Van Nostrand Reinhold Co.

A nonlinear triatomic molecule has three rotational degrees of freedom, as indicated in Fig. 2-4, and, thus, it has three vibrational degrees of freedom, and three normal modes of vibration, as shown in Fig. 2-4,d. Such molecules as SO_2 and H_2O are symmetric nonlinear triatomic molecules. Since they have permanent dipole moments, their rotational motion results in absorption in the far infrared and microwave regions. The three moments of inertia obtained from rotational spectra permit the calculation of the bond lengths and bond angles of the triatomic molecule. Molecular structure data obtained from microwave spectra for several molecules are shown in Table 6-6.

9. G. Herzberg, *Molecular Spectra and Molecular Structure*, Vol. II, *Infrared and Raman Spectra of Polyatomic Molecules*, D. Van Nostrand Company, Princeton, 1945, pp. 154, 172. Henceforth referred to as Vol. II.

Table 6-6. Bond lengths and bond angles obtained from microwave spectra

Molecule	Bond	length/10^{-10}m	Bond	angle
SO_2	S-O,	1.4308 ± 0.0002	OSO,	$119°19' \pm 2'$
H_2O	H-O,	0.95721 ± 0.0003	HOH,	$104.522° \pm 0.05°$
F_2O	F-O,	1.4053 ± 0.0004	FOF,	$103°4' \pm 3'$
Cl_2O	Cl-O,	1.7004 ± 0.0004	OClO,	$117°36' \pm 1°$

W. Gordy and R. L. Cook, *Microwave Molecular Spectra*, Wiley-Interscience, New York, 1970, pp. 688–89

The three vibrational degrees of freedom of such molecules as SO_2 and H_2O are all active in the infrared, so that three rotation-vibration bands are observed. According to the valence-bond force field model, the vibrational potential energy of such a molecule is[10]

$$U = \tfrac{1}{2} [k_1 \, (Q_1{}^2 + Q_2{}^2) + k_\delta \, \delta^2] \qquad (6-58)$$

in which Q_1 is the change in one bond length, Q_2 is the change in the other bond length, δ is the change in bond angle, and k_1 and k_δ are the force constants. The asymmetric stretching frequency depends only on k_1; thus k_1 can be calculated from this frequency alone. The symmetric stretching frequency and the bending frequency depend on both k_1 and k_δ; thus two frequencies must be known to calculate k_δ.

A detailed mathematical treatment of the structural significance of vibrational spectra for more complex molecules can be found in Herzberg. But, even in the absence of a mathematical treatment, the infrared and Raman spectra yield structural information. Although, in principle, every vibration involves the whole molecule, in practice, groups of atoms have been found to exhibit vibrational bands that are sufficiently characteristic of a group to identify the corresponding group as part of the molecule. At the same time, the interaction among groups in the molecule is such that the vibrational spectrum of each molecule is unique and positively identifies it.[11] Figure 6-11 shows a Fourier transform spectrum of an asymmetric stretching band of methane. The position and shape of the band can be used to identify methane in a sample, and the rotational structure can be used to to calculate the moment of inertia and the bond length.

6-3. Electronic spectra

The spacing of the electronic energy levels described in Chapter 5 is such

10. G. Herzberg, Vol. II., p. 168.
11. L. J. Bellamy, *The Infrared Spectra of Complex Molecules*, Methuen, London, 1958.

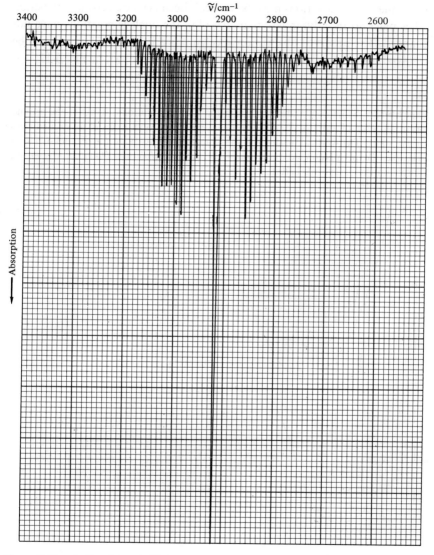

Figure 6-11. A Fourier transform spectrum of an asymmetric stretching vibration-rotation band for CH_4. (Courtesy of J. Marlier, University of Wisconsin-Madison.)

that all molecules absorb radiation in the ultraviolet or far ultraviolet regions. Those substances we perceive as colored absorb in the visible region as well. Since most molecules are at the lowest vibrational level when in the electronic ground state at ordinary temperatures, the transitions shown in Fig. 6-12 are from that level to several vibrational levels in an excited state. On the basis of the Born-Oppenheimer approximation, the Franck-Condon principle states that such transitions must be vertical. Because the electronic transition takes place so rapidly, in less than 10^{-14} s, the nuclear positions have not changed. The observed absorption spectrum is a broad band, which can be resolved into lines corresponding to the transitions shown in Fig. 6-12.

Figure 6-12. Scheme of transitions from the lowest vibrational levels of the ground electronic state of a diatomic molecule to several vibrational levels of an excited electronic state. The vertical nature of the transitions illustrates the Franck-Condon principle. (By permission, adapted from P. W. Atkins, *Molecular Quantum Mechanics*, Clarendon Press, Oxford, 1970.)

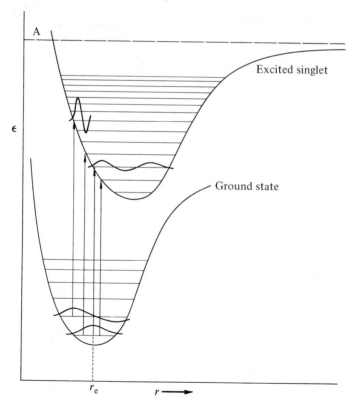

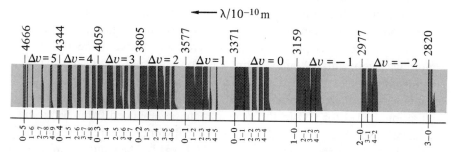

Figure 6-13. Portion of a photographic recording of the emission spectrum of N_2, showing the vibrational bands. (From *Molecular Spectra and Molecular Structure*, Vol. 1, by G. Herzberg. © 1950 by Litton Educational Publishing, Inc. Reprinted by permission of Van Nostrand Reinhold Company.)

Thermal or electrical excitation of molecules makes it possible to observe corresponding emission spectra. An emission spectrum of N_2, with the broad bands obtained at low resolution, is shown in Fig. 6-13. At higher resolution, each band is seen to consist of a large number of closely spaced lines, as seen for the 380.5 nm band of N_2 in Fig. 6-14.

Exercise 6-18. Calculate, in ergs and joules, the energy that corresponds to the transition in N_2 represented by the absorption band at 337.1 nm. Compare your result to the energy calculated previously for some rotational and vibrational transitions.

We have seen that the fine structure of vibrational spectra consists of characteristic lines for simultaneous rotational and vibrational transitions.

Figure 6-14. The 380.5 nm band of N_2 at high resolution. (From *Molecular Spectra and Molecular Structure*, Vol. 1, by G. Herzberg. © 1950 by Litton Educational Publishing, Inc. Reprinted by permission of Van Nostrand Reinhold Company.)

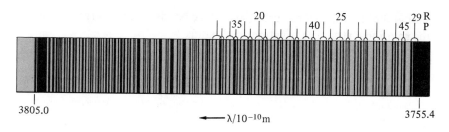

Similarly, the fine structure of electronic band spectra consists of characteristic lines for simultaneous rotational, vibrational, and electronic transitions. Strong support for the wave-mechanical interpretation of spectroscopic data is provided by the agreement between the values of molecular constants obtained from electronic spectra in the ultraviolet region and values obtained in the infrared and microwave regions. For example, the difference in energy between the lowest and the first excited vibrational level for CO in the ground state, as obtained from electronic spectra,[12] is 2143.2 cm^{-1}, and the fundamental vibrational frequency found from the infrared spectrum,[13] is also 2143.2 cm^{-1}. The equilibrium internuclear distance for the ground state of CO calculated from the rotational fine structure of the electronic spectrum[14] is $1.1281_9 \times 10^{-10}$m, whereas that calculated from microwave spectra[15] is 1.1282×10^{-10}m.

When visible and ultraviolet absorption spectra are observed for samples in the liquid or solid state or in solution, the vibrational and rotational structure is blurred due to strong molecular interactions. Nevertheless, such spectra can provide data for estimating the energy of electronic transitions and for qualitative identification and quantitative analysis of chemical species.

As with vibrational and rotational transitions, we shall state without proof the selection rules for electronic transitions for linear molecules. The quantum number for electronic angular momentum, Λ, can change only by

$$\Delta\Lambda = 0, \pm 1 \tag{6-59}$$

For example, the transitions $\Sigma \leftrightarrow \Sigma$ and $\Sigma \leftrightarrow \Pi$ are allowed, but $\Sigma \leftrightarrow \Delta$ is not allowed. Also, transitions can occur only between g and u states, but not between g and g or between u and u. Since the change in the spin quantum number is

$$\Delta S = 0 \tag{6-60}$$

there can be no transitions between singlet and triplet states. In contrast to vibration-rotation transitions,

$$\Delta J = 0, \pm 1 \tag{6-61}$$

so that the rotational structure of electronic bands is very nearly the same as the rotational structure of vibrational bands.

12. G. Herzberg, Vol. I, p. 155.
13. G. Herzberg, Vol. I, p. 62.
14. G. Herzberg, Vol. I, p. 522.
15. W. Gordy and R. L Cook, *Microwave Molecular Spectra*, Wiley-Interscience, New York, 1970, p. 677.

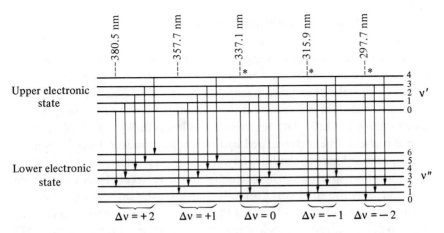

Figure 6-15. Scheme of the vibrational levels of N_2 in the two electronic states between which transitions result in the bands shown in Fig. 6-13.

Because there is no restriction on the change in the vibrational quantum number v in electronic transitions, electronic spectra contain for any electronic transition a group of bands, each group corresponding to a positive or negative value of Δv, as indicated in Fig. 6-13. For $\Delta v = 2$, for example, there is a band at 380.5 nm corresponding to the transition from $v' = 0$ to $v'' = 2$, and, at progressively shorter wavelengths, bands corresponding to the transitions $v' = 1 \to v'' = 3$, $v' = 2 \to v'' = 4$, etc. Rotational fine structure of the 380.5 nm band is shown in Fig. 6-14. From such detailed spectra, information can be obtained on the moment of inertia of the molecule at each vibrational level, and from the vibrational bands, the potential energy curve for each electronic state can be determined.

Figure 6-15 shows the vibrational levels in the upper and lower electronic states and the transitions that account for some of the vibrational bands in Fig. 6-13. An absorption spectrum in the same wavelength range would show only the bands corresponding to the reverse of the transitions marked with an asterisk, since N_2 at ordinary temperatures is in the lowest vibrational level of the ground state.

Exercise 6-19. Table 6-7 gives the wavenumber (in cm^{-1}) of some emission bands of CO, where v' is the vibrational quantum number of the upper electronic state, and v'' is the vibrational quantum number of the lower electronic state. From the data in the table, calculate three independent values for the energy level spacing between (a) $v' = 0$ and $v' = 1$, (b) $v'' = 0$ and $v'' = 1$, (c) $v' = 3$ and $v' = 4$, and (d) $v'' = 3$ and $v'' = 4$.

Polyatomic molecules

We shall not treat the electronic spectra of polyatomic molecules in as detailed a fashion as the spectra of diatomic molecules because the theory is more complex and because the data are more limited. Most polyatomic molecules are not stable at temperatures at which emission spectra are obtained, so only absorption measurements are available. Such measurements provide information about the vibrational levels of the first excited state in gases but little data for vibrational levels in liquids or solutions.

In the case of linear conjugated hydrocarbons, however, there is a simple wave-mechnical model, the *free electron model*, that approximately predicts the absorption spectrum. The geometry of the molecule is assumed to be determined by the sp² hybrid orbitals shared by the carbon and hydrogen atoms. The remaining p orbitals of the planar molecules have a maximum electron density above and below the molecular plane. They form π-LCAO molecular orbitals that span the molecule. In its simplest form, the model neglects interelectronic repulsion, and the assumption is made that all the π electrons are free to move in a one-dimensional potential well of constant potential energy, the length of which is the end-to-end distance of the molecule (particle in a box). The potential energy is assumed to be infinite outside the molecule and is taken to equal zero inside the molecule, as discussed in Sec. 5-1.

As in Eq. 5-12, the energy levels are

$$\varepsilon = \frac{n^2 h^2}{8ml^2}$$

where l is the length of the molecule. If we apply the Pauli principle, only two electrons can occupy an energy level, so that a molecule with N π-electrons has $N/2$ levels filled in the ground state, and the longest wavelength absorbed results in a transition from $n = N/2$ to $n = (N/2) + 1$.

Table 6-7. Wavenumber of emission bands of CO (cm^{-1})

		v''				
		0	1	2	3	4
	0	(64,703)	62,602	60,485	58,393	56,329
	1	66,231	64,088	—	59,882	57,818
	2	67,675	65,533	63,416	61,325	—
v'	3	69,088	66,944	64,828	—	60,675
	4	70,470	68,323	(66,199)	64,117	62,055

Exercise 6-20. Octatetraene has an absorption peak at 304 nm. On the basis of a free-electron model, calculate the length of the molecule.

6-4. Photoelectron spectra

Quanta of visible and ultraviolet light have sufficient energy to release electrons from the illuminated surfaces of some metals. The quantitative study of this phenomenon, the photoelectric effect, led Einstein to postulate the quantized nature of electromagnetic radiation, as we saw in Sec. 3-2.

Photons in the ultraviolet, with wavelengths from 50 nm to 150 nm, have sufficient energy to ionize valence electrons from gaseous atoms and molecules, and X-ray photons have sufficient energy to ionize inner shell electrons. The quantitative study of the kinetic energy spectrum of the electrons produced by photoionization is called photoelectron spectroscopy. The kinetic energy of each electron produced is given by the Einstein equation

$$K = h\nu - I - W \tag{6-62}$$

in which ν is the frequency of the ionizing radiation, I is the ionization energy of the electron, and W is a correction for the additional work required to remove an electron from a solid. The experimental determination of photoelectron spectra requires a monochromatic radiation source of the appropriate frequency, a device for sorting the emitted electrons according to their kinetic energy, and a detector to count the number of electrons as a function of kinetic energy. By means of Eq. 6-62, the electron count can be expressed as a function of ionization energy (with W assumed to be known or constant for a given system).

Ultraviolet photoelectron spectroscopy[16]
The most common source of ionizing radiation for ultraviolet photoelectron spectroscopy is the resonance radiation from a He discharge tube at 58.4 nm, with a photon energy of 3.40×10^{-18} J. Magnetic and electrostatic deflection as well as variable retarding potentials have been used to determine the kinetic energy of emitted electrons.

A typical photoelectron spectrum, that of Xe, is shown in Fig. 6-16. The spectrum shows two peaks because the product Xe^+ ion has two possible energy states, $^2P_{\frac{1}{2}}$ and $^2P_{\frac{3}{2}}$.

Exercise 6-21. Show that $^2P_{\frac{1}{2}}$ and $^2P_{\frac{3}{2}}$ are the correct terms to represent the possible Russell-Saunders coupling states of Xe^+ with the valence electron configuration $(4s)^2 (4p)^5$.

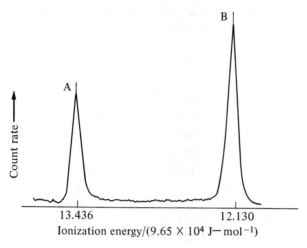

B

A

Count rate

13.436   12.130
Ionization energy/$(9.65 \times 10^4$ J$-$mol$^{-1})$

Figure 6-16. The He 58.4 nm photoelectron spectrum of Xe, showing peaks corresponding to the $^2P_{\frac{1}{2}}$ and $^2P_{\frac{3}{2}}$ state of the product ion. Peak A, Xe + $hv \rightarrow$ Xe$^+$ ($^2P_{\frac{1}{2}}$) + e^-; peak B Xe + $hv \rightarrow$ Xe$^+$ ($^2P_{\frac{3}{2}}$) + e^-. [Reprinted with permission from A. D. Baker, *Accounts Chem. Res.*, *3*, 17 (1970). Copyright by the American Chemical Society.]

The fine structure of photoelectron spectra for diatomic molecules is determined by the vibrational energy levels of the product ion. From the spectrum of H_2 in Fig. 6-17, we can calculate the force constant and dissociation energy of $H_2{}^+$. The spectrum of N_2 in Fig. 6-18 contains several peaks that can be related to the highest energy, occupied molecular orbitals. The vibrational spacing of each peak indicates whether the orbital is bonding or antibonding; an electron removed from a bonding orbital leaves an ion with a lower vibrational frequency than the parent molecule, whereas an electron removed from an antibonding orbital leaves an ion with a higher vibrational frequency than the parent molecule.

X-ray photoelectron spectroscopy[17]
Electrons produced in X-ray photoelectron spectroscopy are analyzed for kinetic energy by focusing techniques similar to those used in ultraviolet photoelectron spectroscopy. The ionization energies of inner electrons depend on the chemical environment of the atom, usually represented by the effective charge or oxidation state of the atom. Figure 6-19 shows the four peaks found in the carbon 1s spectrum of $CF_3COOCH_2CH_3$ (ethyltri-

16. A. D. Baker *Accounts Chem. Res.*, *3*, 17–25 (1970).
17. J. M. Hollander and W. L. Jolly, *Accounts Chem. Res.*, *3*, 193–200 (1970).

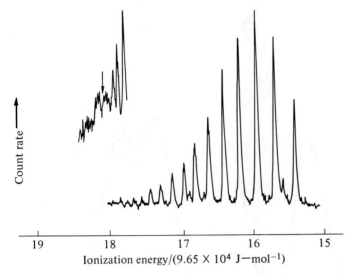

Figure 6-17. The He 58.4 nm photoelectron spectrum of H_2, showing the vibrational levels and the dissociation limit (18.2, arrow) of the H_2^+ ion. [By permission, from D. W. Turner, *Proc. Roy. Soc. London, A307*, 15(1968).]

Figure 6-18. The He 58.4 nm photoelectron spectrum of N_2, showing peaks corresponding to the $^2\Sigma_u$, $^2\Pi_u$, and $^2\Sigma_g$ states of N_2^+ and vibrational structure for the latter two states. [Reprinted with permission, from A. D. Baker, *Accounts Chem. Res., 3*, 17 (1970). Copyright by the American Chemical Society.]

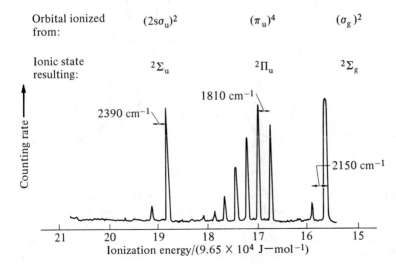

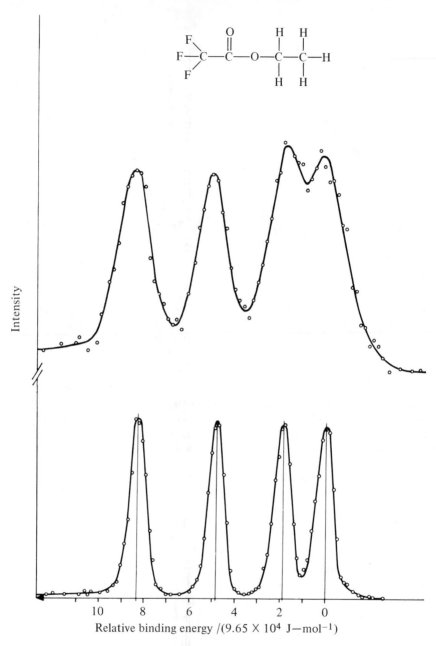

Figure 6-19. X-ray photoelectron spectrum of $CF_3COOCH_2CH_3$ (ethyl trifluoroacetate), showing peaks corresponding to the binding energies of the 1s electrons of the four distinct groups of C atoms in the molecule, relative to the value for the methyl carbon. Upper curve, lower resolution; lower curve, higher resolution. [By permission, from K. Siegbahn, *J. Electron Spectroscopy and Related Phenomena*, **5**, 3-97 (1974).]

216

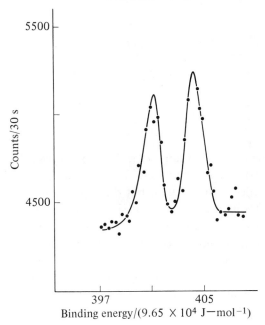

Figure 6-20. X-ray photoelectron spectrum of $Na_2N_2O_3$, the 1s electrons of nitrogen, showing that the two nitrogens in $N_2O_3{}^{2-}$ are chemically distinct. [Reprinted with permission from J. M. Hollander and W. Jolly, *Accounts Chem. Res.*, 3, 193 (1970). Copyright by the American Chemical Society.]

fluoroacetate), corresponding to the four structurally distinct C atoms in the molecule. The nitrogen 1s spectrum of $Na_2N_2O_3$ in Fig. 6-20 shows that the structure of the $N_2O_3{}^{-2}$ ion is asymmetrical.

6-5. Electron paramagnetic resonance spectra

The splitting of the electronic energy levels of an atom in an external magnetic field (Zeeman effect) is illustrated schematically in Fig. 4-18. Most molecules have closed shell electron configurations, with zero resultant orbital and spin angular momentum, and do not show such splitting. But, free radicals, with one or more unpaired electrons, and transition metal complexes, some of which also have unpaired electrons, do exhibit splitting of electronic levels in a magnetic field.

According to Eq. 4-101, the splitting between Zeeman levels for atoms is

$$\Delta\varepsilon = g\mu_B BM_J$$

in which the Landé g factor is

$$g = 1 + \frac{J(J + 1) + S(S + 1) - L(L + 1)}{2J(J + 1)} \tag{6-63}$$

the Bohr magneton is

$$\mu_B = \frac{eh}{4\pi m} \tag{6-64}$$

M_j is the component of net angular momentum in the direction of the field, and B is the magnitude of the external magnetic field. The form of Eq. 4-101 can be used for the Zeeman effect in molecules, but the expression for g depends on the electronic states involved in the transition. For most free radicals, g has a value close to two, the value for a free electron, which indicates that orbital angular momentum contributes little to the total angular momentum. The calculated value of $\Delta\varepsilon$ for such a free radical in a magnetic field of 0.34 webers/m² (3400 gauss) is 6.3×10^{-24}J, and the frequency difference between Zeeman lines is about 10^{10}s^{-1}. At a wavelength of 260 nm, the wavelength difference would be about 0.003 nm, a difference very difficult to detect spectroscopically.

Exercise 6-22. Verify the numerical values for $\Delta\varepsilon$, $\Delta\nu$, and $\Delta\lambda$ given in the preceding paragraph.

Rabi and his collaborators developed a method for determining Zeeman splitting directly, rather than as a small change in the position of a spectral line. The frequency of radiation required to excite an electron from one Zeeman level to the next higher one is

$$\nu = \frac{\Delta\varepsilon}{h}$$

$$= \frac{6.3 \times 10^{-24}\text{J}}{6.6 \times 10^{-34}\text{J-s}}$$

$$= 9.5 \times 10^{9}\text{s}^{-1} \ (= 9.5 \text{ GHz})$$

This frequency is in the microwave region. Rabi and his students developed procedures to measure the absorption of radiation in the microwave region while working with radar during World War II. This method was used for molecules with a magnetic moment observed in an external magnetic field. (Earlier experiments in the radiofrequency region were performed on nuclei rather than electrons.) The experiment is now usually carried out by irradiating the sample with a fixed frequency of microwave radiation and sweeping the magnetic field to find the field value at which absorption (resonance) occurs.

The value of g for free radicals is close to the free electron value of 2.0023 because the coupling between spin angular momentum and orbital angular momentum is small. The differences that do occur can be measured precisely and used to identify different radicals. The value of g for transition metal complexes with unpaired spins is strongly dependent on the electronic state of the complex and provides a rigorous test for models of the electronic state.

The fine structure of the resonance spectrum of free radicals and transition metal ions provides information on the interaction between the magnetic moment of the unpaired spin and the magnetic moment of nuclei in the molecule. One mechanism for this interaction is direct *dipole-dipole coupling* between the electron magnetic moment and the nuclear magnetic moment. In a magnetic field, when the magnetic moments of the electron and the nucleus are parallel, the interaction energy is proportional to r^{-3} $(1 - 3 \cos^2 \theta)$,[18] where r is the distance between dipoles and θ is the angle between the direction of the magnetic field and the vector from the nucleus to the electron. This interaction averages to zero for an electron in an s orbital because all possible orientations of the electron magnetic moment are equally represented. An electron in a p orbital on the magnetic nucleus does not sample all orientations equally, and the net interaction is not zero.

The second mechanism for interaction between the electron magnetic moment and the nuclear magnetic moment is a nonclassical one, the *Fermi contact interaction*. The energy of the interaction is proportional to the value of ψ^2 for the unpaired electron at the nucleus. Thus, the interaction energy is zero except for electrons in s atomic orbitals or molecular orbitals that have nonzero values of ψ^2 at the nucleus. The Fermi contact interaction is isotropic, that is, independent of orientation.

In a single crystal, the dipole interaction may have different values depending on the orientation of the crystal with respect to the magnetic field. Such data can provide information on the orbital that describes the unpaired spin, if the crystal structure is known. In a fluid medium, the tumbling of molecules with respect to the magnetic field averages the dipole interaction to zero, so that only the Fermi contact interaction is observed.

18. P. W. Atkins, *Molecular Quantum Mechanics*, Clarendon Press, Oxford, 1970, p. 449; *Quanta*, Clarendon Press, Oxford, 1974, p. 111.

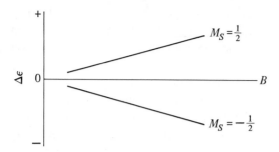

Figure 6-21. The effect of magnetic field strength on the difference in energy between electrons with spin quantum number $M_S = \frac{1}{2}$ and those with $M_S = -\frac{1}{2}$.

For a species with one unpaired electron, the value of M_s is either $\frac{1}{2}$ or $-\frac{1}{2}$. If there is no orbital contribution to the net angular momentum, the energy of the unpaired electron varies with the field strength as

$$\Delta\varepsilon = gM_s \mu_B B \tag{6-65}$$

neglecting the effect of the magnetic nucleus. The variation is shown graphically in Fig. 6-21. The electron-nucleus magnetic interaction, whether dipole or Fermi contact, is proportional to $M_I M_s$ in a sufficiently strong magnetic field. The quantum number for the z component of the nuclear spin angular momentum, M_I, can take on the $2I + 1$ values $I, I - 1, \ldots, -I$, where I is the nuclear spin quantum number. The interaction energy is small compared to the effect of, and independent of the strength of, the magnetic field. Thus,

Figure 6-22. The effect of magnetic field strength on the difference in energy between electrons with $M_S = \frac{1}{2}$ and those with $M_S = -\frac{1}{2}$, in the presence of hydrogen nuclei with $M_I = \frac{1}{2}$ and $M_I = -\frac{1}{2}$. The frequency of radiation is ν, and a and c are the values of the field at which the electrons come into resonance. If $I = 0$, resonance occurs at b.

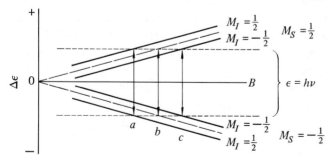

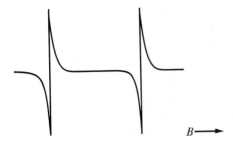

Figure 6-23. A sketch of an electron spin resonance peak in which the unpaired electron is coupled to one H nucleus.

for each value of M_I, there is a line parallel to each line in Fig. 6-21. The value of I for the hydrogen nucleus is $\frac{1}{2}$, so that M_I can be $\frac{1}{2}$ or $-\frac{1}{2}$. The resulting possible variations of $\Delta\varepsilon$ with magnetic field for an unpaired electron about a hydrogen nucleus are shown schematically in Fig. 6-22. If v is the frequency of the microwave radiation, then a, b, and c represent the values of magnetic field at which the radiation will cause a transition from $M_s = -\frac{1}{2}$ to $M_s = \frac{1}{2}$ when $M_I = \frac{1}{2}$, $I = 0$, and $M_I = -\frac{1}{2}$. The direct interaction of the nucleus with the magnetic field is small on the scale of the electron spin resonance experiment, so that we can treat M_I as unchanged during the electron spin transition. Thus, the resonance is observed at a and c, equally spaced about b, rather than at b as would be expected if there were no interaction with the nucleus. The spectrum would look like that shown in Fig. 6-23, in which the position of the center of the spectrum is a measure of g, and the spacing between the peaks is a measure of the nucleus-electron interaction. The form of the curves in Fig. 6-23, the first derivative of the more familiar absorption curves, is produced by the electronic circuits used in the detection and recording system.

For a given value of I, there are $2I + 1$ values of M_I; thus, there will be $2I + 1$ peaks in the spectrum if the unpaired electron interacts with a single nucleus of spin I. The fine structure of the spectrum can thus help to identify the nucleus with which the unpaired electron is interacting. In solution, where only the Fermi contact interaction is measured, the splitting between peaks is a measure of the wave function of the unpaired electron at the nucleus. Electron paramagnetic resonance spectra thus can be used to check wave-mechanical calculations.

An interesting example of such a test is seen in the spectra of aromatic radical ions. The simple Hückel model described in Sec. 5-4 predicts an unpaired electron in a π molecular orbital, for which the probability density is zero in the plane of the molecule, where all the nuclei are located. The spectra of such radicals show fine structure that can be most easily interpreted by assuming interaction with the hydrogen nuclei. The fallacy of the simple Hückel description is the fundamental one of assigning the unpaired spin to a particular one-electron function. The wave function for the molecule is a many-electron function, and the spins of all the electrons are coupled. Thus, it is not possible to assign the unpaired electron to a single orbital; the unpaired spin is distributed among all the orbitals used to describe the molecule, including the σ orbitals of the carbon and hydrogen atoms.

6-6.   Nuclear magnetic resonance spectra

Nuclei, like electrons, have an intrinsic angular momentum I of magnitude $[I(I + 1)]^{\frac{1}{2}} h/2 \pi$, where I is the nuclear spin quantum number. For nuclei with odd mass numbers, I is an odd integral multiple of $\frac{1}{2}$. Nuclei with even mass numbers have $I = 0$ if the atomic number is even and integral values of I if the atomic number is odd.

There is a magnetic moment μ related to the angular momentum by

$$\mu = \gamma_n I \tag{6-66}$$

where γ_n is called the *magnetogyric ratio*. The magnitude of the magnetic moment is given by

$$\mu = \gamma_n [I(I + 1)]^{\frac{1}{2}} \frac{h}{2 \pi} \tag{6-67}$$

By analogy to Eq. 4-98 for the magnetic moment of an electron, μ is frequently given by the expression

$$\mu = g_n \mu_n [I(I + 1)]^{\frac{1}{2}} \tag{6-68}$$

where g_n is the nuclear g-factor, and μ_n is the *nuclear magneton*.

$$\mu_n = \frac{eh}{4 \pi m_p} \tag{6-69}$$

where m_p is the mass of the proton. The energy of a nuclear magnetic moment in an external magnetic field of magnitude B is given by an equation of the same form as Eq. 4-101.

$$\Delta \varepsilon = -g_n M_I \mu_n B \tag{6-70}$$

where M_I is the quantum number for the component of the spin angular momentum in the direction of the magnetic field, which can take on the $2I + 1$ values $-I, -I + 1, \ldots, I$.

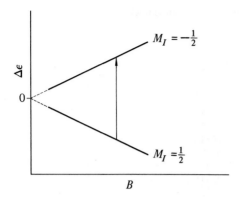

Figure 6-24. The effect of magnetic field on the difference in energy between a proton with $M_1 = \frac{1}{2}$ and $M_1 = -\frac{1}{2}$.

The selection rule for transitions between the resulting levels is $\Delta M_I = \pm 1$. Thus only one frequency can be absorbed, and its value is

$$v = \frac{g_n \mu_n B}{h} \qquad (6\text{-}71)$$

For nuclear magnetic resonance, the frequencies required are in the radio-frequency range. The most common technique is to radiate a sample with a fixed frequency and vary the magnetic field to find the value of the field at which the resonance condition in Eq. 6-71 is satisfied and absorption occurs. The quantities g_n and I distinguish one nucleus from another. As a result of the differences between nuclei, a different radiofrequency oscillator must be used for each nucleus studied with a given magnet. Values of g_n and I for several nuclei are given in Table 6-8.

The proton, with $I = \frac{1}{2}$, is the nucleus most studied in magnetic resonance. The possible values of M_I are $\frac{1}{2}$ and $-\frac{1}{2}$, and the energy of a proton in an external magnetic field varies with field strength as shown schematically in Fig. 6-24. At a field strength of 1.41 W-m^{-2}, the resonance frequency of the proton is 60 MHz, in the radiofrequency region. The frequency for electron paramagnetic resonance in a field of the same magnitude is much larger because each resonance frequency varies inversely with the mass of the particle being observed.

Exercise 6-23. Verify the proton resonance frequency at a field of 1.41 W-m^{-2}.

The value of the magnetic field at which nuclei in molecules resonate with a particular radiofrequency is different from the value of the magnetic field

Table 6-8.   Nuclear *g*-factors and nuclear spin for some nuclei[a]

Nucleus	g_n	I
^{1}H	3.22473	$\frac{1}{2}$
^{2}H	0.60626	1
^{13}C	0.81078	$\frac{1}{2}$
^{14}N	0.28537	1
^{15}N	−0.32683	$\frac{1}{2}$
^{19}F	3.03375	$\frac{1}{2}$
^{31}P	1.3054	$\frac{1}{2}$
^{17}O	−0.6400	$\frac{5}{2}$
^{23}Na	1.1444	$\frac{3}{2}$
^{35}Cl	0.42391	$\frac{3}{2}$

[a] Calculated from data published by Varian Associates.

at which free nuclei of the same kind resonate. The external magnetic field affects the motion of electrons surrounding the nuclei, thus inducing local magnetic fields. These local fields are determined by the electron distribution around the nucleus, which is described by the wave function. The difference between the fields at which free and molecular nuclei resonate is called the *chemical shift*.

The free nucleus is not a convenient reference for chemical purposes, so chemical compounds that exhibit only a single sharp resonance line are used as references for chemical shift measurements. The most common scales for chemical shift define as zero the field at which the protons in $(CH_3)_4Si$ (tetramethylsilane) resonate. The local field of a nucleus is related to the external field through the *screening constant* σ, by the relation

$$B_{loc} = B(1 - \sigma) \tag{6-72}$$

The field-independent chemical shift δ is defined in terms of σ as

$$\delta = (\sigma_{ref} - \sigma) \times 10^6 \tag{6-73}$$

where the factor of 10^6 is included to give δ a convenient order of magnitude, with σ of the order of 10^{-5} for the hydrogen nucleus. From Eq. 6-71 we can see that the field at which a free nucleus resonates with a radiation of frequency v is

$$B_0 = \frac{hv}{(g_n\mu_n)}$$

For a nucleus in an atom or molecule, the corresponding equation is

$$B_{loc} = B(1 - \sigma) = \frac{hv}{(g_n\mu_n)} \tag{6-74}$$

and for the reference nucleus the resonance field is

$$B_{ref} (1 - \sigma_{ref}) = \frac{h\nu}{(g_n \mu_n)} \qquad (6\text{-}75)$$

We can relate the definition of δ to the experimentally observed resonance fields by equating the left sides of Eq. 6-75 and Eq. 6-74. Then

$$\frac{B}{B_{ref}} = \frac{1 - \sigma_{ref}}{1 - \sigma}$$

and

$$\frac{B_{ref} - B}{B_{ref}} = \frac{\sigma_{ref} - \sigma}{1 - \sigma}$$

$$\approx \sigma_{ref} - \sigma \qquad (6\text{-}76)$$

since $1 - \sigma \approx 1$. Thus,

$$\delta = \left(\frac{B_{ref} - B}{B_{ref}} \right) \times 10^6 \qquad (6\text{-}77)$$

The result is expressed in parts per million. Several convenient references and their chemical shifts are shown in Table 6-9.

Table 6-9. Chemical shifts of some convenient references for proton magnetic resonance[a]

Substance	Chemical shift (ppm)
Tetramethylsilane	0.0
Cyclohexane	1.6
Dioxane	3.8
Water	5.2
Methylene chloride	5.8
Benzene	6.9
Chloroform	7.7
Sulfuric acid	11.6

[a] Adapted from data in J. A. Pople, W. G. Schneider, and H. J. Bernstein, *High-resolution Nuclear Magnetic Resonance*, McGraw-Hill Book Co., New York, 1959.

The phenomenon of chemical shift makes nuclear magnetic resonance a powerful method for the determination of molecular structure. Early experiments with CH_3CH_2OH (ethanol) showed three peaks, as indicated in Fig. 6-25. The relative areas of $1:2:3$ suggest that the peak at lowest field represents the OH proton, the next peak CH_2 protons, and the peak at highest field CH_3 protons. This result is consistent with the electron density

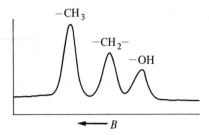

Figure 6-25. The nuclear magnetic resonance spectrum of CH_3CH_2OH (ethanol) at low resolution. [By permission, from J. T. Arnold, *Phys. Rev.*, *102*, 136 (1956).]

and, hence, the shielding predicted for each kind of proton. Even at low resolution, nuclear magnetic resonance provides information about the number of nuclei in each of the possible chemical environments in a molecule and determines how many different such environments exist.

At a somewhat higher resolution, a fine structure appears that suggests which groups of protons are adjacent in the molecule. The CH_3CH_2OH (ethanol) spectrum at higher resolution is shown in Fig. 6-26, in which the CH_2 peak is split into a quartet, and the CH_3 peak is split into a triplet. The OH peak remains, as a singlet. The splitting can be explained most easily in terms of the effect neighboring magnetic nuclei have on the local magnetic field at a nucleus. Nuclear spin-spin coupling occurs through the intervening

Figure 6-26. The nuclear magnetic resonance spectrum of CH_3CH_2OH (ethanol) at resolution high enough to display spin-spin splitting. Splitting of and by the OH proton does not occur because of an exchange catalyzed by a trace of acid. [By permission, from J. T. Arnold, *Phys. Rev.*, *102*, 136 (1956).]

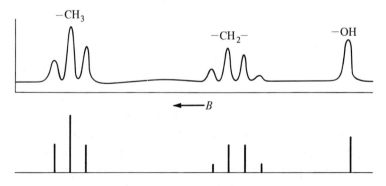

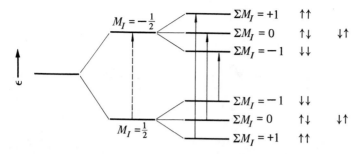

Figure 6-27. The effect of a set of two perturbing protons on the energy levels of a set of equivalent protons in a magnetic field. A single peak is replaced by a triplet as a result of the perturbation.

electrons by three different mechanisms.[19] The first involves the direct effect of the nuclear magnetic moment on the orbital magnetic moments of the electrons. The second is a dipolar coupling between the nuclear magnetic moment and the electron spin magnetic moment, as described in Sec. 6-5. The third is the Fermi contact interaction, also discussed in Sec. 6-5. In each case, the effect is then transmitted to the neighboring nucleus. Since the interaction is entirely internal to the molecule, its magnitude, measured by the *spin-spin coupling constant J*, is independent of the magnetic field. It has the dimensions of reciprocal time and is usually expressed in Hertz.

The difference between the values of the field at which two coupled sets of nuclei resonate can be expressed as an equivalent frequency difference through Eq. 6-71. When this frequency difference is much greater than the coupling constant, the pattern of splitting can be explained simply. In the absence of interacting nuclei, there are two $(2I + 1)$ orientations of a hydrogen nucleus of spin $\frac{1}{2}$ and two energy levels in the magnetic field, with a single transition as shown in Fig. 6-24. If there are n neighboring protons perturbing the protons under observation, there are $n + 1$ different sums of M_I for the perturbing protons and $n + 1$ perturbed energy levels for each of the unperturbed levels. A schematic diagram of the levels for two perturbing protons is shown in Fig. 6-27. The resonance peak would be split into a triplet, with relative intensities $1:2:1$ corresponding to the number of ways each ΣM_I could be attained. The distance between members of a multiplet is the coupling constant. Thus, the CH_3 peak in Fig. 6-26 is a triplet because

19. A. Carrington and A. D. McLachlan, *Introduction to Magnetic Resonance*, Harper & Row, New York, 1967, pp. 64-66.

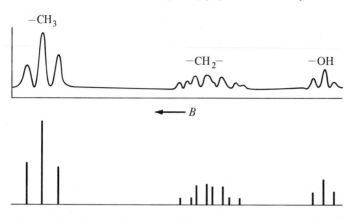

Figure 6-28. The nuclear magnetic resonance spectrum of purified CH_3CH_2OH (ethanol) at high resolution, showing the expected splitting of the methylene and hydroxyl protons. [By permission, from J. T. Arnold, *Phys. Rev.*, *102*, 136 (1956).]

it is coupled to two $-CH_2-$ protons, and the $-CH_2-$ peak is a quartet because it is coupled to three CH_3 protons.

The hydroxyl peak showed no splitting in this experiment. This was attributed to very rapid exchange of protons between CH_3CH_2OH molecules so that only an "average" of their electronic environment was observed. Very carefully purified samples of CH_3CH_2OH do, however, show the expected splitting, as shown in Fig. 6-28. The rapid exchange has been shown to result from contamination of the alcohol with traces of acid. Although equivalent nuclei do couple with each other, no splitting is observed, and the coupling constant cannot be calculated from ordinary spectra.[20]

Nuclear magnetic resonance spectra can also provide information on rates and equilibrium in certain chemical processes in which magnetic nuclei move between two different chemical environments. The nature of the observed resonance curves depends on the relative magnitudes of δv and v_e, where δv is the interval between the resonance peaks expressed in frequency units ($\delta v = g_n \mu_n \delta B/h$), and v_e is the frequency of exchange between the two sites. If v_e is very large compared to δv, only a single line is observed at a frequency between the two resonances. If the proportion of the two sites vary with some parameter, such as temperature or pH, the position of the intermediate can be used to calculate the equilibrium proportion of the two

20. J. A. Pople, W. G. Schneider, and H. J. Bernstein, *High-resolution Nuclear Magnetic Resonance*, McGraw-Hill Book Co., New York, 1959, pp. 115–16.

sites. Examples include *cis-trans* isomers, two rotational conformers with a potential energy barrier hindering free rotation, or two sites in a proton-exchange reaction. If v_e is very small compared to δv, two peaks are seen, but both peaks are broadened, with a width approximately equal to v_e. A study of peak widths as a function of concentration and temperature provides information about rate constants and energies of activation.

Problems

6-1. The following wavenumbers were observed [R. L. Hansler and R. A. Oetjen, *J. Chem. Phys.*, *21*, 1340 (1953).] for the indicated rotational transitions in $H^{35}Cl$ and $D^{35}Cl$. What are the internuclear distances for the two isotopic species? Are they the same within the experimental uncertainty?

Transition	$\tilde{v}(HCl)/cm^{-1}$	Transition	$\tilde{v}(DCl)/cm^{-1}$
$J = 3 \rightarrow J = 4$	83.32	$J = 6 \rightarrow J = 7$	75.15
$J = 4 \rightarrow J = 5$	104.13	$J = 7 \rightarrow J = 8$	85.90
$J = 5 \rightarrow J = 6$	124.73	$J = 8 \rightarrow J = 9$	96.51
$J = 6 \rightarrow J = 7$	145.37	$J = 9 \rightarrow J = 10$	107.14
$J = 7 \rightarrow J = 8$	165.89	$J = 10 \rightarrow J = 11$	117.81
$J = 8 \rightarrow J = 9$	186.23	$J = 11 \rightarrow J = 12$	128.26
$J = 9 \rightarrow J = 10$	206.60	$J = 12 \rightarrow J = 13$	138.80
$J = 10 \rightarrow J = 11$	226.86	$J = 13 \rightarrow J = 14$	149.26

6-2. Calculate the internuclear distance of each of the molecules listed in Table 6-1 from the reported values of the frequency. Are they the same as those listed as r_e? For HCl and DCl, compare your results with the values in Table 6-4.

6-3. From the results given in Table 6-2, we can see that the force constant of a diatomic molecule is independent of isotopic substitution. Comment on this independence.

6-4. The wavenumbers of the lines in the fundamental vibration-rotation band of $H^{35}Cl$ are given below. [E. K. Plyler, A. Danti, L. R. Blaine, and E. D. Tidwell, *J. Res. Natl. Bur. Standards U.S.*, *A64*, 29 (1960).] Calculate k_0, r_0, r_1, and r_e from the data.

Line	$\tilde{v}/cm^{-1}$	Line	$\tilde{v}/cm^{-1}$
P(10)	2625.98	P(1)	2865.10
P(9)	2677.74	R(0)	2906.25
P(8)	2703.02	R(1)	2925.91
P(7)	2727.79	R(2)	2944.92
P(6)	2752.05	R(3)	2963.30
P(5)	2775.77	R(4)	2981.02
P(4)	2798.95	R(5)	2998.07
P(3)	2821.58	R(6)	3014.44
P(2)	2843.63	R(7)	3030.10
		R(8)	3045.07

6-5. Calculate the force constant k and the internuclear distance r of O_2 from the data in Fig. 6-10. Assume the rigid-rotor, harmonic-oscillator model.

6-6. Methane is a spherical top molecule, and its vibration-rotation band, shown in Fig. 6-11, has the same mathematical structure as that of a diatomic molecule. Read the wavenumbers from the graph and calculate the length of the C-H bond and the force constant for the vibration, based on the rigid-rotor, harmonic-oscillator approximation.

6-7. The $VO_2{}^+$ ion has been used as an electron spin resonance probe in the study of metalloenzymes. [R. DeKoch, D. J. West, J. C. Cannon, and N. D. Chasteen, *Biochem*, *13*, 4347 (1974).] The electron spin resonance spectrum of V shows a fine structure with seven peaks. What is the value of I for the V nucleus?

6-8. The radical ion $[(SO_3)_2 NO]^{-2}$ exhibits an electron spin resonance spectrum with three peaks of equal intensity. [P. B. Ayscough, *Electron Spin Resonance in Chemistry*, Methuen, London, pp. 62–65.] $I = 1$ for ^{14}N, 0 for ^{16}O, and 0 for ^{32}S. What can you conclude from these data?

6-9. How might one design an experiment to determine the coupling constant between the CH_3 protons of CH_3CH_2OH (ethanol)? Hint: The coupling constant is an electronic property and independent of the nucleus involved.

6-10. Estimate the dissociation energy of H_2^+ from the data in Fig. 6-17 and compare with the values found in Chapter 5.

7 · Determination of molecular structure

The need for more information about molecules than is available from the molecular formula was recognized when isomers began to be identified. The Van't Hoff-Le Bel model of the tetrahedral carbon atom and the Kekulé structure of benzene were developed in response to this need. They systematized the chemical properties of organic compounds satisfactorily, but there was no way in which results could be verified physically. Quantitative measures of bond lengths and bond angles and of the internal motion of atoms within molecules awaited more sophisticated tools. The modern observer must nevertheless admire the qualitative and semiquantitative models built in the nineteenth and the early twentieth centuries.

Now that the concepts of quantum mechanics have verified the earlier predictions from chemical observation and have extended them, through the logic and precision of mathematics, the details of bond length, bond angle, and internal motion can be described. A number of techniques for structure determination are in use. The mass spectrometer can be used to identity the masses of fragments into which molecules are shattered by energetic electrons. The infrared spectrophotometer can be used to identify specific functional groups in molecules from their characteristic vibrational patterns. The nuclear magnetic resonance spectrometer allows the identification of the different chemical environments in which magnetic nuclei, most frequently, protons, are found, and the determination of the numbers of nuclei in each environment and their relative locations in the molecule. These instrumental methods provide the information on connectivity that is obtained from chemical data, but more quickly, more precisely, and with much smaller samples.

Wave-mechanical interpretation of spectroscopic data allows us to obtain precise values for the equilibrium internuclear distance of diatomic molecules. A complete description of the vibrational motion can also be obtained. For somewhat larger molecules, precise interatomic distances can be calculated from infrared, Raman, and microwave spectra, but the relationship of the vibrational motion to these values is obscure. Neutron, electron, and X-ray diffraction data provide precise interatomic distances and bond angles for a substantial range of molecules.

7-1. Mass spectroscopic methods

Aston invented the mass spectrograph in 1919 to test the hypothesis that elements with nonintegral atomic masses are mixtures of isotopes of different integral mass. Modern versions, now called spectrometers because they use ion detectors and an electrical recording device instead of a photographic plate, permit the determination of isotopic masses and isotope ratios to a precision much greater than that for the chemical determination of atomic mass.

More recently, the mass spectrometer has been used to identify the molecular mass of organic molecules and the masses of the fragments produced by bombardment of the molecules by an electron beam. The type of apparatus used is shown schematically in Fig. 7-1. When the stream of gas

Figure 7-1.   Scheme of a magnetic deflection mass spectrometer.

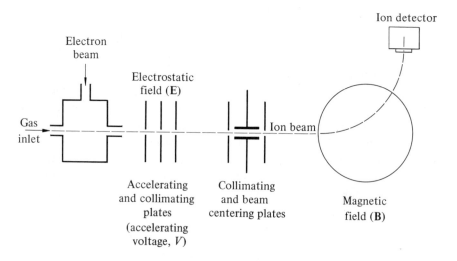

molecules is bombarded by the electron beam, ions of the parent molecule and of the fragments of the parent molecule are produced. These ions are accelerated by the electrostatic field **E**, collimated, and then passed through a magnetic field that is at right angles to their path. The work done on an ion of charge q and mass m by the electrostatic field **E** is equal to Vq, where V is the accelerating potential. Thus, the resulting kinetic energy of the ion is

$$\tfrac{1}{2}mv^2 = Vq \tag{7-1}$$

The magnitude of the force exerted by the magnetic field is Bqv, and

$$Bqv = \frac{mv^2}{r} \tag{7-2}$$

in which r is the radius of curvature produced by the magnetic field, and v^2/r is the centripetal acceleration. Eliminating v from Eqs. 7-1 and 7-2, we obtain

$$\frac{m}{q} = \frac{B^2 r^2}{2V} \tag{7-3}$$

Exercise 7-1.   Verify Eq. 7-3.

For a fixed magnetic field, and with r fixed by the position of the detector, variation of the accelerating voltage V will bring beams of particles of different values of m/q into the detector, and a plot of ion current against V can be calibrated as a plot of relative abundance against m/q. Since the most prominent ion of each species will be the ion resulting from the removal of one electron, the abscissa will, in effect, be a mass scale. Alternatively, the magnetic field can be varied at a constant accelerating potential.

Exercise 7-2.   How does m/q vary with position on a variable potential scale? On a variable magnetic field scale?

The electron beam is accelerated by potentials of 70–90 V, which provide sufficient energy to ionize the molecules in the sample and to break chemical bonds and produce molecular fragments, which are also ionized. The pressure in a mass spectrometer is of the order of 10^{-5} Pa (10^{-10} atm); thus, collisions are rare, and the excited species produced by electron bombardment decompose by unimolecular mechanisms. The accumulation of mass spectrometric results provides a data base for hypotheses about the relative stability of fragments produced in the decomposition reactions and the most probable paths of decomposition of the excited species.

The parent ion, the one produced by ionization of the sample without further decomposition, is frequently, though not always, the species of greatest mass (including isotopic substitution) in the spectrum. The masses of the *nuclides*, the isotopes of all the elements, are known to a very high precision; therefore, the mass of a parent ion and the elemental composition of a parent ion can be determined directly from a high resolution mass spectrum. With spectra obtained routinely on an analytic mass spectrometer, the intensity of isotopic peaks frequently provides the key to interpretation of the spectrum. The relative abundance of the common isotopes of C, H, N, O, S, Cl, and Br are listed in Table 7-1.

Exercise 7-3. Given the following exact isotopic masses, calculate the value of m/e for the parent peaks in the mass spectra of CO, N_2, and C_2H_4, all with nearest integral mass 28. ^{12}C, 12.00000; ^{14}N, 14.00307; ^{16}O, 15.99491; ^{1}H, 1.007825.

Table 7-1. Relative abundance of common isotopes of several elements

Element	Mass number	Abundance	Mass number	Abundance
H	1	100	2	0.015
C	12	100	13	1.12
N	14	100	15	0.37
O	16	100	18	0.204
	17	0.037		
S	32	100	33	0.80
	34	4.44		
Cl	35	100	37	32.40
Br	79	100	81	97.86

Handbook of Chemistry and Physics, 54th ed., R. C. Weast, ed., © CRC Press, 1973. Used by permission of CRC Press.

The chance of having a single isotopic substitution in a molecule depends on the number of atoms of an element in the molecule and the abundance of the isotope that is substituted. Thus, for a single substitution, the intensity of a peak corresponding to isotopic substitutions compared to that of the common isotopic composition is given approximately by (abundance of isotopic substituent) × (number of atoms of element in molecule). For example, if the mass number of the common molecule is 28, the molecule may be CO, C_2H_4, or N_2. For each molecule, if M is the mass of the parent ion, the relative intensities of the $(M + 1)$ and $(M + 2)$ peaks are as in the following table.

	$M + 1$		
$^{12}C^{17}O$	(0.037)(1)	=	0.037
$^{13}C^{16}O$	(1.12)(1)	=	1.12
			1.16
$^{14}N^{15}N$	(0.37)(2)	=	0.74
$^{12}C^{13}C^{1}H_4$	(1.12)(2)	=	2.24
$^{12}C_2{}^{2}H^{1}H_3$	(0.015)(4)	=	0.06
			2.30
	$M + 2$		
$^{12}C^{18}O$	(0.204)(1)	=	0.204

Because we are considering uncommon isotopes the probability of double substitution is so small that it is not thought to be significant.

Table 7-2. Mass spectral data

Mass number	Relative abundance	Mass number	Relative abundance
1	1.11	37	1.01
12	0.13	38	1.89
13	0.26	39	12.5
14	0.96	40	1.63
15	5.30	41	27.8
16	0.12	42	12.2
25	0.46	43	100
26	6.17	44	3.33
27	37.1	50	1.29
28	32.6	51	1.05
29	44.2	57	2.42
30	0.98	58	12.3
		59	0.54

B. J. Zwolinski *et. al., Catalogue of Selected Mass Spectral Data*, Serial No's 4, 63, 113, 1583, 1587, and 1597, American Petroleum Institute Research Project 44, Thermodynamics Research Center, Texas A. & M. University, College Station, Texas (loose-leaf data sheets, extant, 1975).

Consider the mass spectrum indicated in Table 7-2. The parent ion is probably of mass number 58. Since the $(M + 1)$ peak is 4.4% of the M peak, it is likely that the parent ion has four C atoms. The hydrocarbon butane, C_4H_{10}, has mass number 58. The molecular formula is confirmed by the presence of a large peak at 43, corresponding to C_3H_7, and an $M + 1$ peak at 44, with 3.3% abundance relative to M. Other peaks consistent with this assignment are those at 29, corresponding to C_2H_5, and 15, corresponding to CH_3.

Exercise 7-4. Use the data in Table 7-2 to eliminate CHNO, C_2H_3O, CH_3N_2, and C_2H_5N as possible elemental compositions for the peak at 43.

The distinction between $n\text{-}C_4H_{10}$ and $i\text{-}C_4H_{10}$ can be made on the basis of the relative abundance of the fragments. That there is a greater abundance of C_2H_5, 29, than CH_3, 15, indicates a straight chain rather than a branched chain. In compounds with longer chains and a greater number of isomers, comparisons of the parent peak and M-15, M-29, and M-43 provide additional information for eliminating structures. A detailed discussion of a large number of compounds can be found in McLafferty.[1]

Exercise 7-5. Figure 7-2 and Table 7-3 show the mass spectra of $n\text{-}C_8H_{18}$ and three isomeric octanes. Determine the structures of the isomeric octanes from the spectra. (Reproduced from *Interpretation of Mass Spectra*, written by F. W. McLafferty, with permission of publishers, W. A. Benjamin, Inc., Advanced Book Program, Reading, Mass., 1967.)

7-2. Diffraction methods

It was pointed out in Sec. 6-2 that when a molecule is exposed to electromagnetic radiation an oscillating dipole is induced in the molecule, resulting in the emission in all directions of electromagnetic radiation of the same frequency as the incident radiation. Radiation scattered from different parts of the molecule exhibits interference if the wavelength of the incident radiation is the same order of magnitude as the dimensions of the molecule. The angular distribution of the scattered radiation provides information about the shape of the molecule, with macromolecules in solution being studied with visible light at large scattering angles and X-rays at small angles.

Radiation scattered from different atoms in a molecule exhibits interference if the wavelength of the incident radiation is of the same order of magnitude as the interatomic distances. The angular distribution of scattered radiation provides information about the interatomic distances in a molecule. The required wavelengths, about 1×10^{-10}m, can be obtained with electrons, X-rays, and neutrons, all of which are used to study structures.

Electron diffraction

Electrons that are accelerated through a potential difference of approximately 40 kV have a wavelength of about 0.06×10^{-10}m.

1. F. W. McLafferty, *Interpretation of Mass Spectra*, W. A. Benjamin, New York, 1967.

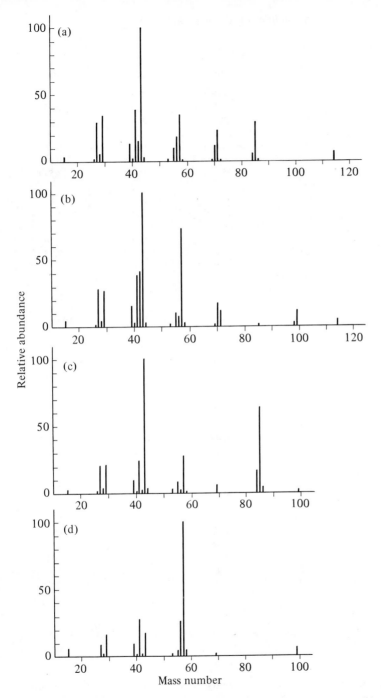

Figure 7-2. Mass spectra of *n*-octane (a) and three isomeric octanes (b), (c), and (d). (Reproduced from *Interpretation of Mass Spectra*, written by F. W. McLafferty, with permission of the publisher, W. A. Benjamin, Inc., Advanced Book Program, Reading, Mass. 1967.)

Table 7-3. Relative abundance

Mass number	$n\text{-}C_8H_{18}$ (a)	(b)	(c)	(d)
15	3.0	4.1	2.3	5.6
26	1.9	1.5	1.3	0.43
27	29.	28.	20.	8.3
28	6.2	4.5	3.2	1.6
29	34.	27.	21.	16.
30	0.74	0.60	0.49	0.31
39	13.	15.	9.4	8.5
40	2.4	2.8	1.2	1.4
41	38.	38.	24.	27.
42	15.	41.	2.3	1.5
43	100.	100.	100.	17.
44	3.3	3.2	3.4	0.60
45	0.05	0.06	0.05	0.03
53	1.7	1.9	2.6	1.3
54	0.77	0.66	0.59	0.28
55	10.	10.	7.7	4.1
56	18.	7.9	2.5	26.
57	34.	73.	27.	100.
58	1.4	3.1	1.2	4.3
59	0.03	0.04	0.03	0.06
69	1.2	1.2	5.5	1.4
70	12.	17.	0.65	0.12
71	23.	12.	0.46	0.33
72	1.2	0.69	0.03	0.03
73	0.02	0.03		
83	0.11	0.09	0.23	0.77
84	5.9	0.79	16.	0.33
85	29.	1.7	63.	0.03
86	1.9	0.12	4.1	
87	0.05		0.10	
98	0.06	3.4	0.15	0.04
99	0.07	12.	1.8	6.1
100	0.01	0.93	0.14	0.47
101	0.01	0.03	0.03	0.03
113		0.08		
114	6.7	4.9	0.00	0.02
115	0.55	0.42		
116	0.03	0.01		

Exercise 7-6. Verify the preceding statement, using Eq. 7-1 and Eq. 3-47. Neglect relativistic effects in your calculation.

Electron beams are scattered strongly in the electrical field associated with the charges of atomic electrons and nuclei. Thus, an incident beam loses a large fraction of the forward-scattered radiation at each successive layer of a thick sample. A thin sample is required to obtain information about

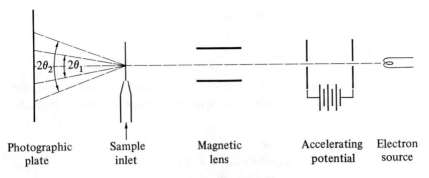

Photographic plate — Sample inlet — Magnetic lens — Accelerating potential — Electron source

Figure 7-3. Scheme of an electron-diffraction apparatus.

interatomic distances from electron diffraction; a jet of dilute gas injected into an electron beam at high vacuum is often used. A scheme of the experimental arrangement is shown in Fig. 7-3. The interference resulting from scattering by the atoms of a diatomic molecule AB is indicated in Fig. 7-4. The incident wavefront is parallel to the x-z plane, and the diffracted wavefront leaves at an angle θ with respect to the y-axis (the direction of the incident beam). The path difference between that part of the diffracted wavefront scattered by A and that scattered by B is $AE - DB = \delta$. If we use the complex exponential notation for the waves scattered at A and B (as in the footnote for Eq. 3-33, p. 79), and if we assume equal scattering from A and B, then the scattered waves are

$$\Phi_A = \Phi_0 \exp\left[-i(2\pi vt - 2\pi x/\lambda)\right] \tag{7-4}$$

and

$$\Phi_B = \Phi_0 \exp\left\{-i[2\pi vt - 2\pi(x + \delta)/\lambda]\right\} \tag{7-5}$$

The resultant scattered wave is the sum of Φ_A and Φ_B, or

$$\Phi = \Phi_A + \Phi_B$$

$$= \Phi_0 \left\{1 + \exp\left[i(2\pi\delta/\lambda)\right]\right\} \exp\left[-i(2\pi vt - 2\pi x/\lambda)\right] \tag{7-6}$$

Exercise 7-7. Verify Eq. 7-6.

The intensity of the scattered radiation is given by

$$I = \Phi\Phi^*$$

$$= \Phi_0{}^2 \left[2 + 2\cos\left(\frac{2\pi\delta}{\lambda}\right)\right]$$

$$= 4\Phi_0{}^2 \cos^2\left(\frac{\pi\delta}{\lambda}\right) \qquad (7\text{-}7)$$

Exercise 7-8. Verify Eq. 7-7.

The path difference, δ and, therefore, the intensity, I, depend on the orientation of the molecule with respect to the incident wavefront. It can be shown that

$$\begin{aligned}
\delta &= AE - DB \\
&= r \sin \alpha \cos [90 - (\beta + \theta)] - r \sin \alpha \sin \beta \\
&= r \sin \alpha [\sin (\beta + \theta) - \sin \beta] \\
&= 2r \sin \alpha \cos\left[\beta + \left(\frac{\theta}{2}\right)\right] \sin\left(\frac{\theta}{2}\right) \qquad (7\text{-}8)
\end{aligned}$$

A schematic representation of this phase relation appears in Fig. 7-4. The quantity $r \sin \alpha$ is the projection of r vertically onto the x-y plane. Multiplication by $\cos [\beta + (\theta/2)]$, an angle in the x-y plane, projects $r \sin \alpha$ onto a

Figure 7-4. Diagram of the phase relation of the electron waves scattered by two atoms, A + B, of a diatomic molecule.

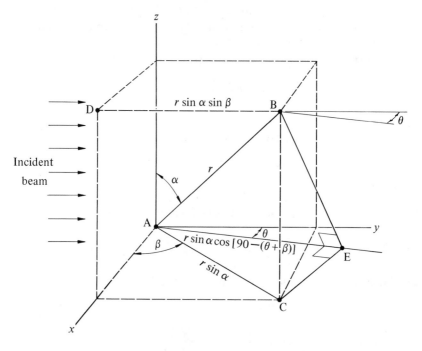

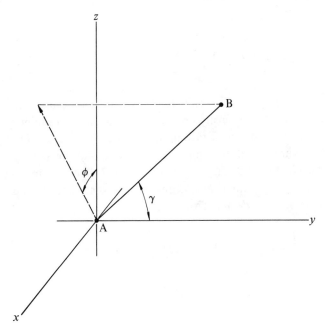

Figure 7-5. The angles γ and ϕ over which the scattered intensity is integrated to include all possible orientations of the scattering molecule with respect to the incident beam.

line at an angle of $\beta + (\theta/2)$ to the line AC. The result is the same as a single projection of r onto that line through an angle γ, so that

$$r \sin \alpha \cos\left[\beta + \left(\frac{\theta}{2}\right)\right] = r \cos \gamma$$

thus,

$$\delta = 2r \cos \gamma \sin\left(\frac{\theta}{2}\right) \tag{7-9}$$

and

$$I = 4\,\Phi_0{}^2 \cos^2\left[\left(\frac{\pi}{\lambda}\right) 2r \cos \gamma \sin\left(\frac{\theta}{2}\right)\right] \tag{7-10}$$

It is common to substitute the variable

$$k = \frac{4\pi}{\lambda} \sin\left(\frac{\theta}{2}\right)$$

so that Eq. 7-10 becomes

$$I = 4\,\Phi_0{}^2\,\cos^2\left(\frac{kr\cos\gamma}{2}\right) \qquad (7\text{-}11)$$

Since the molecules in the gas being studied are oriented randomly with respect to the incident beam, the intensity observed at an angle θ is obtained by averaging I over all possible orientations of the gas molecules. The average is taken over a variation in the angle γ from 0 to π, and the angle ϕ from 0 to 2π, as shown in Fig. 7-5. The quantity I in Eq. 7-11 is multiplied by the infinitesimal solid angle $\sin\gamma\,d\gamma\,d\phi$, integrated over γ and ϕ, and divided by the integral of $\sin\gamma\,d\gamma\,d\phi$. Thus,

$$\bar{I}(\theta) = \frac{\displaystyle\int_0^\pi\int_0^{2\pi} 4\,\Phi_0{}^2\,[\cos^2\,{}^1\!/_2\,(kr\cos\gamma)]\,\sin\gamma\,d\gamma\,d\phi}{\displaystyle\int_0^\pi\int_0^{2\pi}\sin\gamma\,d\gamma\,d\phi}$$

Figure 7-6. The variation of scattering intensity function with the variable $k = (4\pi/\lambda)$ $\sin(\theta/2)$ for CH_4 (curve A) and CD_4 (curve B). [By permission, from L. S. Bartell, K. Kuchitsu, and R. J. de Neui, *J. Chem. Phys.*, *35*, 1211 (1961).]

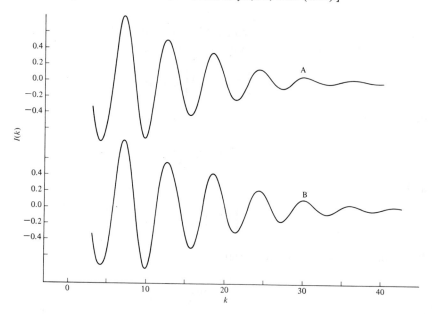

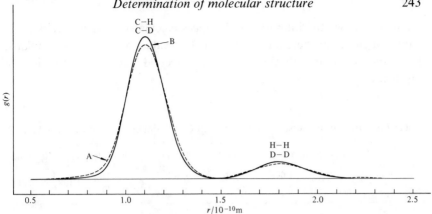

Figure 7-7. The radial distribution curve for CH_4 (——) and CD_4 (– – –) determined from electron-diffraction data. [By permission, from L. S. Bartell, K. Kuchitsu, and R. J. de Neui, *J. Chem. Phys.*, **35**, 1211 (1961).]

and the result is[2]

$$\bar{I}(\theta) = \frac{\Phi_0{}^2}{\pi} \left(1 + \frac{\sin kr}{kr} \right) \tag{7-12}$$

A graph of the intensity function as a function of k for CH_4 and CD_4 is shown in Fig. 7-6; the photographic plate on which the results are recorded shows a group of concentric rings.

In order to use Eq. 7-12, a value of r is assumed, and a function $\bar{I}(\theta)$ is calculated for comparison with the experimental results. Such a procedure is feasible for diatomic molecules, but it becomes rather difficult to use for more complex molecules, since a number of interatomic distances must be found by trial and error. The procedure is much more direct when a Fourier transform method is used to convert the intensity as a function of the angle to a radial distribution function, electron density as a function of interatomic distance.[3] When such a function is plotted, peaks are obtained at each interatomic distance, and a knowledge of approximate bond lengths enables us to identify the peaks of particular pairs of atoms.

The radial distribution curves for CH_4 and CD_4 are shown in Fig. 7-7.[4]

2. L. V. Azaroff, *Elements of X-Ray Crystallography*, McGraw-Hill Book Co., New York, 1968, p. 175.
3. L. V. Azaroff, *op. cit.*, pp. 576–79; L. Pauling and L. O. Brockway, *J. Am. Chem. Soc.*, **57**, 2684–92 (1935).
4. L. S. Bartell, K. Kuchitsu, and R. J. de Neui, *J. Chem. Phys.*, **35**, 1211 (1961).

Table 7-4 shows the interatomic distances calculated from the radial distribution curve for these molecules. The first values are averages calculated from the radial distribution curve, the second values are equilibrium distances, r_e.

Table 7-4. Interatomic distances in CH_4 and CD_4 from electron diffraction data

	C–H	C–D	H–H	H–D	Bond lengths, $r_e/10^{-10}$ m
$r/10^{-10}$ m	1.101_4	1.098_6	1.803	1.801	C–H } C–D } $1.108_2 \pm 0.004$

X-ray diffraction

In 1912, von Laue suggested that X-rays might be characterized by their interaction with the regularly spaced atoms of a crystal, surmising that they were electromagnetic radiation of a wavelength short enough that interference would result. Acting upon his suggestion, Friedrich and Knipping carried out the experiment, passing collimated beams of X-rays through a variety of single crystals to obtain well-defined interference patterns.

We must consider the geometry of crystals and the systems used to describe the arrangements of atoms in crystals before we discuss the diffraction of X-rays in more detail. It was observed early in the study of crystals that, however the relative sizes of the faces of a crystal might vary, the angles between faces are constant. The Abbé René Haüy had suggested, in 1784, that this external order is a reflection of the ordered internal arrangements of atoms in a crystal.

The concept of a space lattice was advanced to describe the possible positions of atoms in a crystal. Every lattice point can be reached from an arbitrarily chosen origin by successive translations along each of three axes spanning the space of the lattice. Every translation is an integral multiple of the characteristic unit of the axis along which the translation is made. Also, every lattice point has the same environment of neighboring points. Using this definition, Bravais showed in 1850 that there are only fourteen possible arrangements of points in a space lattice, grouped conveniently in seven lattice systems.

Figure 7-8 shows the Bravais lattices, and Table 7-5 defines the characteristic features of the lattice systems. The axes of translation are the edges of the unit cell for the simple lattices; other axes are more convenient for the face-centered and body-centered lattices.

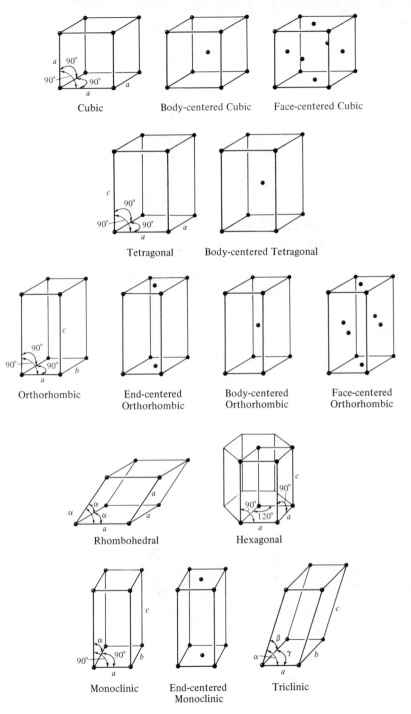

Figure 7-8. The 14 Bravais lattices.

Table 7-5.   Characteristics of the seven crystal systems

Lattice system	Unit cell axes	Unit cell angles
Cubic	$a = b = c$	$\alpha = \beta = \gamma = 90°$
Tetragonal	$a = b \neq c$	$\alpha = \beta = \gamma = 90°$
Orthorhombic	$a \neq b \neq c$	$\alpha = \beta = \gamma = 90°$
Rhombohedral	$a = b = c$	$\alpha = \beta = \gamma < 120°, \neq 90°$
Hexagonal	$a = b \neq c$	$\alpha = \beta = 90°, \gamma = 120°$
Monoclinic	$a \neq b \neq c$	$\beta = \gamma = 90°, \alpha > 90°$
Triclinic	$a \neq b \neq c$	$\alpha \neq \beta \neq \gamma \neq 90°$

Exercise 7-9.   For each lattice shown in Fig. 7-8, determine the number of neighbors and next nearest neighbors and their relative distances.

In order to understand the conventional notation for describing the planes of lattice points, let us consider a lattice in which the three axes defined by

Figure 7-9.   Several highly populated planes of (a) a simple cubic lattice, (b) a face-centered cubic lattice, and (c) a body-centered cubic lattice.

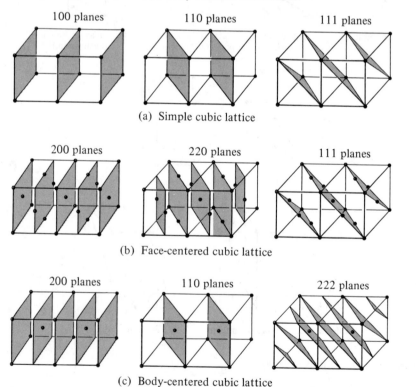

100 planes   110 planes   111 planes

(a) Simple cubic lattice

200 planes   220 planes   111 planes

(b) Face-centered cubic lattice

200 planes   110 planes   222 planes

(c) Body-centered cubic lattice

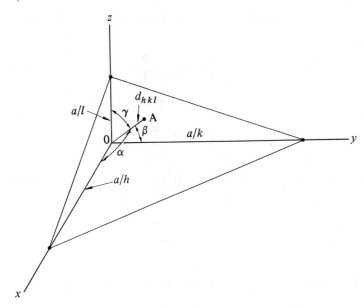

Figure 7-10. Relation of the Miller indices h, k, and l to the direction cosines of the hkl plane in a crystal.

the unit cell are intersected by a set of parallel lattice planes. The intercepts along the axes are spaced a/h, b/k, and c/l, respectively, where h, k, and l are integers. The integers h, k, and l are called the *Miller indices* of the planes, and the planes are called hkl planes. When fractions occur, they are multiplied by the smallest common factor that will convert all three to integers.

Exercise 7-10. A set of parallel lattice planes in a cubic lattice intercepts the x axis at $3a/3$, the y axis at $a/2$, and the z axis at $3a/4$. Calculate the Miller indices of the planes.

Figure 7-9 shows several of the most closely spaced planes in cubic, body-centered cubic, and face-centered cubic lattices.

Exercise 7-11. Calculate the ratio of interplanar distances of the closest planes parallel to the 100, 110, and 111 faces for cubic, body-centered cubic, and face-centered cubic lattices of the same cell size; see Fig. 7-9.

The general relationship for the distance d_{hkl} between parallel planes hkl can be deduced from the geometrical relationships shown in Fig. 7-10. The line OA, of length d_{hkl}, is the line from the origin perpendicular to the nearest plane hkl, α is the angle between OA and the x axis, β is the angle between OA and the y axis, and γ is the angle between OA and the z axis. Thus,

$$d_{hkl} = \frac{a}{h} \cos \alpha = \frac{a}{k} \cos \beta = \frac{a}{l} \cos \gamma \qquad (7\text{-}13)$$

or

$$\cos \alpha = \frac{h}{a} (d_{hkl})$$

$$\cos \beta = \frac{k}{a} (d_{hkl}) \qquad (7\text{-}14)$$

and

$$\cos \gamma = \frac{l}{a} (d_{hkl})$$

Since $\cos \alpha$, $\cos \beta$, and $\cos \gamma$ are the direction cosines of the line OA,

$$\cos^2 \alpha + \cos^2 \beta + \cos^2 \gamma = 1$$

and

$$d_{hkl} = \frac{a}{(h^2 + k^2 + l^2)^{\frac{1}{2}}} \qquad (7\text{-}15)$$

Although it can usually be determined that a crystal is cubic from its external geometry, the distinction between simple cubic, body-centered cubic, and face-centered cubic cannot be made except by X-ray diffraction methods. These methods were developed by W. H. Bragg and W. L. Bragg, father and son, only a year after Van Laue's work. The X-ray spectrometer they used is shown schematically in Fig. 7-11. As the crystal is rotated an angle θ from the incident beam, the ionization chamber used to detect the scattered X-rays is rotated by an angle 2θ so that the angle of reflection equals the angle of incidence.

The Braggs assumed that X-rays are reflected from planes of atoms in the crystal. For a given wavelength, constructive interference occurs only if the difference in path for reflections from parallel planes is an integral number of wavelengths. It can be seen from Fig. 7-12 that the path difference between beams from successive planes is $AB + BC$ or $2d \sin \theta$. Thus, the condition for constructive interference, now known as *Bragg's law*, is

$$n\lambda = 2d \sin \theta \quad n = 1, 2, 3 \ldots \qquad (7\text{-}16)$$

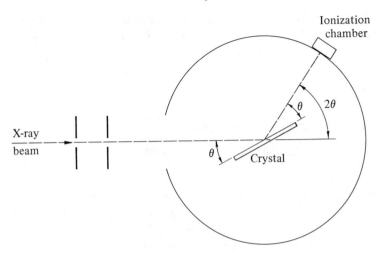

Figure 7-11.   Scheme of a Bragg X-ray reflection apparatus.

In their initial experiments, the Braggs had to determine the wavelength of the X-rays used as well as the spacing in the crystal. They were able to do this from an examination of the angles at which X-rays were reflected from various faces of the NaCl crystal. They found that the sines of the smallest angles at which reflections were found from planes parallel to the 100, 110, and 111 faces of the crystal are in the ratio $1 : \sqrt{2} : (\sqrt{3})/2$, so that the corresponding distances are in the ratio $1 : 1/\sqrt{2} : 2/\sqrt{3}$. From the results of Exercise 7-10, it can be seen that these interplanar distances correspond to

Figure 7-12.   Geometric construction for the Bragg reflection law.

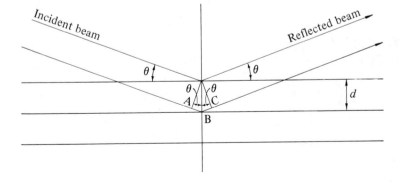

the 200, 220, and 111 planes of a face-centered cubic lattice. The reflections that might be expected from the 100 and 110 planes do not appear because, at the angle at which reflections from the 100 planes interfere constructively, reflections from the 200 planes interfere destructively; similar destructive interference occurs between the 110 and 220 planes.

Exercise 7-12.   Show that, for the angle for which

$$2d_{100} \sin \theta = \lambda, \qquad 2d_{200} \sin \theta = \frac{\lambda}{2}$$

If the units of the NaCl lattice were NaCl molecules at each lattice point, reflections of successively higher order from each face should decrease regularly in intensity. The Braggs observed that this relationship was followed for the 200 and 400 reflections and for the 220 and 440 reflections, but the 111 reflection was much weaker than the 222 reflection. They concluded from these data that the structure of the NaCl crystal must be that shown in Fig. 7-13, in which Na^+ ions occupy the points of a face-centered cubic lattice and Cl^- ions occupy the center of the edges and the body center. Each Na^+ is surrounded by six Cl^- ions. The crystal can also be considered, of course, as one in which the Cl^- ions occupy a face-centered cubic lattice and the Na^+ ions occupy the center of each edge and the body center.

The 200 and 220 reflections are normal because each of the planes involved have equal numbers of Na^+ and Cl^- ions. The 111 reflections are weak because the planes parallel to this face are alternately all Na^+ ions or all Cl^- ions at an interplanar distance one-half d_{111}. Thus, reflections from successive planes interfere destructively but do not cancel entirely, since Na^+ and

Figure 7-13.   The NaCl lattice; Na ($\bigcirc$), Cl ($\bullet$).

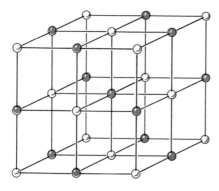

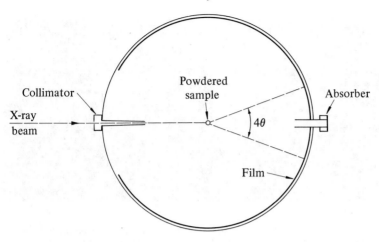

Figure 7-14. Scheme of a powder X-ray camera.

Cl^- do not scatter X-rays equally well. There are no planes between the 222 planes, so the 222 planes reflect strongly.

Since there are four formula units of NaCl in the unit cell shown in Fig. 7-13, the size of the unit cell can be obtained from the density of the crystal, and the wavelength of the X-ray can be calculated from the angle of reflection at which constructive interference occurs.

Exercise 7-13. The density of NaCl crystals is 2.16 g-cm^{-3}. For Cu Kα radiation, the smallest angle for which reflection from the 111 face occurs is 13° 41′. Calculate the wavelength of the X-rays and the length of the unit cell edge.

Once the wavelength of the X-rays is determined, the size and shape of the unit cell can be determined from the angles at which constructive interference occurs. As in the example of the NaCl crystal, the relative intensities of the reflections must be used to determine the arrangement of the atoms in the unit cell. Even without further analysis, information on the angles of reflection and the intensities of these reflections allow us to identify a crystal.

The Bragg method requires the use of single crystals, but Hull in the United States and Debye and Scherrer in Germany discovered almost simultaneously, in 1917, that a polycrystalline powdered sample can be used to obtain sufficient information for identification and to calculate the size and shape of the unit cell. Their experimental apparatus is shown schematically in Fig. 7-14. The X-ray beam impinges on the powdered

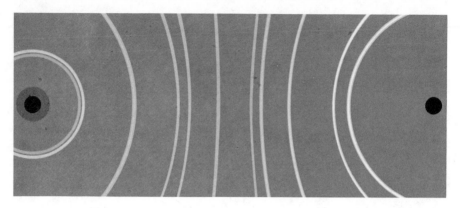

Figure 7-15. A "powder" X-ray photograph obtained with a polycrystalline gold film. [By permission, from T. J. Louzon and T. H. Spencer, *The Western Electric Engineer*, *18*, 10–16 (1974).]

sample, usually contained in a fine glass capillary placed on the axis of the cylindrical camera. Since the crystals are randomly arranged in the sample, reflections must be obtained simultaneously for lattice planes parallel to all the possible crystal faces. The reflections from a given plane form a cone and strike the photographic film at the circumference of the camera in symmetric arcs, as shown in Fig. 7-15. The angle θ can be calculated from the radius of the camera and the distance on the film between the two portions of the arc resulting from a given reflection.

Exercise 7-14. In the powder X-ray diagram of KCl, taken with Cu Kα radiation of wavelength 154 pm, lines closest to the incident beam are found at positions corresponding to $\theta = 14°\ 11'$, $20°\ 15'$, and $25°\ 6'$. The density of KCl is 1.99 g-cm^{-3}. Determine the positions of the K$^+$ and Cl$^-$ ions in a KCl unit cell from these data. Hint: The scattered radiation from K$^+$ and Cl$^-$ have essentially the same intensity because the two ions are isoelectronic.

The determination of the structure of more complicated molecules in crystals requires the measurement of the intensity and position of a large number of reflections from a single crystal. The intensity of each reflection can be expressed as a Fourier series with respect to the electron density in the plane responsible for the reflection. By a Fourier transform, the electron density in the crystal can be expressed as a Fourier series with respect to the intensities of the X-ray reflections. Usually, a trial structure, from which intensities can be calculated for comparison with the experimental results must be obtained. Successive refinements then lead to a structure that fits the observed intensities within experimental precision. Modern diffractometers with recording counters directly linked to high-speed digital computers have made it possible to determine the structure of such complex molecules as vitamin B-12 and the protein myoglobin. Electron density contour maps obtained from X-ray data for several molecules are shown in Fig. 7-16.

Neutron Diffraction

The positions of hydrogen atoms in a crystal are very difficult to determine by X-ray diffraction because X-ray scattering intensity is determined by electron density. Since neutron scattering is determined by the magnetic interaction between neutrons and the sample nuclei, neutron diffraction methods can be used to determine the positions of hydrogen atoms in crystals. Neutron diffraction can also be used to study the loci of magnetic moments in crystals, since neutrons have spin and interact with magnetic moments.

Neutron diffraction methods require the use of a nuclear pile, and the beam of neutrons from the pile is monochromatized by selecting a reflected beam from a single crystal of known spacing. The average thermal energy of neutrons at room temperature is such that their De Broglie wavelength is about 1.4×10^{-10}m. Neutrons of longer wavelength can be obtained by cooling the neutron source.

Exercise 7-15. Calculate the wavelength of a neutron for which the kinetic energy is the average thermal energy at 300 K. To what temperature does a neutron source have to be cooled to provide neutrons with a wavelength of 10×10^{-10}m?

In 1957, Peterson and Levey[5] carried out a neutron diffraction study on D_2O crystals that confirmed Pauling's hypothesis on the position of hydrogen atoms in the crystal. Each oxygen atom in the crystal is bonded covalently to two hydrogens and hydrogen-bonded to two others, with the four hydrogens arranged tetrahedrally about the oxygen. There are six

5. S. W. Peterson and H. A. Levy, *Acta Cryst.*, *10*, 70–76 (1957).

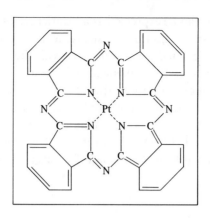

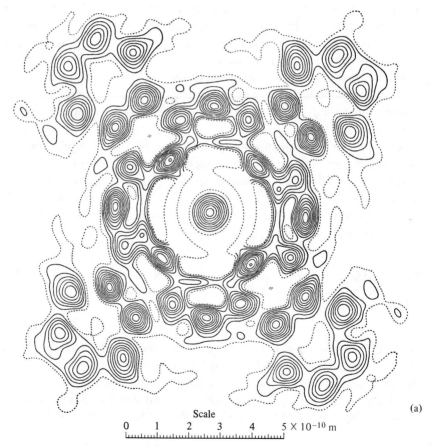

Scale

0   1   2   3   4   5 × 10⁻¹⁰ m

(a)

Figure 7-16. Electron density maps derived from X-ray diffraction data. (a) Platinum phthalocyanine. (By permission, from J. M. Robertson and I. Woodward, *J. Chem. Soc., 1940,* 36.) (b) *Trans*-stilbene. [By permission, from J. M. Robertson and I. Woodward, *Proc. Roy. Soc., A162,* 568 (1937).]

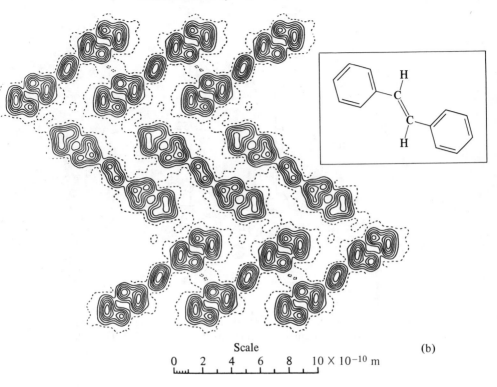

Scale   (b)

$0 \quad 2 \quad 4 \quad 6 \quad 8 \quad 10 \times 10^{-10}$ m

different, equally probable arrangements of the two types of bonding to hydrogen about each oxygen [the number of ways in which four distinguishable objects can be divided into two groups of two is $4!/(2!)(2!) = 6$]. The average structure, which determines the diffraction pattern, has one-half a hydrogen atom at each of two sites symmetrically distributed between oxygen atoms. The results obtained by Peterson and Levy are shown in Fig. 7-17, in which the two $\frac{1}{2}$D atoms are 0.752×10^{-10}m apart, and each $\frac{1}{2}$D is approximately 1.003×10^{-10}m from an oxygen.

Benoit and his collaborators[6] used small angle diffraction of neutrons, with wavelengths ranging from 1×10^{-10}m to 12×10^{-10}m, to determine the configuration of perdeuterated polystyrene in amorphous solid H-polystyrene. The properties of such high polymers as H-polystyrene in

6. J. P. Cotton, D. Decker, H. Benoit, B. Farnoux, J. Higgins, G. Jannink, R. Ober, C. Picot, and J. des Cloizeaux, *Macromolecules*, **7**, 863–72 (1974).

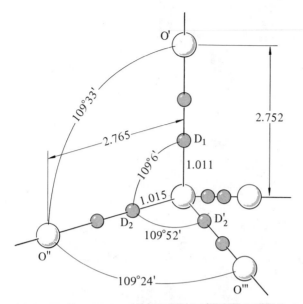

Figure 7-17. The position of D atoms in crystals of solid D_2O at $-50°C$. Oxygen ($\bigcirc$); deuterium ($\frac{1}{2}D$), ($\bullet$). [By permission, from S. W. Peterson and H. A. Levy, *Acta Cryst.*, *10*, 70–76 (1957).]

solution have been explained by assuming that they exist in a random coil configuration (see Chapter 18). It was believed that this configuration is retained in the amorphous solid, but no experimental evidence was obtained. The scattering of neutrons by D and H nuclei is different enough so that the deuterated polymer can be observed in the presence of the normal H-polymer. The results indicated clearly that the polymer configuration is that of a random coil in the amorphous solid state.

Problems

7-1. Compare the energy of an electron accelerated by a potential of 80 V to the ionization energies of atoms in Table 4-4, the ionization energies of molecules shown in Figs. 6-16 through 6-20, and the dissociation energies of molecules given in Chapter 5.

7-2. Calculate the average time between collisions for a species of molar mass equal to 28×10^{-3} kg-mol^{-1} at 298 K and a pressure of 10^{-5} Pa. Assume a molecular diameter of 5×10^{-10} m. Compare to the results of Exercise 2-5.

7-3. The following mass numbers and relative abundances were observed in a mass

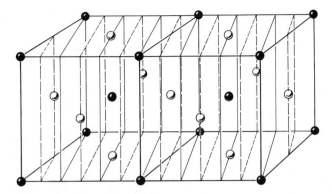

Figure 7-18. Two unit cells of a cubic lattice. Lattice points of the simple cubic lattice (●); lattice points in the face-centered cubic lattice (● and ○); lattice points in the body-centered cubic lattice (● and ☉). The planes shown are 310 (–––) and 620 (——). (Courtesy of Julie Sedgwick, Lawrence University.)

spectrum. [B. J. Zwolinski *et. al.*, *Catalogue of Selected Mass Spectral Data*, Serial No. 97, American Petroleum Institute Research Project 44, Thermodynamics Research Center, Texas A. & M. University, College Station, Texas (loose-leaf data sheets extant, 1975).] Identify the sample.

Mass number	Abundance
16	5.2
32	11.
33	0.10
34	0.42
48	49.
49	0.41
50	2.3
64	100.
65	0.88
66	4.9

7-4. Figure 7-18 shows the 310 and 620 planes for two unit cells of a cubic lattice: simple, body-centered, and face-centered. In which of the three lattices will reflections from each of these planes appear in the X-ray pattern? Explain your answer. Calculate the interplanar distance for each set of planes in terms of the length of a cube edge a.

7-5. The unit cell of CaF_2 (fluorite) is a cube ($a = 5.45 \times 10^{-10}$m). The Ca^{2+} ions are located at the corners and the face centers of the cube. The F^- ions are located at the following locations:

$(a/4, a/4, a/4)$, $(a/4, a/4, 3a/4)$, $(a/4, 3a/4, 3a/4)$, $(3a/4, a/4, a/4)$,
$(3a/4, 3a/4, a/4)$, $(3a/4, a/4, 3a/4)$, $(3a/4, 3a/4, 3a/4)$, $(a/4, 3a/4, a/4)$.

(a) Draw a diagram of the unit cell. (b) Show the arrangement of ions in the 100, 110, and 111 planes. Calculate the nearest Ca-Ca, F-F, and Ca-F distances in terms of a. (c) Calculate the number of formula units of CaF_2 in the unit cell. (d) Calculate the Bragg angles at which diffraction will be observed for the 100, 110, and 111 planes for $n = 1$ and $n = 2$, if $\lambda = 1.5 \times 10^{-10}$m.

8 · *Molecular structure and equilibrium*

In Chapter 2 we discussed the kinetic theory of gases, a simple statistical treatment of the motion of a collection of gas molecules based on classical mechanics. The simple empirical gas laws and the high temperature heat capacity of gases were derived from this theory, but the temperature dependence of the heat capacity and the behavior of systems in chemical equilibrium could not be treated. In Chapter 3 we followed Einstein's application of Planck's quantum hypothesis to the temperature dependence of the heat capacity of crystalline solids. Now, using the results of the quantum theory of molecular structure, we can reconsider the statistical treatment of a collection of molecules and extend it to systems in chemical equilibrium.

8-1. Averages over quantum states

To describe the macroscopic state of an isolated system, it is necessary to specify the energy E, the volume V, and the number of molecules in the system N. To describe the microscopic state of such a system classically, it is necessary to specify the position and momentum of each molecule in the system. Many microscopic states are consistent with each macroscopic state, and simple kinetic theory predicts macroscopic variables from an average over suitable microscopic states, as we saw in Chapter 2.

A microscopic state is described in quantum mechanics by specifying the wave function of the system, one of the many wave functions consistent with the given energy and volume. For a system in which molecules can be considered *independent* of one another, like the molecules of an ideal gas, a microscopic state can be described by specifying the values of the electronic, vibrational, rotational, and translational quantum numbers for each molecule; that is, the wave function ψ and the energy eigenvalue ε must be specified for each molecule. In Chapter 5, electrons in a single molecule were described by using all possible permutations of the appropriate wave function. Since molecules in a gas, like electrons in a molecule, are indistinguishable, their quantum state must be described by allowing all the permutations of wave functions among the molecules. Many combinations of microscopic states are consistent with a given macroscopic state, so macroscopic variables are obtained by averaging over microscopic states. This procedure does not differ from the procedure based on the classical description. In the classical description, molecules interchange kinetic energy upon collision; in the quantum-mechanical description, molecules interchange energy and change quantum numbers upon collision. At *equilibrium*, while microscopic changes continue, macroscopic observed values and averages over microscopic states do not change with time.

A necessary first step in describing a system is to enumerate the quantum states that are *accessible* to the system. The description of a collection of ammonia molecules, for example, ordinarily need only include quantum levels of ammonia; but, in the presence of a suitable catalyst, a dissociation equilibrium with nitrogen and hydrogen is established, and quantum levels of nitrogen and hydrogen molecules must also be included. We observed earlier that insofar as electrons or other particles with half-integral spin are concerned, only quantum states that are antisymmetric with respect to particle interchange are accessible. Similarly, for photons and for particles with integral spin, only quantum states that are symmetric with respect to particle interchange are accessible. The former situation is described by *Fermi-Dirac* statistics, and the latter by *Bose-Einstein* statistics.

A probability must then be assigned to each accessible quantum state of a system so that a weighted statistical average can be obtained. In Sec. 2-5, the postulate of random motion was used to derive the probability function for the classical Maxwell distribution function. The corresponding postulate here is that *all accessible quantum states of the system corresponding to a given value of the energy of the system are equally probable.*

It is convenient to assume that a system in a given macroscopic state passes through all the accessible quantum states many times during the time it takes to make a macroscopic measurement and that equal time is

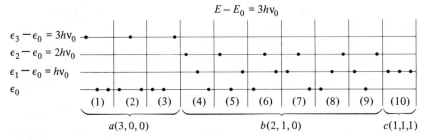

Figure 8-1. A representation of the ten possible *quantum states* and three possible *distributions* of a system of three *distinguishable* oscillators that share a thermal energy $E - E_0 = 3\,hv_0$.

spent in each of the equally probable states. Accordingly, the averages obtained are *averages over time.*[1]

Small Systems[2]

Let us consider a system in which the quantum states can be enumerated, and the averaging can be carried out simply. Harmonic oscillators in a one-dimensional crystalline lattice comprise such a system. The oscillators, being localized in space, are distinguishable, in contrast to the molecules in a gas or a liquid.[3] The eigenvalues of the energy of each oscillator are given by an expression of the same form as the vibrational energy of a diatomic molecule (Eq. 6-33),

$$\varepsilon_v = (v + {}^1\!/_2)\,hv_0 \qquad (8\text{-}1)$$

where v is the quantum number of the oscillator, and v_0 is the classical fundamental frequency of the oscillator.

All oscillators with the same fundamental frequency have the same energy level spacing. Figure 8-1 is a representation of a system of three oscillators that share an energy above the zero-point energy $E - E_0 = 3\,hv_0$. A *quantum state* of the system is described by specifying the position of each dis-

---

1. An alternative approach uses an average over an *ensemble* of systems, one that includes all the accessible quantum states consistent with the macroscopic state, rather than a time average. A clear account of this approach can be found in K. Denbigh, *The Principles of Chemical Equilibrium*, Cambridge University Press, 1961, Part III.
2. R. W. Gurney, *Introduction to Statistical Mechanics*, McGraw-Hill Book Co., New York, 1949.
3. Strictly speaking, the normal modes of vibration, equal in number to the number of oscillators, are distinguishable.

tinguishable oscillator on a set of equally spaced energy levels. A *distribution* is described by specifying the *number* of oscillators at each energy level without regard to their identity. Since there are ten quantum states consistent with the total energy $E - E_0$, distribution *a* has a probability of 0.3, distribution *b* has a probability of 0.6, and distribution *c* has a probability of 0.1. The sum of the probabilities is necessarily one.

The energy levels of an oscillator are determined by the nature of the solid, and they are independent of temperature. The number of possible quantum states of the system depends on the energy $E-E_0$; this quantity increases with increasing temperature, as does the number of quantum states.

The average number of oscillators at any particular energy level $\bar{N}(\varepsilon_j)$ is defined as the sum of the number of oscillators occupying that level in each possible quantum state divided by the number of quantum states. In the example in Fig. 8-1, they are $\bar{N}(\varepsilon_0) = 1.2$, $\bar{N}(\varepsilon_1) = 0.9$, $\bar{N}(\varepsilon_2) = 0.6$, and $\bar{N}(\varepsilon_3) = 0.3$. The *average distribution* is described by specifying all the numbers $\bar{N}(\varepsilon_j)$ for a system. An alternative method of obtaining the *average distribution* is to multiply the population of each level in each distribution by the probability of that distribution and to sum the product over all distributions. Thus $\bar{N}(\varepsilon_0) = (2)(0.3) + (1)(0.6) + (0)(0.1) = 1.2$, etc. Each level shown has an average population that is less than that of any lower energy level. This is a *graded distribution*. All average distributions are graded, even when no single distribution is graded.

Exercise 8-1. Enumerate the quantum states of a system of three oscillators consistent with $E - E_0 = $ (a) $4hv_0$, and (b) $5hv_0$. Describe the average distribution and the most probable distribution. What is the effect of increasing energy on the shape of the average distribution?

Large Systems

It would be cumbersome to enumerate all the quantum states of a system containing a large number of molecules. Mathematical expressions are derived instead to describe the number of quantum states for a given macroscopic state of a system and to describe the number of quantum states for a given *distribution* of a system. If a system of N oscillators has a distribution characterized by N_0 oscillators with energy ε_0, N_1 oscillators with energy ε_1, etc., the number of quantum states of the distribution is equal to the number of ways N distinguishable objects can be partitioned into groups of $N_0, N_1, N_2, \cdots$ objects; that is

$$\Omega_D = \frac{N!}{(N_0!)(N_1!)\cdots(N_n!)\cdots} \tag{8-2}$$

where Ω_D is the number of quantum states in a given distribution.

Exercise 8-2. Verify Eq. 8-2 for the distributions in Figure 8-1. By definition, $0! = 1$.

To calculate the total number of quantum states for a system we use the following representation[4] of the quantum states of a system of three oscillators with $E - E_0 = 3hv_0$, with each line representing one quantum state. The number on each line refers to the corresponding quantum state in Fig. 8-1.

$$
\begin{array}{rl}
(3) & \cdot \mid \cdot \; \cdot \; \varepsilon \; \varepsilon \; \varepsilon \\
(8) & \cdot \mid \cdot \; \varepsilon \; \cdot \; \varepsilon \; \varepsilon \\
(6) & \cdot \mid \cdot \; \varepsilon \; \varepsilon \; \cdot \; \varepsilon \\
(2) & \cdot \mid \cdot \; \varepsilon \; \varepsilon \; \varepsilon \; \cdot \\
(9) & \cdot \mid \varepsilon \; \cdot \; \cdot \; \varepsilon \; \varepsilon \\
(10) & \cdot \mid \varepsilon \; \cdot \; \varepsilon \; \cdot \; \varepsilon \\
(7) & \cdot \mid \varepsilon \; \cdot \; \varepsilon \; \varepsilon \; \cdot \\
(5) & \cdot \mid \varepsilon \; \varepsilon \; \cdot \; \cdot \; \varepsilon \\
(4) & \cdot \mid \varepsilon \; \varepsilon \; \cdot \; \varepsilon \; \cdot \\
(1) & \cdot \mid \varepsilon \; \varepsilon \; \varepsilon \; \cdot \; \cdot
\end{array}
$$

In this representation, the oscillators are indicated by dots. Dots followed on the right by one, two, or three ε's represent oscillators with energies ε_1, ε_2, and ε_3, respectively. Dots followed by another dot or dots at the right end of the diagram represent oscillators with energy ε_0. Thus, the first line represents a quantum state with the first two oscillators at ε_0 and the third at ε_3. The vertical line in the diagram separates the single constant symbol on each line, the dot representing the first oscillator, from the symbols that are permuted from one line to another. If n is the number of units (hv_0) of energy in the system, and N is the number of oscillators, the number of quantum states Ω is just the number of distinguishable permutations of $(N - 1)$ dots and n ε's or the number of ways of arranging n *identical* objects in $(N + n - 1)$ distinguishable positions; that is

$$
\Omega = \frac{(N + n - 1)!}{n! \, (N - 1)!} \tag{8-3}
$$

where Ω is the total number of quantum states.

Exercise 8-3. Verify Eq. 8-3 for the system shown in Fig. 8-1.

The distributions possible for a small system, such as that represented in Fig. 8-1 or those systems in Exercise 8-1, all have probabilities of comparable

4. H. A. Bent, *The Second Law*, Oxford University Press, New York, 1965, p. 155.

magnitude, none can be neglected in calculating the average distribution, and the most probable distribution is not a graded distribution.

Consider a relatively small system of 20 oscillators with $E - E_0 = 20\,hv_0$. The total number of quantum states consistent with this energy is

$$\Omega = \frac{39!}{(19!)(20!)}$$

$$= 6.89 \times 10^{10}$$

For a distribution with $N_1 = 20$ and $N_j = 0, j \neq 1, \Omega_D = 1$. For the graded distribution with $N_3 = 2, N_2 = 4, N_1 = 6$, and $N_0 = 8$,

$$\Omega_D = \frac{20!}{2!\,4!\,6!\,8!} = 1.75 \times 10^9$$

If $N_2 = 5, N_1 = 10$, and $N_0 = 5$,

$$\Omega_D = \frac{20!}{5!\,10!\,5!} = 4.66 \times 10^7$$

Clearly, a distribution with $\Omega_D = 1$ can be ignored in calculating the average distribution; a distribution with Ω_D of the order of 10^7 is much less important than one with Ω_D of the order of 10^9. Thus, even for a moderate number of oscillators, the average distribution is very little different from a distribution of high probability. The difference becomes infinitesimal for a number of oscillators representing a small fraction of a mole (10^{-6} mole $= 6.02 \times 10^{17}$ oscillators), as shown in Table 8-1, where $\Omega_{D,\,max}$ is the number of quantum states in the most probable distribution.

Table 8-1. A comparison of the total number of quantum states and the number of quantum states of the most probable distribution for large systems[a]

Number of oscillators	$(E - E_0)/hv_0$	$\log \Omega$	$\log \Omega_{D,\,max}$
1×10^{15}	5.8198×10^{14}	4.5195×10^{14}	4.5195×10^{14}
1×10^{18}	5.8198×10^{17}	4.5195×10^{17}	4.5195×10^{17}
1×10^{20}	5.8198×10^{19}	4.5195×10^{19}	4.5195×10^{19}

[a] The data are expressed as log Ω because the numbers involved are very large and because the approximation used in calculating the factorial, as discussed below, leads to the logarithm.

There are many distributions of high probability besides the most probable distribution, but they are all close to the most probable distribution. (A schematic graphical representation is shown in Fig. 8-2.) The latter can

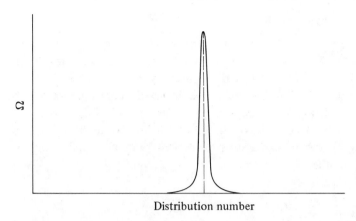

Figure 8-2. A schematic representation of the variation of Ω for different distributions, showing very low values of Ω for any distribution except those very close to the most probable distribution (– – –).

therefore be used to represent the average distribution. As can be seen from the most probable distribution, in Table 8-2, the average population of a level could change by as many as 10^6 oscillators without changing the shape of the distribution significantly.

Table 8-2. Most probable distribution of a system of 10^{20} oscillators with $E - E_0 = 0.58198 \times 10^{20} \, h\nu_0$

Level	Population
0	6.3212×10^{19}
1	2.3254×10^{19}
2	8.5548×10^{18}
3	3.1471×10^{18}
4	1.1578×10^{18}
5	4.2592×10^{17}
6	1.5669×10^{17}
7	5.7642×10^{16}
8	2.1205×10^{16}
9	7.8010×10^{15}
10	2.8698×10^{15}
11	1.0558×10^{15}
12	3.8839×10^{14}
13	1.4288×10^{14}
14	5.2563×10^{13}
15	1.9338×10^{13}
16	7.1136×10^{12}
17	2.6169×10^{12}

8-2. The most probable distribution

For a distribution of N oscillators, such that N_0 have energy ε_0, N_1 have energy ε_1, etc., Eq. 8-2 gives the number of quantum states consistent with the distribution. The most probable distribution is the one for which Ω_D is a maximum, subject to the macroscopic conditions of a constant number of oscillators and a constant total energy. The maximum for $\ln \Omega_D$ and that for Ω_D correspond to the same values of N_j. It is convenient to find the maximum value for $\ln \Omega_D$ by using Stirling's approximation to the factorial, which is valid for large numbers.[5]

Stirling's formula is

$$n! \approx (2\pi)^{\frac{1}{2}} n^{n+\frac{1}{2}} \exp(-n)$$

The error for 10! is only 0.8% and for 100! is 0.08%. The formula for $\ln n!$ is

$$\ln n! \approx \tfrac{1}{2} \ln(2\pi) + (n + \tfrac{1}{2}) \ln n - n$$

which reduces, for n sufficiently large, to

$$\ln n! \approx n \ln n - n \tag{8-4}$$

Since $\ln \Omega_D$ is a function of all the N_j, the condition for a maximum is

$$\delta \ln \Omega_D = \left(\frac{\partial \ln \Omega_D}{\partial N_0} \right) \delta N_0 + \left(\frac{\partial \ln \Omega_D}{\partial N_1} \right) \delta N_1 + \cdots$$

$$= \sum_{j=0}^{\infty} \left(\frac{\partial \ln \Omega_D}{\partial N_j} \right) \delta N_j = 0 \tag{8-5}$$

From Eq. 8-2 and Stirling's approximation, we obtain

$$\ln \Omega_D = \ln N! - \sum_{j=0}^{\infty} (N_j \ln N_j - N_j) \tag{8-6}$$

so that

$$\frac{\partial \ln \Omega_D}{\partial N_j} = -\ln N_j$$

and

$$\delta \ln \Omega_D = -\sum_{j=0}^{\infty} \ln N_j \, \delta N_j = 0 \tag{8-7}$$

5. W. Feller, *An Introduction to Probability Theory and Its Applications*, 2nd ed., vol. 1, John Wiley & Sons, New York, 1957, pp. 50–52.

Exercise 8-4. Verify Eq. 8-7, starting with Eq. 8-2.

For arbitrary values of all the δN_j, Eq. 8-7 requires that $\ln N_j = 0$ for all j. This is an unrealistic condition, since it implies that $N_j = 1$ for all j. We shall see that this corresponds to a condition of infinite temperature. If the conditions that N and E remain constant are imposed, that is

$$\sum_{j=0}^{\infty} \delta N_j = 0 \tag{8-8}$$

$$\sum_{j=0}^{\infty} (\varepsilon_j - \varepsilon_0) \, \delta N_j = 0 \tag{8-9}$$

then the values of all but two of N_j can be chosen arbitrarily, and the remaining two depend on them. Equations 8-8 and 8-9 can be multiplied by the undetermined constants μ and β, respectively, and Eq. 8-7 subtracted, to give

$$\sum_{j=0}^{\infty} [\mu + \beta(\varepsilon_j - \varepsilon_0) + \ln N_j] \, \delta N_j = 0 \tag{8-10}$$

Note that μ must be a pure number and β must have the dimensions of an inverse energy.

Let μ and β be chosen so that, for the most probable distribution,

$$\mu + \beta(\varepsilon_0 - \varepsilon_0) + \ln N_0 = 0$$

and

$$\mu + \beta(\varepsilon_1 - \varepsilon_0) + \ln N_1 = 0$$

The remaining values of δN_j are independent, so the coefficient of δN_j is zero for all j, and

$$\mu + \beta(\varepsilon_j - \varepsilon_0) + \ln N_j = 0$$

or

$$N_j = \exp(-\mu) \exp[-\beta(\varepsilon_j - \varepsilon_0)] \tag{8-11}$$

Substituting from Eq. 8-11 in the expression for the total number of oscillators, we obtain

$$N = \sum_{j=0}^{\infty} N_j = \exp(-\mu) \sum_{j=0}^{\infty} \exp[-\beta(\varepsilon_j - \varepsilon_0)] \tag{8-12}$$

We obtain the fundamental equation of the Boltzmann distribution,

$$\frac{N_j}{N} = \frac{\exp[-\beta(\varepsilon_j - \varepsilon_0)]}{\sum_{j=0}^{\infty} \exp[-\beta(\varepsilon_j - \varepsilon_0)]} \tag{8-13}$$

by dividing Eq. 8-11 by Eq. 8-12. The summation in Eq. 8-13 occurs frequently; it is given the symbol z, and it is called the *molecular partition function*. It describes the partition of the oscillators among the possible molecular quantum states.

If we write Eq. 8-13 as

$$N_j = \frac{N}{z} \exp\left(-\beta(\varepsilon_j - \varepsilon_0)\right] \tag{8-14}$$

it is observed that β must be positive, since the system has a finite energy. A macroscopic state for which $E - E_0$ is known is used to evaluate β, just as empirical observations were used to complete the equation of simple kinetic theory. The thermal energy of N one-dimensional oscillators is known to be NkT at sufficiently high temperatures, from classical equipartition theory and from the empirical law of Dulong and Petit (Sec. 2-3). From the quantum-statistical model, the energy of a system above the ground state (equal to the thermal energy) is

$$E - E_0 = \sum_{j=0}^{\infty} N_j (\varepsilon_j - \varepsilon_0)$$

$$= \frac{N}{z} \sum_{j=0}^{\infty} (\varepsilon_j - \varepsilon_0) \exp\left[-\beta(\varepsilon_j - \varepsilon_0)\right] \tag{8-15}$$

For a quantum oscillator, $(\varepsilon_j - \varepsilon_0)$ equals $jh\nu_0$, so

$$E - E_0 = \frac{N}{z} \sum_{j=0}^{\infty} jh\nu_0 \exp\left(-j\beta h\nu_0\right)$$

$$= \frac{Nh\nu_0 \sum_{j=0}^{\infty} j \exp\left(-j\beta h\nu_0\right)}{\sum_{j=0}^{\infty} \exp\left(-j\beta h\nu_0\right)} \tag{8-16}$$

If we let $\exp\left(-\beta h\nu_0\right) = x$,

$$\sum_{j=0}^{\infty} \exp\left(-j\beta h\nu_0\right) = 1 + x + x^2 + \ldots = s_1 \tag{8-17}$$

and

$$\sum_{j=0}^{\infty} j \exp\left(-j\beta h\nu_0\right) = x(1 + 2x + 3x^2 + \ldots) = s_2 \tag{8-18}$$

Clearly,

$$s_2 = x \frac{ds_1}{dx}$$

Since

$$x s_1 = x + x^2 + \ldots = s_1 - 1$$

$$s_1 = \frac{1}{1 - x}$$

Therefore,

$$s_2 = x \frac{ds_1}{dx}$$

$$= \frac{x}{(1 - x)^2}$$

Returning to Eq. 8-17 and 8-18, we now have

$$\sum_{j=0}^{\infty} \exp(-j\beta h\nu_0) = [1 - \exp(-\beta h\nu_0)]^{-1} = z \qquad (8\text{-}19)$$

and

$$\sum_{j=0}^{\infty} j \exp(-j\beta h\nu_0) = \exp(-\beta h\nu_0)[1 - \exp(-\beta h\nu_0)]^{-2}$$

so that Eq. 8-16 becomes

$$E - E_0 = \frac{N h\nu_0 \exp(-\beta h\nu_0)}{1 - \exp(-\beta h\nu_0)}$$

$$= \frac{N h\nu_0}{(\exp \beta h\nu_0) - 1} \qquad (8\text{-}20)$$

At high temperatures, where $h\nu_0 <<< kT$, we find the classical result $E - E_0 = NkT$ by expanding the exponential in a Taylor's series about zero, and, if the higher terms are neglected,

$$\exp \beta h\nu_0 = 1 + \beta h\nu_0 + \frac{1}{2!}(\beta h\nu_0)^2 + \cdots$$

$$\simeq 1 + \beta h\nu_0 \qquad (8\text{-}21)$$

Substituting from Eq. 8-21 in Eq. 8-20, we obtain

$$E - E_0 \frac{N h\nu_0}{1 + (\beta h\nu_0 - 1)}$$

$$= \frac{N}{\beta} = NkT$$

and

$$\beta = \frac{1}{kT} \qquad (8\text{-}22)$$

Now the Boltzmann distribution can be expressed as

$$N_j = \frac{N}{z} \exp\left(-\frac{\varepsilon_j - \varepsilon_0}{kT}\right) \qquad (8\text{-}23)$$

and

$$z = \sum_{j=0}^{\infty} \exp\left(-\frac{\varepsilon_j - \varepsilon_0}{kT}\right) = \left[1 - \exp\left(-\frac{h v_0}{kT}\right)\right]^{-1} \quad (8\text{-}24)$$

The average thermal energy per oscillator is (from Eq. 8.20 and Eq. 8.22)

$$\bar{\varepsilon} - \varepsilon_0 = \frac{E - E_0}{N}$$

$$= \frac{h v_0}{(\exp h v_0 / kT) - 1} \quad (8\text{-}25)$$

The latter expression was used by Planck to derive the spectral distribution of black-body radiation.

8-3. Energy and heat capacity of crystals and the partition function

Given an expression for the *most probable distribution*, which we use as a very good approximation to the *average distribution*, we can calculate some of the macroscopic properties of simple systems. In this section, the energy and heat capacity of crystals are calculated by using the partition function calculated from the Boltzmann distribution.

From the Boltzmann distribution (Eq. 8-23), the population of all energy levels other than the lowest level is zero at $T = 0$ K. As energy is added to the system, and the temperature increases, the energy of the system above ground level is given by Eq. 8-15, with $\beta = 1/kT$, as

$$E - E_0 = \frac{N}{z} \sum_{j=0}^{\infty} (\varepsilon_j - \varepsilon_0) \exp\left[\frac{-(\varepsilon_j - \varepsilon_0)}{kT}\right] \quad (8\text{-}26)$$

The relation between the energy of the system and the partition function can be made clear if we differentiate z with respect to temperature at constant volume. (The condition of constant volume is imposed because the ε_j are functions of the volume.)

$$z = \sum_{j=0}^{\infty} \exp\left[\frac{-(\varepsilon_j - \varepsilon_0)}{kT}\right]$$

and

$$\left(\frac{\partial z}{\partial T}\right)_V = \sum_{j=0}^{\infty} \frac{\varepsilon_j - \varepsilon_0}{kT^2} \exp\left[\frac{-(\varepsilon_j - \varepsilon_0)}{kT}\right] \quad (8\text{-}27)$$

As T approaches the limit of 0 K, the expression for z approaches one, since the first term in the summation is equal to one for all T, and the remaining terms go to zero at 0 K. As T approaches infinity, every term becomes one, and z equals the number of levels.

From Eqs. 8-26 and 8-27, we can see that

$$E - E_0 = \frac{NkT^2}{z}\left(\frac{\partial z}{\partial T}\right)_V$$

$$= NkT^2\left(\frac{\partial \ln z}{\partial T}\right)_V \tag{8-28}$$

Exercise 8-5. Verify Eq. 8-28.

Thus, knowledge of the partition function and its temperature dependence permits us to calculate the energy of a crystal as a function of the temperature. The value of the heat capacity at constant volume, (Eq. 2-31)

$$C_v = \left(\frac{\partial E}{\partial T}\right)_V$$

can be obtained by differentiating Eq. 8-25 with respect to T.

$$E - E_0 = \frac{Nh\nu_0}{\exp{(h\nu_0/kT)} - 1} \tag{8-29}$$

and

$$C_v = Nk\left(\frac{h\nu_0}{kT}\right)^2 \frac{\exp{(h\nu_0/kT)}}{[\exp{(h\nu_0/kT)} - 1]^2} \tag{8-30}$$

As shown in Sec. 3-3, Eq. 8-30 reduces at high temperature to

$$C_v = Nk$$

or

$$\mathscr{C}_v = 3Lk$$

for a mole of three-dimensional oscillators, in agreement with the law of Dulong and Petit. At low temperatures, at which

$$\exp\left(\frac{h\nu_0}{kT}\right) \gg 1$$

the corresponding expression is

$$C_v = 3Lk\left(\frac{h\nu_0}{kT}\right)^2 \exp\left(\frac{-h\nu_0}{kT}\right) \tag{8-31}$$

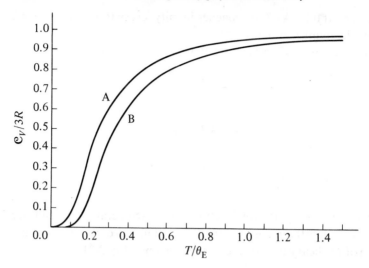

Figure 8-3. A plot of $\mathscr{C}_v/3R$ against T/θ_E for several monatomic solids, showing the fit of experimental heat capacity data to the Einstein model. Curve A, a summary of experimental results. Curve B, the theoretical curve. [E. Schrödinger, *Physik, Z., 20,* 420 (1919).]

Since v_0 is the parameter in the heat capacity equation that characterizes a particular crystal, and since hv_0/k has the dimensions of a temperature, usually designated θ_E, the Einstein model for the heat capacity of a crystal is frequently expressed as

$$\mathscr{C}_v = 3Lk\left(\frac{\theta_E}{T}\right)^2 \frac{\exp(\theta_E/T)}{[\exp(\theta_E/T) - 1]^2} \tag{8-32}$$

so that the heat capacity is a universal function of θ_E/T. Once a value of θ_E has been chosen for each solid, to give the best fit to Eq. 8-32, the heat capacity of all monatomic solids should fall on the same curve when plotted against T/θ_E. Figure 8-3 shows that the Einstein model can be used to account for the decrease in $\mathscr{C}_v$ with decreasing temperature, but it also deviates significantly from the experimental results at very low temperature.

The limiting factor in Einstein's treatment is the assumption that all the oscillators have the same fundamental frequency. Debye[6] and Born and von Kármán[7] developed models that treated a crystalline solid as an elastic

6. P. Debye, *Ann. Physik, 39,* 789 (1912).
7. M. Born and T. von Kármán, *Physik Z., 13,* 297 (1912); *14,* 15, 65 (1913).

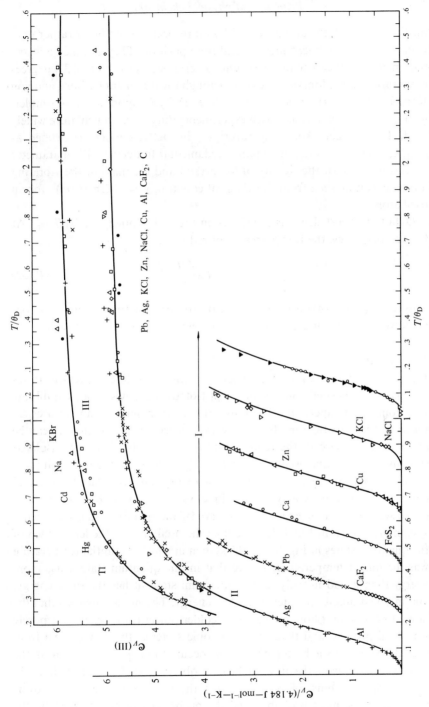

Figure 8-4. A plot of $\mathscr{C}_v$ against T/θ_D for several monatomic solids, showing the fit of the experimental heat capacity data to the Debye model. Curves I and III have displaced coordinates, to display all the data clearly. [E. Schrödinger, *Physik. Z.*, **20**, 450 (1919).]

273

medium in which the atoms vibrated with respect to each other rather than with respect to an independent equilibrium position. They assumed a distribution of frequencies up to a maximum determined by the limit of $3N$ degrees of vibrational freedom of $3N$ atoms. Though the treatment by Born and von Kármán is more rigorous and complete, Debye's approach was simpler, giving very close agreement with experiment; thus, it was much more widely used. The characteristic temperature θ_D in Debye's model is defined as $h v_M/k$, where v_M is the maximum fundamental frequency. The parameter θ_D is a measure of the rigidity of the crystal and the mass of the vibrating atom, as can be seen from the classical equation $v = (1/2\pi)(k/m)^{\frac{1}{2}}$ for an oscillator.

Data for several elements are shown in Fig. 8-4, plotted against T/θ_D. At low temperatures, the Debye treatment reduces to

$$\mathscr{C}_v = \frac{12\pi^4}{5} Lk\left(\frac{T}{\theta_D}\right)^3 \tag{8-33}$$

This result permits extrapolation of heat capacity data to absolute zero from the lowest temperature at which experimental measurement is feasible.

Flow of heat.

It is a familiar observation that when two objects at different temperatures are placed in thermal contact, but kept isolated otherwise, they equilibrate to a common temperature that falls between their initial temperatures. In terms of the statistical model, the average energy of the oscillators can be taken as a measure of the temperature (Eq. 8-25). Consider two isolated systems, one with one oscillator and $E - E_0 = 2h v_0$ and the other with two oscillators and $E - E_0 = h v_0$. When the two systems are placed in thermal contact, they comprise a system of three oscillators, with $E - E_0 = 3h v_0$ (since we assume that energy is conserved); the quantum states of such a system are shown in Fig. 8-1. If we take the oscillator on the left in each of the quantum states in Fig. 8-1 as the system in which the initial temperature was the higher temperature, we see that in one quantum state it has more energy than it had initially, in two quantum states it has the same energy, and in seven quantum states it has less energy. The model suggests that the flow of heat from a hotter object to a colder object is the result of a purely statistical process, and that, in microscopic systems, the flow of heat from a colder object to a hotter object can occur, though not as often as the reverse. The reason the flow of heat is observed to be unidirectional in macroscopic systems, of course, is that the most probable state is overwhelmingly the most probable, and the probability of heat transfer in the opposite direction is negligible.

8-4. The second law of thermodynamics

The unidirectional nature of change in the universe, one example of which is the flow of heat from a hotter system to a colder system, is expressed in the *second law of thermodynamics*. Clausius stated the second law, in 1850, in the form "the entropy of the universe tends to a maximum". From the point of view of classical thermodynamics, the entropy of a system is a function for which the changes summarize concisely the possible, complicated ways in which the variables that describe a system (temperature, pressure, composition, etc.) can change in natural processes. By 1900, Boltzmann and Planck had developed a statistical definition of the entropy that related the abstract thermodynamic concept to a molecular model. The Boltzmann-Planck definition is

$$S = k \ln \Omega \tag{8-34}$$

where S is the entropy of a system, k is the Boltzmann constant, and Ω is the number of quantum states consistent with the macroscopic state of the system. Since, for molecular systems, $\ln \Omega$ for the most probable distribution is not significantly different from $\ln \Omega_{tot}$, we shall use the most probable distribution in all computations.

Entropy of crystals.

As we have seen, the energy and the partition function of a system of oscillators increase continuously with increasing temperature. The shape of the most probable distribution also changes with temperature. The fraction of oscillators in the lowest level is

$$\frac{N_0}{N} = \frac{\exp\left[-(\varepsilon_0 - \varepsilon_0)/kT\right]}{z}$$

$$= \frac{1}{z}$$

Thus, from Eq. 8-24,

$$\frac{N_0}{N} = 1 - \exp\left(\frac{-hv_0}{kT}\right) \tag{8-35}$$

Exercise 8-6.   Verify Eq. 8-35.

The fraction of oscillators in the lowest level is 1 at $T = 0$ K, and it decreases continuously to a value of (the number of accessible quantum levels)$^{-1}$ as the temperature increases. The fraction of oscillators in the jth level ($j \neq 0$) is

$$\frac{N_j}{N} = \frac{\exp\left[-(\varepsilon_j - \varepsilon_0)/kT\right]}{z}$$

$$= \exp\left(\frac{-jhv_0}{kT}\right)\left[1 - \exp\left(\frac{-hv_0}{kT}\right)\right]$$

and increases to a maximum value as the temperature increases. Beyond the temperature at which maximum occupancy is attained, the fraction of oscillators in the jth level decreases with increasing temperature to a limiting value of $1/z$ as T approaches ∞. That is, the oscillators become evenly distributed as T approaches ∞.

Exercise 8-7. Find an expression for the ratio hv_0/kT for which N_j/N is a maximum.

As the temperature increases and the distribution of oscillators among energy levels equalizes, the number of quantum states also increases. (See the results of Exercise 8-1.) This increase in the number of quantum states consistent with a given macroscopic state has been characterized as an increase in "disorder" or a decrease in the "information" about the microscopic state of the system. Thus, the Boltzmann-Planck definition of entropy relates the abstract thermodynamic concept to the "disorder" of a molecular system or the "information" about the system.

According to Eq. 8-6 (applying Stirling's approximation), we obtain

$$\ln \Omega = \ln N! - \sum_{j=0}^{\infty} (N_j \ln N_j - N_j)$$

$$= N \ln N - N - \sum_{j=0}^{\infty} (N_j \ln N_j - N_j)$$

$$= N \ln N - \sum_{j=0}^{\infty} (N_j \ln N_j) \tag{8-36}$$

for any distribution. Substituting for N_j of the most probable distribution from Eq. 8-23 in Eq. 8-36, we obtain

$$\ln \Omega = N \ln N - \sum_{j=0}^{\infty} \left(\frac{N}{z} \exp\left[\frac{-(\varepsilon_j - \varepsilon_0)}{kT}\right]\right) \ln\left\{\frac{N}{z} \exp\left[\frac{-(\varepsilon_j - \varepsilon_0)}{kT}\right]\right\}$$

$$= N \ln z + \frac{E - E_0}{kT} \tag{8-37}$$

Exercise 8-8. Verify Eq. 8-37.

Thus, the entropy of a system of oscillators is

$$S = k N \ln z + \frac{E - E_0}{T} \tag{8-38}$$

At $T = 0$ K, all the oscillators are in the ground state, $\Omega = 1$ (if the ground state is not degenerate), and $S = 0$. These relations led to Planck's formulation of the *third law of thermodynamics* in the form "as the temperature diminishes indefinitely, the entropy of a chemically homogeneous body of finite density approaches indefinitely near to the value zero".[8] Lewis and Gibson[9] modified this statement to limit its application to "perfect crystalline substances", since a solid solution would still have a finite entropy of mixing (see below), and one would expect a liquid to be in a disordered state at $T = 0$ K.

If an amount of heat Q is absorbed by a crystal at constant volume, so that no mechanical work is done, the heat absorbed is equal to the increase in energy of the system. As the energy increases, the temperature increases, and the amount of heat absorbed per unit temperature interval is just the heat capacity. Thus,

$$\left(\frac{\partial Q}{\partial T}\right)_V = \left[\frac{\partial (E - E_0)}{\partial T}\right]_V = C_V \tag{8-39}$$

From Eq. 8-38, the corresponding derivative of the entropy is

$$\left(\frac{\partial S}{\partial T}\right)_V = kN\left(\frac{\partial \ln z}{\partial T}\right)_V - \frac{E - E_0}{T^2} + \frac{1}{T}\left[\frac{\partial (E - E_0)}{\partial T}\right]_V \tag{8-40}$$

From Eq. 8-28, we see that the first term on the right in Eq. 8-40 is equal to $(1/T^2)(E - E_0)$, so that

$$\left(\frac{\partial S}{\partial T}\right)_V = \frac{1}{T}\left[\frac{\partial (E - E_0)}{\partial T}\right]_V \tag{8-41}$$

$$= \frac{C_V}{T} \tag{8-42}$$

Since C_V and T are both positive, the entropy of a crystal is a monotone increasing function of the temperature, and the entropy at any temperature is given by

$$S - S_0 = \int_0^T \left(\frac{\partial S}{\partial T}\right)_V dT$$

$$= \int_0^T \frac{C_V}{T} dT \tag{8-43}$$

8. M. Planck, *Treatise on Thermodynamics*, Dover Publications, Inc., New York, N.D., p. 274.
9. G. N. Lewis and G. E. Gibson, *J. Am. Chem. Soc.*, 42, 1529 (1920); G. N. Lewis and M. Randall, *Thermodynamics*, McGraw-Hill Book Co., New York, 1923, p. 448. Helium, the only substance that is liquid at 0 K, has a standard entropy of zero at 0 K (see Fig. 15-12).

Since heat capacity measurements cannot be made at a temperature close to $T = 0$ K, the Debye model (Eq. 8-33) is used to extrapolate from the lowest temperature at which C_V can be measured to O K. Data on the heat capacity as a function of the temperature, treated according to Eq. 8-43, lead to the calculation of the entropy of crystals even when the simple Einstein and Debye models are not applicable.

One can also conclude from Eq. 8-41 that

$$\left[\frac{\partial S}{\partial (E - E_0)} \right]_V = \frac{\left(\dfrac{\partial S}{\partial T} \right)_V}{\left[\dfrac{\partial (E - E_0)}{\partial T} \right]_V} = \frac{1}{T} \tag{8-44}$$

or

$$(\delta S)_V = \frac{[\delta (E - E_0)]_V}{T} \tag{8-45}$$

If a quantity of heat $-\delta Q = \delta Q_A = \delta (E_A - E_{A,0})$ is lost from a set of oscillators A at temperature T_A and transferred to a set of oscillators B at temperature T_B, so that $\delta Q_B = \delta Q = \delta (E_B - E_{B,0})$, then

$$\delta S_A = \frac{-\delta Q}{T_A}$$

$$\delta S_B = \frac{\delta Q}{T_B}$$

and

$$\delta S = \delta S_A + \delta S_B$$

$$= \delta Q \left[\frac{1}{T_A} + \frac{1}{T_B} \right] \tag{8-46}$$

Thus, the entropy change δS is positive only if $T_A > T_B$. The statistical definitions of entropy and energy lead us to the conclusion that a given change in energy results in a larger change of entropy at low temperature than at high temperature and, thus, relate the Clausius statement of the second law to the spontaneous direction of the flow of heat.

Entropy Change on Mixing

The occurrence of spontaneous mixing at a constant temperature is common, spontaneous unmixing is not. A consideration of a statistical model of a mixing process can bring out the relation of this observation to the second law of thermodynamics. If a set of N_A oscillators A and a set of N_B oscillators B are kept separate, there is only one distinguishable spatial configuration of the combined set of oscillators. (Since the temperature is fixed, and since we assume that the fundamental frequencies of the oscillators are not affected by the chemical identity of their neighbors, the mixing process does not change the *thermal entropy* of the system, and we do not need to consider the many *thermal* quantum states of the system.) Any interchange of identical oscillators among lattice sites in the crystal does not constitute a distinguishable configuration. When the restraint that keeps them separate is removed and the oscillators are arranged randomly on the lattice, there are many distinguishable configurations such that

$$\Omega = \frac{(N_A + N_B)!}{(N_A!)(N_B!)} \tag{8-47}$$

which is the number of distinguishable arrangements of N_A objects of one kind and N_B objects of a second kind on $N_A + N_B$ sites. Applying Eq. 8-34, and using the Stirling approximation, we obtain for the configurational entropy of the randomly mixed state

$$\begin{aligned} S_C &= -kN_A \ln \frac{N_A}{N_A + N_B} - kN_B \ln \frac{N_B}{N_A + N_B} \\ &= -kN_A \ln X_A - kN_B \ln X_B \end{aligned} \tag{8-48}$$

where X_A and X_B are the mole fractions of A and B. Statistically, then, the mixed state has a higher entropy because it has many more arrangements, or more disorder, and the observer has less information about the positions of the oscillators.

Exercise 8-9. Verify Eq. 8-48, starting from Eq. 8-47.

The existence of an entropy of mixing led Lewis and Gibson to suggest that solutions should not be included in the statement of the third law. Even in the case of several pure substances, the possibility of alternative arrangements in the crystalline lattice can lead to a residual entropy at $0\,K$. Carbon monoxide molecules, for example, can be frozen into a crystalline lattice in either of two positions because the C and O atoms are so similar in size and because the molecule is very slightly polar. As the crystal is cooled further, not enough thermal energy is available to "anneal" the crystal to its single most stable state. If we assume that both arrangements are equally probable, the entropy of a mole of CO molecules at $0\,K$ should be

$$S_0 = -\left(\frac{kL}{2}\right) \ln (0.5) - \left(\frac{kL}{2}\right) \ln (0.5)$$

$$= R \ln 2$$

$$= (8.31 \text{ J-K}^{-1})(0.693)$$

$$= 5.76 \text{ J-K}^{-1}$$

The entropy of CO calculated from heat capacity measurements leads to a value of S_0 equal to 4.6 J-K^{-1}, which tells us that the crystal is not in a completely random state.

8-5. Macroscopic properties of gases and the partition function

The statistical treatment of gases differs from that of crystals in two major respects. Since gas molecules move freely and randomly in a container, individual gas molecules cannot be identified as the location of oscillators in a crystal can be. Thus, the number of quantum states of the system must be calculated on the basis of the indistinguishability of molecules. Also, the number of energy levels and molecular quantum states accessible to gas molecules is much greater than the number accessible to oscillators in a crystal. According to Eq. 5-13, the energy of a particle in an infinite, three-dimensional potential well is

$$\varepsilon - \varepsilon_0 = \frac{(n_x^2 + n_y^2 + n_z^2)h^2}{8 m a^2} \tag{8-49}$$

where n_x, n_y, and n_z are the translational quantum numbers, and a is the length of one side of a cubic container. It is clear that the energy level spacing is inversely related to the size of the container, so that the energy levels for gas molecules in a container that is very large compared to the size of the molecule are much more closely spaced than the levels of an oscillator confined to a lattice site that is not much larger than the molecule. In contrast to the distribution of oscillators given in Table 8-2, where the number of oscillators in every level that is significantly populated is very large, the number of levels accessible to gas molecules is larger, by many orders of magnitude, than the number of molecules, and each level corresponds to a number of molecular quantum states.

We can obtain the number of molecular quantum states corresponding to energies between zero and any value $(\varepsilon - \varepsilon_0)$ from a consideration of a Cartesian coordinate system in which the three axes represent the quantum numbers n_x, n_y, and n_z.[10] Each point for which the quantum numbers have positive integral values represents a different quantum state, and each point corresponds to a unit cube in one octant of the coordinate system. All states of a given energy $(\varepsilon - \varepsilon_0)$ correspond to points on a sphere of radius r such that

$$r = (n_x^2 + n_y^2 + n_z^2)^{\frac{1}{2}} = \left[\frac{8ma^2(\varepsilon - \varepsilon_0)}{h^2}\right]^{\frac{1}{2}} \tag{8-50}$$

At sufficiently high values of energy, ω_ε, the number of molecular quantum states with energy less than $(\varepsilon - \varepsilon_0)$, is equal to the volume of the octant enclosed by the sphere of radius r and the three coordinate planes, since each point corresponds to a unit cube. Thus,

$$\omega_\varepsilon = \frac{1}{8}\left(\frac{4}{3}\pi r^3\right)$$

$$= \frac{\pi}{6}\left[\frac{8ma^2(\varepsilon - \varepsilon_0)}{h^2}\right]^{\frac{3}{2}} \tag{8-51}$$

Similarly, the number of molecular states with energy between $(\varepsilon - \varepsilon_0)$ and $(\varepsilon - \varepsilon_0) + \delta(\varepsilon - \varepsilon_0)$ is

$$\delta\omega_\varepsilon = \frac{\pi}{4}\left(\frac{8ma^2}{h^2}\right)^{\frac{3}{2}}(\varepsilon - \varepsilon_0)^{\frac{1}{2}}\,\delta(\varepsilon - \varepsilon_0) \tag{8-52}$$

Exercise 8-10. For a gas of molecular mass equal to 20 atomic mass units and enclosed in a cubic container, for which $a = 10$ cm, calculate the value of ω_ε for a value of $(\varepsilon - \varepsilon_0)$ equal to the thermal energy of the molecules at 300 K; calculate the number of molecules in the container at a pressure of 1×10^5 Pa.

Since the number of molecular quantum states is much larger than the number of molecules, the average population of each state will be less than one. Thus, we cannot use Stirling's approximation to evaluate the number of quantum states of the system when we consider individual molecular states.

Exercise 8-11. Calculate the energy difference between the state with $n_x = n_y = n_z = 10$ and the state with $n_x = n_y = 10$ and $n_z = 11$ for the gas in Exercise 8-10.

10. R. C. Tolman, *The Principles of Statistical Mechanics*, Oxford University Press, London, 1938, p. 290.

The molecular quantum states are so dense on an energy scale that one can find a very large number of states within a small enough interval about any energy $(\varepsilon - \varepsilon_0)$ so that they are all essentially at the same energy. Let g_j be the number of molecular quantum states at energy $(\varepsilon_j - \varepsilon_0)$, and let N_j be the population of that group of states in a particular distribution. The number of distinguishable quantum states of a group of N_j indistinguishable molecules among g_j distinguishable molecular states, with no limitations on the population of any molecular state, is

$$\Omega_j = \frac{(N_j + g_j - 1)!}{(N_j!)(g_j - 1)!} \tag{8-53}$$

$$\simeq \frac{(N_j + g_j)!}{N_j! g_j!} \tag{8-54}$$

since N_j and g_j are very large compared to 1. This is the number of arrangements of N_j indistinguishable objects into groups by means of $(g_j - 1)$ partitions.

Exercise 8-12. Verify Eq. 8-53 by enumerating the possible arrangements of three indistinguishable particles among three states. Hint: See Eq. 8-3 and the discussion preceding it.

The total number of quantum states of the distribution is

$$\Omega = \prod_{j=0}^{\infty} (\Omega_j) \tag{8-55}$$

where the symbol $\prod$ indicates a product of all the Ω_j, since the number of arrangements at each $(\varepsilon_j - \varepsilon_0)$ is independent of the number of arrangements at other energies. Thus,

$$\ln \Omega = \sum_{j=0}^{\infty} \ln \Omega_j$$

$$= \sum_{j=0}^{\infty} \ln \frac{(N_j + g_j)!}{N_j! g_j!} \tag{8-56}$$

By application of Stirling's approximation, Eq. 8-56 reduces to

$$\ln \Omega = \sum_{j=0}^{\infty} N_j \ln (N_j + g_j) + \sum_{j=0}^{\infty} g_j \ln (N_j + g_j)$$

$$- \sum_{j=0}^{\infty} N_j \ln N_j - \sum_{j=0}^{\infty} g_j \ln g_j \tag{8-57}$$

$$= \sum_{j=0}^{\infty} N_j \ln \left(1 + \frac{g_j}{N_j} \right) + \sum_{j=0}^{\infty} g_j \ln \left(1 + \frac{N_j}{g_j} \right) \tag{8-58}$$

Exercise 8-13. Derive Eq. 8-57 from Eq. 8-56.

The most probable distribution here, as for the crystal, is obtained by the method of undetermined multipliers, with the conditions that the total energy and the total number of molecules are constant. The result is

$$1 + \frac{g_j}{N_j} = \exp \mu \exp [\beta(\varepsilon_j - \varepsilon_0)] \qquad (8\text{-}59)$$

Equation 8-59 represents a *Bose-Einstein distribution*, in which more than one particle can be described by the same molecular quantum state.

Exercise 8-14. Verify Eq. 8-59. Hint: Use Eq. 8-57 to find $\partial \ln \Omega / \partial N_j$.

Since, at the temperatures and pressures at which a gas behaves ideally, $g_j/N_j \gg 1$, Eq. 8-59 can be approximated for the case of an ideal gas as

$$\frac{g_j}{N_j} = \exp \mu \exp [\beta(\varepsilon_j - \varepsilon_0)]$$

or

$$N_j = g_j \exp (-\mu) \exp [-\beta(\varepsilon_j - \varepsilon_0)] \qquad (8\text{-}60)$$

Similarly, Eq. 8-58 can be approximated by

$$\ln \Omega = \sum_{j=0}^{\infty} N_j \ln \frac{g_j}{N_j} + \sum_{j=0}^{\infty} N_j$$

$$= N + \sum_{j=0}^{\infty} N_j \ln \frac{g_j}{N_j} \qquad (8\text{-}61)$$

since $g_j/N_j \gg 1$ and $\ln (1 + N_j/g_j) \simeq (N_j/g_j)$ for $N_j/g_j \ll 1$.

Exercise 8-15. Verify Eq. 8-61. Hint: Expand $\ln [1 + N_j/g_j)]$ in a Taylor's series about zero.

The total number of molecules is

$$N = \sum_{j=0}^{\infty} N_j$$
$$= \exp (-\mu) \sum_{j=0}^{\infty} g_j \exp [-\beta(\varepsilon_j - \varepsilon_0)] \qquad (8\text{-}62)$$

We obtain the *Boltzmann approximation* to the Bose-Einstein distribution from the division of Eq. 8-60 by Eq. 8-62 as

$$\frac{N_j}{N} = \frac{g_j \exp\left[-\beta(\varepsilon_j - \varepsilon_0)\right]}{\sum\limits_{j=0}^{\infty} g_j \exp\left[-\beta(\varepsilon_j - \varepsilon_0)\right]} \tag{8-63}$$

or

$$N_j = \frac{N}{z} g_j \exp\left[-\beta(\varepsilon_j - \varepsilon_0)\right] \tag{8-64}$$

where z is the *molecular partition function*. This is to be distinguished from the ensemble partition function, obtained from the ensemble average.

The thermal energy of monatomic gases

The parameter β can be evaluated by calculating the statistical value for the thermal energy of a monatomic gas and equating it to the classical equipartition value.

$$E - E_0 = \sum_{j=0}^{\infty} N_j (\varepsilon_j - \varepsilon_0)$$

$$= \frac{N \sum\limits_{j=0}^{\infty} g_j (\varepsilon_j - \varepsilon_0) \exp\left[-\beta(\varepsilon_j - \varepsilon_0)\right]}{\sum\limits_{j=0}^{\infty} g_j \exp\left[-\beta(\varepsilon_j - \varepsilon_0)\right]} \tag{8-65}$$

Since the energy levels are very closely spaced, the summations in Eq. 8-65 can be replaced by integrals to a very good approximation. In these integrals, g_j is taken equal to $\delta\omega_\varepsilon$ of Eq. 8-52, since the only energy to be considered is translational energy. Thus,

$$E - E_0 = \frac{N \int_0^\infty \frac{\pi}{4}\left(\frac{8\,ma^2}{h^2}\right)^{\frac{3}{2}} (\varepsilon - \varepsilon_0)^{\frac{3}{2}} \exp\left[-\beta(\varepsilon - \varepsilon_0)\right] d(\varepsilon - \varepsilon_0)}{\int_0^\infty \frac{\pi}{4}\left(\frac{8\,ma^2}{h^2}\right)^{\frac{3}{2}} (\varepsilon - \varepsilon_0)^{\frac{1}{2}} \exp\left[-\beta(\varepsilon - \varepsilon_0)\right] d(\varepsilon - \varepsilon_0)}$$

$$= \frac{N \int_0^\infty (\varepsilon - \varepsilon_0)^{\frac{3}{2}} \exp\left[-\beta(\varepsilon - \varepsilon_0)\right] d(\varepsilon - \varepsilon_0)}{\int_0^\infty (\varepsilon - \varepsilon_0)^{\frac{1}{2}} \exp\left[-\beta(\varepsilon - \varepsilon_0)\right] d(\varepsilon - \varepsilon_0)}$$

From integral tables, we obtain the results

$$E - E_0 = \frac{N \left[\frac{3}{2\beta}\left(\frac{1}{2\beta}\right)\left(\frac{\Pi}{\beta}\right)^{\frac{1}{2}}\right]}{\left[\left(\frac{1}{2\beta}\right)\left(\frac{\Pi}{\beta}\right)^{\frac{1}{2}}\right]}$$

$$= N \frac{3}{2\beta} \tag{8-66}$$

In Sec. 2-3, it was shown that $E - E_0$ for a monatomic ideal gas is $(3/2) NkT$, so that,

$$\beta = \frac{1}{kT} \tag{8-67}$$

(Henceforth, we shall take $\beta = 1/kT$ without proving it in each specific case.) Thus, we can write for the thermal energy of an ideal gas,

$$E - E_0 = \frac{N}{z} \sum_{j=0}^{\infty} g_j (\varepsilon_j - \varepsilon_0) \exp \left[- \frac{(\varepsilon_j - \varepsilon_0)}{kT} \right] \tag{8-68}$$

where

$$z = \sum_{j=0}^{\infty} g_j \exp \left[- \frac{(\varepsilon_j - \varepsilon_0)}{kT} \right] \tag{8-69}$$

As for the oscillator (Eq. 8-28)

$$E - E_0 = NkT^2 \left(\frac{\partial \ln z}{\partial T} \right)_V \tag{8-70}$$

Exercise 8-16. Verify Eq. 8-70.

Evaluation of the partition function
We shall evaluate the partition function of an ideal gas for the simple model in which the energy of a molecule is the sum of the electronic, vibrational, rotational, and translational energies, with each type of energy independent of the others. Thus,

$$\varepsilon_j - \varepsilon_0 = (\varepsilon_{n, \text{elec}} - \varepsilon_{0, \text{elec}}) + (\varepsilon_{k, \text{vib}} - \varepsilon_{0, \text{vib}})$$
$$+ (\varepsilon_{l, \text{rot}} - \varepsilon_{0, \text{rot}}) + (\varepsilon_{m, \text{trans}} - \varepsilon_{0, \text{trans}}) \tag{8-71}$$

and

$$\exp \left[- \frac{(\varepsilon_j - \varepsilon_0)}{kT} \right]$$

$$= \exp \left[- \frac{(\varepsilon_{n, \text{elec}} - \varepsilon_{0, \text{elec}}) + (\varepsilon_{k, \text{vib}} - \varepsilon_{0, \text{vib}}) + (\varepsilon_{l, \text{rot}} - \varepsilon_{0\, \text{rot}}) + (\varepsilon_{m, \text{trans}} - \varepsilon_{0, \text{trans}})}{kT} \right]$$

$$= \left\{ \exp \left[- \frac{(\varepsilon_{n, \text{elec}} - \varepsilon_{0, \text{elec}})}{kT} \right] \right\} \left\{ \exp \left[- \frac{(\varepsilon_{k, \text{vib}} - \varepsilon_{0, \text{vib}})}{kT} \right] \right\} \left\{ \exp \left[- \frac{(\varepsilon_{l, \text{rot}} - \varepsilon_{0, \text{rot}})}{kT} \right] \right\}$$

$$\left\{ \exp \left[- \frac{(\varepsilon_{m, \text{trans}} - \varepsilon_{0, \text{trans}})}{kT} \right] \right\}$$

Similarly,

$$g_j = g_{n, \text{elec}}\, g_{k, \text{vib}}\, g_{l, \text{rot}}\, g_{m, \text{trans}}$$

The partition function is, then,

$$z = \sum_{j=0}^{\infty} g_j \exp\left[-\frac{(\varepsilon_j - \varepsilon_0)}{KT}\right]$$

$$= \sum_{n=0}^{\infty}\sum_{k=0}^{\infty}\sum_{l=0}^{\infty}\sum_{m=0}^{\infty}\left\{g_{n,\,elec}\exp\left[-\frac{(\varepsilon_{n,\,elec} - \varepsilon_{0,elec})}{kT}\right]\right\}$$

$$\left\{g_{k,\,vib}\exp\left[-\frac{(\varepsilon_{k,\,vib} - \varepsilon_{0,\,vib})}{kT}\right]\right\}$$

$$\left\{g_{l,\,rot}\exp\left[-\frac{(\varepsilon_{l,\,rot} - \varepsilon_{0,\,rot})}{kT}\right]\right\}$$

$$\left\{g_{m,\,trans}\exp\left[-\frac{(\varepsilon_{m,\,trans} - \varepsilon_{0,\,trans})}{kT}\right]\right\}$$

$$= \left\{\sum_{n=0}^{\infty}g_{n,\,elec}\exp\left[-\frac{(\varepsilon_{n,\,elec} - \varepsilon_{0,elec})}{kT}\right]\right\}$$

$$\left\{\sum_{k=0}^{\infty}g_{k,\,vib}\exp\left[-\frac{(\varepsilon_{k,\,vib} - \varepsilon_{0,\,vib})}{kT}\right]\right\}$$

$$\left\{\sum_{l=0}^{\infty}g_{l,\,rot}\exp\left[-\frac{(\varepsilon_{l,\,rot} - \varepsilon_{0,\,rot})}{kT}\right]\right\}$$

$$\left\{\sum_{m=0}^{\infty}g_{m,\,trans}\exp\left[-\frac{(\varepsilon_{m,\,trans} - \varepsilon_{0,\,trans})}{kT}\right]\right\}$$

$$= z_{elec}\,z_{vib}\,z_{rot}\,z_{trans} \tag{8-72}$$

Exercise 8-17. Show that $\sum_{i=0}^{2}\sum_{j=0}^{2} a_i b_j$ is equal to $\left(\sum_{i=0}^{2} a_i\right)\left(\sum_{j=0}^{2} b_j\right)$

Thus, whenever the energies are additive, the partition function can be expressed as a product.

The electronic partition function

Because electronic energy levels are so widely spaced, the electronic partition function

$$z_{elec} = \sum_{n=0}^{\infty} g_{n,\,elec}\,\exp\left[-\frac{(\varepsilon_{n,\,elec} - \varepsilon_{0,\,elec})}{kT}\right] \tag{8-73}$$

has only one term of appreciable magnitude at ordinary temperatures and is equal to 1 unless the ground electronic state is degenerate, as in a free radical like NO.

Exercise 8-18. Carbon monoxide exhibits a strong absorption line at 1546×10^{-10}m as a result of an electronic transition. Calculate the fraction of CO molecules in the excited state for which this line represents a transition from the ground state, at 300 K. At what temperature would $(\varepsilon - \varepsilon_0)$ for this excited state become equal to kT? What is the value of z_{elec} at 300 K?

The vibrational partition function

The form of the vibrational partition function for a diatomic molecule is the same as that for a one-dimensional oscillator. That is (see Eq. 8-24),

$$z_{\text{vib}} = \sum_{v=0}^{\infty} g_{\text{vib}} \exp\left(\frac{-vhv_0}{kT}\right)$$

$$= \left[1 - \exp\left(\frac{-hv_0}{kT}\right)\right]^{-1} \tag{8-74}$$

since g_{vib} is equal to 1 for one vibrational degree of freedom. At sufficiently high temperatures, the exponential in Eq. 8-74 can be expanded in a Taylor's series about zero, and higher terms can be neglected. The high temperature-limiting form of z_{vib}, corresponding to classical behavior, is

$$z_{\text{vib}} = \frac{kT}{hv_0} \tag{8-75}$$

Exercise 8-19. Verify Eq. 8-75.

Exercise 8-20. Carbon monoxide has a strong absorption line in the infrared at a wavelength of 4.65 μm. Calculate the vibrational partition function for CO at 300 K. Compute the temperature at which the vibrational motion of gaseous CO would be classical, i.e., the temperature at which $(kT/hv_0) > 10$.

For a polyatomic molecule, there are a number of vibrational degrees of freedom, as discussed in Sec. 6-2. The vibrational partition function is a product over the vibrational degrees of freedom of terms such as that in Eq. 8-74. In principle, the value of v_0 for one vibrational degree of freedom can be calculated by solution of the Schrödinger equation for the molecule, but, in practice, it is obtained from observation of molecular spectra.

The rotational partition function

The rotational partition function for a diatomic molecule is (see Eq. 6-9)

$$z_{\text{rot}} = \sum_{J=0}^{\infty} g_J \exp\left[-\frac{(\varepsilon_J - \varepsilon_0)}{kT}\right]$$

$$= \sum_{J=0}^{\infty} (2J + 1) \exp \left[-J(J + 1) \frac{h^2}{8\pi^2 IkT} \right] \qquad (8\text{-}76)$$

where $2J + 1$ represents the number of possible orientations of the angular momentum vector with respect to an external reference axis for each value of the rotational quantum number J. Data from which the rotational partition function is calculated are obtained from rotational spectra or the rotational fine structure of vibrational or electronic spectra in the form of the rotational constant B, where

$$B = \frac{h}{8\pi^2 I} \qquad (8\text{-}77)$$

and the rotational spacing is (see Secs. 6-1 and 6-2)

$$\Delta v = 2B \qquad (8\text{-}78)$$

In our discussion of vibration-rotation spectra in Sec. 6-2, we did not consider the relative intensity of the lines of the rotational fine structure shown in Fig. 6-6. Because of the independence of the separate partition functions one should be able to describe the distribution of molecules among *rotational* levels as

$$\frac{N_{J,\,rot}}{N_{0,\,rot}} = g_J \exp \left[-\frac{(\varepsilon_{J,\,rot} - \varepsilon_{0,\,rot})}{kT} \right]$$

$$= (2J + 1) \exp \left[-J(J + 1) \frac{h^2}{8\pi^2 IkT} \right] \qquad (8\text{-}79)$$

Since absorbance is proportional to the number of absorbing molecules, Eq. 8-79 should also approximate the relative absorbances of lines corresponding to transitions *from* levels with quantum number J.

There is no expression in closed form for the summation in Eq. 8-76, but the summation can be carried out directly at low temperatures. At high temperatures, the energy levels are sufficiently close together, compared to the thermal energy, that the sum can be replaced by an integral

$$z_{rot} = \int_0^{\infty} \exp \left[-J(J + 1) \frac{h^2}{8\pi^2 IkT} \right] (2J + 1) \, dJ \qquad (8\text{-}80)$$

Since $(2J + 1)\,dJ = d[J(J + 1)]$

and

$$\int_0^{\infty} \exp(-ax) \, dx = \frac{1}{a}$$

$$z_{rot} = \frac{8\pi^2 I k T}{h^2} = \frac{k T}{B h} \tag{8-81}$$

Equation 8-81 can be applied to rotational degrees of freedom in which identical configurations of the molecule occur only after a rotation of 2π. If the molecule has elements of symmetry, such that identical configurations occur after a rotation of $2\pi/\sigma$,

$$z_{rot} = \frac{8\pi^2 I k T}{\sigma h^2} \tag{8-82}$$

For a homonuclear diatomic molecule, $\sigma = 2$. We will not consider the effects of symmetry on the low-temperature rotational partition function.

Exercise 8-21. Carbon monoxide exhibits a rotational spectrum in the far infrared with lines essentially equally spaced, and $\Delta v = 1.15 \times 10^{11} s^{-1}$. Calculate the temperature at which $k T = 10 h^2 / 8\pi^2 I$ and the value of z_{rot} at 300 K from Eq. 8-81.

The translational partition function
The translational partition function of an ideal gas is

$$z_{trans} = \sum_{j=0}^{\infty} g_j \exp\left[-\frac{(\varepsilon_{j,\, trans} - \varepsilon_{0,\, trans})}{k T} \right] \tag{8-83}$$

At all temperatures, translational energy level spacing is small enough, compared to the thermal energy, that the summation in Eq. 8-83 can be replaced by an integral. As before, we shall take g_j equal to $\delta\omega_\varepsilon$ of Eq. 8-52. The expression for z_{trans} is then

$$
\begin{aligned}
z_{trans} &= \int_0^\infty \frac{\pi}{4} \left(\frac{8 m a^2}{h^2} \right)^{\frac{3}{2}} (\varepsilon - \varepsilon_0)^{\frac{1}{2}} \exp\left[-\frac{(\varepsilon - \varepsilon_0)}{k T} \right] d(\varepsilon - \varepsilon_0) \\
&= \frac{\pi}{4} \left(\frac{8 m a^2}{h^2} \right)^{\frac{3}{2}} \int_0^\infty (\varepsilon - \varepsilon_0)^{\frac{1}{2}} \exp\left[-\frac{(\varepsilon - \varepsilon_0)}{k T} \right] d(\varepsilon - \varepsilon_0) \quad (8\text{-}84)
\end{aligned}
$$

From tables of integrals we learn that the integral in Eq. 8-84 is equal to $(k T/2)(\pi k T)^{\frac{1}{2}}$, so that

$$z_{trans} = \left(\frac{2\pi m k T}{h^2} \right)^{\frac{3}{2}} V \tag{8-85}$$

where V, the volume of the system, is equal to a^3.

Exercise 8-22. Calculate the translational partition function for CO at a temperature of 300 K in a volume of 1.00 dm³.

Exercise 8-23. Calculate the temperature at which kT becomes equal to ten times the energy level spacing calculated for translational levels in Exercise 8-11.

The thermal energy of an ideal gas

The thermal energy of an ideal gas is (Eq. 8-70)

$$E - E_0 = NkT^2 \left(\frac{\partial \ln z}{\partial T} \right)$$

Substituting for z in Eq. 8-70 from Eq. 8-72, we obtain

$$E - E_0 = NkT^2 \left[\left(\frac{\partial \ln z_{elec}}{\partial T} \right)_V + \left(\frac{\partial \ln z_{vib}}{\partial T} \right)_V + \left(\frac{\partial \ln z_{rot}}{\partial T} \right)_V + \left(\frac{\partial \ln z_{trans}}{\partial T} \right)_V \right]$$

$$= (E - E_0)_{elec} + (E - E_0)_{vib} + (E - E_0)_{rot} + (E - E_0)_{trans} \quad (8\text{-}86)$$

This equation confirms our calculation to be consistent with the initial approximation that the energy of the molecule is the sum of the electronic, vibrational, rotational, and translational energy, and the energy of the system is then the sum of the energies of the molecules.

Since, at ordinary temperatures, $z_{elec} = 1$, the electronic thermal energy and the electronic heat capacity are equal to zero, in agreement with simple kinetic theory. Since the partition function for a vibrational degree of freedom of a molecule of an ideal gas is of the same form as that for a one-dimensional oscillator in a crystal, the thermal energy of vibration for a vibrational degree of freedom and the vibrational heat capacity are also of the same form (Eqs. 8-29 and 8-30). The high temperature form of z_{vib} (Eq. 8-75) leads to a vibrational thermal energy

$$(E - E_0)_{vib} = NkT \quad (8\text{-}87)$$

in agreement with the equipartition value of RT per mole.

Although there is no expression in closed form for z_{rot}, the temperature derivative is of such a form that the rotational thermal energy and the rotational heat capacity are temperature dependent. The high temperature form of the rotational partition function (Eq. 8-81) leads to a rotational thermal energy

$$(E - E_0)_{rot} = NkT \quad (8\text{-}88)$$

in agreement with the equipartition value for the two rotational degrees of freedom of a diatomic molecule.

Exercise 8-24. Verify Eq. 8-88.

The translational partition function for an ideal gas (Eq. 8-85) leads to a translational thermal energy

$$(E - E_0)_{trans} = \tfrac{3}{2} NkT \tag{8-89}$$

consistent with the equipartition value for three translational degrees of freedom, since translational motion is classical at all temperatures for macroscopic containers. Figure 2-2 shows the experimental values of C_v for several gases as a function of temperature. This temperature dependence can be calculated from the partition function.

Exercise 8-25. Verify Eq. 8-89.

The entropy of an ideal gas
The Boltzmann-Planck definition of entropy (Eq. 8-34), together with the expression for $\ln \Omega$ for an ideal gas (Eq. 8-61), leads to the result

$$S = k\left[N + \sum_{j=0}^{\infty} N_j \ln\left(\frac{g_j}{N_j}\right)\right] \tag{8-90}$$

From Eq. 8-64, with $\beta = \dfrac{1}{kT}$,

$$\frac{g_j}{N_j} = \frac{z}{N} \exp\left[\frac{(\varepsilon_j - \varepsilon_0)}{kT}\right]$$

and

$$\ln\left(\frac{g_j}{N_j}\right) = \ln\left(\frac{z}{N}\right) + \frac{(\varepsilon_j - \varepsilon_0)}{kT} \tag{8-91}$$

Substituting from Eq. 8-91 in Eq. 8-90, we obtain

$$S = k\left[N + \sum_{j=0}^{\infty} N_j \ln\left(\frac{z}{N}\right) + \sum_{j=0}^{\infty} N_j \frac{(\varepsilon_j - \varepsilon_0)}{kT}\right]$$

$$= kN\left[1 + \ln\left(\frac{z}{N}\right)\right] + \frac{E - E_0}{T} \tag{8-92}$$

Exercise 8-26. Verify Eq. 8-92.

Because gas molecules must be treated statistically as indistinguishable molecules, whereas oscillators in a crystal can be treated as distinguishable, the expressions for the entropy for the two are different. The difference can be seen clearly if Eq. 8-38 and Eq. 8-92 are rearranged to give

$$S_{crys} = kN \ln z + \frac{E - E_0}{T}$$

$$= k \ln z^N + \frac{E - E_0}{T} \qquad (8\text{-}93)$$

and

$$S_{\text{gas}} = kN \ln z + \frac{E - E_0}{T} - k[N \ln N - N]$$

$$= k \ln \left(\frac{z^N}{N!} \right) + \frac{E - E_0}{T} \qquad (8\text{-}94)$$

Exercise 8-27. Verify Eq. 8-94.

Equation 8-94 can also be written in the form

$$S_{\text{gas}} = kN \ln z + \frac{E - E_0}{T} - k \ln (N!) \qquad (8\text{-}95)$$

We have seen that $(E - E_0)$ can be expressed as a sum of the electronic, vibrational, rotational, and translational contributions to the thermal energy. Similarly, it is clear from Eq. 8-72 that the first term in Eq. 8-95 can be expressed as a similar sum. Since the electronic contribution to the entropy is expected to be zero for a nondegenerate ground state, and since the third term in Eq. 8-95 is applicable to monatomic gases that do not have rotational and vibrational degrees of freedom, this term is assigned to the translational contribution to the entropy.

Thus, the contributions of the various degrees of freedom of a diatomic molecule to the entropy are:

$$S_{\text{elec}} = 0$$

$$S_{\text{vib}} = kN \ln z_{\text{vib}} + \frac{(E - E_0)_{\text{vib}}}{T}$$

$$= kN \ln \left[1 - \exp \left(\frac{-h\nu_0}{kT} \right) \right]^{-1} + N h\nu_0 / T \left[\exp \left(\frac{h\nu_0}{kT} \right) - 1 \right] \quad (8\text{-}96)$$

$$S_{\text{rot}} = kN \ln z_{\text{rot}} + \frac{(E - E_0)_{\text{rot}}}{T} \qquad (8\text{-}97)$$

and

$$S_{\text{trans}} = kN \ln z_{\text{trans}} + \frac{(E - E_0)_{\text{trans}}}{T} - k \ln (N!)$$

$$= kN \ln \left[\left(\frac{2\pi mkt}{h^2} \right)^{\frac{3}{2}} V \right] + \frac{3}{2} Nk - k \ln (N!) \qquad (8\text{-}98)$$

The high temperature values for the vibrational and rotational entropies are:

$$S_{vib} = kN \ln \frac{kT}{hv_0} + Nk \qquad (8\text{-}99)$$

and

$$S_{rot} = kN \ln \frac{8\pi^2 I k T}{h^2} + Nk \qquad (8\text{-}100)$$

A more convenient form of Eq. 8-98 can be obtained by applying Stirling's approximation, and the result is

$$S_{trans} = kN \ln \left[\left(\frac{2\pi mkT}{h^2} \right)^{\frac{3}{2}} \frac{V}{N} \right] + \frac{5}{2} Nk \qquad (8\text{-}101)$$

Exercise 8-28. Calculate the vibrational, rotational, and translational contributions to the entropy of one mole of CO contained in a volume of 22.0 dm^3 at a temperature of 300 K, using Eqs. 8-96, 8-100, and 8-101.

The total entropy of one mole of CO at 298.1 K and 1.01×10^5 Pa pressure, calculated from Eqs. 8-96, 8-100, and 8-101, was found[11] to be 197.9 J-K^{-1}. This value is called the *spectroscopic entropy* because spectroscopic data are used in the calculation. The calorimetric value, obtained by integration of Eq. 8-43, using heat capacity data[12] was found to be 193.3 J-K^{-1}. The discrepancy between these two values led to the conclusion that crystalline CO has a residual molar entropy of 4.6 J-K^{-1} at 0K, as noted in Sec. 8-4.

8-6. Isomerization equilibrium in gases

Consider the isomerization reaction

$$A(g) \rightleftarrows B(g) \qquad (8\text{-}102)$$

The concentration equilibrium constant for such a reaction is

$$K_c = \frac{[B]}{[A]} \qquad (8\text{-}103)$$

11. J. O. Clayton and W. F. Giauque, *J. Am. Chem. Soc.*, **54**, 2610 (1932).
12. *Ibid.*

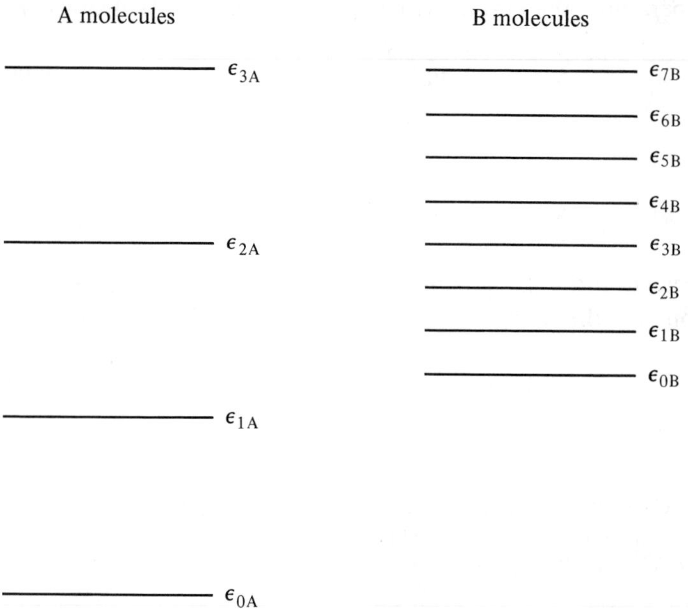

A molecules   B molecules

Figure 8-5. Scheme of the energy levels of the isomeric molecules A and B, showing the differences in ε_0 and in energy level spacing.

where the square brackets indicate concentration in mol-dm^{-3}. Since both species occupy the same volume, one can also write

$$K_c = \frac{n_B}{n_A} = \frac{N_B}{N_A} \tag{8-104}$$

If the chemical reaction in Eq. 8-102 were not possible, the equilibrium state of either A(g) or B(g) would be represented by the most probable distribution of A molecules or B molecules among the corresponding energy levels, which are shown schematically in Fig. 8-5. The two distributions would be

$$N_{j,\,A} = \frac{N_A}{z_A}\, g_{j,\,A} \exp\left[-\frac{(\varepsilon_{j,\,A} - \varepsilon_{0,\,A})}{kT}\right] \tag{8-105}$$

and

$$N_{j,\,B} = \frac{N_B}{z_B}\, g_{j,\,B} \exp\left[-\frac{(\varepsilon_{j,\,B} - \varepsilon_{0,\,B})}{kT}\right] \tag{8-106}$$

In statistical terms, the possibility of the isomerization reaction means that all the energy levels in Fig. 8-5 are accessible to the molecules and that all the quantum states of the system represented by distributions of molecules among the two sets of levels are accessible to the system. We designate as B molecules all those in the right-hand set of levels and as A molecules all those in the left-hand set of levels in any distribution. The equilibrium state of the system in which reaction is possible is represented by the most probable distribution of molecules among both sets of levels, with $\varepsilon_{0, A}$ taken as the zero of energy. Thus, the population of any A level is

$$N_{j, A} = \frac{N_A + N_B}{z_{A+B}} g_{j, A} \exp -\left[\frac{(\varepsilon_{j, A} - \varepsilon_{0, A})}{kT}\right] \qquad (8\text{-}107)$$

and the population of any B level is

$$N_{j, B} = \frac{N_A + N_B}{z_{A+B}} g_{j, B} \exp \left[-\frac{(\varepsilon_{j, B} - \varepsilon_{0, A})}{kT}\right]$$

$$= \frac{N_A + N_B}{z_{A+B}} g_{j, B} \exp \left[-\frac{(\varepsilon_{j, B} - \varepsilon_{0, B})}{kT}\right] \exp \left[-\frac{(\varepsilon_{0, B} - \varepsilon_{0, A})}{kT}\right] \qquad (8\text{-}108)$$

where

$$z_{A+B} = \sum_{j=0}^{\infty} \left\{ g_{j, A} \exp \left[-\frac{(\varepsilon_{j, A} - \varepsilon_{0, A})}{kT}\right] + g_{j, B} \exp \left[-\frac{(\varepsilon_{j, B} - \varepsilon_{0, A})}{kT}\right] \right\} \qquad (8\text{-}109)$$

The number of A molecules at equilibrium is, then,

$$N_A = \sum_{j=0}^{\infty} N_{j, A}$$

$$= \frac{N_A + N_B}{z_{A+B}} \sum_{j=0}^{\infty} g_{j, A} \exp \left[-\frac{(\varepsilon_{j, A} - \varepsilon_{0, A})}{kT}\right]$$

$$= \frac{N_A + N_B}{z_{A+B}} z_A \qquad (8\text{-}110)$$

and the number of B molecules at equilibrium is

$$N_B = \sum_{j=0}^{\infty} N_{j, B}$$

$$= \frac{N_A + N_B}{z_{A+B}} \exp \left[-\frac{(\varepsilon_{0, B} - \varepsilon_{0, A})}{kT}\right] \sum_{j=0}^{\infty} g_{j, B} \exp \left[-\frac{(\varepsilon_{j, B} - \varepsilon_{0, B})}{kT}\right]$$

$$= \frac{N_A + N_B}{z_{A+B}} z_B \exp \left[-\frac{(\varepsilon_{0, B} - \varepsilon_{0, A})}{kT}\right] \qquad (8\text{-}111)$$

where z_B is defined with respect to $\varepsilon_{0, B}$.

We can obtain a statistical expression for the equilibrium constant by substituting from Eqs. 8-110 and 8-111 in Eq. 8-104, so that

$$K_c = \frac{N_B}{N_A} = \frac{z_B}{z_A} \exp\left[-\frac{(\varepsilon_{0,B} - \varepsilon_{0,A})}{kT}\right] \qquad (8\text{-}112)$$

It can be seen from Eq. 8-112 that there are two factors determining the relative number of A molecules and B molecules at equilibrium. The first factor is the ratio of partition functions, which is determined by the relative density of energy levels of the two molecules. The second factor is the difference in energy between the lowest levels of the two molecules.

We have seen that all partition functions decrease in magnitude with decreasing temperature and approach a value of 1 at 0 K. The energy difference $(\varepsilon_{0,B} - \varepsilon_{0,A})$ is temperature independent, so that the exponential factor is dominant at low temperature. As the temperature increases, the exponential factor decreases in magnitude, and the ratio of partition functions becomes the dominant term.

Exercise 8-29. If, in Fig. 8-5, $(\varepsilon_{0,B} - \varepsilon_{0,A})$ is equal to 1.00×10^{-20} J and if the energy levels are equally spaced, with spacings of the A levels equal to 1.00×10^{-20} J and spacings of the B levels equal to 2.00×10^{-21} J, calculate the value of K_c at 100, 500, and 1000 K (assume $g_j = 1$ for all levels).

Since the zero of energy for a molecule has been chosen as the lowest translational, rotational, vibrational, and electronic energy, the ground electronic states of the two molecules determine $\varepsilon_{0,B} - \varepsilon_{0,A}$. The masses of the two isomers are the same, so that their translational energy levels are identical. The differences in energy-level spacing must occur as differences in vibrational and rotational energy-level spacing. Thus, the A levels might represent a more compact, tightly bound isomer with a smaller moment of inertia and greater force constants, whereas the B levels might represent a looser, more extended isomer.

8-7. Phase equilibria

The statistical arguments that enable us to predict the effect of temperature on the chemical equilibrium between two isomers can also lead to an understanding of the effect of temperature on the stability of the solid, liquid, and gaseous states of matter. At low temperatures, the phase at lowest energy is stable regardless of energy-level spacing; at higher temperatures, the phase with the more dense energy-level spacing is stable.

Crystalline solids exist, of course, as a result of strong intermolecular forces and are, therefore, at a lower potential energy than are liquids or gases. The empirical evidence for this statement is in the observation of a *latent heat of sublimation* that is absorbed when a solid is converted to a gas and a *latent heat of fusion* that is absorbed when a solid is converted to a liquid. Similarly, a liquid is at a lower potential energy than a gas, due to intermolecular forces, but the molecules interact less strongly in a liquid than in a solid. A *latent heat of vaporization* is observed when a liquid is converted to a gas.

The energy levels of a crystal are relatively widely spaced, being those of an oscillator bound strongly to its nearest neighbors in the crystalline lattice. The energy levels of a liquid are somewhat more closely spaced than those of a crystal because the molecules, though bound in the liquid, are free to move in a broader and shallower potential well. In the gas phase, in which

Figure 8-6. Scheme of the energy levels of solid, liquid, and gaseous phases of the same substance, showing the differences in ε_0 and in energy level spacing.

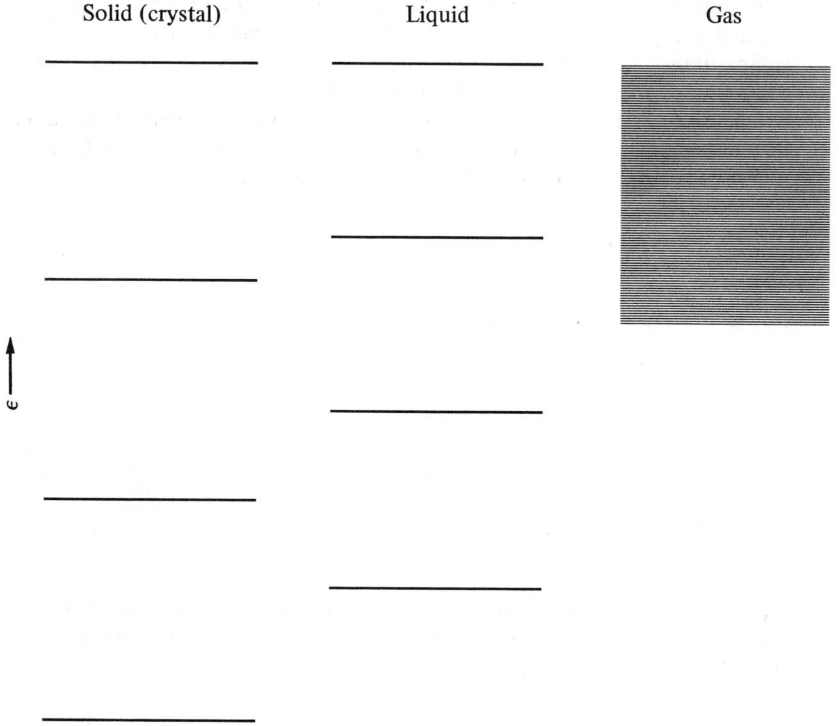

Solid (crystal)   Liquid   Gas

the molecules interact with one another only by collisions, and in which they are free to move through a very large volume, the energy levels are very closely spaced. A schematic representation of energy levels of solid, liquid, and gaseous phases are shown in Fig. 8-6.

On the basis of Fig. 8-6 and the preceding discussion, it is clear that the crystal is the stable phase at lower temperature and that the liquid and gas are the stable phases at higher temperatures. No simple statistical model, however, can yet account for the discontinuity in slope of the isotherm for phase transitions. Model calculations lead to smooth curves analogous to the van der Waals curve shown in Fig. 1-8.[13]

Problems

8-1. Calculate the translational, rotational, and vibrational contributions to the partition function of $H^{35}Cl$ and to the molar thermal energy and molar entropy of $H^{35}Cl$ at 298 K and at 1000 K at 1×10^5 Pa. Use the spectroscopic data given in Chapter 6 (p. 229). Compare your results for the thermal energy with the equipartition value. The entropy at 298 K is 186.80 $J\text{-mol}^{-1}\text{-K}^{-1}$.

8-2. The value of v_0 for copper in the Einstein model is $6.25 \times 10^{12} s^{-1}$. Calculate the molecular partition function for copper at 298 and 1000 K. Calculate the molar thermal energy of copper at 298 and 1000 K. Compare your results with the equipartition value. Calculate the molar entropy of copper at 298 and 1000 K. The molar entropy at 298 K is 33.05 $J\text{-mol}^{-1}\text{-K}^{-1}$.

8-3. Calculate the partition function and the molar entropy for a hypothetical copper monatomic gas at a temperature of 1000 K and a pressure of 1×10^5 Pa. Compare it to the value obtained for the solid at the same temperature.

13. T. L. Hill, *Introduction to Statistical Thermodynamics*, Addison-Wesley, Reading, 1960, Chap. 14; H. E. Stanley, *Introduction to Phase Transitions and Critical Phenomena*, Clarendon Press, Oxford, 1971.

9 · The first law of thermodynamics. Energy and enthalpy

In Chapter 8 a statistical model was developed from which energy and heat capacity for simple crystals and ideal gases could be calculated, using data on molecular structure and molecular spectra. It was clear, however, that this model is not easily applicable to complex crystals and gases or to imperfect gases and liquids—to the first group because the calculations are too complex, to the second because intermolecular forces make it impossible to consider the molecules as independent of one another.

It is useful to have available a method by which all systems can be treated without regard to molecular models. Classical thermodynamics, developed in the 19th century by Carnot, Joule, Mayer, Kelvin, Clausius, Helmholtz, Gibbs, and Planck, provides such a treatment. The thermodynamic description of material systems is entirely in terms of macroscopic properties, but only relationships among macroscopic properties are described, not the values of particular properties. A few thermodynamic laws, generalizations from experience, form the basis for derivations of these relationships in terms of functions of the state of the system. *Energy* and *entropy* are usually chosen as fundamental functions, and other functions can be expressed in terms of them. The resulting relationships cannot provide information about the microscopic structure of matter, but they do serve as criteria of the validity of statistical models like those discussed in Chapter 8. Energy is discussed in this chapter, entropy in Chapter 10.

9-1. Thermodynamic systems and variables

A thermodynamic *system* is any part of the universe that is set apart for study and consideration. Its boundaries may be material, or they may be convenient imaginary surfaces. If matter can pass in and out of the system it is said to be *open*; otherwise, it is *closed*. Material boundaries of a system can be of two kinds. If a system cannot be influenced by its surroundings without a change in its volume, the boundary is said to be *adiabatic* (we shall see that this corresponds to an insulating wall). If a system can be influenced by its surroundings without a change in its volume, the boundary is said to be *thermally conducting*. (This statement must be modified when the system is influenced by force fields—electric, magnetic, gravitational, or centrifugal.)

Of the many variables that can be used to describe thermodynamic systems, two groups are particularly important. One group, the *intensive variables*, do not change in value when two or more identical systems are joined to form a large system. They are characteristic of the state of the system but independent of the quantity of matter. Examples of intensive variables include the temperature T, pressure P, density ρ, viscosity η, electrical conductivity κ, and such quantities as the molar energy $\mathscr{E}$ and molar heat capacity $\mathscr{C}$. The other group, the *extensive variables*, are multiplied by n when n identical systems are joined. Examples are mass m, volume V, energy E, and heat capacity C.

State of the system

If the chemical composition of a system does not change, we can specify the values of *two* intensive variables to fix the values of all other intensive variables and, thus, define the state of the system. A functional relationship between any other intensive variable and the two arbitrarily chosen independent variables constitutes an *equation of state*. Equation 1-13, the ideal gas law, is an example of such an equation. Thermodynamics describes only *equilibrium states*, that is, those states that do not change with time. Nevertheless, thermodynamic analysis does allow us to predict the way in which changes in the determining variables alter the equilibrium state.

The zeroth law of thermodynamics

If the adiabatic wall separating two systems, each of which are at equilibrium, is replaced by a thermally conducting wall, and no change takes place, the two systems are said to have been in *thermal equilibrium* initially. In the other possible case, that some change takes place, the systems are said to have reached thermal equilibrium after some time has passed. It has been observed that thermal equilibrium is a *transitive property*; that is, if A and B are in thermal equilibrium, and A and C are in thermal equilibrium, then B and C are in thermal equilibrium. Thus, all systems in thermal equilibrium with one another have one *intensive property* in common. This statement is the *zeroth law of thermodynamics*, so named because it was a logical prerequisite to the first and second laws, but formulated later.

The characteristic intensive property of thermal equilibrium is the *temperature*. As pointed out in Sec. 1-2, the temperature is determined from measurements of the pressure and molar volume of the system. In systems other than ideal gases, though the functional relationship would be different, as for a van der Waals gas, the temperature would still be determined by pressure and molar volume measurements. From the zeroth law, the equation of state of a complex system need not be known to measure the temperature; an ideal gas thermometer, or some system calibrated in terms of an ideal gas thermometer, can be used to measure the temperature of any system with which it is in thermal equilibrium.

9-2. Work, energy, and heat. The first law

There is an explicit statistical model for calculating the thermal energy of a system from the number and structure of its component molecules and the temperature for a few systems, like gases at low pressure, as shown in Sec. 8-5. When there is no such model, an empirical method can be used to determine the energy of a system. The basis for such a determination is a series of experiments reported by Joule in 1845 and 1850[1].

Joule found that when work is done on a system with *adiabatic* walls, the change of state of the system is dependent only on the amount of work done and independent of the way in which the work is done, whether the work is frictional or electric or the work of expansion of gases. From these observations, it is reasonable to deduce that there is some function of the state of the system in which the difference is equal to the adiabatic work done. Joule's experiments provide the basis for defining the *energy E* as a *state function*, a thermodynamic property of the system. Then

$$\Delta E = E_2 - E_1 = W \quad \text{(adiabatic)} \tag{9-1}$$

1. J. P. Joule, *Phil. Mag.*, 27, 205 (1845); J. P. Joule, *Phil. Trans. Roy. Soc.*, 140, 61–82 (1850); H. A. Boorse and L. Motz, *The World of the Atom*, Basic Books, Inc., New York, 1966, pp. 236–59.

For an infinitesimal change, the corresponding expression is

$$\delta E = \delta W \quad \text{(adiabatic)} \tag{9-2}$$

One should note that the thermodynamic definition of energy applies only to differences in energy and does not specify an absolute value of E for any state of the system.

Heat

The energy of a system can also be changed without work being done on the system. The change in energy of a system in thermal contact with its surroundings without work being done is a result of the transfer of *heat*. The heat transferred Q is defined by the relation

$$Q = E_2 - E_1 \quad \text{(no work)} \tag{9-3}$$

or

$$\delta Q = \delta E \quad \text{(no work)} \tag{9-4}$$

It is understood that an amount of heat Q can be transferred *to* one system only if the same amount of heat is transferred *from* another system. In thermodynamics, the term *heat* is used only to describe this transfer of energy between systems in thermal contact. Thus, Q, like W, is not a state function; the thermal energy is a characteristic property of the system; Q and W are not.

The first law

The values of ΔE, Q, and W can be independently determined when work is done on a system in thermal contact with another system. The work W on going from state 1 to state 2 over a given path is obtained from changes in the surroundings, using the definition[2]

$$W = \int_{1 \atop \text{Path}}^{2} \mathbf{F}_{ex} \cdot d\mathbf{S} \tag{9-5}$$

The quantity $\mathbf{F}_{ex} \cdot d\mathbf{S}$ is the scalar product of the external force acting on the system and the displacement through which it acts. The integral is evaluated over the path taken in doing work on the system, and the value of the integral depends on the path. According to Eq. 9-5, W is positive when $\mathbf{F}$ and $\mathbf{S}$ are in the same direction, that is when work is done by the surroundings

---

2. An alternative definition in which W has the opposite sign is also used. Appendix C-3 provides a brief introduction to the scalar product.

on the system; W is negative when the system does work on the surroundings. The energy change ΔE is obtained from experiments in which the same change of state is carried out adiabatically. The heat Q is obtained from the change of state of the heat source, on which no work is done. The heat Q is positive when it is transferred from the surroundings to the system and negative when it is transferred from the system to the surroundings.

However Q and W vary, the sum $Q + W$ always equals ΔE. Thus, for a finite change,

$$\Delta E = E_2 - E_1 = Q + W \tag{9-6}$$

and for an infinitesimal change

$$\delta E = \delta Q + \delta W \tag{9-7}$$

The *first law* of *thermodynamics* can be summarized in the following statements:

$$\Delta E = E_2 - E_1 = W \quad \text{(adiabatic)}$$

$$Q = \Delta E \quad \text{(no work)}$$

$$\Delta E = Q + W$$

$$\oint dE = 0 \tag{9-8}$$

That is, E is a function of the state of the system. Since there is no change in energy when a system returns to its original state, Eq. 9-8 is a statement of the conservation of energy. It may be stated alternatively as: $\int_A^B dE$ is independent of the path taken between A and B.

Exercise 9-1. Show that Eq. 9-8 can be derived from the statement given in the preceding sentence. Hint: Consider a system that goes from state A to state B on one path and returns from state B to state A by another path.

9-3. Application of the first law to gases

It is convenient to describe the molar energy of a pure gas as a function of the two intensive variables, temperature and molar volume,

$$\mathscr{E} = f(T, \mathscr{V}) \tag{9-9}$$

As the independent variables change, the energy change is

$$\delta \mathscr{E} = \left(\frac{\partial \mathscr{E}}{\partial T}\right)_{\mathscr{V}} \delta T + \left(\frac{\partial \mathscr{E}}{\partial \mathscr{V}}\right)_T \delta \mathscr{V} \tag{9-10}$$

Exercise 9-2. Given that $E = n\mathscr{E}$ and that $V = n\mathscr{V}$, show from Eq. 9-10 that, at a constant number of moles n,

$$\delta E = \left(\frac{\partial E}{\partial T}\right)_V \delta T + \left(\frac{\partial E}{\partial V}\right)_T \delta V \tag{9-11}$$

Similarly, Exercise 9-2 indicates that

$$E = g(T, V) \quad \text{(constant } n) \tag{9-12}$$

When the volume is constant, $\delta V = 0$, so

$$\delta E_V = \left(\frac{\delta E}{\partial T}\right)_V \delta T \tag{9-13}$$

No work can be done on a system of constant volume in the absence of a nonmechanical force, and

$$\delta E_V = \delta E \quad \text{(no work)} = \delta Q_V \tag{9-14}$$

therefore,

$$\delta Q_V = \left(\frac{\partial E}{\partial T}\right)_V \delta T \tag{9-15}$$

and

$$\left(\frac{\partial E}{\partial T}\right)_V = \left(\frac{\partial Q}{\partial T}\right)_V = C_V \tag{9-16}$$

The molar heat capacity at constant volume $\mathscr{C}_V$ is a characteristic of the gas and must be determined experimentally or from a molecular model.

In 1845, Joule reported experiments designed to determine $(\partial E/\partial V)_T$. He used pressures low enough for the gases to be considered ideal; his aparatus is shown schematically in Fig. 9-1. Vessel A was filled with air to a pressure of about 20×10^5 Pa, and vessel B was evacuated. He let the system come to thermal equilibrium in a well-insulated water bath and then opened the stopcock connecting the two vessels. He found no change in the temperature of the water. Since the gas expanded into a vacuum, no work was done; since the temperature of the water did not change, no heat was exchanged; therefore $\Delta E = 0$ for the gas. Thus, in Eq. 9-11, δE and δT are equal to zero, and $\delta V \neq 0$. Therefore

$$\left(\frac{\partial E}{\partial V}\right)_T = 0 \tag{9-17}$$

Joule concluded from these results that the energy of a gas is independent of the volume, and, therefore, that no forces act between gas molecules.

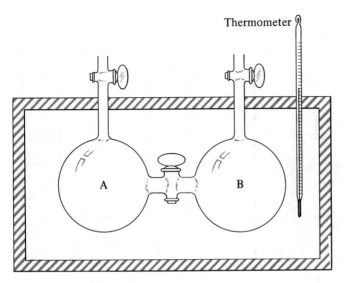

Figure 9-1. Diagram of the Joule gas expansion apparatus.

His inability to detect a temperature change was a result of the large heat capacity of the water. A later, more sensitive, experiment that Joule and Thomson devised did show an energy change upon expansion. But, an ideal gas can be characterized as one for which the ideal gas law (Eq. 1-14) and Eq. 9-17 are valid.

From Eq. 9-11 and Eq. 9-16, we can write

$$\delta E = C_V \, \delta T + \left(\frac{\partial E}{\partial V}\right)_T \delta V \tag{9-18}$$

Since E is a state function, δE for a given change of state is the same whether T is changed first and then V, or V is changed first and then T. Thus we would expect that

$$\frac{\partial^2 E}{\partial T \partial V} = \left[\frac{\partial}{\partial T}\left(\frac{\partial E}{\partial V}\right)_T\right]_V = \left[\frac{\partial}{\partial V}\left(\frac{\partial E}{\partial T}\right)_V\right]_T = \frac{\partial^2 E}{\partial V \partial T} \tag{9-19}$$

For an ideal gas, $(\partial E/\partial V)_T = 0$, so that

$$\left[\frac{\partial}{\partial V}\left(\frac{\partial E}{\partial T}\right)_V\right]_T = \left(\frac{\partial C_V}{\partial V}\right)_T = 0 \tag{9-20}$$

and C_V is a function of the temperature only.

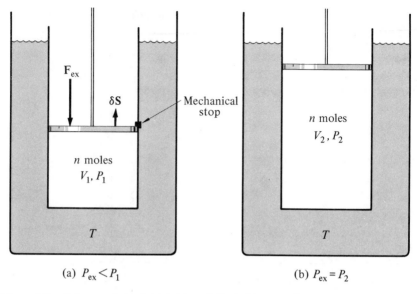

Figure 9-2. Scheme of the initial (a) and final (b) states in the constant-pressure, isothermal expansion of a gas.

Isothermal expansion and compression of gases

Consider a sample of n moles of gas in a cylinder of volume V_1, immersed in a thermal reservoir sufficiently large that heat can be transferred to or from it without significantly changing the temperature T. The pressure of the gas is P_1, and the external pressure is $P_{ex} < P_1$ (the piston is held at V_1 by a mechanical stop, as indicated in Fig. 9-2.)

When the mechanical stop is removed, the gas expands until its pressure is equal to P_{ex}, and a new equilibrium state is attained, at V_2, $P_2 = P_{ex}$, and T. The expansion is called *isothermal* because the initial and final temperatures are the same. During the course of the expansion, there is *no* thermodynamic description of the state of the system because it is changing with time, and there is no uniform density or pressure, or thermodynamic temperature, until the final equilibrium state is reached (after some oscillation).

The work done in the expansion is given by Eq. 9-5 as

$$W = \int_{1 \atop \text{Path}}^{2} \mathbf{F}_{ex} \cdot d\mathbf{S}$$

In evaluating the scalar product, we always take the differential displacement δS to be in the positive direction; thus, from Fig. 9-2,

$$\mathbf{F}_{ex} \cdot \delta \mathbf{S} = |\mathbf{F}_{ex}| \, |\delta \mathbf{S}| \cos \pi$$

$$= -|\mathbf{F}_{ex}| \, |\delta \mathbf{S}| \qquad (9\text{-}21)$$

But

$$|\mathbf{F}_{ex}| = P_{ex} \, A \qquad (9\text{-}22)$$

where A is the area of the piston. Thus,

$$\mathbf{F}_{ex} \cdot \delta \mathbf{S} = -P_{ex} \, A \, |\delta \mathbf{S}|$$

$$= -P_{ex} \, \delta V \qquad (9\text{-}23)$$

and

$$W = -\int_{V_1}^{V_2} P_{ex} \, dV \qquad (9\text{-}24)$$

Equation 9-24 is the fundamental equation to be used in calculating the work done in any expansion or compression.

Since $P_{ex} = P_2$ is constant for the expansion illustrated in Fig. 9-2, the work done is

$$W = -P_2 \int_{V_1}^{V_2} dV$$

$$= -P_2 (V_2 - V_1) \qquad (9\text{-}25)$$

As we might expect, W is negative for $V_2 > V_1$, representing a loss of energy to the surroundings. If the gas in the cylinder is ideal, $\Delta E = 0$, since the temperature does not change and E depends only on the temperature for an ideal gas. Therefore,

$$Q = \Delta E - W$$

$$= -W$$

$$= P_2 (V_2 - V_1) \qquad (9\text{-}26)$$

Since the energy of the gas does not change in the expansion, the energy required for the performance of work comes from the thermal energy of the bath.

Reversible expansion. A special case of gaseous expansion, and one that can only be approached and not exactly realized, is that in which the external pressure is adjusted continually so that it differs only infinitesimally from the pressure of the gas. By an infinitesimal change in the external pressure, the direction of the process can be reversed, hence the designation, *reversible*. Such an expansion must be carried out extremely slowly, so that the gas has a definite equilibrium pressure at every point, and the entire expansion process can be described thermodynamically.

If a reversible expansion is carried out between the initial state and the final state as shown in Fig. 9-2, the work done is

$$W = -\int_{V_1}^{V_2} P \, dV \tag{9-27}$$

since $P \approx P_{ex}$. To evaluate the integral, the equation of state is used to express P as a function of V. For an ideal gas,

$$P = \frac{nRT}{V}$$

so that

$$W = -\int_{V_1}^{V_2} nRT \frac{dV}{V}$$

$$= -nRT \ln \frac{V_2}{V_1} \tag{9-28}$$

Again,

$$\Delta E = 0$$

and

$$Q = -W$$

$$= nRT \ln \frac{V_2}{V_1} \tag{9-29}$$

Figure 9-3 shows the isotherm of an ideal gas at temperature T, which represents the path of the reversible expansion. The area under the curve from V_1 to V_2 is positive, and the work done is negative. The corresponding area for the work done at a constant P_{ex} equal to P_2 is the area under the horizontal line at P_2. It can be seen that the reversible work is greater in magnitude than the work done at constant pressure. We show in Chapter 11 that the magnitude of the reversible work is always the maximum obtainable from any change of state.

Consider now the reversal of both the constant pressure expansion and the reversible expansion. In order to return the system to its initial state at

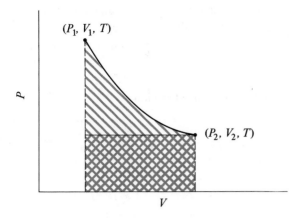

Figure 9-3. The isotherm for the reversible expansion of an ideal gas. The hatched area (total) represents the magnitude of the work of a reversible expansion or compression. The cross-hatched area (below) represents the magnitude of the work of a constant-pressure expansion at P_2.

constant pressure, the external pressure would have to be equal to P_1. Thus, the work done in the reversal of the constant pressure expansion is

$$W = - \int_{V_2}^{V_1} P_1 \, dV$$

$$= -P_1 (V_1 - V_2)$$

$$= P_1 (V_2 - V_1) \tag{9-30}$$

The corresponding value of Q is

$$Q = -W$$
$$= -P_1 (V_2 - V_1) \tag{9-31}$$

The values of W and Q for the reversal of the reversible expansion are equal and opposite in sign to those for the expansion because the path taken is the same, but the direction taken is opposite. The values of Q, W, and ΔE for the expansion, the reversal, and the complete cycle for the reversible and constant pressure processes are collected in Table 9-1. It can be seen from this table that for the reversible case the system is returned to its initial state and the surroundings are returned to their initial state because the heat and work quantities in the expansion and the reversal are equal but opposite. In contrast, in the constant-pressure expansion and compression, though the system is returned to its initial state, the surroundings are not, since a net amount of work from the surroundings has been converted into heat. This *irreversibility* is characteristic of all *real* processes in nature and is an example of the content of the second law of thermodynamics, discussed in Chapter 10.

Table 9-1. Cycles for expansion and compression of an ideal gas

	ΔE	Q	W
Reversible expansion	0	$n R T \ln V_2/V_1$	$-n R T \ln V_2/V_1$
Reversible compression	0	$-n R T \ln V_2/V_1$	$n R T \ln V_2/V_1$
Complete cycle	0	0	0
Constant pressure expansion	0	$P_2 (V_2 - V_1)$	$-P_2 (V_2 - V_1)$
Constant pressure compression	0	$-P_1 (V_2 - V_1)$	$P_1 (V_2 - V_1)$
Complete cycle	0	$(P_2 - P_1)(V_2 - V_1) < 0$	$(P_1 - P_2)(V_2 - V_1) > 0$

Exercise 9-3. (a) Calculate W, ΔE, and Q in joules for the reversible, isothermal expansion of 0.0500 mole of an ideal gas from an initial pressure of 10.0×10^5 Pa to a final pressure of 1.00×10^5 Pa at a temperature of 300 K. (b) Calculate the corresponding quantities for the reversal of the expansion. (c) Calculate W, ΔE, and Q for the expansion of the gas from the same initial state to the same final state at a constant external pressure of 1.00×10^5 Pa. (d) Calculate the corresponding quantities for the reversal of the constant-pressure expansion at an external pressure of 10.0×10^5 Pa.

Adiabatic expansion and compression of gases

By definition, $\delta Q = 0$ in an adiabatic process and (Eq. 9-2)

$$\delta E = \delta W$$

or (Eq. 9-1),

$$W = \Delta E = E_2 - E_1$$

Thus, in an adiabatic expansion, any work done is at the expense of the energy of the gas, and there should be a decrease in the temperature of the gas. According to Eqs. 9-1 and 9-2, the magnitude of the change in energy, and therefore the magnitude of the change in temperature, depend on how much work is done. Since the external pressure in an irreversible expansion is less by a finite amount than the gas pressure, whereas the external pressure differs only infinitesimally from that of the gas in a reversible expansion, the work done in an irreversible expansion is smaller in magnitude than that in a reversible expansion, and the temperature change is correspondingly smaller. Joule's experiment, illustrated in Fig. 9-1, is the limiting example of an adiabatic expansion in which no work is done and in which there is no temperature change.

A quantitative description of the reversible adiabatic expansion can be obtained if we substitute

$$\delta E = C_V \delta T \tag{9-32}$$

and

$$\delta W = -P \delta V \tag{9-33}$$

in Eq. 9-2. These substitutions are valid for an ideal gas and a reversible expansion, respectively.

Exercise 9-4. Verify the preceding statement.

Thus,

$$C_V \delta T = -P \delta V$$

$$= -nRT \frac{\delta V}{V}$$

or

$$\frac{C_V}{T} \delta T = -nR \frac{\delta V}{V} \tag{9-34}$$

If the temperature range is small, so that C_V can be considered constant, Eq. 9-34 can be integrated as follows from an initial state T_1, V_1 to a final state T_2, V_2. Thus,

$$\int_{T_1}^{T_2} \frac{dT}{T} = -\frac{nR}{C_V} \int_{V_1}^{V_2} \frac{dV}{V}$$

and

$$\ln \frac{T_2}{T_1} = -\frac{nR}{C_V} \ln \frac{V_2}{V_1} \tag{9-35}$$

Taking antilogarithms of both sides of Eq. 9-35, we have

$$\frac{T_2}{T_1} = \left(\frac{V_1}{V_2} \right)^{nR/C_v} = \left(\frac{V_1}{V_2} \right)^{R/\mathscr{C}_v} \tag{9-36}$$

Thus, the temperature decreases in the reversible adiabatic expansion of an ideal gas and increases in the corresponding compression.

Exercise 9-5. Calculate the work done and the final temperature in the reversible adiabatic expansion of an ideal gas from the initial state of Exercise 9-3 to the same final volume, and calculate the final temperature. Assume that $\mathscr{C}_V = \frac{3}{2} R$ for an ideal gas. Compare W for the reversible isothermal expansion. (Note that the final states are not the same.)

9-4. The enthalpy function

We can deduce from Eq. 9-14 that the heat exchanged at constant volume is equal to the change in a state function; that is

$$Q_V = \Delta E_V \tag{9-37}$$

Since many processes in chemical, biological, and geological systems occur at constant pressure, rather than constant volume, it is of interest to see whether the heat exchanged at constant pressure is also equal to the change in a state function. From Eq. 9-7 we have

$$\delta Q = \delta E - \delta W \tag{9-38}$$

For a constant pressure process in which the only work is against the pressure of the environment

$$\delta W = -P\,\delta V$$

and

$$\delta Q_P = \delta E + P\,\delta V \tag{9-39}$$

Since the pressure is constant, $V\,\delta P$ can be added to the right side of Eq. 9-39 without changing its value. Thus,

$$\begin{aligned} \delta Q_P &= \delta E + P\,\delta V + V\,\delta P \\ &= \delta E + \delta(PV) \\ &= \delta(E + PV) \end{aligned} \tag{9-40}$$

Since E, P, and V are state functions, $E + PV$ is a state function. Thus, we find that the heat exchanged at constant pressure is equal to the change in a state function if the only work done is against the pressure of the surroundings. The quantity $(E + PV)$ is given the symbol H; it is denoted the *enthalpy*.

$$H = E + PV \tag{9-41}$$

Thus,

$$\delta Q_P = \delta H_P \tag{9-42}$$

and

$$Q_P = H_2 - H_1 = \Delta H_P \tag{9-43}$$

Heat capacity at constant pressure
The heat capacity at constant pressure, C_p, is, by definition,

$$C_p = \left(\frac{\partial Q}{\partial T}\right)_P \tag{9-44}$$

and, from Eq. 9-42,

$$C_p = \left(\frac{\partial H}{\partial T}\right)_P \tag{9-45}$$

Since Q is not a state function, the heat capacity $(\partial Q/\partial T)$ is also not generally equal to a state function; it can vary from zero to infinity depending on the relative values of Q and W in any process. It is only under certain conditions, such as *constant volume* and *constant pressure*, that the heat capacity is equal to a state function.

The relationship between C_V and C_p can be obtained from the definition of H (Eq. 9-41) and the definitions of C_p and C_V (Eqs. 9-45 and 9-16). Since

$$H = E + PV$$

$$\left(\frac{\partial H}{\partial T}\right)_P = \left(\frac{\partial E}{\partial T}\right)_P + P\left(\frac{\partial V}{\partial T}\right)_P \tag{9-46}$$

From the definitions of the heat capacity

$$C_p - C_V = \left(\frac{\partial H}{\partial T}\right)_P - \left(\frac{\partial E}{\partial T}\right)_V$$

$$= \left(\frac{\partial E}{\partial T}\right)_P + P\left(\frac{\partial V}{\partial T}\right)_P - \left(\frac{\partial E}{\partial T}\right)_V \tag{9-47}$$

If we consider E as a function of P and T,

$$\delta E = \left(\frac{\partial E}{\partial P}\right)_T \delta P + \left(\frac{\partial E}{\partial T}\right)_P \delta T \tag{9-48}$$

whereas, for E as a function of V and T (Eq. 9-11),

$$\delta E = \left(\frac{\partial E}{\partial V}\right)_T \delta V + \left(\frac{\partial E}{\partial T}\right)_V \delta T$$

We can relate the two by expressing δV for V as a function of T and P.

$$\delta V = \left(\frac{\partial V}{\partial T}\right)_P \delta T + \left(\frac{\partial V}{\partial P}\right)_T \delta P \tag{9-49}$$

Substituting from Eq. 9-49 in Eq. 9-11, we obtain

$$\delta E = \left(\frac{\partial E}{\partial V}\right)_T \left(\frac{\partial V}{\partial P}\right)_T \delta P + \left(\frac{\partial E}{\partial V}\right)_T \left(\frac{\partial V}{\partial T}\right)_P \delta T + \left(\frac{\partial E}{\partial T}\right)_V \delta T \quad (9\text{-}50)$$

The value of δE for a given change of state is the same whether it is expressed as a function of V and T or as a function of P and T, so we can equate coefficients of like terms in Eq. 9-50 and Eq. 9-48. Thus,

$$\left(\frac{\partial E}{\partial T}\right)_P = \left(\frac{\partial E}{\partial V}\right)_T \left(\frac{\partial V}{\partial T}\right)_P + \left(\frac{\partial E}{\partial T}\right)_V \quad (9\text{-}51)$$

Substituting from Eq. 9-51 in Eq. 9-47, we obtain

$$C_p - C_V = \left(\frac{\partial E}{\partial V}\right)_T \left(\frac{\partial V}{\partial T}\right)_P + P\left(\frac{\partial V}{\partial T}\right)_P \quad (9\text{-}52)$$

For an ideal gas, $(\partial E/\partial V)_T = 0$ and $V = nRT/P$, so that

$$C_p - C_V = nR \quad (9\text{-}53)$$

Exercise 9-6. For a monatomic ideal gas and a diatomic ideal gas, determine the value of $\mathscr{C}_p$ at a temperature for which the equipartition limit is applicable.

Exercise 9-7. From Eqs. 9-36 and 9-53, derive the following relation between P and V in a reversible adiabatic expansion of an ideal gas.

$$\frac{P_2}{P_1} = \left(\frac{V_1}{V_2}\right)^{\gamma} \quad (9\text{-}54)$$

where

$$\gamma = \mathscr{C}_p/\mathscr{C}_V \quad (9\text{-}55)$$

The Joule-Thomson effect

Joule and Thomson (later Lord Kelvin) developed a more sensitive method to detect intermolecular interactions in gases than Joule's original experiment. In their apparatus, shown schematically in Fig. 9-4, a gas in a cylinder at some constant pressure P_1 was forced through a porous diaphragm into a second cylinder in which the pressure was maintained at a constant lower pressure P_2. The diffusion of the gas through the porous diaphragm was slow enough so that the gas reached equilibrium at the pressure P_2 by the time it reached the other side. Since the apparatus was well insulated, there was no heat exchange with the surroundings. The temperature in each cylinder was recorded.

Consider the initial state to be that with all the gas in cylinder 1, at pressure

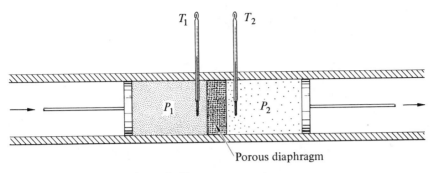

Figure 9-4. Diagram of the Joule-Thomson apparatus.

P_1, volume V_1, and temperature T_1. At the end of the experiment, all the gas has been forced through the diaphragm into cylinder 2. Since $Q = 0$,

$$\Delta E = E_2 - E_1 = W$$
$$= -P_1(0 - V_1) - P_2(V_2 - 0)$$
$$= P_1 V_1 - P_2 V_2$$

Rearranging, we have

$$E_2 + P_2 V_2 = E_1 + P_1 V_1$$

or

$$H_2 = H_1$$

Thus, the Joule-Thomson experiment is performed at constant enthalpy.

If the pressure difference is sufficiently small, the result of the experiment can be expressed as

$$\mu_{JT} = \left(\frac{\partial T}{\partial P}\right)_H \tag{9-56}$$

where μ_{JT} is the Joule-Thomson coefficient. If the gas cools upon expansion, as a result of doing work against attractive intermolecular forces, μ_{JT} is positive. If the gas heats upon expansion, as a result of expanding when repulsive forces predominate, μ_{JT} is negative.

Figure 9-5 shows how the sign of the Joule-Thomson coefficient varies with temperature and pressure for N_2. At sufficiently high pressure, only heating occurs upon expansion. At lower pressures, heating occurs at a high temperature, cooling over a range of intermediate temperatures, and heating at a very low temperature. The temperature at which the coefficient changes

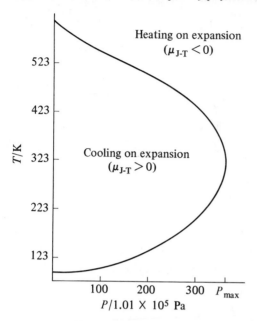

Figure 9-5. The temperature and pressure at which the Joule-Thomson coefficient of N_2 equals zero; that is, the dependence of the Joule-Thomson inversion temperature on pressure. [By permission, from J. R. Roebuck and H. Osterberg, *Phys. Rev., 48,* 450 (1935).]

sign for a given gas is called the Joule-Thomson *inversion temperature.* Early efforts to liquefy gases by cooling on expansion had mixed results because the experimenters were working above the upper inversion temperature for some gases. As a result of the discovery of the Joule-Thomson effect, liquefaction of gases through the cooling effect of expansion between the upper and lower inversion temperatures became a routine industrial process and made the study of phenomena at temperatures approaching 0 K possible.[3]

The relation of the Joule-Thomson coefficient to intermolecular forces can be obtained from a consideration of the total differential of H as a function of T and P.

$$\delta H = \left(\frac{\partial H}{\partial T}\right)_P \delta T + \left(\frac{\partial H}{\partial P}\right)_T \delta P \tag{9-57}$$

At constant H, $\delta H = 0$ and

$$\frac{\delta T}{\delta P} = \left(\frac{\partial T}{\partial P}\right)_H = -\frac{(\partial H/\partial P)_T}{(\partial H/\partial T)_P} = -\frac{(\partial H/\partial P)_T}{C_p} \tag{9-58}$$

3. K. Mendelssohn, *The Quest for Absolute Zero,* McGraw-Hill Book Co., New York, 1966.

Like $(\partial E/\partial V)_T$, $(\partial H/\partial P)_T$ is a measure of the intermolecular forces in the gas. The latter is negative when attractive forces predominate and positive when repulsive forces predominate. Using the second law of thermodynamics in Chapter 10, we shall obtain explicit expressions for $(\partial H/\partial P)_T$ for real gases from the equation of state.

9-5. Thermochemistry

In the early days of thermodynamics, it was suggested that the sign of the heat of reaction was a criterion for determining whether a reaction is spontaneous. Though it can now be seen that we must also know the change in entropy to predict spontaneity, measurement of thermal effects is still an important part of chemical thermodynamics.

Calorimetry

The most common instrument for measuring the thermal effects of chemical reactions is the adiabatic *bomb calorimeter*, which is shown schematically in Fig. 9-6. The sample is placed in cup A in the bomb, in contact with the ignition wire, in an atmosphere of O_2 at a pressure of about 25×10^5 Pa. The stirrer is started, and the temperature is recorded as a function of time to provide a basis for corrections for heat loss. After ignition, the temperature is recorded again as a function of time until the temperature change is again

Figure 9-6. Diagram of a bomb calorimeter.

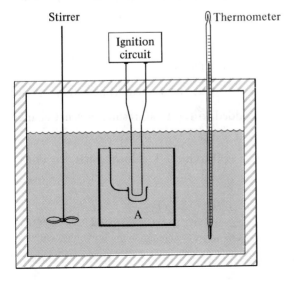

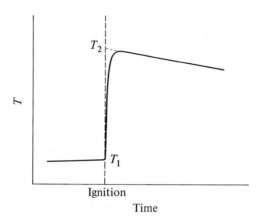

Figure 9-7. Representation of a temperature-time curve for a combustion carried out in a bomb calorimeter.

steady. A typical experimental curve is shown schematically in Fig. 9-7. The almost linear portion of the curve after ignition is extrapolated back to the time of ignition to determine what temperature would have been reached if there had been no leakage of heat from the calorimeter. The quantity $T_2 - T_1$ then represents the corrected temperature change of the calorimeter and its contents as a result of the chemical reaction.

After a correction has been made for heat loss to the surroundings, the heat Q can be taken equal to zero for the calorimeter and its contents. Since the volume is constant, and no non-mechanical work is done, W is also zero. Therefore, $\Delta E = 0$. The change of state to which this value of ΔE corresponds is

$$[\text{Reactants} + \text{calorimeter}](T_1) \longrightarrow [\text{Products} + \text{calorimeter}](T_2) \quad (9\text{-}59)$$

The change of state for which ΔE is desired is

$$[\text{Reactants} + \text{calorimeter}](T_1) \longrightarrow [\text{Products} + \text{calorimeter}](T_1) \quad (9\text{-}60)$$

The change of state that, when added to Eq. 9-60, results in a net change equal to Eq. 9-59 is

$$[\text{Products} + \text{calorimeter}](T_1) \longrightarrow [\text{Products} + \text{calorimeter}](T_2) \quad (9\text{-}61)$$

Since E is a state function,

$$\Delta E_{59} = 0 = \Delta E_{60} + \Delta E_{61}$$

and

$$\Delta E_{60} = -\Delta E_{61} \quad (9\text{-}62)$$

Thus, the desired ΔE can be obtained by measuring ΔE for Eq. 9-61.

Two methods are commonly used to measure ΔE_{61}. The more direct method is to measure the amount of electrical work required to produce a temperature change of $T_2 - T_1$ in the calorimeter plus products. Since $Q = 0$, $W = \Delta E_{61}$. Alternatively, since $\Delta E = C_V \Delta T$ over a small temperature range in which C_V is constant, a reaction of known ΔE can be carried out in the same system and a value of C_V obtained. The implicit assumption is made that the products of the two reactions are sufficiently similar, and that the heat capacity of the products is a sufficiently small fraction of the total heat capacity, that C_V is the same for the two experiments.

Exercise 9-8. When 450.74 mg of jet-engine fuel was burned in a bomb calorimeter under a pressure of 25×10^5 Pa of O_2, the temperature change was found to be 1.7656 K. In order to produce the same temperature change in the calorimeter and products, it was necessary to operate a 10,000-W heater for 2061.9 s. A correction of 104.57 J was made for the energy liberated by the burning of the ignition wire. Calculate the value of ΔE per gram for combustion of the fuel.

Frequently, the value of ΔH must be calculated from experimental values of ΔE. Since

$$H = E + PV$$
$$\Delta H = \Delta E + \Delta(PV) \tag{9-63}$$

The quantity $\Delta(PV)$ is sufficiently small for reactions with solid and liquid reactants and products that only the value of $\Delta(PV)$ for gaseous reactants and products need be considered. If the gases are approximated as being ideal,

$$\Delta(PV) = \Delta(nRT)$$
$$= \Delta n(RT)$$

so that

$$\Delta H = \Delta E + \Delta n(RT) \tag{9-64}$$

where Δn is the change in the number of moles of gas as a result of the reaction. Constant pressure calorimeters, of course, lead directly to values of ΔH.

If ΔH is *positive*, and heat is *absorbed* upon reaction, the reaction is termed *endothermic*. If ΔH is *negative*, and heat is *released* upon reaction, the reaction is termed *exothermic*.

Exercise 9-9. If the fuel in Exercise 9-8 is a hydrocarbon with the formula C_8H_{18}, calculate $\Delta \mathscr{E}$ and $\Delta \mathscr{H}$ for the combustion of 1 mole of fuel.

Table 9-2. Standard states for the enthalpy of pure substances

Phase or substance	Standard state
Pure solid	The most stable form at 1.01×10^5 Pa pressure and the specified temperature
Pure Liquid	The most stable form at 1.01×10^5 Pa pressure and the specified temperature
Pure gas	Zero pressure and the specified temperature. The hypothetical ideal gas at 1.01×10^5 Pa has the same enthalpy as the real gas at zero pressure
Carbon	Graphite

I. M. Klotz and R. M. Rosenberg, *Chemical Thermodynamics*, 3rd ed., W. A. Benjamin, Menlo Park, 1972, pp. 336–37.

Standard states. Standard enthalpy of formation

Since it is not possible to obtain absolute values of energy and enthalpy, it is useful to define an arbitrary reference state, called the *standard state*. Standard states for the enthalpy are shown in Table 9-2. When $\Delta \mathcal{H}$ refers to a reaction in which the reactants and products are in their standard states, it is called the *standard enthalpy change* and given the symbol $\Delta \mathcal{H}^{\ominus}$. The chemical equation corresponding to $\Delta \mathcal{H}^{\ominus}$ and the phases of reactants and products must be specified. For example, at 298.15 K,

$$H_2(g) + \tfrac{1}{2} O_2(g) = H_2O(l) \qquad \Delta \mathcal{H}^{\ominus} = -285.83 \text{ kJ-mol}^{-1}$$

$$2 H_2(g) + O_2(g) = 2 H_2O(l) \qquad \Delta \mathcal{H}^{\ominus} = -571.66 \text{ kJ-mol}^{-1}$$

Rather than tabulate standard enthalpy changes for all the reactions for which they have been measured, it is more economical to choose fewer reactions and calculate the standard enthalpy changes for all other reactions from their values. Such a set includes the reactions by which one mole of a compound in its standard state is formed from the elements in their standard states. The enthalpy of such a reaction is the *standard enthalpy of formation*, written $\Delta \mathcal{H} f^{\ominus}$. Then, for any reaction,

$$\Delta \mathcal{H}^{\ominus} = \sum_{\text{prod}} v_j \, \Delta \mathcal{H} f^{\ominus}{}_j - \sum_{\text{react}} v_j \, \Delta \mathcal{H} f^{\ominus}{}_j \qquad (9\text{-}65)$$

where the v_j are the stoichiometric coefficients in the chemical equation. Equation 9-65 is valid because H is a state function. Its content was stated as an empirical generalization before it was deduced from the first law, and it was known as Hess's law of constant heat summation.

Exercise 9-10. For the reaction

$$CO_2(g) + H_2(g) = CO(g) + H_2O(g)$$

at 1500 K, verify Eq. 9-65 by showing that the equations corresponding to $\Delta \mathscr{H}f^{\,\ominus}$ can be added and subtracted appropriately to obtain the desired reaction.

Table 9-3 contains some standard enthalpies of formation at $T = 298.15\,K$.

Table 9-3. Standard enthalpy of formation at 298.15 K

Compound	$\Delta \mathscr{H}f^{\,\ominus}/(kJ\text{-}mol^{-1})$	Compound	$\Delta \mathscr{H}f^{\,\ominus}/(kJ\text{-}mol^{-1})$
$O_3(g)$	143.	$H(g)$	217.97
$HD(g)$	0.32	$H_2O(g)$	−241.82
$H_2O(l)$	−285.83	$H_2O_2(l)$	−187.7
$HF(l)$	−299.7	$HF(g)$	−271.
$HCl(g)$	−92.307	$HClO_4(l)$	−40.6
$HClO_4 \cdot H_2O(c)$	−382.2	$HBr(g)$	−36.4
$HI(g)$	26.5	$S(monoclinic)$	0.3
$S_8(g)$	102.3	$SO_2(g)$	−296.83
$SO_3(g)$	−395.7	$H_2SO_4(l)$	−813.988
$NO(g)$	90.24	$NO_2(g)$	33.2
$N_2O_4(g)$	9.16	$NH_3(g)$	−46.11
$HNO_3(l)$	−174.1	$NH_4NO_3(c)$	−365.6
$NH_4Cl(c)$	−314.4	$H_3PO_4(c)$	−1279.
$P(g)$	314.6	$PCl_5(g)$	−375.
$PCl_3(g)$	−287.	$CO(g)$	−110.52
$C(diamond)$	1.897	$C(g)$	716.682
$CO_2(g)$	−393.51	$CH_4(g)$	−74.81
$CH_3OH(l)$	−238.7	$C_2H_2(g)$	226.7
$C_2H_4(g)$	52.26	$C_2H_6(g)$	84.68
$SiO_2(quartz)$	−910.94	$SiO_2(cristobalite)$	−909.48
$SiO_2(tridymite)$	−909.06	$BH_3(g)$	100.
$B_2H_6(g)$	36.	$Al_2O_3(corundum)$	−1676.
$Al_2O_3 \cdot 3H_2O(gibbsite)$	−2563.	$AgCl(c)$	−127.07
$Hg_2SO_4(c)$	−743.12	$NaOH(c)$	−425.60

NBS Technical Note 270–3, "Selected Values of Chemical Thermodynamic Properties", 1968.

Exercise 9-11. Calculate $\Delta \mathscr{H}^{\,\ominus}$ at 298.15 K for the following reactions, using values from Table 9-3.

$$PCl_5(g) = PCl_3(g) + Cl_2(g)$$
$$N_2O_4(g) = 2\,NO_2(g)$$
$$CO(g) + \tfrac{1}{2}\,O_2(g) = CO_2(g)$$
$$2\,CO(g) + O_2(g) = 2\,CO_2(g)$$

Bond energies

When thermodynamic data are not available for a particular compound, a reasonable estimate of the enthalpy of formation can be calculated from tables of *bond energies*. The bond energy for a pair of atoms is the *average value* of the molar enthalpy change for the dissociation of the bond in a variety of compounds.

In the hypothetical dissociation of CH_4 into gaseous atoms, for example,

$$CH_4(g) = C(g) + 4H(g) \qquad (9\text{-}66)$$

one could define the bond energy ε_{C-H} as being equal to $\Delta \mathcal{H}/4$. This value would be different from any of the actual bond dissociation enthalpies in the steps

$$CH_4(g) = CH_3(g) + H(g) \quad \Delta \mathcal{H}_1$$

$$CH_3(g) = CH_2(g) + H(g) \quad \Delta \mathcal{H}_2$$

$$CH_2(g) = CH(g) \ + H(g) \quad \Delta \mathcal{H}_3$$

$$CH(g) \ \ = C(g) \ \ \ + H(g) \quad \Delta \mathcal{H}_4$$

Values of $\Delta \mathcal{H}_1, \Delta \mathcal{H}_2, \Delta \mathcal{H}_3$, and $\Delta \mathcal{H}_4$ are difficult to obtain experimentally. Possible sources include mass spectrometric data for appearance potentials of particular radicals and spectroscopic data for systems in which these radicals are present. Theoretical calculations can also be used to estimate these values.

Since $\mathcal{H}$ is a state function, however,

$$\Delta \mathcal{H} = \Delta \mathcal{H}_1 + \Delta \mathcal{H}_2 + \Delta \mathcal{H}_3 + \Delta \mathcal{H}_4$$

The value of $\Delta \mathcal{H}$ for Eq. 9-66 can be obtained from the values of $\Delta \mathcal{H}$ for the following reactions:

$$CH_4(g) = C\,(graphite) + 2\,H_2(g) \quad \Delta \mathcal{H} = -\Delta \mathcal{H} f^{\ominus}[CH_4]$$

$$C\,(graphite) = C(g) \qquad\qquad\qquad \Delta \mathcal{H} = \Delta \mathcal{H} f^{\ominus}[C(g)]$$

$$2\,H_2(g) = 4\,H(g) \qquad\qquad\qquad\; \Delta \mathcal{H} = 4\,\Delta \mathcal{H} f^{\ominus}[H(g)]$$

Exercise 9-12. Calculate ε_{C-H} from the data in Table 9-3.

Exercise 9-13. Use the bond energies from Table 9-4 and any additional data necessary from Table 9-3 to calculate $\Delta \mathcal{H}$ for the reaction

$$P(c) + {}^5\!/_2\,Cl_2(g) = PCl_5(g)$$

Compare with the value of $\Delta Hf^{\ominus}$ for PCl_5 (g) in Table 9-3.

Exercise 9-14. Use bond energies from Table 9-4 and any necessary additional data from Table 9-3 to obtain $\Delta Hf^{\ominus}$ for $C_6H_6(g)$, assuming a single Kekulé structure for benzene. Calculate the resonance (or delocalization) energy of benzene by comparing this value to the experimental value of $\Delta Hf^{\ominus} = 80.668$ kJ-mol^{-1}.

Table 9-4 lists some values of bond energies at 298.15 K.

Table 9-4.   Bond energies at 298.15 K[a]

Bond	$\varepsilon/(\text{kJ-mol}^{-1})$
H−H	435.9
C−C	346.
C=C	610.0
C≡C	835.1
N−N	160.
N≡N	944.7
O−O	150
O=O	498.3
S−S	230.
Cl−Cl	242.1
C−H	413.
N−H	391.
O−H	462.8
S−H	350.
C−Cl	340.
P−Cl	330.

[a]T. L. Cottrell, *The Strengths of Chemical Bonds*, 2nd ed., Butterworths, London 1958, pp. 270–89; A. G. Gaydon, *Dissociation Energies*, Chapman and Hall, Ltd., London, 1968.

Temperature dependence of $\Delta \mathscr{H}$

Tables such as 9-3 and 9-4 provide data only at 298.15 K. Data are needed also for other temperatures at which reactions can occur. Equation 9-45 can be used to calculate $\Delta \mathscr{H}$ at one temperature from data on $\Delta \mathscr{H}$ at another temperature and values for the heat capacity of reactions and products as a function of temperature over the desired range of temperature.

Consider the reaction

$$N_2(g) + 3\,H_2(g) = 2\,NH_3(g) \tag{9-67}$$

One can express the enthalpy change as

$$\Delta \mathscr{H} = 2\,\mathscr{H}(NH_3) - \mathscr{H}(N_2) - 3\,\mathscr{H}(H_2) \tag{9-68}$$

Differentiating Eq. 9-68 with respect to temperature, we obtain

$$\left[\frac{\partial(\Delta \mathscr{H})}{\partial T}\right]_P = 2\left[\frac{\partial \mathscr{H}(NH_3)}{\partial T}\right]_P - \left[\frac{\partial \mathscr{H}(N_2)}{\partial T}\right]_P - 3\left[\frac{\partial \mathscr{H}(H_2)}{\partial T}\right]_P$$

$$= 2\,\mathscr{C}_p(NH_3) - \mathscr{C}_p(N_2) - 3\,\mathscr{C}_p(H_2)$$

$$= \Delta \mathscr{C}_p \tag{9-69}$$

If we integrate Eq. 9-69 between the limits T_1 and T_2, we obtain

$$\int_{T_1}^{T_2} \left[\frac{\partial(\Delta \mathcal{H})}{\partial T} \right]_P dT = \Delta \mathcal{H}_{T_2} - \Delta \mathcal{H}_{T_1}$$

$$= \int_{T_1}^{T_2} \Delta \mathscr{C}_p \, dT \qquad (9\text{-}70)$$

or

$$\Delta \mathcal{H}_{T_2} = \Delta \mathcal{H}_{T_1} + \int_{T_1}^{T_2} \Delta \mathscr{C}_p \, dT \qquad (9\text{-}71)$$

The values of the heat capacity of reactants and products as a function of temperature must be known in order to evaluate the integral in Eq. 9-71. These heat capacity data are usually expressed as an empirical equation of the form

$$\mathscr{C}_p = a + bT + cT^2 + dT^3 \qquad (9\text{-}72)$$

or

$$\mathscr{C}_p = a + bT + \frac{c'}{T^2} \qquad (9\text{-}73)$$

Table 9-5 lists values of the empirical constants of Eq. 9-72 or Eq. 9-73 for several substances of interest. The values given are for $\mathscr{C}_p$ ($J\text{-}K^{-1}\text{-}mol^{-1}$) and have been determined from values calculated from a statistical model.

Exercise 9-15.   Calculate $\Delta \mathcal{H}$ at 298.15 K for the reaction

$$2\,SO_2(g) + O_2(g) = 2\,SO_3(g) \qquad (9\text{-}74)$$

from Table 9-3. Using the empirical constants in Table 9-5, calculate $\Delta \mathcal{H}$ for the same reaction at 500 K.

Alternatively, Eq. 9-69 can be integrated indefinitely, and the result is

$$\Delta \mathcal{H} = \Delta \mathcal{H}_I + \int \Delta \mathscr{C}_p \, dT$$

$$= \Delta \mathcal{H}_I + \int [\Delta a + (\Delta b)T + (\Delta c)T^2 + (\Delta d)T^3]\, dT$$

$$= \Delta \mathcal{H}_I + (\Delta a)T + \left(\frac{\Delta b}{2}\right)T^2 + \left(\frac{\Delta c}{3}\right)T^3 + \left(\frac{\Delta d}{4}\right)T^4 \quad (9\text{-}75)$$

where $\Delta \mathcal{H}_I$ is a constant of integration. When a known value of $\Delta \mathcal{H}$ at some value of T is inserted in Eq. 9-75, the value of $\Delta \mathcal{H}_I$ can be calculated. When this value is substituted in Eq. 9-75, the result is an explicit equation for $\Delta \mathcal{H}$ as a function of T, valid in the temperature range for which the heat capacity data are valid.

Exercise 9-16. Obtain an equation for $\Delta\mathscr{H}$ of Eq. 9-74 as a function of T over the range 298.15 K to 1000 K.

Table 9-5. Molar heat capacity coefficients at constant pressure[a]

Substance	Temperature range (K)	$a/(\text{J-mol}^{-1}\,\text{K}^{-1})$	$b/(10^{-3}$ $\text{J-mol}^{-1}\,\text{K}^{-2})$	$c/(10^{-7}$ $\text{J-mol}^{-1}\,\text{K}^{-3})$	$d/(10^{-9}$ $\text{J-mol}^{-1}\,\text{K}^{-4})$
$H_2(g)$	300–1500	29.066	−0.8364	20.12	
$O_2(g)$	300–1500	25.72	12.98	−38.6	
$N_2(g)$	300–1500	27.30	5.230	−0.04	
$Cl_2(g)$	300–1500	31.696	10.144	−40.38	
$H_2O(g)$	300–1500	30.36	9.615	11.84	
$CO_2(g)$	300–1500	21.56	63.697	−405.1	9.678
$CO(g)$	300–1500	26.86	6.966	−8.20	
$D_2(g)$	300–1500	28.58	0.879	19.6	
$DH(g)$	300–1500	29.25	−1.15	25.0	
$SO_2(g)$	300–1800	25.72	57.923	−380.9	8.606
$SO_3(g)$	300–1200	15.07	151.92	−1206.2	36.19
$CH_4(g)$	300–1500	17.45	60.459	11.2	−7.205
$C_2H_4(g)$	300–1500	11.32	122.01	379.0	
$C_2H_6(g)$	300–1500	5.351	177.67	−687.01	8.514
$C_6H_6(g)$	300–1500	−39.66	501.787	−3376.6	85.462
$NH_3(g)$	300–1000	25.89	33.00	−30.5	
$PCl_3(g)$	300–1000	83.965	−1.21	−11.32[b]	
$PCl_5(g)$	300–500	19.83	449.065	−4987.33	

[a]Reprinted with permission from H. M. Spencer and J. L. Justice, *J. Am. Chem. Soc.*, **56**, 2311 (1934); H. M. Spencer and G. N. Flanagan. *J. Am. Chem. Soc.*, **64**, 2511 (1942); H. M. Spencer, *J. Am. Chem. Soc.*, **67**, 1859 (1945). Copyright by the American Chemical Society.
[b]This is the constant of Eq. 10-73, $c'/(\text{J-mol}^{-1}\text{-K})$.

The statistical model developed in Chapter 8 provides another approach to the temperature dependence of the enthalpy. This approach is limited to molecules for which adequate spectroscopic data are available to permit calculation of the partition function. From Eq. 8-70,

$$E - E_0 = NkT^2 \left(\frac{\partial \ln z}{\partial T} \right)_V$$

for an ideal gas. For one mole of gas, $N = L$, and

$$\mathscr{E} - \mathscr{E}_0 = RT^2 \left(\frac{\partial \ln z'}{\partial T} \right)_V \qquad (9\text{-}76)$$

where z' is the partition function calculated using the volume per mole. Since $H = E + PV$,

$$\mathscr{H} - \mathscr{E}_0 = RT^2 \left(\frac{\partial \ln z'}{\partial T} \right)_V + P\mathscr{V}$$

$$= RT^2 \left(\frac{\partial \ln z'}{\partial T} \right)_V + RT \qquad (9\text{-}77)$$

Thus, the statistical model leads directly to a calculation of the enthalpy function $\mathscr{H} - \mathscr{E}_0$.

Since the calculations are for ideal gases, the results are those for the standard state, $\mathscr{H}^\ominus - \mathscr{E}_0{}^\ominus$. For any chemical reaction

$$\Delta \mathscr{H}^\ominus = \sum_{\text{prod}} v_j (\mathscr{H}^\ominus - \mathscr{E}_{0,j}^\ominus) - \sum_{\text{react}} v_j (\mathscr{H}^\ominus - \mathscr{E}_{0,j}^\ominus)$$

$$+ \sum_{\text{prod}} v_j \mathscr{E}_{0,j}^\ominus - \sum_{\text{react}} v_j \mathscr{E}_{0,j}^\ominus \qquad (9\text{-}78)$$

$$= \sum_{\text{prod}} v_j (\mathscr{H}^\ominus - \mathscr{E}_{0,j}^\ominus) - \sum_{\text{react}} v_j (\mathscr{H}^\ominus - \mathscr{E}_{0,j}^\ominus) + \Delta \mathscr{E}_0{}^\ominus \qquad (9\text{-}79)$$

The quantities in the summations are obtained from the statistical calculations. A value of $\Delta \mathscr{H}$ at one temperature must be determined from experimental data in order to calculate $\Delta \mathscr{E}_0{}^\ominus$.

Tables of values of $\mathscr{H}^\ominus - \mathscr{E}_0{}^\ominus$ at various temperatures have been prepared to avoid repetitious calculations of the partition function. Since $\mathscr{H}_0{}^\ominus = \mathscr{E}_0{}^\ominus$, the functions calculated are sometimes given as $\mathscr{H}^\ominus - \mathscr{H}_0{}^\ominus$. The tables in which these functions are tabulated also list values of $\Delta \mathscr{H} f_0{}^\ominus$ for each substance, from which $\Delta \mathscr{E}_0{}^\ominus$ can be calculated. Such tables use results from spectroscopic and from calorimetric measurements.

Problems

9-1. The enthalpy of combustion of graphite is an essential quantity in the calculation of the standard enthalpy of formation of organic compounds. Given the value of $\Delta \mathscr{H}^\ominus$ for the combustion of one mole of C_8H_{18} from Exercise 9-9, show how $\Delta \mathscr{H}^\ominus$ of combustion of graphite and one additional datum are required to calculate $\Delta \mathscr{H} f^\ominus$ of C_8H_{18}.

9-2. In a classic example of precise calorimetry, P. Hawtin, L. B. Lewis, N. Moul, and R. H. Phillips [*Proc. Roy. Soc., London*, A261, 67–95 (1966)], redetermined the value of $\Delta \mathscr{H}^\ominus$ for the combustion of graphite and diamond. Their results at 298.16 K were

$$C(\text{graphite}) + O_2(g) = CO_2(g) \quad \Delta \mathscr{H}^\ominus = -3.93475 \times 10^5 \text{ J-mol}^{-1} \pm 25$$

$$C(\text{diamond}) + O_2(g) = CO_2(g) \quad \Delta \mathscr{H}^\ominus = -3.95347 \times 10^5 \text{ J-mol}^{-1} \pm 70$$

Calculate $\Delta \mathscr{H} f^\ominus$ for C(diamond) and compare with the value in Table 9-3, which is based on earlier experiments. Assuming that the uncertainty in the two calculated values of $\Delta \mathscr{H} f^\ominus$ are approximately equal, decide whether there is a significant difference between them.

9-3. G. B. Kistiakowsky, Ruhoff, Smith, and Vaughn [*J. Am. Chem. Soc.*, 58, 137, 146

(1936)] determined $\Delta\mathscr{H}^{\ominus}$ of hydrogenation of benzene and cyclohexene at 355 K as follows:

$$C_6H_6(g) + 3H_2(g) = C_6H_{12}(g) \quad \Delta\mathscr{H}^{\ominus} = -208.4 \text{ J-mol}^{-1}$$

$$C_6H_{10}(g) + H_2(g) = C_6H_{12}(g) \quad \Delta\mathscr{H}^{\ominus} = -119.7 \text{ J-mol}^{-1}$$

Estimate the "resonance energy" of benzene from these data.

9-4. An ideal gas is changed from state T_1, P_1, to state T_2, P_2. Find two alternative paths between the two states and show that Q and W are path- dependent. Hint: Plot the hypothetical state points on a T, P coordinate system. Equation 9-49 may be useful.

10 · The second law of thermodynamics. Entropy and free energy

The first law of thermodynamics expresses the impossibility of creating energy from nothing, the impossibility of a perpetual motion machine "of the first kind". But, the first law and the zeroth law together are inadequate to describe completely the thermodynamic behavior of a system. Many events can be imagined, but they are never observed, and their occurrence would not contradict the first law. The spontaneous freezing of water at a pressure of 1.0×10^5 Pa at a temperature above $0\,°C$, with the consequent *heating* of the surroundings, does not occur, but its occurrence would not violate the first law. Similarly, the spontaneous decomposition of water into hydrogen and oxygen, with consequent *cooling*, does not occur, but its occurrence would not violate the first law. The spontaneous transport of heat from a colder object to a warmer object, with no other change, does not occur, but its occurrence would not violate the first law. These three examples illustrate the essential unidirectionality of natural processes, which is expressed in the *second law of thermodynamics*.

Planck suggested a form of the second law that emphasizes the irreversibility of nature: *It is impossible to reverse completely any natural process without a compensating change in another system.* Thus, any one of the changes mentioned above, which are the reverse of spontaneous changes, can be made to occur, but only when it is coupled with a spontaneous change in another system.

The *entropy S* is the thermodynamic function that, together with the energy, permits the quantitative description of this irreversibility. G. N. Lewis based his operational definition of entropy directly on the irrevers-

328

ibility of nature by stating: *The increase in entropy of a system that results from an irreversible change is equal to the entropy change in a standard system that is required to restore the system being measured to its initial state by a reversible process.*[1]

The second law of thermodynamics, like the first, is a summary of empirical observations that cannot be derived from some more basic principle. It is possible, however, to derive from one statement of the second law other equivalent forms that may be more useful in the quantitative description of thermodynamic systems.[2]

10-1. The Carnot engine

The entropy S, like the wave function ψ, is defined rather than derived. Entropy can be better understood, however, by considering the properties of a reversible heat engine, following the discussion of Sadi Carnot. Carnot's work, *The Motive Power of Heat*, appeared in 1824, before he was 30 years old, and it provided the basis for both the first and second laws of thermodynamics.

He devised the concept of an ideal, reversible heat engine, which operates at maximum efficiency. The engine exchanges work with its surroundings through the expansion and compression of some working substance, steam, for example. The walls of the engine can be made adiabatic, or they can be made thermally conducting and in thermal contact with either of two thermal reservoirs at different temperatures. In a heat engine, heat is absorbed from the high temperature reservoir, some of the heat is converted to work, and the rest is given off to the low temperature reservoir. In a reverse heat engine, such as a heat pump or refrigerator, heat is transported from the low temperature reservoir to the high temperature reservoir, while work is done on the engine by the surroundings. In all cases, the working substance is returned to its initial state at the end of a cycle of operations. Carnot was concerned with determining whether the efficiency of an engine is limited and how the engine's efficiency is related to the nature of the working substance. The two functions of the engine are illustrated schematically in Fig. 10-1. The temperatures t_2 and t_1 can be measured on any arbitrary thermometric scale.

1. G. N. Lewis and M. Randall, *Thermodynamics*, McGraw-Hill Book Co., New York, 1923, pp. 113–16.
2. An enjoyable and informative treatment of the second law can be found in H. Bent, *The Second Law*, Oxford University Press, New York, 1965.

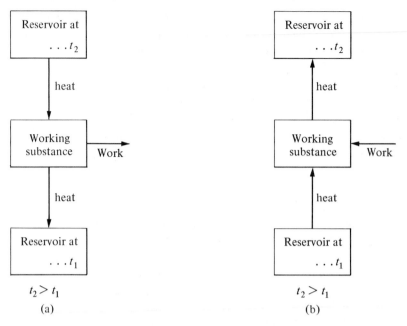

Figure 10-1. Scheme of a Carnot engine as (a) a heat engine, and (b) a refrigerator or heat pump.

Consider a heat engine cycle in which an amount of heat $Q_2 > 0$ is exchanged with the heat reservoir at t_2, an amount of work $W < 0$ is done, and an amount of heat $Q_1 < 0$ is exchanged with the reservoir at t_1. Since the working substance is returned to its initial state,

$$\Delta E = Q_2 + Q_1 + W = 0 \tag{10-1}$$

The *conversion factor* ε of the engine is defined as the maximum fraction of the heat absorbed from the high temperature reservoir that can be converted to work in an ideal engine; that is,

$$\varepsilon = \frac{|W|}{|Q_2|} \tag{10-2}$$

where the absolute values are used to ensure that ε is a positive quantity. Substituting from Eq. 10-1 in Eq. 10-2, we obtain the alternate form

$$\varepsilon = \frac{|Q_2 + Q_1|}{|Q_2|} \tag{10-3}$$

Since Q_2 and Q_1 are of opposite sign, and since $|Q_2|$ is always greater than $|Q_1|$, we can rewrite Eq. 10-3 as

$$\varepsilon = \frac{|Q_2| - |Q_1|}{|Q_2|}$$

$$= 1 - \frac{|Q_1|}{|Q_2|} \tag{10-4}$$

Carnot considered two reversible engines that utilized two different working substances but were coupled to the same thermal reservoirs. One acted as a heat engine and the other as a refrigerator, as illustrated in Fig. 10-2. If one assumes that the conversion factor of the heat engine is greater than that of the refrigerator, then

$$\varepsilon' = \frac{|W'|}{|Q_2'|} > \varepsilon = \frac{|W|}{|Q_2|} \tag{10-5}$$

If the work output of the heat engine is the work input for the refrigerator, $|W| = |W'|$, and

$$\frac{1}{|Q_2'|} > \frac{1}{|Q_2|}$$

or

$$|Q_2| > |Q_2'| \tag{10-6}$$

If the magnitude of the work is equal for the two engines,

$$|Q_1| > |Q_1'| \tag{10-7}$$

Figure 10-2. Scheme of two coupled Carnot engines, one acting as a heat engine, the other as a refrigerator or heat pump.

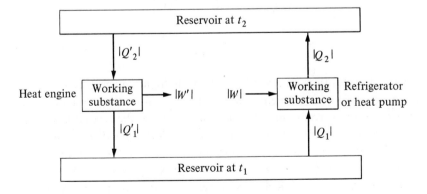

since

$$|Q_2| - |Q_2'| = |Q_1| - |Q_1'| > 0$$

The net result is seen to be the transport of a net amount of heat, $|Q_1| - |Q_1'| = |Q_2| - |Q_2'|$, from the reservoir at low temperature to the reservoir at high temperature without any other change in the system or surroundings. This is contrary to experience, as summarized in Clausius's statement of the second law: *It is impossible to construct a machine that operates in a cycle whose sole effect is the transport of heat from a reservoir at a low temperature to one at a higher temperature.* Carnot thus concluded that ε is *independent of the working substance and dependent only on the temperatures of the reservoirs.*

If we assume that the heat delivered to the low temperature reservoir by the heat engine is just transferred to the refrigerator,

$$|Q_1'| = |Q_1|$$

Then

$$|W| - |W'| = |Q_2| - |Q_2'|$$

We have assumed that $\varepsilon' > \varepsilon$, that is

$$\left(1 - \frac{|Q_1'|}{|Q_2'|}\right) > \left(1 - \frac{|Q_1|}{|Q_2|}\right)$$

implying that

$$\frac{|Q_1'|}{|Q_2'|} < \frac{|Q_1|}{|Q_2|}$$

Because

$$|Q_1'| = |Q_1|, |Q_2'| > |Q_2|$$

and

$$|W'| > |W|$$

This pair of cycles thus removes heat from the high temperature reservoir and converts it to an equivalent amount of work without any other change in the system or surroundings. Kelvin's statement of the second law summarizes the observation that this cannot occur: *It is impossible to construct a machine that operates in a cycle whose sole effect is to remove heat from a reservoir at constant temperature and do an equivalent amount of work.*

Both the Clausius and Kelvin statements of the second law describe examples of perpetual motion machines "of the second kind", which are considered impossible according to the second law. The Kelvin statement forms the basis of concern for the thermal pollution of our environment as a result of the increasing conversion of thermal energy to electrical and mechanical work.

Exercise 10-1.   Show, through combinations of reversible cycles, such as those used above, that ε cannot be less than that of the most efficient reversible cycle without contradicting one of the statements of the second law.

Having demonstrated that the assumption that ε depended on the nature of the working substance leads to a contradiction, Carnot expressed the result as *Carnot's theorem*: *The conversion factor of a reversible engine depends only on the temperatures of the reservoirs and is independent of the nature of the working substance.*

10-2.   The thermodynamic temperature scale[3]

In Sec. 1-3, we derived an ideal gas temperature scale that is independent of the unique properties of any particular gas but is dependent on the common properties of all gases in the limit of zero pressure. Kelvin showed, in 1850, that a thermodynamic temperature scale could be derived from Carnot's theorem, using only the properties of heat engines in the limit of reversible operation. He also showed that the ideal gas and thermodynamic temperature scales coincide if a common reference point is chosen.

Consider three thermal reservoirs at temperatures t_1, t_2, and t_3, such that $t_3 > t_2 > t_1$, and consider three Carnot engines, the first operating between t_2 and t_1, the second between t_3 and t_1, and the third between t_3 and t_2. Conversion factors of the three engines are:

$$1 - \frac{|Q_1|}{|Q_2|} = f(t_2, t_1)$$

$$1 - \frac{|Q_1|}{|Q_3|} = f(t_3, t_1)$$

and

$$1 - \frac{|Q_2|}{|Q_3|} = f(t_3, t_2)$$

3. K. Denbigh, *The Principles of Chemical Equilibrium*, Cambridge University Press, Cambridge, 1961, pp. 26–32.

or

$$\frac{|Q_1|}{|Q_2|} = g(t_2, t_1) \tag{10-8}$$

$$\frac{|Q_1|}{|Q_3|} = g(t_3, t_1) \tag{10-9}$$

$$\frac{|Q_2|}{|Q_3|} = g(t_3, t_2) \tag{10-10}$$

where $1 - f(t_i, t_j) = g(t_i, t_j)$. Dividing Eq. 10-9 by Eq. 10-10, we obtain

$$\frac{|Q_1|}{|Q_2|} = \frac{g(t_3, t_1)}{g(t_3, t_2)} = g(t_2, t_1) \tag{10-11}$$

and we see that t_3 must cancel, on taking the ratio. Therefore, we can write

$$g(t_2, t_1) = \frac{h(t_1)}{h(t_2)} \tag{10-12}$$

and

$$\frac{|Q_1|}{|Q_2|} = \frac{h(t_1)}{h(t_2)} \tag{10-13}$$

Then, $|Q|$ is a thermometric property, and the simplest temperature scale that can be defined using this result is one in which $|Q|$ is directly proportional to the temperature, so that

$$\frac{|Q_1|}{|Q_2|} = \frac{T_1}{T_2} \tag{10-14}$$

(We shall use the symbol T for the thermodynamic temperature scale in anticipation of the proof that it can be made coincident with the ideal gas scale.) Equation 10-4 can now be written

$$\varepsilon = 1 - \frac{T_1}{T_2}$$

$$= \frac{T_2 - T_1}{T_2} \tag{10-15}$$

Exercise 10-2. Calculate the conversion factor for a Carnot engine operating between (a) the normal boiling point of water and 25 °C, (b) high pressure steam at 200 °C and 25 °C, and (c) mercury vapor at 400 °C and 25 °C.

Exercise 10-3. A reversible ideal heat pump operating between the interior of a house at 25 °C and the exterior at -10 °C is used to heat the house. For each 1000 J of elec-

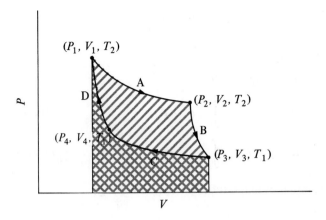

Figure 10-3.   The steps of a Carnot cycle plotted on a *P-V* diagram. Curve A, isothermal expansion at T_2; B, adiabatic expansion; C, isothermal compression at T_1; D, adiabatic compression.

trical energy used to operate the heat pump, how much heat is transported into the house? If the same amount of electrical energy were used to heat the house directly by means of a resistance heater, how much heat would be produced?

Relation to the ideal gas scale

Consider a reversible Carnot engine in which the working substance is an ideal gas. The cycle operates in four steps: A, the isothermal reversible expansion of the gas from P_1, V_1 to P_2, V_2 while in thermal contact with a high temperature reservoir at (ideal gas) temperature T_2; B, the adiabatic reversible expansion of the gas from P_2, V_2, T_2 to P_3, V_3, T_1, where T_1 is the temperature of the low temperature reservoir; C, the isothermal reversible compression of the gas from P_3, V_3, to P_4, V_4, while in thermal contact with the reservoir at temperature T_1; and D, the adiabatic reversible compression of the gas from P_4, V_4, T_1 to the initial state, P_1, V_1, T_2. The cycle is indicated schematically in Fig. 10-3. The entire shaded area in Fig. 10-3 represents the magnitude of the work of expansion, the cross-hatched area represents the magnitude of the work of compression, and the area enclosed by the four steps represents the magnitude of the net work of the cycle.

The thermodynamic changes for the steps in the cycle can be calculated as follows. For steps A and C, which are isothermal,

$$\Delta E_A = \Delta E_C = 0$$

$$-Q_2 = W_A = -nRT_2 \ln \frac{V_2}{V_1} \qquad (10\text{-}16)$$

$$-Q_1 = W_C = -nRT_1 \ln \frac{V_4}{V_3} \tag{10-17}$$

For steps B and D, which are adiabatic,

$$Q_B = Q_D = 0$$

$$W_B = \Delta E_B = \int_{T_2}^{T_1} C_V \, dT \tag{10-18}$$

$$W_D = \Delta E_D = \int_{T_1}^{T_2} C_V \, dT \tag{10-19}$$

Since W_B and W_D are equal in magnitude but opposite in sign,

$$W = W_A + W_C$$

$$= -nRT_2 \ln \left(\frac{V_2}{V_1}\right) - nRT_1 \ln \left(\frac{V_4}{V_3}\right) \tag{10-20}$$

and

$$\varepsilon = \frac{|W|}{|Q_2|}$$

$$= \frac{nRT_2 \ln (V_2/V_1) + nRT_1 \ln (V_4/V_3)}{nRT_2 \ln (V_2/V_1)} \tag{10-21}$$

Equation 9-34 can be used to write equations for steps B and D as

$$\frac{1}{nR} \int_{T_2}^{T_1} C_V \frac{dT}{T} = -\int_{V_2}^{V_3} \frac{dV}{V} = -\ln \frac{V_3}{V_2} \tag{10-22}$$

and

$$\frac{1}{nR} \int_{T_1}^{T_2} C_V \frac{dT}{T} = -\int_{V_4}^{V_1} \frac{dV}{V} = -\ln \frac{V_1}{V_4} \tag{10-23}$$

The left sides of Eqs. 10-22 and 10-23 are equal in magnitude but opposite in sign, so

$$\ln \frac{V_3}{V_2} = -\ln \frac{V_1}{V_4} = \ln \frac{V_4}{V_1}$$

and

$$\ln \frac{V_4}{V_3} = -\ln \frac{V_2}{V_1} \tag{10-24}$$

Equation 10-21, when we rewrite it using Eq. 10-24, becomes

$$\varepsilon = \frac{T_2 \ln (V_2/V_1) - T_1 \ln (V_2/V_1)}{T_2 \ln (V_2/V_1)}$$

$$= \frac{T_2 - T_1}{T_2}$$

$$= 1 - \frac{T_1}{T_2} \tag{10-25}$$

According to Carnot's theorem, the conversion factor of a Carnot engine is independent of the working substance, so

$$\left(1 - \frac{T_1}{T_2}\right)_{\text{thermodynamic}} = \left(1 - \frac{T_1}{T_2}\right)_{\text{ideal gas}}$$

and

$$\left(\frac{T_1}{T_2}\right)_{\text{thermodynamic}} = \left(\frac{T_1}{T_2}\right)_{\text{ideal gas}}$$

Since the two temperature scales are proportional, they may be made to coincide by assigning the same value on both scales to a common fixed point. The point chosen, by international agreement, is the triple point of water at 273.16 K, the temperature at which solid, liquid, and gaseous H_2O are in equilibrium.

Exercise 10-4. Draw diagrams of a Carnot cycle for an ideal gas using the following coordinates: (a) T, V; (b) E, V; (c) P, T; (d) H, P. Indicate clearly each step and its direction.

10-3. The entropy function

Equation 10-14 can be rearranged to read

$$\frac{|Q_1|}{T_1} = \frac{|Q_2|}{T_2}$$

Since Q_1 and Q_2 are opposite in sign,

$$\frac{Q_1}{T_1} = -\frac{Q_2}{T_2}$$

and

$$\frac{Q_1}{T_1} + \frac{Q_2}{T_2} = 0 \tag{10-26}$$

Thus, though Q is not a state function, and $Q_{\text{cyc}} \neq 0$, $(Q/T)_{\text{cyc}} = 0$ for a reversible Carnot cycle. It is clearly useful to define a function that is the ratio of the heat to the temperature. Clausius did so, in 1850, and named it the *entropy* S. Entropy is an extensive variable, and it is defined only in terms of differences, as follows:

$$\delta S = \frac{\delta Q_{rev}}{T} \tag{10-27}$$

For a macroscopic change of state,

$$\Delta S = S_2 - S_1 = \int_1^2 \frac{dQ_{rev}}{T} \tag{10-28}$$

For an isothermal change of state,

$$\Delta S = \frac{1}{T} \int_1^2 dQ_{rev}$$

$$= \frac{Q_{rev}}{T} \tag{10-29}$$

The definition of the entropy specifies that it be calculated from the heat exchanged when the system goes from state 1 to state 2 by a *reversible* path. Since ΔS is independent of the particular reversible path chosen, S is a *state function*, and we write

$$\oint dS = 0 \tag{10-30}$$

That S is a state function also follows directly from G. N. Lewis's operational definition: *The entropy change in the standard system upon reversibly returning the system being measured to its original state is independent of the path taken by that system in its irreversible change.*

Exercise 10-5. Consider the change of state in which n moles of an ideal gas go from P_1, V_1, T_1 to P_2, V_2, T_2. Two of the possible reversible paths for this change of state are (a) reversible heating from T_1 to T_2 at constant pressure P_1 to volume V', followed by isothermal reversible change of pressure at T_2 to the final state P_2, V_2, T_2; (b) isothermal reversible change of pressure at T_1 to the intermediate state P_2, V'', T_1, followed by reversible heating from T_1 to T_2 at constant pressure P_2. Calculate ΔS for the change of state along the two paths and show that they are equal. Hint: Divide the integral into two parts for each path, one at constant pressure and one at constant temperature.

Exercise 10-6. Plot a Carnot cycle on a graph with vertical coordinate T and horizontal coordinate S. What is the significance of the area (a) under the upper curve? (b) under the lower curve? (c) enclosed by the cycle?

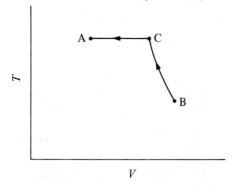

Figure 10-4. Representation on a *T-V* diagram of the reversible steps used to restore a system to state A after an irreversible, adiabatic change from state A to state B.

Irreversible processes

Consider an irreversible change from state A to state B while the system is isolated from its surroundings. As shown in Fig. 10-4, A and B are two points on the *T, V* coordinates. The path is not indicated because intermediate states of an irreversible process are not defined thermodynamically. Since *S* is a state function,

$$\Delta S = S_B - S_A \qquad (10\text{-}31)$$

and ΔS is independent of the path.

In order to evaluate ΔS, let us return the system to state A by the reversible path shown in Fig. 10-4. The path from B to C represents a reversible adiabatic step, and the path from C to A represents a reversible isothermal step. Since *S* is a state function,

$$S_B - S_A = -[(S_C - S_B) + (S_A - S_C)] \qquad (10\text{-}32)$$

The value of $(S_C - S_B)$ for the adiabatic reversible path is

$$S_C - S_B = \int_B^C \frac{dQ_{rev}\,(BC)}{T} = 0 \qquad (10\text{-}33)$$

The value of $S_A - S_C$ for the isothermal reversible path is, from Eq. 10-29,

$$S_A - S_C = \frac{Q_{rev}\,(CA)}{T} \qquad (10\text{-}34)$$

Substituting from Eq. 10-33 and Eq. 10-34 in Eq. 10-32, we have

$$S_B - S_A = -\frac{Q_{rev}(CA)}{T}$$

Since the step from C to A is the only nonadiabatic step, Q for that step is equal to Q for the cycle. Since $\Delta E = 0$ for the cycle, $Q = -W$, and

$$S_B - S_A = -\frac{Q}{T} = \frac{W}{T} \tag{10-36}$$

But Q cannot be positive. If it were, the net result of the complete cycle would be the removal of heat from a thermal reservoir at constant temperature and its conversion to work without any other change in the system or surroundings. This result would contradict the Kelvin statement of the second law and is not seen. Therefore Q must be negative, and

$$\Delta S = S_B - S_A \geqslant 0 \quad \text{(adiabatic, or isolated system)} \tag{10-37}$$

where the equality applies to the reversible case. For an infinitesimal change, the corresponding equation is

$$\delta S \geqslant 0 \quad \text{(adiabatic, or isolated system)} \tag{10-38}$$

When an irreversible change occurs in a system in contact with a thermal reservoir, the system and reservoir together can be considered an isolated system. Then Eq. 10-38 can be written as

$$\delta S = \delta S_s + \delta S_r \geqslant 0 \tag{10-39}$$

where δS_s is the entropy change of the system and δS_r is the entropy change of the reservoir. Though the change in the system is irreversible, we can consider the absorption of an infinitesimal amount of heat by the reservoir to be reversible, so that

$$\delta S_r = \frac{\delta Q_r}{T_r}$$

The system and reservoir are at the same temperature, and $\delta Q_r = -\delta Q_s$, so

$$\delta S_s - \frac{\delta Q_s}{T} \geqslant 0$$

or

$$\delta S \geqslant \frac{\delta Q}{T} \tag{10-40}$$

Equation 10-40 is a general mathematical statement of the *second law of thermodynamics*, of which Eq. 10-38 is a special case. It is important, when applying the second law to systems in thermal contact with their surroundings, to use Eq. 10-40 and not Eq. 10-38. The second law does not state that the entropy of a system cannot decrease, only that the magnitude of the entropy increase in the surroundings must be greater than the magnitude of the entropy decrease in the system. This consideration removes the apparent paradox of a living organism, which decreases its entropy as it develops into a highly organized system at the expense of a much larger increase in the entropy of the surroundings due to release of heat by the organism and the conversion of complex food stuffs to simpler molecules like H_2O and CO_2. When crystals form spontaneously, the entropy decrease due to ordering of the crystalline units is smaller than the concomitant entropy increase in the surroundings, due to the heat of formation of the crystals.

10-4.   Entropy changes in some simple systems

The exchange of heat
Consider two subsystems A and B at temperatures T_A and T_B, respectively. If the two are isolated from the surroundings but in thermal contact with each other, there will be a flow of heat if T_A and T_B are not equal. If we assume that an infinitesmal amount of heat flows from A to B and that the gain or loss of heat for each subsystem can be treated as reversible, then

$$\delta S_A = \frac{\delta Q_A}{T_A} = -\frac{|\delta Q|}{T_A}$$

and

$$\delta S_B = \frac{\delta Q_B}{T_B} = \frac{|\delta Q|}{T_B}$$

Thus, the total entropy change is

$$\delta S = \delta S_A + \delta S_B$$

$$= -\frac{|\delta Q|}{T_A} + \frac{|\delta Q|}{T_B}$$

$$= |\delta Q|\left(\frac{1}{T_B} - \frac{1}{T_A}\right)$$

The second law (Eq. 10-38) states that $\Delta S \geqslant 0$ for an isolated system (A + B). The equality holds if $T_B = T_A$, in which case the heat exchange is reversible. If $T_B < T_A$, $\Delta S > 0$, in agreement with the common observation that heat flows spontaneously from a system at higher temperature to one at lower temperature. This is really a statement of probability rather than one of certainty, as was pointed out in Sec. 8-3.

Isothermal expansion of gases
By definition (Eq. 10-27),

$$\delta S = \frac{\delta Q_{rev}}{T}$$

From the first law (Eq. 9-7),

$$\delta Q_{rev} = \delta E - \delta W_{rev}$$

so that

$$\delta S = \frac{\delta E}{T} - \frac{\delta W_{rev}}{T}$$

If the only work done is pressure-volume work (Eq. 9-33),

$$\delta W_{rev} = -P \, \delta V$$

and

$$\delta S = \frac{1}{T} \delta E + \frac{P}{T} \delta V \tag{10-41}$$

If we consider E as a function of T and V (Eq. 9-11),

$$\delta E = \left(\frac{\partial E}{\partial T}\right)_V \delta T + \left(\frac{\partial E}{\partial V}\right)_T \delta V$$

Substituting from Eq. 9-11 in Eq. 10-41, we have

$$\delta S = \frac{1}{T}\left(\frac{\partial E}{\partial T}\right)_V \delta T + \left[\frac{1}{T}\left(\frac{\partial E}{\partial V}\right)_T + \frac{P}{T}\right]\delta V \tag{10-42}$$

If we consider S as a function of T and V,

$$\delta S = \left(\frac{\partial S}{\partial T}\right)_V \delta T + \left(\frac{\partial S}{\partial V}\right)_T \delta V \tag{10-43}$$

Equating coefficients of like terms in Eq. 10-42 and Eq. 10-43, we obtain

$$\left(\frac{\partial S}{\partial T}\right)_V = \frac{1}{T}\left(\frac{\partial E}{\partial T}\right)_V$$

$$= \frac{1}{T} C_V \tag{10-44}$$

and

$$\left(\frac{\partial S}{\partial V}\right)_T = \frac{1}{T}\left(\frac{\partial E}{\partial V}\right)_T + \frac{P}{T} \tag{10-45}$$

Since S is a state function, the second derivative with respect to T and V has the same value independent of the order of differentiation, Thus,

$$\frac{\partial^2 S}{\partial T \partial V} = \left[\frac{\partial}{\partial T}\left(\frac{\partial S}{\partial V}\right)_T\right]_V = \left[\frac{\partial}{\partial V}\left(\frac{\partial S}{\partial T}\right)_V\right]_T = \frac{\partial^2 S}{\partial V \partial T}$$

and

$$\frac{1}{T}\frac{\partial^2 E}{\partial T \partial V} - \frac{1}{T^2}\left(\frac{\partial E}{\partial V}\right)_T - \frac{P}{T^2} + \frac{1}{T}\left(\frac{\partial P}{\partial T}\right)_V = \frac{1}{T}\frac{\partial^2 E}{\partial V \partial T}$$

From Eq. 9-19,

$$\frac{\partial^2 E}{\partial T \partial V} = \frac{\partial^2 E}{\partial V \partial T}$$

so that

$$\left(\frac{\partial E}{\partial V}\right)_T = T\left(\frac{\partial P}{\partial T}\right)_V - P \tag{10-46}$$

Exercise 10-6. By considering the relation between δS and δH, show that $(\partial H/\partial P)_T$ is equal to $V - T\,(\partial V/\partial T)_P$.

For an ideal gas, $P = nRT/V$,

$$\left(\frac{\partial P}{\partial T}\right)_V = \frac{nR}{V}$$

and

$$\left(\frac{\partial E}{\partial V}\right)_T = 0$$

Equation 10-45 thus reduces to

$$\left(\frac{\partial S}{\partial V}\right)_T = \frac{P}{T} = \frac{nR}{V} \tag{10-47}$$

In an isothermal expansion, from Eq. 10-43,

$$\delta S = \left(\frac{\partial S}{\partial V}\right)_T \delta V$$

$$= \frac{nR}{V}\delta V \tag{10-48}$$

and

$$\Delta S = S_2 - S_1$$

$$= \int_{V_1}^{V_2} nR \frac{dV}{V}$$

$$= n R \ln \frac{V_2}{V_1} \qquad (10\text{-}49)$$

Exercise 10-7. Calculate the entropy change in the isothermal reversible expansion of 0.0350 mole of ideal gas from a volume of 0.890 dm³ to a volume of 2.98 dm³. What is the entropy change if the same change of state is carried out irreversibly at a constant external pressure equal to the final pressure of the system ($T = 300$ K)? Calculate the entropy change in the surroundings for each case, assuming that heat transfer between the system and surroundings is reversible. Calculate the total entropy change for the system and surroundings in each case.

In terms of the statistical model described in Sec. 8-5, only the translational partition function changes in an isothermal expansion. Since the values of the translational energy levels are (Eq. 8-49) inversely proportional to $V^{\frac{2}{3}}$, the expansion of a gas results in more closely spaced energy levels, an increase in g_j, in Eq. 8-83, and an increase in the number of energy levels in the system and, thus, in the value of the partition function z. In terms of the Boltzmann-Planck statistical definition of S (Eq. 8-34), there are many more possible quantum states for an expanded gas and therefore greater entropy.

Exercise 10-8. Show that Eq. 8-101 leads to Eq. 10-49 for the entropy change in the isothermal expansion of an ideal gas.

Isochoric temperature change
When the temperature of a system of constant composition changes at constant volume, it follows from Eq. 10-43 that

$$\delta S = \left(\frac{\partial S}{\partial T} \right)_V \delta T$$

$$= \frac{C_V}{T} \delta T \qquad (10\text{-}50)$$

For a finite change of temperature,

$$\Delta S = S_2 - S_1$$

$$= \int_{T_1}^{T_2} (C_V/T) \, dT \tag{10-51}$$

Equation 10-51 is identical with Eq. 8-43, which was derived from a statistical model.

In the general case, C_V is a function of T and V. Thus, the value of the integral in Eq. 10-51 depends on V as well as the limits T_1 and T_2. For an ideal gas, C_V is a function only of T (Eq. 9-20) and the value of ΔS depends only on the limits of integration.

Exercise 10-9. Calculate the entropy change for the heating of 0.0350 mole of a monatomic ideal gas from 300 K to 400 K.

Isobaric temperature change
When the temperature of a system is changed at constant pressure, from Eq. 10-27 and Eq. 9-42,

$$\delta S = \frac{\delta Q_{\text{rev}}}{T}$$

$$= \frac{\delta H}{T} \tag{10-52}$$

From Eq. 9-57, if the pressure is constant,

$$\delta H = \left(\frac{\partial H}{\partial T} \right)_P \delta T$$

$$= C_p \, \delta T \tag{10-53}$$

so that

$$\delta S = \frac{C_p}{T} \delta T \tag{10-54}$$

For a finite change of temperature,

$$\Delta S = S_2 - S_1 = \int_{T_1}^{T_2} \frac{C_p}{T} \, dT \tag{10-55}$$

Adiabatic expansion of gases
Since S is a state function, it may seem strange to have $\Delta S = 0$ for the reversible adiabatic expansion between state A and state B, whereas $\Delta S > 0$ for the irreversible adiabatic expansion from the same initial state to the same final volume. It was pointed out in Sec. 9-3 that the final temperature

in an adiabatic expansion depends on the amount of work done. In particular, the *temperature decrease is less* for an irreversible expansion than for a reversible reaction, and the *entropy increase is therefore greater*.

Changes of phase

Consider the conversion of liquid water to ice at $0°C$ and 1.01×10^5 Pa pressure.

$$H_2O \quad (l, 1.01 \times 10^5 \text{ Pa}, 298.15 \text{ K}) = H_2O \quad (c, 1.01 \times 10^5 \text{ Pa}, 298.15 \text{ K})$$
$$(10\text{-}56)$$

Since the system is at equilibrium, the change of phase can be carried out reversibly, and since it is at constant pressure, δQ is equal to δH. Thus,

$$\delta S = \frac{\delta Q_{rev}}{T}$$

$$= \frac{\delta H}{T}$$

For a finite change,

$$\Delta S = \int_1^2 \frac{dH}{T}$$

$$= \frac{1}{T} \int_1^2 dH$$

$$= \frac{\Delta H}{T} \tag{10-57}$$

Since the heat lost by the system is equal to that gained by the surroundings, the entropy change of the surroundings is equal in magnitude and opposite in sign to that of the system, in accord with the second law.

The conversion of supercooled liquid water to ice at $-10°C$ is a spontaneous, irreversible change of state, though the entropy change of the system is negative. In order to calculate ΔS, it is necessary to find a reversible path between the initial and final states. One such path is

$$H_2O(l, 263 \text{ K}, 1.01 \times 10^5 \text{ Pa}) = H_2O(l, 273 \text{ K}, 1.01 \times 10^5 \text{ Pa}) \qquad \Delta \mathscr{S}_1$$

$$H_2O(l, 273 \text{ K}, 1.01 \times 10^5 \text{ Pa}) = H_2O(c, 273 \text{ K}, 1.01 \times 10^5 \text{ Pa}) \qquad \Delta \mathscr{S}_2$$

$$H_2O(c, 273 \text{ K}, 1.01 \times 10^5 \text{ Pa}) = H_2O(c, 263 \text{ K}, 1.01 \times 10^5 \text{ Pa}) \qquad \Delta \mathscr{S}_3$$

$$\Delta \mathscr{S}_1 = \int_{263 \text{ K}}^{273 \text{ K}} \mathscr{C}_p (l) \frac{dT}{T} \approx \mathscr{C}_p (l) \ln \frac{273}{263}$$

$$\Delta \mathscr{S}_2 = \frac{\Delta \mathscr{H}_{273}}{273 \text{ K}}$$

and

$$\Delta \mathscr{S}_3 = \int_{273 \text{ K}}^{263 \text{ K}} \mathscr{C}_p \text{ (c)} \frac{dT}{T} \approx \mathscr{C}_p \text{ (c)} \ln \frac{263}{273}$$

if $\mathscr{C}_p$ is constant over this temperature range. Thus,

$$\Delta \mathscr{S}_{263} = \frac{\Delta \mathscr{H}_{273}}{273} + [\mathscr{C}_p \text{ (c)} - \mathscr{C}_p \text{ (l)}] \ln \frac{263}{273}$$

$$= \Delta \mathscr{S}_{273} + \Delta \mathscr{C}_p \ln \frac{263}{273} \qquad (10\text{-}58)$$

It is easier to obtain the same equation by analogy to Eq. 9-71 for ΔH.

$$\Delta \mathscr{S}_{263} = \Delta \mathscr{S}_{273} + \int_{273 \text{ K}}^{263 \text{ K}} (\Delta \mathscr{C}_p) \frac{dT}{T} \qquad (10\text{-}59)$$

If the temperature change is small enough that $\Delta \mathscr{C}_p$ can be considered a constant,

$$\Delta \mathscr{S}_{263} \approx \Delta \mathscr{S}_{273} + \Delta \mathscr{C}_p \ln \frac{263}{273}$$

To compare $\Delta \mathscr{S}_{263}$ to $Q_{263}/263$, we can use Eq. 9-71 to calculate $\Delta \mathscr{H}_{263}$.

Exercise 10-10. The $\Delta \mathscr{H}_{273}$ for the crystallization of water is 6008 J-mol^{-1}. Assume that the molar heat capacities of liquid water and crystalline water (ice) are constant and have the respective values 75.4 J-mol^{-1}-K^{-1} and 37.1 J-mol^{-1}-K^{-1}. Calculate $\Delta \mathscr{S}$ and Q/T for the crystallization of 20 g of H_2O at 273 K and 263 K, and compare the results with Eq. 10-40.

10-5. The calculation of standard entropies

We now have a thermodynamic basis for calculating the *standard entropy*, $\mathscr{S}_T^\circ - \mathscr{S}_0^\circ$, for any substance. For simple substances, we can compare the thermodynamic result with the statistical calculation of entropy described in Chapter 8. The standard state for a gas is zero pressure, or the hypothetical ideal gas state at a pressure of 1.01×10^5 Pa. The standard state for pure solids and pure liquids is the most stable form at a given temperature and at a pressure of 1.01×10^5 Pa.

For the thermodynamic calculation, we can use Eq. 10-55 and Eq. 10-57. The data required are the heat capacity of the substance from a temperature near 0 K to the temperature of interest and the values of ΔH for any phase transitions that occur in this temperature range. The Debye equation in the form

$$\mathscr{C}_p \approx \mathscr{C}_V = (1910 \text{ J-mol}^{-1} \text{ K}^{-1})\left(\frac{T}{\theta_D}\right)^3 \qquad (10\text{-}60)$$

is used to evaluate the constant θ_D from an experimental value of $\mathscr{C}_p$ at a temperature below 20 K. Then

$$\Delta\mathscr{S}_1 = S^\ominus_{20} - S^\ominus_0$$

$$= \int_{0\,\text{K}}^{20\,\text{K}}\left(\frac{1910}{\theta_D{}^3}\right) T^2\, dT$$

$$= \left(\frac{1910}{\theta_D{}^3}\right)\left[\frac{(20)^3}{3}\right]$$

From 20 K to the melting point T_M,

$$\Delta\mathscr{S}_2 = \mathscr{S}^\ominus_{T_M} - \mathscr{S}^\ominus_{20}$$

$$= \int_{20\,\text{K}}^{T_M} \mathscr{C}_p\,(c)\,\frac{dT}{T}$$

This integral is evaluated graphically or numerically from heat capacity data for the solid over the temperature range (see Appendix C-2). Either $\mathscr{C}_p/T$ as a function of T, or $\mathscr{C}_p$ as a function of $\ln T$, can be used as the variable for the integration. If solid-solid transitions occur, the entropy of transition, as calculated from Eq. 10-57, must be included.

The entropy change on melting is

$$\Delta\mathscr{S}_3 = \frac{\Delta\mathscr{H}_M}{T_M}$$

From the melting point to the boiling point, the entropy change over the liquid range is

$$\Delta\mathscr{S}_4 = \int_{T_M}^{T_B} \mathscr{C}_p\,(l)\,\frac{dT}{T}$$

The entropy of vaporization is

$$\Delta\mathscr{S}_5 = \frac{\Delta\mathscr{H}_v}{T_B}$$

The entropy change of the gas from T_B to the temperature T at which the standard entropy is desired is

$$\Delta \mathscr{S}_6 = \int_{T_B}^{T} \mathscr{C}_p \, (\text{g}) \, \frac{dT}{T}$$

Since the standard state for the entropy of a gas is *the ideal gas* at 1.01×10^5 Pa pressure, a small correction is made for imperfection of the gas.[4]

If the crystal is in a single quantum state at 0 K, the entropy calculated from the statistical model should agree with the value of $\mathscr{S}_T^{\ominus} - \mathscr{S}_0^{\ominus}$ calculated thermodynamically. Possible reasons for lack of agreement have been discussed in Sec. 8-4.

Exercise 10-11. J. O. Clayton and W. F. Giauque [*J. Am. Chem. Soc.*, *54*, 2610, (1932)] made careful measurements of the heat capacity of solid and liquid CO and of the heats of phase transitions over the temperature range from 11.70 K to the normal boiling point at 81.61 K. Figure 10-5, showing the heat capacity data of two crystalline forms and the liquid, is taken from the original paper. Calculate the standard entropy at 298.15 K of 1 mole of CO in the ideal gas state at 1.01×10^5 Pa. (All measurements were at 1.01×10^5 Pa.) The molar heat capacity at 11.70 K is 6.088 J-mol^{-1}-K^{-1}.

$$\int_{11.70 \text{ K}}^{61.55 \text{ K}} \mathscr{C}_p \, \frac{dT}{T} = 40.30 \text{ J-mol}^{-1}\text{-K}^{-1}$$

The value of $\Delta \mathscr{H}^{\ominus}$ for the solid to solid transition at 61.55 K is 633.0 J-mol^{-1}.

$$\int_{61.55 \text{ K}}^{68.09 \text{ K}} \mathscr{C}_p \, \frac{dT}{T} = 5.138 \text{ J-mol}^{-1}\text{-K}^{-1}$$

The heat of fusion at 68.09 K is 815.6 J-mol^{-1}.

$$\int_{68.09 \text{ K}}^{81.61 \text{ K}} \mathscr{C}_p \, \frac{dT}{T} = 10.92 \text{ J-mol}^{-1}\text{-K}^{-1}$$

The heat of vaporization at 81.61 K is 6040.0 J-mol^{-1}. The correction for imperfection of the gas at the boiling point is 0.88 J-mol^{-1}-K^{-1}. The $\Delta \mathscr{S}$ for the ideal gas from 81.61 K to 298.15 K was calculated from the statistical model as 37.64 J-mol^{-1}-K^{-1}.

Exercise 10-12. Explain qualitatively, on the basis of differences in structure, the relative values of $\mathscr{S}^{\ominus} - \mathscr{S}_0^{\ominus}$ for (a) H_2 and D_2; (b) $H_2O(g)$ and $H_2O(l)$; (c) SO_2 and SO_3; (d) O_2 and O_3.

Table 10-1 lists the standard entropy of several substances at 298.15 K.

4. I. M. Klotz and R. M. Rosenberg, *Chemical Thermodynamics*, 3rd ed., W. A. Benjamin, Menlo Park, 1972, pp. 190–93.

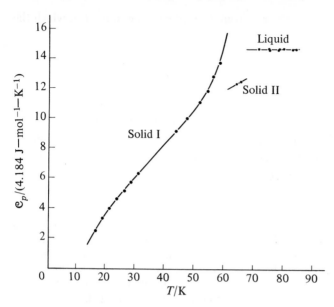

Figure 10-5. Plot of $\mathscr{C}_p$ for carbon monoxide as a function of temperature from 11.70 K to the normal boiling point at 81.61 K, showing values for two crystalline forms and the liquid. [By permission, from J. O. Clayton and W. F. Giauque, *J. Am. Chem. Soc., 54,* 2610 (1932).]

Table 10-1. Standard entropy at 298.15 K

Substance	$(\mathscr{S}^{\ominus} - \mathscr{S}^{\ominus}_0)/$ $(\text{J-mol}^{-1}\text{-K}^{-1})$	Substance	$(\mathscr{S}^{\ominus} - \mathscr{S}^{\ominus}_0)/$ $(\text{J-mol}^{-1}\text{-K}^{-1})$
$O_2(g)$	205.03	$CH_4(g)$	186.15
$O_3(g)$	238.8	$C(graphite)$	5.740
$H_2(g)$	130.57	$C(diamond)$	2.38
$HD(g)$	143.69	$C_2H_2(g)$	200.8
$D_2(g)$	144.85	$C_2H_4(g)$	219.5
$H_2O(g)$	188.72	$SiO_2(quartz)$	41.84
$H_2O(l)$	69.91	$SiO_2(cristobalite)$	42.68
$HF(g)$	173.67	$SiO_2(tridymite)$	43.5
$Cl_2(g)$	222.96	$B_2H_6(g)$	232.0
$HCl(g)$	186.80	$Al_2O_3(corundum)$	50.92
$SO_2(g)$	248.1	$Ag(c)$	42.68
$SO_3(g)$	256.6	$AgCl(c)$	96.2
$N_2(g)$	191.5	$C_6H_6(g)$	269.2
$NH_3(g)$	192.3	$NH_4Cl(c)$	94.6
$PCl_3(g)$	311.7	$PCl_5(g)$	364.5
$NO_2(g)$	240.0	$N_2O_4(g)$	304.2
$CO(g)$	197.56	$CO_2(g)$	213.6

NBS Technical Note 270–3, "Selected Values of Chemical Thermodynamic Properties", 1968.

10-6.   Irreversibility, spontaneity, and equilibrium. The free energy functions

The mathematical statement of the second law in Eq. 10-40 can be applied as a criterion to distinguish between reversible and irreversible processes. If the system is *adiabatic*, Eq. 10-38 is applicable. If the system is of constant volume, so that no mechanical work can be done, and if no other external forces, such as electrical, magnetic, or gravitational forces, exchange work with the system, then any *irreversible* process must correspond to a *spontaneous* change of state of the system. A reversible process occurs only if such a system is *at equilibrium*, that is, if the system is in a state such that any fluctuation is followed by a return to the initial state. In terms of the statistical model, a fluctuation away from the most probable distribution is followed by a return to the most probable distribution.

Since Q and W are both equal to zero for the system just described, $\Delta E = 0$. Thus, *the criterion for spontaneity and equilibrium for a system at constant E and V is*

$$\delta S \geqslant 0$$

or, *for finite change,*

$$\Delta S \geqslant 0$$

where the equality refers to an equilibrium state and the inequality to a spontaneous change of state.

Systems at constant temperature and volume
For a system at constant volume, if no work other than pressure-volume work is done,

$$\delta Q = \delta E$$

and Eq. 10-40 becomes

$$\delta S \geqslant \frac{\delta E}{T}$$

or

$$\delta E - T \delta S \leqslant 0 \tag{10-61}$$

When T is constant, $- S \delta T$ can be added to the left side of Eq. 10-61 without changing its value. Thus,

$$\delta E - T \delta S - S \delta T \leqslant 0$$

$$\delta E - \delta(TS) \leqslant 0$$

and

$$\delta(E - TS) \leqslant 0 \qquad (10\text{-}62)$$

The equality in Eq. 10-62 refers to an *equilibrium state*, since a reversible process can take place only in a system at equilibrium if the temperature and volume are fixed and no work other than pressure-volume work is done. The inequality refers to a *spontaneous change of state*, which is irreversible in the absence of an opposing force.

Since E, T, and S are state functions, the quantity in parentheses in Eq. 10-62 is also a state function, designated the *Helmholtz free energy*, A, where

$$A = E - TS \qquad (10\text{-}63)$$

The criterion of spontaneity and equilibrium at constant temperature and volume then is

$$\delta A \leqslant 0 \qquad (10\text{-}64)$$

or, *for a finite change,*

$$\Delta A = (A_2 - A_1) \leqslant 0 \qquad (10\text{-}65)$$

Systems at constant temperature and pressure
Most chemical, biological, and geological processes occur under conditions of constant temperature and pressure. It is therefore useful to derive a criterion for spontaneity and equilibrium applicable to these conditions. For a system at constant pressure, in which the only work done is that against the constant pressure of the environment,

$$\begin{aligned}\delta Q &= \delta E - \delta W \\ &= \delta E + P\,\delta V \end{aligned} \qquad (10\text{-}66)$$

Substitution from Eq. 10-66 in Eq. 10-40 leads to

$$\delta S \geqslant \frac{\delta E + P\,\delta V}{T}$$

or

$$\delta E + P\,\delta V - T\,\delta S \leqslant 0 \qquad (10\text{-}67)$$

Since the temperature and pressure are constant, we can add $V\,\delta P$ and $-S\,\delta T$ to the left side of Eq. 10-67 without changing its value. Thus,

$$\delta E + P\,\delta V + V\,\delta P - T\,\delta S - S\,\delta T \leqslant 0$$
$$\delta E + \delta(PV) - \delta(TS) \leqslant 0$$

$$\delta(E + PV - TS) \leqslant 0$$
$$\delta(H - TS) \leqslant 0 \qquad (10\text{-}68)$$

Since the only work done is that against the constant pressure of the environment, the inequality in Eq. 10-68 refers to a *spontaneous change of state* as well as an irreversible process, and the equality refers to an *equilibrium state* as well as a reversible process. Since H, T, and S are state functions, the quantity in parentheses in Eq. 10-68 is also a state function, designated the *Gibbs free energy*,

$$G = H - TS \qquad (10\text{-}69)$$

The criterion for spontaneity and equilibrium at constant temperature and pressure then is

$$\delta G \leqslant 0 \qquad (10\text{-}70)$$

or, *for a finite change of state,*

$$\Delta G = (G_2 - G_1) \leqslant 0 \qquad (10\text{-}71)$$

Spontaneous changes of state, for which $\Delta G < 0$, are termed *exergonic*, whereas nonspontaneous changes of state, for which $\Delta G > 0$, are termed *endergonic*.

Exercise 10-13. Determine the criterion for spontaneity and equilibrium at (a) constant S, V, and (b) constant S, P, if no work other than pressure-volume work is done.

10-7. Work and the free energy functions

When the first law expression for δQ (Eq. 9-7) is substituted in Eq. 10-40, the result is

$$\delta S \geqslant \frac{\delta E - \delta W}{T}$$

or

$$\delta E - T\,\delta S \leqslant \delta W \qquad (10\text{-}72)$$

Equation 10-72 is a general statement of the *combined first and second laws of thermodynamics.*

For a change of state at constant temperature, $-S\,\delta T$ can be added to the left side of Eq. 11-72 without a change of value. Thus,

$$\delta E - T\delta S - S\delta T \leqslant \delta W$$
$$\delta A \leqslant \delta W \qquad (10\text{-}73)$$

or, for a finite change

$$\Delta A = (A_2 - A_1) \leqslant W \tag{10-74}$$

Because $\delta A = \delta W_{rev}$ for a reversible, isothermal process, A is sometimes called the *work function*. For an irreversible process, $\delta A < \delta W_{irr}$. When the change of state is spontaneous, $\delta A < 0$ and $\delta W_{irr} < 0$. Thus

$$|\delta A| > |\delta W_{irr}| \quad \text{and} \quad |\delta W_{irr}| < |\delta W_{rev}| \quad \text{(spontaneous)} \tag{10-75}$$

The magnitude of the reversible work obtained from a given spontaneous change of state is a *maximum* for that change.

When an external agent does work on a system and thereby produces a nonspontaneous change of state in the system, $\delta A > 0$ and $\delta W > 0$. The equality $\delta A = \delta W_{rev}$ and the inequality $\delta A < \delta W_{irr}$ still apply. Since δA and δW are both positive in this instance,

$$|\delta W_{irr}| > |\delta W_{rev}| \quad \text{(nonspontaneous)} \tag{10-76}$$

Thus, in producing a nonspontaneous change of state in a system, the *minimum* work is done by a reversible process. One of the most common examples of the application of Eqs. 10-75 and 10-76 is to the charging and discharging of a galvanic cell. The limiting value of the work obtained from the discharge of a galvanic cell is calculated from the voltage when no current is drawn, the reversible potential. Similarly, it takes more work to charge a galvanic cell at a finite rate, irreversibly, than it does to charge it reversibly. Of course, the latter process requires an infinite time, since the system must be allowed to come to equilibrium after each infinitesimal change.

It is sometimes convenient to express δW in Eq. 10-72 as the sum of the pressure-volume work and all other work δW_{net}. Equation 10-72 then becomes

$$\delta E - T\delta S \leqslant \delta W_{net} - P_{ex} \, \delta V \tag{10-77}$$

For a system at constant temperature and pressure $P_{ex} = P$, and $-V\delta P$ and $-S\,\delta T$ can be added without changing the equation. Thus,

$$\delta E - T\delta S - S\delta T + P\delta V + V\delta P \leqslant \delta W_{net}$$

That is,

$$\delta G \leqslant \delta W_{net} \tag{10-78}$$

or for a finite change,

$$\Delta G = (G_2 - G_1) \leqslant W_{net} \tag{10-79}$$

The change in the Gibbs free energy, then, is a limiting value of the work other than pressure-volume work at constant pressure and temperature. For the reasons given above for the Helmholtz free energy,

$$\left| \delta W_{\text{net, rev}} \right| > \left| \delta W_{\text{net, irr}} \right| \quad \text{(spontaneous)} \tag{10-80}$$

and

$$\left| \delta W_{\text{net, rev}} \right| < \left| \delta W_{\text{net, irr}} \right| \quad \text{(nonspontaneous)} \tag{10-81}$$

Exercise 10-14. Verify Eqs. 10-80 and 10-81.

Equation 10-78 is particularly important in determining the maximum amount of non-pressure–volume work that can be obtained from chemical reactions in biological systems. It is applicable to osmotic work of transport against concentration gradients, to mechanical work of such contractile tissue as muscle and flagella, and to electrical work in the transmission of nerve impulses and in discharges in the electric eel.[5]

Exercise 10-15. Find the state function whose change is equal to $\delta W_{\text{net, rev}}$ under conditions of (a) constant E, V; (b) constant S, V; (c) constant S, P.

10-8. Properties of the free energy functions

Equation 10-69 defines the Gibbs free energy as

$$G = H - TS$$

$$= E + PV - TS \tag{10-82}$$

and for an infinitesimal change of state,

$$\delta G = \delta E + P\,\delta V + V\,\delta P - T\,\delta S - S\,\delta T \tag{10-83}$$

From the first law (Eq. 9-7),

$$\delta E = \delta Q + \delta W$$

For convenience in calculating δG for a given change of state, we choose to carry out that change of state reversibly, and with only pressure-volume work, so that $\delta Q = T\,\delta S$ and $\delta W = -P\,\delta V$. Substitution of these relations in Eq. 10-83 gives us

$$\delta G = V\,\delta P - S\,\delta T \tag{10-84}$$

5. A. L. Lehninger, *Bioenergetics*, W. A. Benjamin, New York, 1965; I. M. Klotz, *Energy Changes in Biochemical Reactions*, 2nd ed., Academic Press, New York, 1967; W. H. Cropper, *J. Chem. Educ.*, **48**, 182 (1971).

Exercise 10-16. Verify Eq. 10-84.

If we consider G as a function of T and P, we can write

$$\delta G = \left(\frac{\partial G}{\partial T}\right)_P \delta T + \left(\frac{\partial G}{\partial P}\right)_T \delta P \tag{10-85}$$

Comparison of Eq. 10-84 and Eq. 10-85 leads to the results

$$\left(\frac{\partial G}{\partial T}\right)_P = -S \tag{10-86}$$

and

$$\left(\frac{\partial G}{\partial P}\right)_T = V \tag{10-87}$$

Exercise 10-17. By similar reasoning, show that

$$\left(\frac{\partial A}{\partial T}\right)_V = -S \tag{10-88}$$

and

$$\left(\frac{\partial A}{\partial V}\right)_T = -P \tag{10-89}$$

Since G is a state function,

$$\frac{\partial^2 G}{\partial T \,\partial V} = \left[\frac{\partial}{\partial T}\left(\frac{\partial G}{\partial P}\right)_T\right]_P = \left[\frac{\partial}{\partial P}\left(\frac{\partial G}{\partial T}\right)_P\right]_T = \frac{\partial^2 G}{\partial P \,\partial T}$$

so that

$$\left(\frac{\partial V}{\partial T}\right)_P = -\left(\frac{\partial S}{\partial P}\right)_T \tag{10-90}$$

Exercise 10-18. From the properties of A, show that

$$\left(\frac{\partial P}{\partial T}\right)_V = \left(\frac{\partial S}{\partial V}\right)_T \tag{10-91}$$

Since, for a finite change of state

$$\Delta G = G_2 - G_1 \tag{10-92}$$

we can write, using Eq. 10-86,

$$\left(\frac{\partial \Delta G}{\partial T}\right)_P = \left(\frac{\partial G_2}{\partial T}\right)_P - \left(\frac{\partial G_1}{\partial T}\right)_P$$

$$= -S_2 + S_1$$

$$= -\Delta S \tag{10-93}$$

and, similarly, using Eq. 10-87,

$$\left(\frac{\partial \Delta G}{\partial P} \right)_T = \Delta V \tag{10-94}$$

ΔG for isothermal changes of state

From the definition of the Gibbs free energy, we can write for a macroscopic, isothermal change in state

$$\Delta G = \Delta H - T\,\Delta S \tag{10-95}$$

At sufficiently low temperatures,

$$|T\,\Delta S| \ll |\Delta H|$$

and

$$\Delta G \approx \Delta H$$

For many reactions, this condition is fulfilled at ordinary temperatures, and this fact accounts for the early misconception that ΔH is a criterion for spontaneity and equilibrium at constant temperature and pressure. At sufficiently high temperatures,

$$|T\,\Delta S| \gg |\Delta H|$$
$$\Delta G \approx T\,\Delta S$$

and the entropy change is the controlling factor in the determination of spontaneity and equilibrium. These thermodynamic conclusions are consistent with the conclusions derived from a statistical model in Sec. 8-6. The thermodynamic quantity ΔH is analogous to the statistical quantity $\Delta \varepsilon_0$; both are the controlling factors in equilibrium at low temperature. The thermodynamic quantity ΔS is analogous to $\ln z_B/z_A$.

The change in the Gibbs free energy for an isothermal change of state can be related to the entropy change in the system and in the surroundings. From Eq. 10-95,

$$\frac{\Delta G}{T} = \frac{\Delta H}{T} - \Delta S$$

where ΔS is the entropy change of the system. At constant pressure, if only mechanical work is done (Eq. 9-43),

$$\Delta H = Q = -Q_{surr}$$

If we assume that the heat exchange with the surroundings, at the same temperature as the system, is reversible, then

$$\Delta S_{\text{surr}} = \frac{Q_{\text{surr}}}{T}$$

Thus,

$$\frac{\Delta G}{T} = -\Delta S_{\text{surr}} - \Delta S$$

and

$$-\frac{\Delta G}{T} = \Delta S \quad (\text{system} + \text{surroundings})$$

The quantity $\Delta G/T$ then provides a measure of the entropy change for system and surroundings in terms of properties of the system only.

10-9. Statistical calculation of free energy functions

Since (Eq. 10-63)

$$A = E - TS$$

a statistical expression for A can be obtained from the corresponding expressions for E and S. For an ideal gas (Eq. 8-70),

$$E - E_0 = NkT^2 \left(\frac{\partial \ln z}{\partial T} \right)_V$$

and (Eq. 8-92)

$$S = kN \ln \frac{z}{N} + \frac{E - E_0}{T} + kN$$

Thus,

$$A = -NkT \ln \frac{z}{N} - NkT + E_0 \tag{10-96}$$

and the molar Helmholtz free energy is

$$\mathscr{A} = -RT \ln \frac{z'}{L} - RT + \mathscr{E}_0 \tag{10-97}$$

or

$$\mathscr{A} - \mathscr{E}_0 = -RT \ln \frac{z'}{L} - RT \tag{10-98}$$

where z' is the partition function calculated using the volume per mole. The standard state value $\mathscr{A}^{\ominus} - \mathscr{E}_0^{\ominus}$ is calculated from Eq. 10-98 by choosing a molar volume at each temperature such that the pressure of the gas is 1.01×10^5 Pa, that is

$$\mathscr{V}^{\ominus} = \frac{RT}{P^{\ominus}}$$

The contributions of translational, rotational, vibrational, and electronic motions to the Helmholtz free energy can be calculated from the corresponding contributions to E and S.

The Gibbs free energy (Eq. 10-69) is

$$G = H - TS$$
$$= A + PV \qquad (10\text{-}99)$$

A statistical expression for P can be obtained through Eq. 10-89,

$$P = -\left(\frac{\partial A}{\partial V}\right)_T$$

$$= NkT\left(\frac{\partial \ln z}{\partial V}\right)_T \qquad (10\text{-}100)$$

Since the translational partition function is the only one that depends on V, and directly proportional to V,

$$\left(\frac{\partial \ln z}{\partial V}\right)_T = \frac{1}{V} \qquad (10\text{-}101)$$

and

$$P = Nk\,\frac{T}{V} \qquad (10\text{-}102)$$

$$= nR\,\frac{T}{V}$$

showing that the ideal gas law can be derived from a statistical model as well as from the kinetic theory. Thus,

$$G = -NkT \ln \frac{z}{N} - NkT + E_0 + NkT$$

$$= -NkT \ln \frac{z}{N} + E_0 \qquad (10\text{-}103)$$

and the molar Gibbs free energy is

$$\mathscr{G} = -RT \ln \frac{z'}{L} + \mathscr{E}_0 \qquad (10\text{-}104)$$

or

$$\mathscr{G} - \mathscr{E}_0 = -RT \ln \frac{z'}{L} \tag{10-105}$$

and

$$\mathscr{G}^\ominus - \mathscr{E}_0^\ominus = -RT \ln \frac{z^\ominus}{L} \tag{10-106}$$

where $\mathscr{V}^\ominus$ is used to calculate $z^\ominus$.

Exercise 10-19. Derive explicit expressions for the translational, rotational, vibrational, and electronic contributions to $\mathscr{G} - \mathscr{E}_0$ for a diatomic molecule. Assume that the high temperature expressions for the rotational and vibrational partition functions apply.

Exercise 10-20. Show that z/N and z'/L are dimensionless quantities and that they are *intensive* quantities.

Let us use Eq. 10-106 to calculate the quantity $\mathscr{G}^\ominus - \mathscr{E}_0^\ominus$ for CO at 300 K. The partition function can be expressed (Eq. 8-72) as

$$z = z_{elec} z_{vib} z_{rot} z_{tran}$$

At 300 K, $kT = (1.38 \times 10^{-23} \text{ J-K}^{-1})(300 \text{ K}) = 4.14 \times 10^{-21}$ J. The strong absorption band of CO at 1546×10^{-10} m (Exercise 8-18) indicates that the first excited electronic state is at an energy

$$\varepsilon - \varepsilon_0 = h\nu = \frac{hc}{\lambda}$$

$$= \frac{(6.63 \times 10^{-34} \text{ J-s})(3.00 \times 10^8 \text{ m-s}^{-1})}{(1546 \times 10^{-10} \text{ m})}$$

$$= 1.29 \times 10^{-18} \text{ J}$$

This value is so large with respect to kT that essentially all the molecules are in the ground electronic state, and $z_{elec} = 1$. The strong vibrational band of CO appears at a wavelength of 4.65×10^{-6} m (Exercise 8-20). The energy spacing of vibrational levels is therefore

$$\Delta\varepsilon = h\nu = \frac{hc}{\lambda}$$

$$= \frac{(6.63 \times 10^{-34} \text{ J-s})(3.00 \times 10^8 \text{ m-s}^{-1})}{(4.65 \times 10^{-6} \text{ m})}$$

$$= 4.28 \times 10^{-20} \text{ J}$$

This value is equal to 10.33 kT, so that

$$z_{vib} = \left[1 - \exp\left(-\frac{\Delta\varepsilon}{kT}\right)\right]^{-1} = (1 - 3.26 \times 10^{-5})^{-1} \approx 1.00$$

From Exercises 6-4 and 18-19, Δv for the rotational spectrum of CO is 1.15×10^{11} Hz. The energy level spacing between $J = 0$ and $J = 1$ is $h\Delta v = 2\,Bh$, so

$$\Delta\varepsilon = (6.63 \times 10^{-34} \text{ J-s})(1.15 \times 10^{11} \text{ s}^{-1})$$
$$= 7.61 \times 10^{-23} \text{ J}$$

This value is sufficiently small with respect to kT that the rotational partition function is given by its classical limit

$$z_{rot} = \frac{kT}{Bh}$$

$$= \frac{(4.14 \times 10^{-21} \text{ J})}{(3.80 \times 10^{-23} \text{ J})}$$

$$= 110$$

The value of $z^{\ominus}$ is obtained by calculating z_t for a value of $\mathscr{V}^{\ominus}$ at 300 K and a pressure of 1.01×10^5 Pa. Thus,

$$\mathscr{V}^{\ominus} = \frac{RT}{P^{\ominus}} = \frac{(8.31 \text{ J-mol}^{-1}\text{-K}^{-1})(300 \text{ K})}{(1.01 \times 10^5 \text{ Pa})}$$

$$= 2.47 \times 10^{-2} \text{ m}^3\text{-mol}^{-1}$$

and

$$z^{\ominus}_{tran} = \left(\frac{2\pi mkT}{h^2}\right)^{\frac{3}{2}} \mathscr{V}^{\ominus}$$

$$= \left[\frac{2\pi(28 \times 10^{-3} \text{ kg-mol}^{-1}/6.02 \times 10^{23} \text{ mol}^{-1})(4.14 \times 10^{-21} \text{ J})}{(6.63 \times 10^{-34} \text{ J-s})^2}\right]^{\frac{3}{2}}$$

$$\times (2.47 \times 10^{-2} \text{ m}^3\text{-mol}^{-1})$$

$$= 3.55 \times 10^{30} \text{ mol}^{-1}$$

Then

$$\frac{z^{\ominus}}{L} = \frac{z_{rot}\,z^{\ominus}_{tran}}{L}$$

$$= \frac{(110)(3.55 \times 10^{30} \text{ mol}^{-1})}{6.02 \times 10^{23} \text{ mol}^{-1}}$$

$$= 6.49 \times 10^8$$

and

$$\ln \frac{z^{\ominus}}{L} = 20.28$$

Thus

$$\mathcal{G}^{\ominus} - \mathcal{E}_0^{\ominus} = -RT \ln \frac{z^{\ominus}}{L}$$

$$= -(8.31 \text{ J-mol}^{-1}\text{-K}^{-1})(300 \text{ K})(20.28)$$

$$= -5.06 \times 10^4 \text{ J-mol}^{-1}$$

as compared to a value of -5.07×10^4 J-mol^{-1} given in the NBS (U.S.) Circular 500.

Problems

10-1. The statement: "You can't unfry an egg" is a paraphrase of the second law. In terms of Planck's statement of the second law, suggest compensating changes in other systems that would be necessary to reverse the frying of an egg.

10-2. Calculate $(\partial E/\partial V)_T$ for (a) a van der Waals gas; (b) a gas that is described by the equation of state: $PV = nRT + na/V$.

10-3. Calculate $(\partial H/\partial P)_T$ for a gas that is described by the equation of state: $PV = nRT + naP$.

10-4. The following data on the heat capacity of solid CO were obtained by Clayton and Giauque [*J. Am. Chem. Soc.*, *54*, 2610 (1932)]. Use numerical or graphical integration, and the Debye T^3 law to calculate $\mathcal{S}^{\ominus}$ (61.55 K) $- \mathcal{S}_0^{\ominus}$. Compare with the value in Exercise 10-11.

T/K	$\mathcal{C}_p/(\text{J-mol}^{-1}\text{-K}^{-1})$	T/K	$\mathcal{C}_p/(\text{J-mol}^{-1}\text{-K}^{-1})$
14.36	6.849	44.21	37.89
16.94	10.28	44.71	38.03
19.37	13.67	47.90	41.37
21.93	16.64	48.34	41.58
24.31	19.13	52.34	46.07
26.64	21.40	55.07	49.08
29.01	23.77	56.82	53.18
31.56	26.24	59.04	56.94
39.85	33.94		

10-5. Comment on the relation of tightness of binding and mass to the values of $\mathcal{S}^{\ominus} - \mathcal{S}_0^{\ominus}$ in Table 10-1, using specific examples to support your generalizations.

10-6. Calculate ΔG for the isothermal expansion described in Exercise 10-7. Does the value depend on whether the expansion is carried out reversibly? How is ΔG related to ΔS? The expansion of a gas is spontaneous as long as $P_{ex} < P$. Which, if any, of the criteria for spontaneity discussed in this chapter is appropriate for this example? What is the value of ΔA for the expansion?

10-7. Find an alternative derivation for Eq. 10-57, starting with Eq. 10-95.

10-8. Derive the following relationships, using any equations in the chapter:

$$[\partial(\Delta G/T)/\partial T]_P = -\Delta H/T^2 \qquad (10\text{-}107)$$

$$(\partial \Delta G/\partial T)_P = (\Delta G - \Delta H)/T \qquad (10\text{-}108)$$

$$[\partial(\Delta G/T)/\partial(1/T)]_P = \Delta H \qquad (10\text{-}109)$$

10.9. Calculate $\mathscr{G}^{\ominus} - \mathscr{E}_0^{\ominus}$ for $H^{35}Cl$ and $D^{35}Cl$ at 300 K, using spectroscopic data from Chapter 6. In the light of your results, comment on the statement: "Isotopes are forms of an element that differ in atomic mass but are identical in chemical properties". Assume $z_{elec} = 1$.

11 · *Free energy and chemical equilibrium in gases*

The most important application of the second law for chemists, biologists, and geologists, is to the problem of chemical equilibrium. That is, given a set of reactants and products of a chemical reaction, what is the composition of the equilibrium mixture? Le Chatelier pointed out the economic importance of this theoretical and experimental problem when he said[1]:

> It is known that in the blast furnace the reduction of iron oxide is produced by carbon monoxide, according to the reaction
>
> $$Fe_2O_3(c) + 3\,CO(g) = 2\,Fe(c) + 3\,CO_2(g)$$
>
> but the gas leaving the chimney contains a considerable proportion of carbon monoxide, which thus carries away a considerable portion of unutilized heat. Because this incomplete reaction was thought to be due to an insufficiently prolonged contact between carbon monoxide and the iron ore, the dimensions of the furnaces have been increased. In England they have been made as high as thirty meters. But the proportion of carbon monoxide escaping has not diminished, thus demonstrating, by an experiment costing several hundred thousand francs, that the reduction of iron oxide by carbon monoxide is a limited reaction. Acquaintance with the laws of chemical equilibrium would have permitted the same conclusion to be reached more rapidly and far more economically.

11-1. Equilibrium in systems of variable composition. The chemical potential

Chemically reacting systems are necessarily systems of changing composi-

1. H. Le Chatelier, *Ann. Mines* (8), *13*, 157 (1888); as translated in G. N. Lewis and M. Randall, *Thermodynamics*, 2nd ed., revised by K. S. Pitzer and L. Brewer, McGraw-Hill Book Co., New York, 1961, p. 2.

tion. Thus, we need to consider the dependence of thermodynamic variables on the composition of the system in order to apply thermodynamic principles to such a system.

Extensive thermodynamic functions depend on the mole numbers n of all the components in a system as well as the two intensive quantities chosen as independent variables (see preceding chapters). For example,

$$G = f(T, P, n_1, n_2, \ldots, n_k) \qquad (11\text{-}1)$$

for a system containing k chemical species, and the variation in G due to infinitesimal changes in the independent variables is

$$\delta G = \left(\frac{\partial G}{\partial T}\right)_{P, n_\omega} \delta T + \left(\frac{\partial G}{\partial P}\right)_{T, n_\omega} \delta P + \sum_{j=1}^{k}\left(\frac{\partial G}{\partial n_j}\right)_{T, P, n_\beta} \delta n_j \qquad (11\text{-}2)$$

where n_ω represents all mole numbers $n_1, n_2, \ldots n_k$, and n_β represents all the mole numbers except n_j. The partial derivative of G with respect to the mole number of component j—with T, P, and all other mole numbers n_β being constant—is called the *partial molar Gibbs free energy* $\mathcal{G}_j$ or the *chemical potential* μ_j of the jth component. The partial derivatives of G with respect to T and P in Eq. 10-86 and Eq. 10-87 were derived for a system in which all the mole numbers are constant, so that they are equal to the corresponding derivatives in Eq. 11-2. We can now rewrite Eq. 11-2 as

$$\delta G = -S(T, P, n_\omega)\,\delta T + V(T, P, n_\omega)\,\delta P + \sum_{j=1}^{k} \mu_j(T, P, n_\omega)\,\delta n_j \qquad (11\text{-}3)$$

in which we indicate explicitly that S, V, and the μ_j are functions of temperature, pressure, and the mole numbers. The μ_j are *intensive variables*, the derivatives of one extensive variable with respect to another. As such they are functions of the composition of the system, which can be expressed in terms of the ratios of mole numbers of the components, but independent of the absolute magnitudes of the n_ω.

At constant temperature and pressure, Eq. 11-3 becomes

$$\delta G = \sum_{j=1}^{k} \mu_j\,\delta n_j \qquad (11\text{-}4)$$

From Eq. 10-99,

$$\delta G = \delta A + P\delta V + V\delta P$$

From this result and Eq. 11-3, we can write

$$\delta A = -S\delta T - P\delta V + \sum_{j=1}^{k} \mu_j\,\delta n_j \qquad (11\text{-}5)$$

Exercise 11-1. Verify Eq. 11-5.

Thus, the chemical potential can also be seen as

$$\mu_j = \left(\frac{\partial A}{\partial n_j}\right)_{T, V, n_\beta} \tag{11-6}$$

At constant temperature and volume, Eq. 11-5 becomes

$$\delta A = \sum_{j=1}^{k} \mu_j \, \delta n_j \tag{11-7}$$

From Eq. 10-64 and Eq. 10-70, valid for any closed system, we can see that

$$\sum_{j=1}^{k} \mu_j \, \delta n_j \leqslant 0 \tag{11-8}$$

is the *criterion for equilibrium and spontaneity both at constant temperature and pressure and at constant temperature and volume.*
One can show similarly that

$$\mu_j = \left(\frac{\partial H}{\partial n_j}\right)_{S, P, n_\beta} \tag{11-9}$$

$$= \left(\frac{\partial E}{\partial n_j}\right)_{S, V, n_\beta} \tag{11-10}$$

$$= -T\left(\frac{\partial S}{\partial n_j}\right)_{E, V, n_\beta} \tag{11-11}$$

$$= -T\left(\frac{\partial S}{\partial n_j}\right)_{H, P, n_\beta} \tag{11-12}$$

Exercise 11-2. Verify Eqs. 11-9 through 11-12.

The quantity $\sum_{j=1}^{k} \mu_j \, \delta n_j$ is thus a *general criterion for equilibrium and spontaneity*, regardless of the external restraints on the system. (See Exercise 10-13.)

The designation, "potential", was suggested by J. Willard Gibbs for the quantity given in Eqs. 11-2, 11-6, 11-9, 11-10, 11-11, and 11-12.[2] The equilibrium state of a chemical system is one at a minimum chemical potential,

2. J. Willard Gibbs, "On the equilibrium of heterogeneous substances", *Trans. Conn. Acad. Sci.*, III, pp. 108–248, Oct. 1875–May 1876, and pp. 343–524, May 1877–July, 1878, *Collected Works of J. Willard Gibbs*, Vol. I, Yale University Press, New Haven, 1928, p. 65.

just as the equilibrium state of an object in a gravitational field is one at a minimum in the gravitational potential. The chemical system proceeds spontaneously toward this equilibrium state just as an object in a gravitational field spontaneously rolls downhill.

For the reaction

$$sS + uU + \ldots = cC + dD + \ldots \qquad (11\text{-}13)$$

Eq. 11-8 becomes

$$\sum_{j=1}^{k} \mu_j \, \delta n_j = \mu_S \, \delta n_S + \mu_U \, \delta n_U + \ldots + \mu_C \, \delta n_C + \mu_D \, \delta n_D + \ldots \leqslant 0 \qquad (11\text{-}14)$$

The changes in the mole numbers of reactants and products in a chemical reaction are not independent but are related by the stoichiometry of the reaction. Thus, for Eq. 11-13, we define a quantity ξ, such that

$$-\frac{\delta n_S}{s} = -\frac{\delta n_U}{u} = - \ldots = \frac{\delta n_C}{c} = \frac{\delta n_D}{d} = \ldots = \delta\xi \qquad (11\text{-}15)$$

where ξ is *the degree of advancement of the reaction.* By definition, the range of values of ξ is from zero to one, and $\delta\xi$ is always positive.

We can rewrite Eqs. 11-4 and 11-14 using this notation as

$$\delta G = (-s\mu_S - u\mu_U - \ldots + c\mu_C + d\mu_D \ldots) \, \delta\xi \qquad (11\text{-}16)$$

and

$$(-s\mu_S - u\mu_U - \ldots + c\mu_C + d\mu_D + \ldots) \, \delta\xi \leqslant 0 \qquad (11\text{-}17)$$

Since $\delta\xi$ is always positive, the quantity in parentheses in Eq. 11-17 is less than or equal to zero. That is,

$$(-s\mu_S - u\mu_U - \ldots + c\mu_C + d\mu_D + \ldots) \leqslant 0$$

or

$$\sum_{\text{prod}} (\nu_j \, \mu_j) - \sum_{\text{react}} (\nu_j \, \mu_j) \leqslant 0 \qquad (11\text{-}18)$$

where ν_j is the stoichiometric coefficient of the jth reactant or product, a dimensionless quantity.

The quantity on the left in Eq. 11-18 is frequently denoted by ΔG, but it is more properly denoted $\Delta\mathscr{G}$, since it is an intensive property, with units (J-mol^{-1}) in SI units. The quantity $\Delta\mathscr{G}$ refers to a reaction in which s moles of S and u moles of U react to form c moles of C and d moles of D in a system that is large enough so the composition of the system and the chemical potentials of the reactants and products are not changed by the reaction. Since this is the equivalent of a differential quantity for a finite system, the alternative notations, $\Delta \tilde{G}$ or $\partial G/\partial \xi$, have been suggested.[3]

11-2. The chemical potential of an ideal gas

A pure ideal gas obeys the equation of state, Eq. 1-14, so that

$$V = \frac{nRT}{P}$$

Equation 10-87 gives us $(\partial G/\partial P)_T = V$, and the change in Gibbs free energy for the isothermal expansion of the gas from the standard pressure $P^{\ominus}$ (equal to 1.01×10^5 Pa) to a pressure P is

$$\Delta G = G - G^{\ominus}$$

$$= \int_{P^{\ominus}}^{P} dG$$

$$= \int_{P^{\ominus}}^{P} \left(\frac{\partial G}{\partial P}\right)_T dP$$

Substituting, we find

$$\Delta G = n\,R\,T \ln \frac{P}{P^{\ominus}} \qquad (11\text{-}19)$$

or

$$G = G^{\ominus} + n\,R\,T \ln \frac{P}{P^{\ominus}} \qquad (11\text{-}20)$$

Since for a pure gas, by definition,

$$\mu = \left(\frac{\partial G}{\partial n}\right)_{T,\,P} = \frac{G}{n}$$

we can write

$$\mu = \frac{G^{\ominus}}{n} + R\,T \ln \frac{P}{P^{\ominus}} = \mu^{\ominus} + R\,T \ln \frac{P}{P^{\ominus}} \qquad (11\text{-}21)$$

where $\mu^{\ominus}$ is the chemical potential of the pure gas when $P = P^{\ominus}$.

3. H. A. Bent, *J. Chem. Educ.*, *50*, 323 (1973); J. N. Spencer, *J. Chem. Educ.*, *51*, 577 (1974).

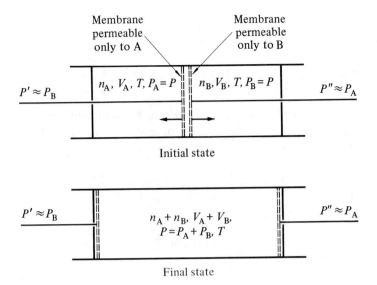

Membrane permeable only to A

Membrane permeable only to B

$P' \approx P_B$ $n_A, V_A, T, P_A = P$ $n_B, V_B, T, P_B = P$ $P'' \approx P_A$

Initial state

$P' \approx P_B$ $n_A + n_B, V_A + V_B,$ $P = P_A + P_B, T$ $P'' \approx P_A$

Final state

Figure 11-1. Scheme of a van't Hoff mixing box, in which a hypothetical reversible mixing process can be carried out. The total volume does not change.

An ideal gas mixture

An ideal gas mixture is defined as one that follows Dalton's law of partial pressures (see Eq. 1-17)

$$P = \sum_{j=1}^{k} \frac{n_j RT}{V} \tag{11-22}$$

and for which the value of the molar energy $\mathscr{E}_j$ for each component is a function only of the temperature and independent of the composition and the pressure.

Consider the system illustrated in Fig. 11-1 (a Van't Hoff mixing box), immersed in a thermal bath at temperature T. Two gases, at the same initial pressure P, are separated by two pistons, the faces of which are semipermeable membranes. The left membrane is permeable only to A, and the right membrane is permeable only to B. If an external pressure P', infinitesimally less than P_B is applied to the left-hand piston, and an external pressure P'' infinitesimally less than P_A, is applied to the right-hand piston, the gases are mixed reversibly without a change in total volume.

The work done in the mixing process is equal to

$$W = W_A + W_B$$

$$= \int_{V_A}^{V_A + V_B} -n_A \, RT \frac{dV}{V} + \int_{V_B}^{V_A + V_B} -n_B \, RT \frac{dV}{V}$$

$$= - n_A \, RT \ln \frac{V_A + V_B}{V_A} - n_B \, RT \ln \frac{V_A + V_B}{V_B} \qquad (11\text{-}23)$$

The initial and final states are at the same temperature and pressure, so that the volumes are proportional to the number of moles of gas. Therefore, Eq. 11-23 can be written as

$$W = - n_A \, RT \ln \frac{n_A + n_B}{n_A} - n_B \, RT \ln \frac{n_A + n_B}{n_B}$$

$$= n_A \, RT \ln X_A + n_B \, RT \ln X_B \qquad (11\text{-}24)$$

where X_A and X_B are the *mole fractions* of A and B. Because ΔE is equal to zero for this isothermal mixing of ideal gases

$$Q_{rev} = - W = - n_A \, RT \ln X_A - n_B \, RT \ln X_B \qquad (11\text{-}25)$$

Here ΔH, like ΔE, is equal to zero for an isothermal process involving ideal gases, and

$$\Delta G_{mix} = \Delta H - T \Delta S$$
$$= - T \Delta S$$

Recalling that $Q_{rev} = T \Delta S$, and using Eq. 11-25, we obtain

$$\Delta S_{mix} = - n_A \, R \ln X_A - n_B \, R \ln X_B \qquad (11\text{-}26)$$

and

$$\Delta G_{mix} = n_A \, RT \ln X_A + n_B \, RT \ln X_B \qquad (11\text{-}27)$$

Exercise 11-3.   Verify Eq. 11-27, starting with Eq. 11-25.

The Gibbs free energy of mixing is also expressed

$$\Delta G_{mix} = G_2 - G_1 \qquad (11\text{-}28)$$

where G_2 is the free energy of the mixture and G_1 is the free energy of the pure gases. The free energy of the mixture can be obtained by integrating Eq. 11-4, in the form

$$\delta G = \mu_A \, \delta n_A + \mu_B \, \delta n_B \qquad (11\text{-}29)$$

at constant composition and therefore constant chemical potential from zero amount of substance to n_A moles of A and n_B moles of B. Thus

$$\Delta G = \int_0^{n_A} \mu_A \, dn_A + \int_0^{n_B} \mu_B \, dn_B$$

$$= \mu_A n_A + \mu_B n_B$$

By definition,

$$\Delta G = \int_1^2 dG$$

$$= G_2 - 0$$

so that

$$G_2 = \mu_A n_A + \mu_B n_B \tag{11-30}$$

It is possible to consider the integration leading to Eq. 11-30 as a mathematical analogue of a physical process in which many infinitesimal systems of fixed composition are joined to form a large system having the same composition.

Using Eq. 11-21 for the pure gases and Eq. 11-30 for the mixture, we now write Eq. 11-28 as

$$\Delta G_{mix} = \mu_A n_A + \mu_B n_B - n_A \mu_A^\ominus - n_A RT \ln \frac{P}{P^\ominus} - n_B \mu_B^\ominus - n_B RT \ln \frac{P}{P^\ominus}$$

$$\tag{11-31}$$

Combining Eq. 11-27 and Eq. 11-31 gives us

$$\left[\mu_A - \mu_A^\ominus - RT \ln \frac{P}{P^\ominus} - RT \ln X_A \right] n_A$$

$$+ \left[\mu_B - \mu_B^\ominus - RT \ln \frac{P}{P^\ominus} - RT \ln X_B \right] n_B = 0 \tag{11-32}$$

Exercise 11-4. Verify Eq. 11-32.

Since Eq. 11-32 is valid for any values of n_A and n_B, both quantities in brackets must be equal to zero, and

$$\mu_A - \mu_A^\ominus - RT \ln \frac{P X_A}{P^\ominus} = 0$$

That is, since $PX_A = p_A$, the partial pressure of A in the mixture,

$$\mu_A = \mu_A^\ominus + RT \ln \frac{p_A}{p^\ominus} \tag{11-33}$$

This result corresponds to Eq. 11-21 for a pure ideal gas and shows that the chemical potential of an ideal gas in an ideal mixture is independent of the presence of other gases. The value of μ_A is equal to $\mu_A^\ominus$ when p_A is equal to $P^\ominus$. Thus, the standard state of an ideal gas in an ideal mixture is the standard state of the pure gas. We shall indicate the relationship to the standard state of the component by writing Eq. 11-33 as

$$\mu_A = \mu_A^\ominus + RT \ln \frac{p_A}{p^\ominus} \tag{11-34}$$

Equation 11-34, together with Eq. 11-30, shows that the *free energy of an ideal mixture of ideal gases is just the sum of the free energies the individual gases would have in the same volume at the same temperature.* Since the energies are also additive, the entropies are additive as well.

As indicated in Eq. 11-3, the chemical potential is a function of the temperature, the pressure, and the composition of the mixture. By definition (Eq. 11-2),

$$\mu_j = \left(\frac{\partial G}{\partial n_j}\right)_{T, P, n_\beta}$$

Then,

$$\left(\frac{\partial \mu_j}{\partial T}\right)_{P, n_\omega} = \left[\frac{\partial}{\partial T}\left(\frac{\partial G}{\partial n_j}\right)_{T, P, n_\beta}\right]_{P, n_\omega}$$

$$= \left[\frac{\partial}{\partial n_j}\left(\frac{\partial G}{\partial T}\right)_{P, n_\omega}\right]_{T, P, n_\beta} \tag{11-35}$$

since the value of the second derivative of a state function is independent of the order of differentiation. Substituting from Eq. 10-86 in Eq. 11-35, we have

$$\left(\frac{\partial \mu_j}{\partial T}\right)_{P, n_\omega} = -\left(\frac{\partial S}{\partial n_j}\right)_{T, P, n_\beta} \tag{11-36}$$

Similarly,

$$\left(\frac{\partial \mu_j}{\partial P}\right)_{T, n_\omega} = \left(\frac{\partial V}{\partial n_j}\right)_{T, P, n_\beta} \tag{11-37}$$

Exercise 11-5. Verify Eq. 11-37.

The derivative of any extensive thermodynamic property B with respect to n_j at constant T, P, and n_β is called the *partial molar* property $\mathscr{B}_j$. Thus,

$$\left(\frac{\partial S}{\partial n_j}\right)_{T, P, n_\beta} = \mathscr{S}_j \tag{11-38}$$

and

$$\left(\frac{\partial V}{\partial n_j}\right)_{T, P, n_\beta} = \mathscr{V}_j \tag{11-39}$$

Any relationship among extensive thermodynamic properties is also valid for the corresponding partial molar properties of a single component of a mixture.

Exercise 11-6. Show that

$$\mu_j = \mathscr{H}_j - T\mathscr{S}_j \tag{11-40}$$

11-3. Chemical equilibrium in ideal gas mixtures. The equilibrium constant

If Eq. 11-34 is used to express the chemical potentials in Eq. 11-18 for the chemical reaction in Eq. 11-13, we have

$$\Delta\mathscr{G} = c\mu_C^\ominus + d\mu_D^\ominus - s\mu_S^\ominus - u\mu_U^\ominus$$

$$+ cRT\ln\frac{p_C}{p^\ominus} + dRT\ln\frac{p_D}{p^\ominus}$$

$$- sRT\ln\frac{p_S}{p^\ominus} - uRT\ln\frac{p_U}{p^\ominus} \leqslant 0 \tag{11-41}$$

If we denote the first four terms in $\Delta\mathscr{G}$ as $\Delta\mathscr{G}^\ominus$, that is, the value of $\Delta\mathscr{G}$ when all the partial pressures are equal to $p^\ominus$, then

$$\Delta\mathscr{G}^\ominus \leqslant -RT\left[c\ln\frac{p_C}{p^\ominus} + d\ln\frac{p_D}{p^\ominus} - s\ln\frac{p_S}{p^\ominus} - u\ln\frac{p_U}{p^\ominus}\right]$$

or

$$\Delta\mathscr{G}^\ominus \leqslant RT\ln\left[\frac{(p_C/p^\ominus)^c\,(p_D/p^\ominus)^d}{(p_S/p^\ominus)^s\,(p_U/p^\ominus)^u}\right] \tag{11-42}$$

and

$$\ln\frac{(p_C/p^\ominus)^c\,(p_D/p^\ominus)^d}{(p_S/p^\ominus)^s\,(p_U/p^\ominus)^u} \leqslant -\frac{\Delta\mathscr{G}^\ominus}{RT} \tag{11-43}$$

The equality in Eq. 11-43 refers to an equilibrium state, and the inequality refers to a spontaneous change of state as written in Eq. 11-13. If the inequality is reversed, the reverse of Eq. 11-13 is spontaneous.

Exercise 11-7. For the reaction,

$$2\ NO_2\,(g) = N_2O_4\,(g)$$

$\Delta \mathscr{G}^{\ominus}$ is equal to -4.770×10^3 J-mol^{-1} at 298.15 K. (a) What is the spontaneous direction of reaction in a mixture in which $p_{NO_2} = 2.67 \times 10^4$ Pa and $p_{N_2O_4} = 1.07 \times 10^5$ Pa? (b) What is the spontaneous direction of reaction in a mixture in which $p_{NO_2} = 1.07 \times 10^5$ Pa and $p_{N_2O_4} = 2.67 \times 10^4$ Pa?

Since $\Delta \mathscr{G}^{\ominus}$, R, and T are constants at any fixed temperature, the ratio on the left in Eq. 11-43 is also a constant, the *equilibrium constant* in terms of partial pressure. It is customary to write

$$K_p = \left[\frac{(p_C/p^{\ominus})^c\,(p_D/p^{\ominus})^d}{(p_S/p^{\ominus})^s\,(p_U/p^{\ominus})^u} \right]_{equil} \tag{11-44}$$

and

$$\ln K_p = -\frac{\Delta \mathscr{G}^{\ominus}}{RT} \tag{11-45}$$

The equilibrium constant can be determined from measurements of the partial pressure of reactants and products at equilibrium, or from values of $\Delta \mathscr{G} f^{\ominus}$ or $\mathscr{G}^{\ominus} - \mathscr{E}_0^{\ominus}$ obtained from equilibrium, calorimetric, or spectroscopic data. The equilibrium constant is dependent on the temperature, but independent of the total pressure and the initial composition of the reaction mixture.

Exercise 11-8. Calculate K_p for the system in Exercise 11-7. What are the equilibrium partial pressures of NO_2 and N_2O_4 if the system is initially pure N_2O_4 at a pressure of 1.01×10^5 Pa and the volume is fixed?

For a reaction in which some reactants and products are gases and others are pure solids, the equilibrium constant involves only the partial pressures of the gases, since all the solids are in their standard states. Thus, the equilibrium constant for the reaction Le Chatelier discussed (p. 364) is

$$K_p = \frac{(p_{CO_2}/p^{\ominus})^3}{(p_{CO}/p^{\ominus})^3}$$

and

$$\Delta \mathscr{G}^{\ominus} = - RT \ln K_p$$

$$= 2\mu_{Fe(c)}^{\ominus} + 3\mu_{CO_2(g)}^{\ominus} - \mu_{Fe_2O_3(c)}^{\ominus} - 3\mu_{CO(g)}^{\ominus}$$

If sufficient time is allowed for the equilibrium ratio of the gases to be reached, no further reaction can be achieved without changing the temperature.

Exercise 11-9. Derive the expression for the equilibrium constant in the preceding paragraph in a way analogous to the derivation of Eq. 11-44.

If mole fractions rather than partial pressures are used to describe the equilibrium composition of a reacting mixture, we have

$$p_j = X_j P$$

and

$$K_p = \left(\frac{X_C^c \cdot X_D^d}{X_S^s \cdot X_U^u} \right) \left(\frac{P}{p^{\ominus}} \right)^{c+d-s-u}$$

$$= K_X \left(\frac{P}{p^{\ominus}} \right)^{\Delta \nu} \tag{11-46}$$

Equation 11-46 makes explicit the dependence of the equilibrium constant K_X on the magnitude of the pressure, in constrast to the independence of K_p.

If concentrations are used to describe the equilibrium composition of a reacting mixture, we have

$$[C] = \frac{n_C}{V}$$

$$= \frac{p_C}{RT} \tag{11-47}$$

and

$$p^{\ominus} = c^{\ominus} RT$$

where $c^{\ominus}$ is the concentration corresponding to the pressure $p^{\ominus}$ at any temperature. Thus,

$$K_p = \frac{([C]/c^{\ominus})^c \, ([D]/c^{\ominus})^d}{([S]/c^{\ominus})^s \, ([U]/c^{\ominus})^u}$$

$$= \frac{[C]^c [D]^d}{[S]^s [U]^u} (c^{\ominus})^{-\Delta v}$$

$$= K_c \qquad (11\text{-}48)$$

Here K_c, like K_p, is independent of the pressure.

11-4. The calculation of equilibrium constants from equilibrium data

Frequently, the density of the equilibrium mixture at a known temperature, pressure, and volume is sufficient information from which to calculate an equilibrium constant. Consider the reaction in which PCl_5 dissociates to form PCl_3 and Cl_2.

$$PCl_5 (g) = PCl_3 (g) + Cl_2 (g) \qquad (11\text{-}49)$$

If n_0 is the number of moles of PCl_5 originally placed in the reaction vessel, and α is the fraction of PCl_5 dissociated at equilibrium, then the number of moles of reactant and products at equilibrium and the corresponding partial pressures are those shown in Table 11-1.

Table 11-1. Calculation of equilibrium partial pressures for reaction in Eq. 11-49

Substance	$PCl_5(g)$	$PCl_3(g)$	$Cl_2(g)$	Total
Moles at equilibrium	$n_0(1 - \alpha)$	$n_0 \alpha$	$n_0 \alpha$	$n_0(1 + \alpha)$
Mole fraction at equilibrium	$\dfrac{1 - \alpha}{1 + \alpha}$	$\dfrac{\alpha}{1 + \alpha}$	$\dfrac{\alpha}{1 + \alpha}$	1
Partial pressure at equilibrium	$\dfrac{1 - \alpha}{1 + \alpha} P$	$\dfrac{\alpha}{1 + \alpha} P$	$\dfrac{\alpha}{1 + \alpha} P$	P

The equilibrium constant is then

$$K_p = \frac{\{[\alpha/(1 + \alpha)](P/p^{\ominus})\}^2}{[(1 - \alpha)/(1 + \alpha)](P/p^{\ominus})}$$

$$= \frac{\alpha^2}{1 - \alpha^2} \frac{P}{p^{\ominus}} \qquad (11\text{-}50)$$

The value of α can be calculated from the ideal gas law applied to the equilibrium mixture.

$$PV = n_0 (1 + \alpha) RT$$

$$= \frac{m_0}{M} (1 + \alpha) RT$$

or

$$\alpha = \frac{MP}{\rho RT} - 1 \qquad (11\text{-}51)$$

where m_0 is the initial mass of the PCl_5, M is the molar mass of PCl_5, and ρ is the density of the equilibrium mixture.

Exercise 11-10.   When 3.40 g of PCl_5 are placed in a 1.00 dm^3 vessel and allowed to equilibrate at 400 K, the equilibrium pressure was found to be 1.01×10^5 Pa. Calculate the degree of dissociation, α, and K_p for Eq. 11-49.

When the number of moles of reactants and products are equal, there is no change in pressure or density, and an analysis for one or more of the species present at equilibrium is required. For the reaction

$$H_2(g) + I_2(g) = 2\,HI(g) \qquad (11\text{-}52)$$

let n_0 be the initial number of moles of H_2 and of I_2 added to the system and let x be the number of moles of HI present at equilibrium. Then the number of moles of H_2 and I_2 at equilibrium are $n_0 - x/2$, and

$$K_p = \frac{x^2}{(n_0 - x/2)^2} \qquad (11\text{-}53)$$

In this case, neither the total number of moles nor the total pressure need be known to calculate K_p. The number of moles of I_2 at equilibrium can be determined by cooling the equilibrium system quickly and carrying out a chemical analysis or by measuring the intensity of light absorption by the I_2 in the equilibrium mixture.

Exercise 11-11.   When 0.0174 mole of H_2 and 0.0174 mole of I_2 are placed in a 1.00 dm^3 reaction vessel at 700 K and 1.01×10^5 Pa, 0.0036 mole of I_2 are found in the equilibrium mixture. Calculate K_p for Eq. 11-52.

Once K_p is found for a reaction, it can be used to calculate the equilibrium composition for any mixture of the same reactants and products at the same temperature.

Exercise 11-12.   Calculate the composition of the equilibrium mixture when 0.00500 mole of H_2 and 0.00500 mole of I_2 are allowed to equilibrate at 700 K.

Exercise 11-13. $\Delta \mathscr{G}^{\ominus}$ for Eq. 9-74 is -1.4177×10^5 J at 298.15 K. Calculate the equilibrium ratio of p_{SO_2}/p_{SO_3} in air, with a partial pressure of O_2 equal to 2.02×10^4 Pa. In view of this result, how do you explain the presence of substantial amounts of SO_2 in the atmosphere?

11-5. The calculation of equilibrium constants from thermal data

According to Eq. 11-45,

$$\ln K_p = - \frac{\Delta \mathscr{G}^{\ominus}}{RT}$$

Thus, K_p can be calculated if $\Delta \mathscr{G}^{\ominus}$ is known for a reaction, even if equilibrium data cannot be obtained. This relationship is particularly important for reactions in which either the reactants or products are present in very low concentration at equilibrium because the concentrations may not be measurable. For any isothermal process (Eq. 10-95),

$$\Delta \mathscr{G}^{\ominus} = \Delta \mathscr{H}^{\ominus} - T \Delta \mathscr{S}^{\ominus}$$

so that a knowledge of $\Delta \mathscr{H}^{\ominus}$ for the reaction and of $\mathscr{S}^{\ominus}$ for each reactant and product is sufficient to calculate $\Delta \mathscr{G}^{\ominus}$ and K_p for the reaction.

Exercise 11-14. Obtain the necessary data from Table 9-3 and Table 10-1 to calculate K_p at 298.15 K for the reaction

$$N_2(g) + 3 H_2(g) = 2 NH_3(g)$$

assuming that $\mathscr{S}_0^{\ominus}$ is equal to zero for reactants and products.

Since the major use of enthalpy and entropy data is in the calculation of free energy data and equilibrium constants, many tabulations of thermodynamic data provide values of $\Delta \mathscr{G} f^{\ominus}$, the *standard Gibbs free energy of formation*, which is the free energy change for the formation of one mole of compound, in its standard state, from the elements, in their standard states. The value of $\Delta \mathscr{G} f^{\ominus}$ for the elements is zero, by definition. The standard states for pure substances are those in Table 9-2. Then, for any reaction,

$$\Delta \mathscr{G}^{\ominus} = \sum_{\text{prod}} (v_j \, \Delta \, \mathscr{G} f_j{}^{\ominus}) - \sum_{\text{react}} (v_j \, \Delta \, \mathscr{G} f_j{}^{\ominus}) \tag{11-54}$$

Values of $\Delta \mathscr{G} f^{\ominus}$ for several substances at 298.15 K are listed in Table 11-2. The units are kilojoules per mole.

Table 11-2. Free energy of formation at 298.15 K[a]

Substance	$\Delta\mathscr{G}f^{\ominus}/(kJ\text{-}mol^{-1})$	Substance	$\Delta\mathscr{G}f^{\ominus}/(kJ\text{-}mol^{-1})$
O(g)	231.75	$N_2O_4(g)$	97.82
$O_3(g)$	163.	$NH_3(g)$	−16.5
HD(g)	−1.46	$PCl_3(g)$	−268.
$H_2O(g)$	−228.59	$PCl_5(g)$	−305.
$H_2O(l)$	−237.18	CO(g)	−137.15
HF(g)	−273	$CO_2(g)$	−394.36
HCl(g)	−95.299	$CH_4(g)$	−50.75
HBr(g)	−53.43	$C_2H_2(g)$	209.2
HI(g)	1.7	$C_2H_4(g)$	68.11
$SO_2(g)$	−300.19	$C_2H_6(g)$	−32.9
$SO_3(g)$	−371.1	$C_6H_6(g)$	−129.66
$H_2S(g)$	−33.6	H(g)	203.26
NO(g)	86.57	D(g)	206.52
$NO_2(g)$	51.30		

[a] U.S. NBS Technical Note 270–3, "Selected Values of Chemical Thermodynamic Properties", 1968, and "Selected Values of Physical and Thermodynamic Properties of Hydrocarbons", Am. Petroleum *Institute Project 44, Carnegie Press, Pittsburgh, 1953.*

11-6. The effect of external variables on equilibrium. Le Chatelier's principle

The principle that describes the effect of changes in external variables on chemical equilibrium was derived by H. Le Chatelier and F. Braun in the latter part of the 19th century.[4] Le Chatelier stated the principle as follows: *Any system in chemical equilibrium, as a result of the variation in one of the factors determining the equilibrium, undergoes a change such that, if this change had occurred by itself, it would have introduced a variation opposite in sign in the factor considered.*[5]

Effect of pressure

The value of K_p is independent of pressure, but the value of K_x, which reflects the chemical composition of the equilibrium system, is not. From Eq. 11-46,

$$K_x = K_p \, (P/p^{\ominus})^{-\Delta v}$$

and

$$\ln K_x = \ln K_p - (\Delta v) \ln (P/p^{\ominus})$$

4. H. Le Chatelier, *Ann. Mines* (8), *13*, 200 (1888); F. Braun, *Z. physik. Chem.*, *1*, 259 (1887).
5. Translated from the quotation in the obituary notice for Le Chatelier, by W. J. Pope, *Nature*, *138*, 711–12 (1936). A detailed discussion can be found in J. de Heer, *J. Chem. Educ.*, *34*, 375 (1957); *35*, 133 (1958).

Thus,

$$\left(\frac{\partial \ln K_X}{\partial P}\right)_T = -\frac{\Delta v}{P} \tag{11-55}$$

$$= -\frac{\Delta \mathscr{V}}{RT} \tag{11-56}$$

where $\Delta \mathscr{V}$ is the partial molar volume change for the reaction as written at a constant temperature and pressure.

Exercise 11-15. Verify Eq. 11-56. Hint: Show that $\mathscr{V}_j = RT/P$ for a component of a mixture of ideal gases.

The value of K_X increases with increasing pressure if Δv and $\Delta \mathscr{V}$ are negative, and decreases with increasing pressure if Δv and $\Delta \mathscr{V}$ are positive. Thus, the equilibrium always shifts with increasing pressure in the direction that results in fewer molecules in the system, a change which, if it had occurred by itself, would have resulted in a decrease in pressure. One can also say that a system in which a chemical equilibrium can shift in response to a change in pressure has a *greater compressibility* than a corresponding system in which no such shift is possible.

Effect of added inert gas
The effect of the addition of inert gas to a system in equilibrium can be seen most simply by writing Eq. 11-46 in the form

$$K_p = \frac{X_C^c \cdot X_D^d}{X_S^s \cdot X_U^u} (P/p^{\ominus})^{\Delta v}$$

$$= \frac{n_C^c \cdot n_D^d}{n_S^s \cdot n_U^u} \frac{(P/p^{\ominus})^{\Delta v}}{n_{\text{tot}}^{\Delta v}} \tag{11-57}$$

where n_{tot} is the total moles of gas in the system.

Exercise 11-16. Verify Eq. 11-57.

If an inert gas is added at *constant volume*, the mole numbers of reactants and products do not change, since P is proportional to n_{tot}, and K_p is constant. If, on the other hand, an inert gas is added at *constant pressure*, the mole numbers of reactants and products change, since P is constant, while n_{tot} increases. The direction of the shift in equilibrium composition depends on the sign of Δv. The effect is the same as that of a *decrease* in total pressure without added inert gas, since there is a decrease in the sum of the partial pressures of reactants and products.

Effect of added reactant or product

A straightforward reading of Le Chatelier's principle suggests that the addition of excess reactant or product to a system at equilibrium should result in a reaction in the direction that decreases the amount of the added reactant or product. Hence, the common practice of adding an excess of a less expensive reactant to increase the conversion at equilibrium of a more expensive reactant. If the addition is made *at constant volume*, P is proportional to n_{tot}, and from Eq. 11-57 we see that the new equilibrium state must be one in which the reaction has occurred as stated above. If the addition is made at constant pressure, the direction of the shift in equilibrium depends on the original equilibrium composition and on the stoichiometric coefficients.

Consider the ammonia synthesis reaction, for which

$$K_p = \frac{(p_{NH_3}/p^{\ominus})^2}{(p_{N_2}/p^{\ominus})(p_{H_2}/p^{\ominus})^3} \tag{11-58}$$

If excess NH_3 is added to a system at equilibrium at constant pressure, p_{NH_3} increases, and p_{N_2} and p_{H_2} decrease. Therefore, the new equilibrium will be reached by conversion of NH_3 to N_2 and H_2. If excess N_2 is added to this system at equilibrium, the initial effect is to decrease p_{NH_3} and p_{H_2}. The value of $p_{H_2}{}^3$ initially can decrease by a greater factor than $p_{NH_3}{}^2/p_{N_2}$ decreases, so that equilibrium is restored by the decomposition of NH_3 to form additional N_2 and H_2, a result obtainable from Eq. 11-58 but not from the usual statement of Le Chatelier's principle.

Exercise 11-17. At 773 K, $\Delta\mathscr{G}f^{\ominus}$ for NH_3 is 3.578×10^4 J-mol^{-1}. [See A. T. Larson and R. L. Dodge, *J. Am. Chem. Soc.*, *46*, 367 (1924).] When 0.909 mole of N_2 and 0.0909 mole of H_2 are allowed to come to equilibrium at a pressure of 1.11×10^6 Pa at 773 K, 7.12×10^{-4} mole of NH_3 are formed. What is the spontaneous direction of reaction when 1.000 mole of N_2 is added to the equilibrium mixture?

Effect of temperature

Differentiation of Eq. 11-45 with respect to temperature at constant pressure gives us

$$\left(\frac{\partial \ln K_p}{\partial T}\right)_P = -\frac{1}{R}\left[\frac{\partial(\Delta\mathscr{G}^{\ominus}/T)}{\partial T}\right]_P \tag{11-59}$$

$$= \frac{\Delta\mathscr{H}^{\ominus}}{RT^2} \tag{11-60}$$

(See Eq. 10-107.) The value of K_p increases with increasing temperature if $\Delta \mathscr{H}^{\ominus}$ is positive, an endothermic reaction, and decreases with increasing temperature if $\Delta \mathscr{H}^{\ominus}$ is negative, an exothermic reaction. Thus, the equilibrium shifts with increasing temperature in a direction such that the temperature would *decrease* if that shift occurred by itself in an isolated system, in the direction that absorbs heat. One can also say that a system in which a chemical equilibrium can shift in response to a change in temperature has a *greater heat capacity* than a corresponding system in which no such shift is possible.

Integration of Eq. 11-60 between two temperatures that are close enough together so $\Delta \mathscr{H}^{\ominus}$ can be considered constant gives us

$$\ln \left[\frac{K_p(T_2)}{K_p(T_1)} \right] = -\frac{\Delta \mathscr{H}^{\ominus}}{R} \left[\frac{1}{T_2} - \frac{1}{T_1} \right] \qquad (11\text{-}61)$$

Equation 11-61 can be used to calculate K_p at one temperature if it is known at another temperature and $\Delta \mathscr{H}^{\ominus}$ is known, or it can be used to calculate $\Delta \mathscr{H}^{\ominus}$ from equilibrium constant data at two temperatures. Indefinite integration of Eq. 11-60 on the assumption that $\Delta \mathscr{H}^{\ominus}$ is a constant over a temperature range yields the *Van't Hoff equation*,

$$\ln K_p = -\frac{\Delta \mathscr{H}^{\ominus}}{RT} + \text{constant}$$

or

$$\log K_p = -\frac{\Delta \mathscr{H}^{\ominus}}{2.303\ RT} + \text{constant}' \qquad (11\text{-}62)$$

The assumption that $\Delta \mathscr{H}^{\ominus}$ is constant can be tested by plotting $\log K_p$ against $1/T$. If $\Delta \mathscr{H}^{\ominus}$ is constant, the experimental points fall on a straight line, and the slope is equal to $-\Delta \mathscr{H}^{\ominus}/(2.303\ R)$. If the plot of $\log K_p$ against $1/T$ yields a curve, the value of $\Delta \mathscr{H}^{\ominus}$ can be calculated by graphical or numerical differentiation of the experimental data. From Eq. 10-109 and Eq. 11-45 we see that

$$\frac{\partial \ln K_p}{\partial (1/T)} = \frac{\Delta \mathscr{H}^{\ominus}}{R} \qquad (11\text{-}63)$$

so the slope of a plot of $\log K_p$ against $1/T$ is equal to $\Delta \mathscr{H}/(2.303\ R)$ at any temperature.

If the expression for $\Delta \mathscr{H}^{\ominus}$ as a function of temperature from Eq. 9-75 is used with Eq. 11-60, integration gives us $\ln K_p$ as a function of temperature.

$$\ln K_p = -\frac{\Delta \mathcal{H}_I}{RT} + \frac{\Delta a}{R} \ln T + \frac{\Delta b}{2R} T$$

$$+ \frac{\Delta c}{6R} T^2 + \frac{\Delta d}{12R} T^3 - \frac{I}{R} \tag{11-64}$$

where $-(I/R)$ is an integration constant and I is the integration constant in the equation for $\Delta \mathcal{G}^{\ominus}/T$ as a function of temperature, since $\ln K_p = -(1/R)$ $(\Delta \mathcal{G}^{\ominus}/T)$.

Exercise 11-18. Verify Eq. 11-64.

Exercise 11-19. Obtain an equation for $\ln K_p$ as a function of T over the range 298.15 K to 1000 K for the reaction in Eq. 9-74. Plot log K_p against $1/T$ over this interval. Over how large a temperature interval can $\Delta \mathcal{H}^{\ominus}$ be assumed constant?

11-7. Calculation of equilibrium constants from spectroscopic data

Since the chemical potential of a component of an ideal gas mixture is independent of the presence of the other gases, the simplest way to relate the chemical potential to spectroscopic data is to equate the chemical potential to the molar free energy as given in Eq. 10-104. That is,

$$\mu_j = \mathcal{E}_{0,j} - RT \ln (z'_j/L) \tag{11-65}$$

Similarly, the standard chemical potential is, from Eq. 10-106,

$$\mu_j^{\ominus} = \mathcal{E}_{0,j}^{\ominus} - RT \ln (z_j^{\ominus}/L) \tag{11-66}$$

From Eq. 11-41, the standard partial molar Gibbs free energy change for Eq. 11-13 is

$$\Delta \mathcal{G}^{\ominus} = c\mu_C^{\ominus} + d\mu_D^{\ominus} - s\mu_S^{\ominus} - u\mu_U^{\ominus}$$

Substituting from Eq. 11-66, we have

$$\Delta \mathcal{G}^{\ominus} = c\mathcal{E}_{0,C}^{\ominus} + d\mathcal{E}_{0,D}^{\ominus} - s\mathcal{E}_{0,S}^{\ominus} - u\mathcal{E}_{0,U}^{\ominus}$$

$$- RT \ln \left[\frac{(z_C^{\ominus})^c (z_D^{\ominus})^d}{(z_S^{\ominus})^s (z_U^{\ominus})^u} L^{-\Delta \nu} \right] \tag{11-67}$$

$$\Delta \mathcal{G}^{\ominus} = \Delta \mathcal{E}_0^{\ominus} - RT \ln \left[\frac{(z_C^{\ominus})^c (z_D^{\ominus})^d}{(z_S^{\ominus})^s (z_U^{\ominus})^u} L^{-\Delta \nu} \right] \tag{11-68}$$

Exercise 11-20. Verify Eq. 11-68.

From Eq. 11-45 and Eq. 11-48

$$\Delta \mathscr{G}^{\ominus} = - R T \ln K_p$$

$$= - R T \ln K_C \qquad (11\text{-}69)$$

Equating the expressions for $\Delta \mathscr{G}^{\ominus}$ from Eq. 11-68 and Eq. 11-69, we have

$$K_p = K_C = \frac{(z_C^{\ominus})^c (z_D^{\ominus})^d}{(z_S^{\ominus})^s (z_U^{\ominus})^u} L^{-\Delta v} \exp\left(\frac{-\Delta \mathscr{E}_0^{\ominus}}{RT}\right) \qquad (11\text{-}70)$$

Exercise 11-21. Verify Eq. 11-70.

In order to avoid repetitious calculation of partition functions, values of $\mathscr{G}^{\ominus} - \mathscr{E}_0^{\ominus}$ are calculated from spectroscopic data and tabulated at convenient temperature intervals. Then, for any reaction,

$$\Delta \mathscr{G}^{\ominus} = \sum_{\text{prod}} [v_j(\mathscr{G}_j^{\ominus} - \mathscr{E}_{0,j}^{\ominus})] - \sum_{\text{react}} [v_j(\mathscr{G}_j^{\ominus} - \mathscr{E}_{0,j}^{\ominus})] + \Delta \mathscr{E}_0^{\ominus} \qquad (11\text{-}71)$$

The value of $\Delta \mathscr{E}_0^{\ominus}$ for a reaction can be calculated if $\Delta \mathscr{G}^{\ominus}$ is known at one temperature from either equilibrium or thermal data or from the use of Eq. 9-79 to calculate $\Delta \mathscr{H}^{\ominus}$. Since $(\mathscr{G}^{\ominus} - \mathscr{E}_0^{\ominus})/T$ varies more slowly with temperature than $\mathscr{G}^{\ominus} - \mathscr{E}_0^{\ominus}$, the former function permits a more precise interpolation between tabulated temperatures and is frequently the function provided in tables of thermodynamic data. Usually, a value of $\Delta \mathscr{H} f_0^{\ominus} = \Delta \mathscr{E} f_0^{\ominus}$ is provided in the same table so that $\Delta \mathscr{E}_0^{\ominus}$ can be obtained.

It is convenient to include, in these tables of data, values for substances for which statistical calculations from partition functions are not feasible. For such substances, 298.15 K is a more convenient reference temperature than 0 K. Many tables of thermodynamic data, therefore, give values of $(\mathscr{G}^{\ominus} - \mathscr{E}_{298}^{\ominus})/T$. Some of the most important sources of thermodynamic data are listed below.

1. "Selected Values of Chemical Thermodynamic Properties," U.S. NBS, Circular 500, 1952; Technical Note 270–73, 1968; Technical Note 270–4, 1969; Technical Note 270–75, 1971.
2. K. K. Kelley, "High temperature heat-content, heat capacity, and entropy data for the elements and inorganic compounds," Bureau of Mines Bulletin 584, 1960.
3. K. K. Kelley and E. G. King, "Entropies of the elements and inorganic compounds," Bureau of Mines Bulletin 592, 1961.
4. R. A. Robie and D. R. Waldbaum, "Thermodynamic properties of min-

erals and related substances at 298.15°K (25.0°C) and one atmosphere (1.013 bars) pressure and at higher temperatures," Geological Survey Bulletin 1259, 1968.

5. "Selected values of physical and thermodynamic properties of hydrocarbons and related compounds," Am. Petroleum Inst. Res. Proj. 44, Carnegie Press, Pittsburgh, 1953.

6. "Selected values of properties of chemical compounds," Manufacturing Chemists' Assoc., Carnegie Inst. of Technology, Pittsburgh, 1955.

7. "JANAF Thermochemical Tables," 2nd ed., National Standard Reference Data System, U.S. NBS, 37, 1971.

8. B. J. Zwolinskie *et al.*, "Selected values of properties of hydrocarbons and related compounds," American Petroleum Institute Research Project 44, Thermodynamics Research Center, Texas A & M University, College Station, Texas (Loose-leaf data sheets).

9. B. J. Zwolinskie *et al.*, "Selected Values of Properties of Chemical Compounds," Thermodynamics Research Center Data Project, Thermodynamics Research Center, Texas A & M University, College Station, Texas (Loose-leaf data sheets).

11-8.   Chemical equilibrium in non-ideal gas mixtures. The fugacity

The equilibrium constant K_p derived in Sec. 11-3 was assumed to be a function of the temperature only and independent of the pressure. Data for the ammonia synthesis over a wide range of pressure are shown in Table 11-3; from these data, it can be seen that a different approach is required to deal with real gases at high pressure.

Table 11-3. $N_2(g) + 3 H_2(g) = 2 NH_3(g)$

$P/1.01 \times 10^5$ Pa	K_p
10	0.0000434
30	0.0000457
50	0.0000476
100	0.0000526
300	0.0000781
600	0.0001674
1000	0.0006230
2000	0.01787
3500	1.1558

L. J. Winchester and B. F. Dodge, *J. Am. Inst. Chem. Eng.*, *2*, 431 (1956).

The fugacity function

The dependence of the chemical potential on the pressure takes the form (Eq. 11-21)

$$\mu = \mu^{\ominus} + RT\ln\left(\frac{P}{P^{\ominus}}\right)$$

as a result of substituting $V = nRT/P$ in the integral,

$$\Delta G = \int_{P^{\ominus}}^{P} V dP \tag{11-72}$$

It would be possible to apply Eq. 11-72 to real gases by substituting for V an empirical expression for V as a function of P, but no simple closed form can be used that is applicable to all gases. A simple form of the equation for the chemical potential and a simple form of the equation for the equilibrium constant that is independent of the gas being discussed are so convenient, however, that G. N. Lewis suggested an alternative procedure. He defined a function, the *fugacity f*, such that, for a pure gas,

$$\mu = \mu^{\ominus} + RT\ln\left(f/f^{\ominus}\right) \tag{11-73}$$

where $\mu^{\ominus}$ is the chemical potential when $f = f^{\ominus}$, and $\mu^{\ominus}$ is a function of the temperature only. It is characteristic of the gas and the standard state chosen. Since all gases approach ideality in the limit of zero pressure, the definition of the fugacity is completed with the limiting relationship

$$\lim_{P \to 0} f/P = 1 \tag{11-74}$$

The fugacity is expressed in the same units as pressure.

The standard state for the fugacity of gases is chosen so that $f^{\ominus}$ equals 1.01×10^5 Pa. But, the standard state is not the real gas with a fugacity of 1.01×10^5 Pa, but the hypothetical state found by extrapolating the low pressure portion of the curve of f against P to a pressure of 1.01×10^5 Pa, as shown in Fig. 11-2. The standard state can also be viewed as the ideal gas state at 1.01×10^5 Pa pressure, a state for which f equals 1.01×10^5 Pa.

Pressure dependence of the fugacity

Having chosen a universal expression of the form of Eq. 11-73, we express the individual differences among gases in terms of the variation of the fugacity with pressure. From Eq. 11-73,

$$\left(\frac{\partial\mu}{\partial P}\right)_T = RT\left[\frac{\partial\ln\left(f/f^{\ominus}\right)}{\partial P}\right]_T \tag{11-75}$$

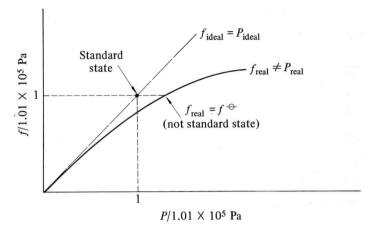

Figure 11-2. A schematic plot of the fugacity of a gas as a function of pressure, showing that the standard state is obtained by extrapolation of low pressure behavior to a pressure of 1.01×10^5 Pa.

From Eq. 11-37 and Eq. 11-39,

$$\left(\frac{\partial \mu}{\partial P}\right)_T = \mathscr{V}$$

so that

$$\left[\frac{\partial \ln (f/f^{\ominus})}{\partial P}\right]_T = \frac{\mathscr{V}}{RT} \qquad (11\text{-}76)$$

We cannot integrate Eq. 11-76 between $P = 0$ and some finite value of P to find the fugacity at P because f goes to zero at $P = 0$, and $\mathscr{V}$ and $\ln f$ become infinite. We know from Eq. 11-74, however, that f/P is finite at $P = 0$, so we shall calculate

$$\left[\frac{\partial \ln (f/P)}{\partial P}\right]_T = \left[\frac{\partial \ln (f/f^{\ominus})}{\partial P}\right]_T - \left[\frac{\partial \ln (P/P^{\ominus})}{\partial P}\right]_T$$

$$= \frac{\mathscr{V}}{RT} - \frac{1}{P}$$

$$= \frac{1}{RT}\left(\mathscr{V} - \frac{RT}{P}\right) \qquad (11\text{-}77)$$

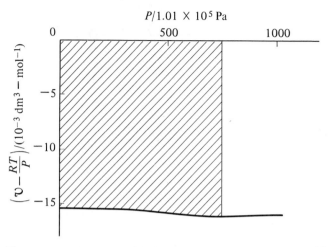

Figure 11-3. A plot of the deviation function, $\mathscr{V}\text{-}(RT/P)$, as a function of P for H_2 at 298.15 K. (By permission, from I. M. Klotz and R. M. Rosenberg, *Chemical Thermodynamics*, W. A. Benjamin, Menlo Park, 1972.)

Although $\mathscr{V}$ and RT/P both go to infinity at $P = 0$, the difference between them remains finite and Eq. 11-77 can be integrated. The result is

$$\ln (f/P) = \frac{1}{RT} \int_0^P \left(\mathscr{V} - \frac{RT}{P}\right) dP \qquad (11\text{-}78)$$

If a table of values of the molar volume $\mathscr{V}$ as a function of P is available, the right side of Eq. 11-78 can be integrated, graphically or numerically. Figure 11-3 shows a plot of $(\mathscr{V} - RT/P)$ against P for hydrogen gas at 298.15 K. The shaded area represents the integral in Eq. 11-78.

If the pressure-volume behavior of a gas can be represented by a virial equation of the form

$$P\mathscr{V} = RT + BP + CP^2 + \ldots \qquad (11\text{-}79)$$

Eq. 11-78 becomes

$$\ln (f/P) = \frac{1}{RT} \int_0^P (B + CP + \ldots) \, dP$$

$$= \frac{1}{RT}\left(BP + \frac{C}{2} P^2 + \ldots\right)$$

The integration can also be carried out using other equations of state, such as the van der Waals, the Dieterici, or the Berthelot equations.

Equation 11-78 can also be expressed in terms of the compressibility factor $Z = PV/RT$ as

$$\ln (f/P) = \int_0^P \frac{Z-1}{P} \, dP \tag{11-81}$$

Exercise 11-22. Verify Eq. 11-81.

To the extent that the law of corresponding states is valid, Z is a universal function of the reduced pressure P_r and the reduced temperature T_r for all gases (Sec. 1-3). Numerical integration of Eq. 11-81 using this universal function provides values of $\ln (f/P)$ as a function of P_r and T_r that are the same for all gases to the same approximation. A graphical representation of some results of this kind is shown in Fig. 11-4.

The quantity f/P is commonly called the *fugacity coefficient*, or the *activity coefficient*, and given the symbol γ;

$$f = \gamma P \tag{11-82}$$

Fugacity in mixtures of gases
The fugacity of gases in mixtures must be obtained to apply these definitions to chemical equilibrium in gases. By analogy to Eq. 11-73 and Eq. 11-74, the corresponding definitions for a component of a mixture of gases are

$$\mu_j = \mu_j^\ominus + RT \ln \frac{f_j}{f^\ominus} \tag{11-83}$$

and

$$\lim_{P \to 0} \left[\frac{f_j}{(X_j P)} \right] = 1 \tag{11-84}$$

Since the standard state is the ideal gas state at 1.01×10^5 Pa pressure, the standard chemical potential and the standard fugacity for a component of a mixture are the same as the corresponding values for the pure gas. This definition is equivalent to the statement that the fugacity of a gas in a mixture is equal to the fugacity of the pure gas in equilibrium with the mixture across a membrane permeable only to that component, since at equilibrium, μ (pure) $= \mu_j$.

Exercise 11-23. Verify the preceding sentence.

For a gas in a mixture, from Eq. 11-37 and Eq. 11-39,

$$\left(\frac{\partial \mu_j}{\partial P} \right) = \mathscr{V}_j$$

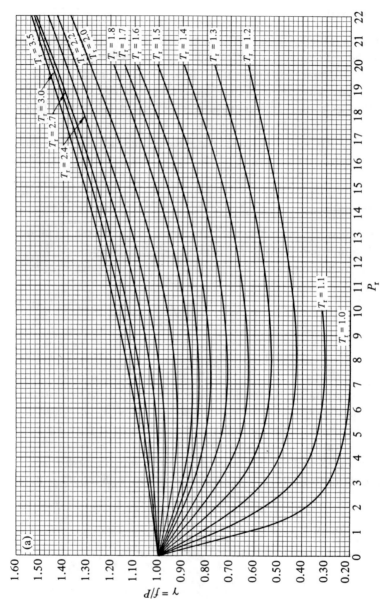

Figure 11-4. Plots of the activity coefficient $\gamma = f/P$ as a function of the reduced pressure P_r at several values of T_r: (a) T_r from 1.0 to 3.5; (b) T_r from 3.5 to 35.0. [By permission, from R. H. Newton, *Ind. Eng. Chem.*, **27**, 302–6 (1935).] (c) Low values of P_r (courtesy of Prof. George Martin Brown, Chemical Engineering Department, Northwestern University). The expressions $T_r = T/(T_c + 8)$ and $P_r = P/(P_c + 8.08 \times 10^5 \text{ pa})$ are used for H_2 and He in (a), (b), and (c).

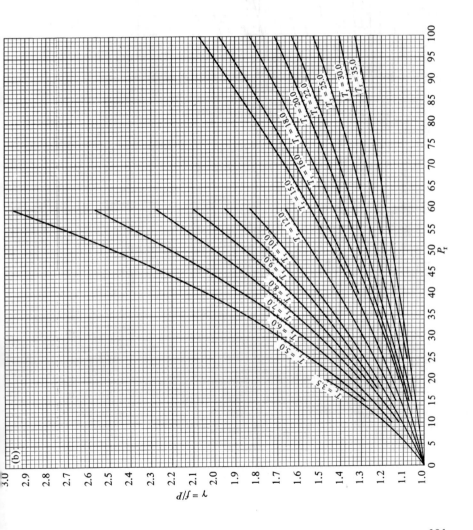

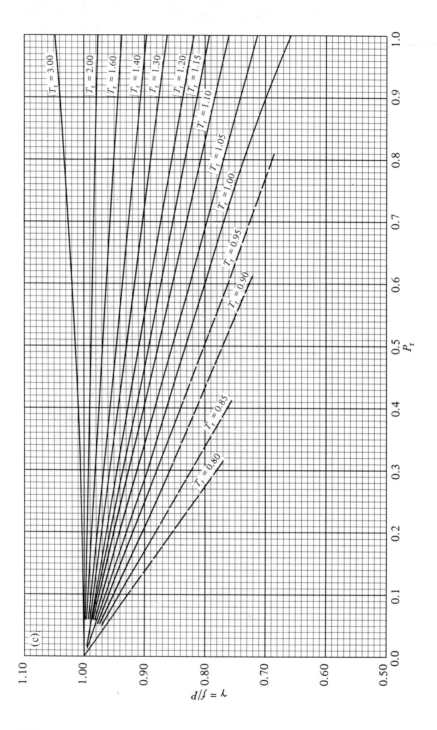

so that

$$\left[\frac{\partial \ln (f_j/f^{\ominus})}{\partial P}\right]_{T,\,n_{\omega}} = \frac{\mathscr{V}_j}{RT} \tag{11-85}$$

As with Eq. 11-76 for the pure gas, Eq. 11-85 cannot be integrated from $P = 0$ to some finite pressure, so we shall evaluate the derivative

$$\left[\frac{\partial \ln (f_j/X_jP)}{\partial P}\right]_{T,\,n_{\omega}} = \left[\frac{\partial \ln (f_j/f^{\ominus})}{\partial P}\right]_{T,\,n_{\omega}} - \left\{\frac{\partial \ln [(X_jP)/p^{\ominus}]}{\partial P}\right\}_{T,\,n_{\omega}}$$

$$= \frac{\mathscr{V}_j}{RT} - \frac{1}{P}$$

$$= \frac{1}{RT}\left(\mathscr{V}_j - \frac{RT}{P}\right) \tag{11-86}$$

If sufficient data are available on the density of a gaseous mixture as a function of temperature, pressure, and composition, $\mathscr{V}_j$ can be calculated for each component as a function of temperature, pressure, and composition, and Eq. 11-86 can be integrated by graphical or numerical methods to obtain values of $\gamma_j = f_j/p_j$. Since such data are available only for a few systems, it is customary to assume that

$$f_j = X_j f_j^* \tag{11-87}$$

where f_j^* is the fugacity of the pure gas at a pressure equal to the total pressure of the mixture. Equation 11-87, which is called the *Lewis and Randall rule*, assumes that deviation from ideality depends only on the total pressure and is independent of the nature of the other gases in the mixture.

To the approximation provided by the Lewis and Randall rule,

$$\gamma_j = \frac{f_j}{p_j} = \frac{X_j f_j^*}{X_j P}$$

$$= \gamma_j^* \tag{11-88}$$

Thus, the fugacity coefficients obtained for the pure gas are used to calculate the fugacity of the component of a mixture.

The fugacity function and the equilibrium constant

One can derive an equilibrium constant in terms of fugacity from Eq. 11-83, together with the criterion for equilibrium in Eq. 11-18 and the expression for $\Delta\mathscr{G}^{\ominus}$ in Eq. 11-42. The result, for Eq. 11-13, is

$$K_f = \frac{(f_C/f^{\ominus})^c (f_D/f^{\ominus})^d}{(f_S/f^{\ominus})^s (f_U/f^{\ominus})^u} \tag{11-89}$$

Exercise 11-24. Verify Eq. 11-89.

If we substitute from Eq. 11-88 for the fugacities in Eq. 11-89 we have

$$K_f = \frac{(p_C/p^{\ominus})^c (p_D/p^{\ominus})^d}{(p_S/p^{\ominus})^s (p_U/p^{\ominus})^u} \cdot \frac{(\gamma_C)^c (\gamma_D)^d}{(\gamma_S)^s (\gamma_U)^u}$$

$$= K_p K_\gamma \qquad\qquad (11\text{-}90)$$

since $\gamma^{\ominus} = 1$. The values of K_γ needed to convert the partial pressure equilibrium constants in Table 11-3 into fugacity equilibrium constants are listed in Table 11-4, together with the resulting K_f.

Table 11-4. $N_2(g) + 3\,H_2(g) = 2\,NH_3(g)$

$P/1.01 \times 10^5$ Pa	K_γ	K_f
10	0.990	0.0000429
30	0.951	0.0000434
50	0.893	0.0000423
100	0.774	0.0000404
300	0.473	0.0000370
600	0.247	0.0000412
1000	0.188	0.000102
2000	0.117	0.00210

L. J. Winchester and B. F. Dodge, *J. Am. Inst. Chem. Eng.*, 2, 431 (1956).

Though the approximation of the Lewis and Randall rule begins to break down at pressures of 1.01×10^8 Pa, the value of K_f is constant within experimental error below that pressure.

The equilibrium constant calculated from tables of $\Delta\mathscr{G}f^{\ominus}$ or tables of $\mathscr{G}^{\ominus} - \mathscr{E}_0^{\ominus}$ is K_f. Calculation of K_γ then permits the calculation of K_p at high pressures, from which equilibrium yields of product can be obtained.

Problems

11-1. Calculate a value of $\Delta\mathscr{G}^{\ominus}$ and a value of K_p at 298 K for the reaction

$$H(g) + D_2(g) = HD(g) + D(g)$$

from data in Table 11-2. Comment on the results as they are related to Problem 10-9.

11-2. Calculate $\Delta\mathscr{G}^{\ominus}$ and K_p at 298 K for the reaction

$$N_2(g) + 3\,H_2(g) = 2\,NH_3(g)$$

from data in Table 11-2 and compare with the results of Exercise 11-14.

11-3. Larson and Dodge [*J. Am. Chem. Soc.*, *46*, 367 (1924).] found the following values of K_p at a pressure of 1.01×10^6 Pa for the reaction in Problem 11-2.

T/K	K_p
623	7.070×10^{-4}
648	3.294×10^{-4}
673	1.669×10^{-4}
698	8.464×10^{-5}
723	4.356×10^{-5}
748	2.663×10^{-5}
773	1.452×10^{-5}

Plot $\log K_p$ against $1/T$. If the plot is linear, calculate $\Delta \mathcal{H}^\ominus$ from the slope. If the plot is not linear, determine the slope graphically at each temperature from 648 to 748 K.

11-4. Use any data in Chapters 9, 10, and 11 and one value of K_p in Problem 11-3 to obtain an analytical expression for $\log K_p$ as a function of T. Compare the values calculated from this analytical expression with the values in Problem 11-3.

11-5. The values of $\Delta \mathcal{G}^\ominus$ obtained from tables of $\Delta \mathcal{G} f^\ominus$ or $\mathcal{G}^\ominus - \mathcal{E}_0^\ominus$ lead to K_f through the relation

$$\Delta \mathcal{G}^\ominus = -RT \ln K_f \qquad (11\text{-}91)$$

where K_f is independent of pressure over a wide range of pressures. Calculate K_f at 723 K for the NH_3 synthesis reaction in Problem 11-2 from $\Delta \mathcal{G} f^\ominus$ and $\Delta \mathcal{H} f^\ominus$ values at 298.15 K and the heat capacity coefficients for the reactants and products. Use the critical constants in Table 1-5 to obtain values of f/P for NH_3, H_2, and N_2 from Fig. 11-4 and calculate K_y at each pressure up to 6.06×10^7 Pa in Table 11-4. Calculate K_p at each pressure and compare with the experimental values of Winchester and Dodge in Table 11-3.

11-6. Given the following molecular properties (G. Herzberg, Vol. I, pp. 532–33), calculate $\Delta \mathcal{G}^\ominus$ and K_p at 298 K for the reaction in Problem 11-1 from a statistical model.

	$\tilde{\nu}_e/cm^{-1}$	$r_e/10^{-10} m$	$\tilde{\nu}_{elec}/cm^{-1}$
H_2	4395.2	0.7417	90,196.1
D_2	3118.5	0.7416	90,634.0
HD	3817.1	0.7414	90,410

12 · *The dynamics of chemical change*

We have examined the structure of molecules, their interaction with electro-magnetic radiation, and the statistical and classical thermodynamic de-scription of chemical equilibrium. We are now prepared to consider the mechanisms of chemical reactions, particularly as they can be elucidated through the study of reaction rates.

12-1. The description of reaction rates

The rate of a chemical reaction can be described by specifying the rate of change of the concentration of any reactant or product. The choice of one reactant or product rather than another will lead to values of the rate of reaction that differ by a factor determined by the stoichiometry of the re-action. Thus, it is convenient, for a reaction

$$aA + bB + \ldots \rightleftarrows cC + dD + \ldots \qquad (12\text{-}1)$$

to define a single rate of reaction r, where

$$r = -\frac{1}{a}\frac{d[A]}{dt} = -\frac{1}{b}\frac{d[B]}{dt} = \frac{1}{c}\frac{d[C]}{dt} = \frac{1}{d}\frac{d[D]}{dt} \qquad (12\text{-}2)$$

Then r is independent of the particular species chosen. It has the dimensions of concentration per time.

Experimental determination of reaction rates
Experimentally, reaction rates are usually determined by measuring the

396

concentration of a particular reactant or product as a function of time. It is essential that the reaction be slow enough so that the concentration does not change appreciably during the time needed to make the measurement. A reaction that can be stopped in the sample used for analysis—by lowering the temperature or removing a catalyst, for example—is equally suitable.

It is frequently more convenient, however, to monitor the change of a physical property of the system with time, and to relate that change to the rate of the reaction, especially if the property can be measured continuously and recorded automatically as a function of time. It is possible to record continuously some chosen property and simultaneously to convert the measurements to a digital form suitable for computer processing.

For example, in the gaseous reaction

$$2 \text{ NO(g)} + \text{Cl}_2 \rightleftarrows 2 \text{ NOCL(g)} \tag{12-3}$$

$$r = -\frac{1}{2}\frac{d[\text{NO}]}{dt} = -\frac{d[\text{Cl}_2]}{dt} = \frac{1}{2}\frac{d[\text{NOCl}]}{dt} \tag{12-4}$$

Since the number of molecules of products and reactants is different, the pressure of the system at constant volume and temperature changes, and the change in pressure with time can be related to r. If the gaseous mixture is assumed to be ideal,

$$P = \{[\text{NO}] + [\text{Cl}_2] + [\text{NOCl}]\} RT \tag{12-5}$$

and

$$\frac{dP}{dt} = RT\left\{\frac{d[\text{NO}]}{dt} + \frac{d[\text{Cl}_2]}{dt} + \frac{d[\text{NOCl}]}{dt}\right\} \tag{12-6}$$

Substituting from Eq. 12-4 in Eq. 12-6, we obtain

$$\frac{dP}{dt} = RT\{-2r - r + 2r\}$$

and

$$r = -\frac{1}{RT}\frac{dP}{dt} \tag{12-7}$$

Exercise 12-1. The inversion of sucrose to form glucose and fructose, represented by the equation

$$S + H_2O \rightleftarrows G + F$$

can be followed by the change in the optical rotation α of the solution. If the optical rotation is a linear function of the concentrations of sucrose, glucose, and fructose, with coefficients characteristic of each sugar, derive a relationship between r and $d\alpha/dt$.

Order of reaction

Chemical kinetics is the branch of chemistry that is concerned with the determination of the relationship between r and the concentrations of reactants and products and with the deductions about mechanism that can be drawn from this information. The rates of *some* reactions are proportional to a product of concentrations raised to a power. For example, if Eq. 12-1 is such a reaction, and if

$$r = k_r[A]^p[B]^q[C]^s[D]^w \tag{12-8}$$

p, q, s, and w are, respectively, the *order of reaction* with respect to A, B, C, and D, and k_r is the *rate constant* of the reaction or the *specific reaction rate*. The sum of p, q, s, and w is the *over-all order of the reaction*. The rate constant does vary with the temperature, but it is independent of concentration or extent of reaction.

Zero-order reactions.   For some reactions of the form

$$A \rightarrow \text{products} \tag{12-9}$$

the rate of reaction can be expressed as

$$r = -\frac{d[A]}{dt} = k_r[A]^n \tag{12-10}$$

When $n = 0$,

$$-\frac{d[A]}{dt} = k_r \tag{12-11}$$

On integration of Eq. 12-11, one obtains

$$[A] = [A]_0 - k_r t \tag{12-12}$$

Reactions that are catalyzed at a solid surface or by an enzyme exhibit zero-order kinetics when the catalyst is saturated with the reactant. (See Sec. 12-6 and Sec. 18-2.)

First-order reactions.   When $n = 1$ in Eq. 12-10,

$$-\frac{d[A]}{dt} = k_r[A] \tag{12-13}$$

Separation of variables leads to

$$\frac{d[A]}{[A]} = -k_r dt \tag{12-14}$$

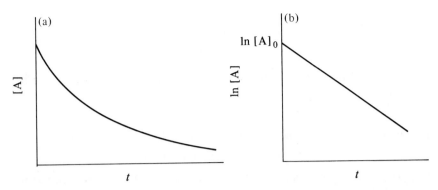

Figure 12-1. (a) An idealized plot of concentration of reactant as a function of time (t) for a first-order reaction. (b) An idealized representation of a linear plot of the natural logarithm of the concentration of reactant as a function of time for a first-order reaction; slope $= -k_r$.

so that, on integrating from $[A]_0$ to $[A]$ and from 0 to t, we obtain

$$\ln \frac{[A]}{[A]_0} = -k_r t \qquad (12\text{-}15)$$

Thus,

$$[A] = [A]_0 \exp(-k_r t) \qquad (12\text{-}16)$$

and a graph of $\ln [A]$ against t is linear, with slope equal to $-k_r$. Another way to check whether a reaction is first order is to calculate k_r from the equation

$$k_r = -\frac{1}{t} \ln \frac{[A]}{[A]_0} \qquad (12\text{-}17)$$

for successive values of t. The values should not vary systematically with t. Confidence in the order assigned to a reaction increases as the degree of completion of the reaction over which the test is valid increases. For reversible reactions, however, the reverse reaction must be taken into account. Schematic plots of $[A]$ against t and of $\ln [A]$ against t are shown in Fig. 12-1. The value of k_r is independent of the concentration units used.

The half-time $t_{\frac{1}{2}}$ of a reaction is defined as the time at which $[A] = [A]_0/2$. Thus, from Eq. 12-15

$$\ln (^1/_2) = -k_r t_{\frac{1}{2}}$$

and

$$t_{\frac{1}{4}} = -\frac{\ln\left(\frac{1}{2}\right)}{k_r} = \frac{\ln 2}{k_r} \tag{12-18}$$

That is, $t_{\frac{1}{4}}$ is independent of the concentration of reactant for a first-order reaction.

Exercise 12-2. The following data were obtained at 318 K for the reaction

$$2 \, N_2O_5 \, (g) = 4 \, NO_2 \, (g) + 0_2 \, (g)$$

by F. Daniels and E. H. Johnston [*J. Am. Chem. Soc.*, *43*, 53 (1921). Test whether the reaction is first order and calculate the rate constant (a) by plotting log $[N_2O_5]$ against t, (b) by calculating a value of k_r for each experimental point from Eq. 12-17, and (c) by plotting $[N_2O_5]$ against t and calculating several values of $t_{\frac{1}{4}}$ from the graph.

Time (min)	$[N_2O_5]/(10^{-3} \text{ mol-dm}^{-3})$
0.0	17.6
10.0	12.4
20.0	9.33
30.0	7.06
40.0	5.31
50.0	3.95
60.0	2.95
70.0	2.22
80.0	1.67
90.0	1.23
100.0	0.94
120.0	0.50
140.0	0.25
160.0	0.14

Exercise 12-3. What are the units of k_r for a first-order reaction?

Second-order reactions. Reactions of the type

$$2 \, A \rightarrow \text{products} \tag{12-19}$$

or

$$A + B \rightarrow \text{products} \tag{12-20}$$

may be second order. If

$$r = -\frac{d[A]}{dt} = k_r[A]^2 \tag{12-21}$$

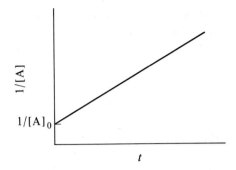

Figure 12-2. An idealized linear plot of the inverse of the concentration of reactant as a function of time (t) for a second-order reaction; slope $= k_r$.

separation of variables and integration as before gives

$$\frac{1}{[A]} = \frac{1}{[A]_0} + k_r t \tag{12-22}$$

A graph of $1/[A]$ against t is a straight line with slope equal to k_r, as shown schematically in Fig. 12-2. Again, an alternative check is to calculate k_r at successive values of t.

If the reaction in Eq. 12-20 is first order with respect to A and first order with respect to B,

$$-\frac{d[A]}{dt} = k_r [A][B] \tag{12-23}$$

It is convenient to define the variable $x = [A]_0 - [A] = [B]_0 - [B]$, so that Eq. 12-23 becomes

$$\frac{dx}{dt} = k_r([A]_0 - x)([B]_0 - x) \tag{12-24}$$

Separation of variables leads to

$$\frac{dx}{([A]_0 - x)([B]_0 - x)} = k_r dt \tag{12-25}$$

Equation 12-25 can be integrated by the method of partial fractions. The result is

$$\frac{1}{[A]_0 - [B]_0} \ln \frac{[A][B]_0}{[B][A]_0} = k_r t \tag{12-26}$$

A plot of ln $([A]/[B])$ against t should be linear, with slope equal to $([A]_0 - [B]_0)k_r$, as shown schematically in Fig. 12-3. Equation 12-26 is not applicable when $[A]_0 = [B]_0$, but Eq. 12-24 then reduces to an equivalent of Eq. 12-21.

Exercise 12-4.   Verify Eq. 12-26 and the sentence following it. Hint: Let

$$\frac{1}{([A]_0 - x)([B]_0 - x)} = \frac{a}{([A]_0 - x)} + \frac{b}{[B]_0 - x}$$

and solve for a and b.

Exercise 12-5.   Test the following data [from H. S. Johnston and D. M. Yost, *J. Chem. Phys.*, *17*, 386 (1949)] for the reaction

$$2\,NO_2(g) + O_3(g) = N_2O_5(g) + O_2(g)$$

for agreement with the rate equation

$$-\frac{d\,[O_3]}{dt} = k_r\,[NO_2]\,[O_3]$$

and calculate the rate constant k_r. These data were among the first obtained by rapid flow methods. Note the time scale.

time/10^{-3} sec	$[NO_2]/(10^{-4}\ mol\text{-}dm^{-3})$	$[O_3]/(10^{-4}\ mol\text{-}dm^{-3})$
0.0	2.50	2.49
3.33	2.38	2.41
6.66	2.31	2.36
10.00	2.23	2.30
13.33	2.16	2.25
16.67	2.11	2.22
33.33	1.82	2.03
50.00	1.60	1.87
66.67	1.36	1.72
83.33	1.17	1.61
100.0	1.01	1.51
116.7	0.86	1.42
133.3	0.78	1.37
150.0	0.71	1.33
166.7	0.64	1.29

Hint: Note the stoichiometry; do not try to use Eq. 12-24 without modification.

Exercise 12-6.   What are the units of k_r for a second-order reaction?

12-2.   Reaction order and reaction mechanism

Complex reactions are assumed to consist of a series of steps, each of which

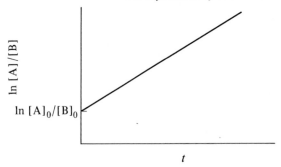

Figure 12-3. An idealized plot of the natural logarithm of the ratio of the concentrations of two reactants as a function of time (t) for a second-order reaction; slope $= \{[A]_0 - [B]_0\}\, k_r$.

is called an *elementary reaction*. Given the simplest chemical equation with integer coefficients that describes an elementary reaction, we assume that the reaction order for each reactant is its coefficient in the chemical equation. In effect, the rate of reaction is considered to be proportional to the frequency of encounters among the reactant molecules, that is, proportional to the concentration of each reactant raised to the integer coefficient just described. Thus, a *unimolecular* reaction, such as the isomerization or decomposition of a single molecule, is first order; a *bimolecular* reaction is second order; and a *termolecular* reaction is third order. In contrast, the number of molecules involved in an over-all reaction cannot always be inferred from the reaction order alone.

When the kinetic order of a reaction corresponds to the stoichiometric equation, it is reasonable to assume that the reaction mechanism is a single elementary reaction that is identical with the stoichiometric equation. The hydrogen-iodine reaction (Eq. 12-27) had been considered a classic example

$$H_2(g) + I_2(g) = 2\,HI(g) \tag{12-27}$$

of a bimolecular gaseous reaction since Bodenstein's kinetic results in the 1890's. Despite the subsequent experimental and theoretical study of the reaction by a number of investigators, it was not until 1967[1] that the mechanism was shown to be

$$
\begin{aligned}
I_2 &\rightleftarrows 2\,I \quad \text{(in rapid equilibrium)} \\
2\,I + H_2 &\rightleftarrows 2\,HI \quad \text{(slow)}
\end{aligned}
\tag{12-28}
$$

1. J. H. Sullivan, *J. Chem. Phys.*, **46**, 73 (1967); *Chem. Eng. News*, Jan. 16, 1971, p. 40.

The rate of the slow reaction is

$$r = \frac{1}{2}\frac{d[\text{HI}]}{dt} = k_r[\text{I}]^2[\text{H}_2] \tag{12-29}$$

One can write an equilibrium constant expression for the rapid dissociation of I_2 as

$$K = \frac{[\text{I}]^2}{[\text{I}_2]} \tag{12-30}$$

Substituting from Eq. 12-30 in Eq. 12-29, we have

$$r = k_r K[\text{I}_2][\text{H}_2] \tag{12-31}$$

This result shows that the termolecular mechanism, Eq. 12-28, is kinetically indistinguishable from the bimolecular mechanism, Eq. 12-27, for which the rate is

$$r = k_r'[\text{H}_2][\text{I}_2] \tag{12-32}$$

where

$$k_r' = k_r K \tag{12-33}$$

The distinction could only be made by comparing the thermal reaction with the photochemical reaction, in which it is known that I_2 is first dissociated into I atoms. Two recent articles indicate that a final decision on the mechanism of this reaction has not yet been reached.[2]

When the reaction order does not correspond to the stoichiometry of the reaction, as in the reactions in Exercises 12-2 and 12-5, the reaction involves more than one elementary reaction. It is then necessary to devise a series of elementary reactions that is consistent with the reaction order found experimentally.

Ogg[3] pointed out in 1947 that the kinetics of the decomposition of nitrogen pentoxide could be explained by the following mechanism:

$$\text{N}_2\text{O}_5 \underset{k_2}{\overset{k_1}{\rightleftarrows}} \text{NO}_2 + \text{NO}_3 \tag{12-34}$$

$$\text{NO}_2 + \text{NO}_3 \xrightarrow{k_3} \text{NO}_2 + \text{NO} + \text{O}_2 \tag{12-35}$$

$$\text{NO} + \text{NO}_3 \xrightarrow{k_4} 2\,\text{NO}_2 \tag{12-36}$$

2. G. G. Hammes and B. Widom, *J. Am. Chem. Soc.*, **96**, 7621–22 (1974); R. M. Noyes, *J. Am. Chem. Soc.*, **96**, 7623–24 (1974).
3. R. A. Ogg, *J. Chem. Phys.*, **15**, 337 (1947).

If Eq. 12-34 is multiplied by two, the mechanism is consistent with the stiochiometry shown in Exercise 12-2. We shall show later in this section that the mechanism is consistent with the experimental first-order kinetics.

Unimolecular reactions

For many years, one of the major problems of chemical kinetics was to find an explanation for the first-order kinetics of unimolecular reactions. The reactant molecules must acquire energy before they can react, and it was shown relatively early that this energy was not obtained by absorption of electromagnetic radiation. The most likely alternative source is through molecular collisions, but a bimolecular collision should result in second-order kinetics.

A first answer to the problem was provided by Lindemann in 1927 when he put forward the *steady-state approximation*. Consider the unimolecular decomposition

$$A \rightarrow \text{products} \tag{12-37}$$

The first step in the proposed mechanism is the production of activated molecules A^* by collision.

$$A + A \underset{k_2}{\overset{k_1}{\rightleftarrows}} A^* + A \tag{12-38}$$

The reaction is indicated as reversible to take into account the possibility of deactivation by collision as an alternative to the reaction to form products,

$$A^* \xrightarrow{k_3} \text{products} \tag{12-39}$$

According to Eq. 12-39, the rate of production of products should be

$$\frac{d[P]}{dt} = r = k_3[A^*] \tag{12-40}$$

But $[A^*]$ is changing according to the equation

$$\frac{d[A^*]}{dt} = k_1[A]^2 - k_2[A][A^*] - k_3[A^*] \tag{12-41}$$

Rather than trying to solve the two differential equations, Eqs. 12-40 and 12-41, simultaneously, Lindemann made the steady-state approximation that $[A^*]$ remains small during the course of the reaction and that $d[A^*]/dt$ is negligible compared to the rates of the processes by which A^* is produced and destroyed. Thus,

$$k_1[A]^2 - k_2[A][A^*] - k_3[A^*] \approx 0$$

$$[A^*] = \frac{k_1 [A]^2}{k_3 + k_2 [A]}$$

and

$$r = k_3 \frac{k_1 [A]^2}{k_3 + k_2 [A]} \tag{12-42}$$

If the pressure is high, so that the rate of deactivation by collision is greater than the rate of production of products, then

$$k_2 [A] \gg k_3$$

and

$$r = \frac{k_3 k_1}{k_2} [A] \tag{12-43}$$

The reaction is then first order. If the pressure is low, so that the rate of production of products is greater than the rate of deactivation by collision, then

$$k_3 \gg k_2 [A]$$

and

$$r = k_1 [A]^2 \tag{12-44}$$

The reaction is then second order. This prediction of the shift in observed kinetics with pressure was a major step in the understanding of unimolecular reactions.

At first glance the steady-state approximation seems to be self-contradictory; from Eq. 12-40, the rate of reaction is proportional to $[A^*]$, which is assumed to be constant, yet from Eq. 12-42 the rate of reaction is a function of $[A]$, which is steadily decreasing as the reaction proceeds. The contradiction is removed by using as the rate of reaction the *initial slope* of the curve of concentration of products against time. (An implicit assumption is that the steady state has been reached before the first measurement.) The initial slope refers to the rate at fixed values of $[A]$ and $[A^*]$. The initial slope is measured for a range of initial concentrations of A, and the order is obtained by plotting log r against log $[A]_0$, as shown in Fig. 12-4. Over a range of concentrations in which Eq. 12-43 is valid, the slope of the line is one, and the intercept is log $(k_3 k_1 / k_2)$. Over a range of lower concentrations for which Eq. 12-44 is valid, the slope of the line is two, and the intercept is log k_1. A plot of data over the entire range of concentrations shows a slope of two at low concentrations and a slope of one at high concentrations.

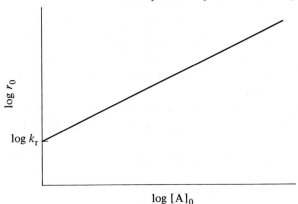

Figure 12-4. An idealized plot of the logarithm of the initial rate against the logarithm of the initial concentration; the order of reaction is equal to the slope.

More complex models have been developed that fit experimental results for first-order gas reactions better than the simple Lindemann model does. The reader is referred to more advanced treatises for treatment of these models.[4]

Complex reactions
The steady-state approximation has also been useful in simplifying the treatment of complex reactions, in which intermediate compounds do not appear in the over-all reaction order. The differential equations that describe the mechanism suggested for the decomposition of N_2O_5 (Eqs. 12-34 to 12-36) are

$$\frac{d[NO_2]}{dt} = k_1[N_2O_5] - k_2[NO_2][NO_3] + 2k_4[NO][NO_3] \qquad (12\text{-}45)$$

for the rate of production of product and

$$\frac{d[NO_3]}{dt} = k_1[N_2O_5] - (k_2 + k_3)[NO_2][NO_3] - k_4[NO][NO_3] \quad (12\text{-}46)$$

and

$$\frac{d[NO]}{dt} = k_3[NO_2][NO_3] - k_4[NO][NO_3] \qquad (12\text{-}47)$$

4. H. S. Johnston, *Gas Phase Reaction Rate Theory*, The Ronald Press Co., New York, 1966, Chap. 15; A. A. Frost and R. G. Pearson, *Kinetics and Mechanism*, 2nd ed., John Wiley & Sons, New York, 1961, pp. 69–75; K. J. Laidler, *Chemical Kinetics*, 2nd ed., McGraw-Hill Book Co., New York, 1965, pp. 143–75.

for the rate of change of concentration of intermediates. An exact solution of the problem would require a simultaneous solution of the three differential equations. The steady-state approximation assumes that the concentrations of the intermediates are small and rapidly reach values that remain constant until the reaction is almost complete.

Thus, the derivatives in Eqs. 12-46 and 12-47 are set equal to zero, and the resulting equations are solved for the concentrations of the intermediates. The results are

$$k_4 [NO][NO_3] = k_3 [NO_2][NO_3] \tag{12-48}$$

and

$$[NO_2][NO_3] = \frac{k_1 [N_2O_5]}{k_2 + 2 k_3} \tag{12-49}$$

Exercise 12-7. Verify Eq. 12-48 and Eq. 12-49.

Substituting from Eq. 12-48 and Eq. 12-49 in Eq. 12-45, we obtain

$$\frac{d[NO_2]}{dt} = \frac{4 k_3 k_1}{k_2 + 2 k_3} [N_2O_5] \tag{12-50}$$

Exercise 12-8. Verify Eq. 12-50.

Since the over-all reaction is

$$2 N_2O_5 (g) = 4 NO_2 (g) + O_2 (g)$$

$$r = \frac{1}{4} \frac{d[NO_2]}{dt} = \frac{k_3 k_1}{k_2 + 2 k_3} [N_2O_5] \tag{12-51}$$

12-3. Temperature and reaction rates

During the 1880's, van't Hoff and Arrhenius observed that the strong temperature dependence of the rate constant for a number of reactions could be described by the equation

$$\frac{d \ln k_r}{dT} = \frac{E_a}{RT^2} \tag{12-52}$$

in which E_a is the *Arrhenius activation energy*. If E_a is independent of temperature, the integrated form of the equation is

$$k_r = B \exp\left(- \frac{E_a}{RT}\right) \tag{12-53}$$

where B is a constant of integration.

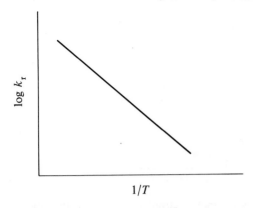

Figure 12-5. An idealized linear plot of log k_r against $1/T$, from which the Arrhenius activation energy is obtained; slope $= -E_a/2.303\ RT$.

Exercise 12-9. Verify Eq. 12-53.

A plot of log k_r against $1/T$ is linear if Eq. 12-53 fits the experimental results for a reaction, and the slope is equal to $-E_a/(2.303\ RT)$, as indicated schematically in Fig. 12-5.

Exercise 12-10. Calculate E_a for the decomposition of N_2O_5 from the accompanying data.

T/K	k_r/min^{-1}
298.0	2.03
308.0	8.09
318.0	29.9
328.0	90.1
338.0	291.5

F. Daniels and E. H. Johnston, *J. Am. Chem. Soc.*, *43*, 53 (1921).

The exponential factor in Eq. 12-53 is reminiscent of the expression obtained from the Maxwell distribution of kinetic energies for the fraction of molecules with energy greater than a given value (See Exercise 2-11). Here, E_a is interpreted as the difference in potential energy between the reactants and a postulated intermediate state called the *activated complex*. The exponential factor in Eq. 12-53 represents the fraction of reactants with sufficient energy to reach the activated-complex state. As indicated in Fig. 12-6, the difference between E_a (forward) and E_a (reverse) is equal to ΔE for an elementary reaction.

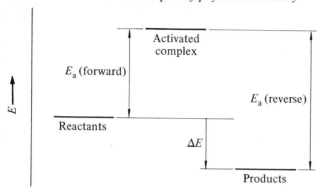

Figure 12-6. The schematic relationship among activation energies for the forward and reverse reactions and ΔE for the elementary reaction; E_a is always a positive quantity, whereas ΔE may be positive or negative.

Collision model

A simple *collision model* leads to an equation of the form of Eq. 12-53. The fundamental assumption of the model is that the rate of a reaction is determined by the rate of collision between reactant molecules. In the case of bimolecular gas reactions, a reasonable estimate of the rate of collisions can be obtained from the kinetic theory of gases. According to Eq. 2-37, the number of collisions between like molecules per unit volume per second is

$$Z = (^1/_2)(2)^{\frac{1}{2}} \pi \sigma_A^2 \bar{u}_A n_A^2$$

where σ_A is the molecular diameter of A in the hard-sphere model, $\bar{u}_A$ is the average molecular speed, and n_A is the number of molecules of A per unit volume.

If it is assumed that every collision results in a reaction, then the rate of reaction is

$$r = Z/L = Z'[A]^2 \tag{12-54}$$

where Z' is known as the frequency factor.

Exercise 12-11. Derive an expression for the frequency factor Z'. Calculate a value for Z' for HI at 556 K, 1×10^5 Pa. Assume $\sigma = 4.1 \times 10^{-10}$ m.

The rate of a bimolecular reaction in which two A react to form products can also be expressed as (Eq. 12-21)

$$r = k_r[A]^2$$

A combination of Eqs. 12-54 and 12-21 leads to the result

$$k_r = (1/2)(2)^{\frac{1}{2}} \pi \sigma_A^2 \bar{u}_A L \tag{12-55}$$

Exercise 12-12. Verify Eq. 12-55.

From Eq. 2-64,

$$\bar{u}_A = (8kT/\pi m)^{\frac{1}{2}}$$
$$= (8RT/\pi M)^{\frac{1}{2}} \tag{12-56}$$

Thus,

$$k_r = 2L\sigma_A^2 (\pi RT/M)^{\frac{1}{2}} \tag{12-57}$$

The value of k_r predicted by Eq. 12-57 for the decomposition of HI at 556 K can be compared with Bodenstein's experimental results. The collision diameter calculated from gas viscosity measurements[5] is 4.1×10^{-10} m. The molecular mass is $.12791$ kg-mol^{-1}. The result of the calculation is that $k_r = 6.8 \times 10^{10}$ dm^3-mol^{-1}-s^{-1}. Since the experimental result[6] is 3.52×10^{-7} dm^3-mol^{-1}-s^{-1}, it is clear that only a small fraction of collisions results in reaction. That fraction should be given by the exponential factor in Eq. 12-53. Since E_a has been found to equal 1.83×10^5 J-mol^{-1} for the HI decomposition,

$$\exp\left(\frac{-E_a}{RT}\right) = 6.3 \times 10^{-18}$$

If B in Eq. 12-53 is taken as equal to the frequency factor Z' of Eq. 12-54,

$$k_r = Z' \exp\left(\frac{-E_a}{RT}\right) \tag{12-58}$$
$$= (6.8 \times 10^{10})(6.3 \times 10^{-18})$$
$$= 4.3 \times 10^{-7} \text{ dm}^3\text{-mol}^{-1}\text{-s}^{-1}$$

The agreement is excellent, considering the large effect of a possible small error in the value of E_a.

When similar calculations are carried out for bimolecular gas reactions of more complex molecules, the agreement is not nearly so good; the calculated

5. J. O. Hirschfelder, C. F. Curtiss, and R. B. Bird, *Molecular Theory of Gases and Liquids*, John Wiley & Sons, Inc., New York, 1954, p. 1112.
6. M. Bodenstein, *Z. physik. Chem.*, *13*, 56 (1894); *22*, 1 (1897); *29*, 295 (1899); G. B. Kistiakowsky, *J. Am. Chem. Soc.*, *50*, 2315 (1928).

frequency factor can be a number of powers of ten greater than the experimental value, as shown in Table 12-1. The discrepancy is usually attributed to a *steric* or *orientation* requirement for a collision to result in reaction. A term, usually designated P, is included in the expression for the rate constant to take this requirement into account, so that

$$k_r = PZ' \exp\left(-\frac{E_a}{RT}\right)$$ (12-59)

Table 12-1.   Frequency factor for some bimolecular gas reactions, $Z'/(dm^3\text{-mol}^{-1}\text{-s}^{-1})$

Reaction	Experimental	Collision theory prediction	Reference[a]
1. Acrolein + butadiene = Tetrahydrobenzaldehyde	1.46×10^6	$\sim 10^{11}$	1
2. Acrolein + isoprene = Methyltetrahydrobenzaldehyde	1.02×10^6	$\sim 10^{11}$	1
3. $NO + O_3 = NO_2 + O_2$	8.0×10^8	4.7×10^{10}	2, 3
4. $NO_2 + O_3 = NO_3 + O_2$	5.9×10^9	6.3×10^{10}	2, 4
5. $NOCl + Cl = NO + Cl_2$	1.14×10^{10}	5.7×10^{10}	2, 5
6. $COCl + Cl = CO + Cl_2$	4.00×10^{11}	6.5×10^{10}	2, 5

[a] Key to references:
1. G. B. Kistiakowsky and J. R. Lacher, *J. Am. Chem. Soc.*, *58*, 123 (1936).
2. D. R. Herschbach, H. S. Johnston, K. S. Pitzer, and R. E. Powell, *J. Chem. Phys.*, *25*, 736 (1956).
3. H. S. Johnston and H. J. Crosby, *J. Chem. Phys.*, *22*, 689 (1954).
4. H. S. Johnston and D. M. Yost, *J. Chem. Phys.*, *17*, 386 (1949).
5. W. G. Burns and F. S. Dainton, *Trans. Farad. Soc.*, *48*, 39, 52 (1952).

Kinetic theory provides no basis for the calculation of the orientation factor. If two molecules must collide in a specific orientation for a reaction to occur, then the molecules must lose their independent translational and rotational degrees of freedom in the formation of the activated complex. As we have seen (Sec. 8-7), this loss of translational and rotational degrees of freedom can be described in terms of a decrease in entropy, and the transition state model provides a method for estimating the change in entropy in the formation of the activated complex.

Transition state model

In the early thirties, London, Wigner, Polanyi, and Eyring developed an alternative model for gaseous reactions based on the assumption that the reactants and the activated complex, or transition state, are in a quasi-equilibrium state. This model also assumes that the energy of the system varies in a continuous fashion from that of the reactants, through the transition state, to the products without a discrete change in electronic state. This latter assumption, the "adiabatic hypothesis," implies that the reaction can be described as if it occurred on a single potential-energy surface.

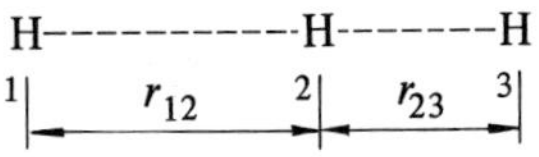

```
H----------H-------H
1|    r₁₂   2|  r₂₃  3|
 |←—————————|←———————|
```

Figure 12-7. The coordinates r_{12} and r_{23} in the linear $H + H_2 = H_2 + H$ reaction.

Potential-energy surface. The computation of the potential-energy[7] surface for a reaction makes possible the theoretical calculation of the activation energy. Consider first the collision of two hydrogen atoms, with electrons of opposite spin. Since the potential energy is a function only of one variable, the internuclear distance, it can be described graphically as a curve, such as the curve for total energy in Fig. 5-4. As the atoms approach one another and the potential energy decreases, the kinetic energy of the atoms increases. As the atoms approach one another more closely than the potential energy minimum, the potential energy increases and the kinetic energy decreases. The atoms continue to approach one another until the potential energy is a positive quantity equal to the initial kinetic energy, and the kinetic energy is then zero. At this distance, the net force acting is repulsive, so the atoms begin to move apart. When they reach the potential energy minimum, they have a kinetic energy that equals the sum of the initial kinetic energy and the depth of the potential well. Thus, they continue to move apart indefinitely, and we must conclude that two atoms that collide cannot form a stable molecule unless a third body, either another molecule or the wall of the container, also collides with them simultaneously and absorbs the excess kinetic energy that prevents their combination. Since the combination results in an increase in the average kinetic energy, the temperature of the surroundings increases. From a consideration of the second law of thermodynamics, the decrease in entropy that results from the substitution of vibrational and rotational degrees of freedom for translational degrees of freedom is more than compensated for by the increase in entropy that results from the increase in temperature.

In order to describe the interaction of *three* hydrogen atoms simply, we shall consider only those interactions that occur with all three atoms in a linear array. The energy of the system then depends on the distances r_{12} and r_{23} shown in Fig. 12-7. A complete graphical representation is three dimensional, with r_{12} and r_{23} the coordinates in a plane and the energy represented

---

7. This function of the nuclear positions is the electronic energy referred to in Secs. 5-1 and 5-2 and serves as the potential energy for the motion of the nuclei. It is the sum of the kinetic and the potential energy of the electrons and the mutual potential energy of the nuclei.

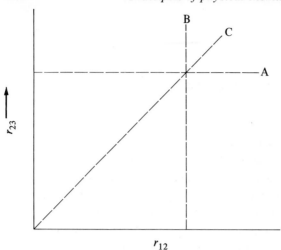

Figure 12-8. The coordinate plane in which r_{12} and r_{23} are the Cartesian coordinates. The line A represents a cross section through the potential energy surface in which r_{12} varies, and r_{23} remains very large. The line B represents a cross section in which r_{23} varies, and r_{12} remains very large. The line C represents a cross section in which r_{12} and r_{23} vary while both remain equal.

by the height above the plane at any point, forming a potential-energy surface.

The coordinate plane is shown in Fig. 12-8. The upper left-hand corner, with r_{12} small and r_{23} large, represents the state.

$$\underset{1 \quad 2}{H - H} + \underset{3}{H}$$

The lower right-hand corner, with r_{12} large and r_{23} small, represents the state.

$$\underset{1}{H} + \underset{2 \quad 3}{H - H}$$

The upper right-hand corner, with both r_{12} and r_{23} large, represents a state in which the three hydrogen atoms are independent.

One way in which we can visualize the nature of the potential-energy surface is to take cross sections through the surface in perpendicular planes through lines A, B, and C in Fig. 12-8. Along line A, r_{23} remains very large, and r_{12} varies. The cross section of the surface, therefore, should closely resemble the potential energy curve of the molecule

$$\underset{1 \quad 2}{H - H}$$

as a function of r_{12}. A schematic representation is shown as curve A, B in Fig. 12-9. The cross section of the surface along line B should be identical with that along line A, since it is the potential energy curve of the molecule

$$\underset{2\qquad 3}{\text{H}-\text{H}}$$

plotted as a function of r_{23}. Along line C, $r_{12} = r_{23}$, and the intersection of the surface with the perpendicular plane through this line represents the energy of the system as a function of $r_{12} = r_{23}$; that is, the symmetrical vibration of the

$$\underset{1\qquad 2\qquad 3}{\text{H}-\text{H}-\text{H}}$$

molecule. A schematic representation is shown as curve C in Fig. 12-9. The minimum in curve C is more shallow than that in curve A and occurs at a greater internuclear distance because two bonds formed by three electrons are weaker than a single bond formed by two electrons.

Exercise 12-13. Explain the statement in the preceding sentence in terms of a simple molecular orbital description of the H_3 molecule, including the number of bonding, nonbonding, and antibonding orbitals and their occupancy.

Figure 12-9. Schematic potential energy curves obtained by taking sections through the potential energy surface for $H + H_2 = H_2 + H$ along lines A, B, and C (Fig. 12-8).

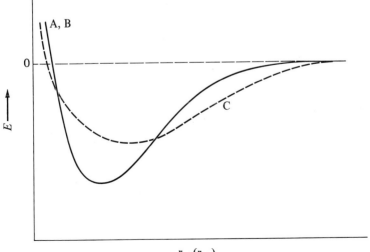

The surface is seen, then, to have a plateau in the region in which both r_{12} and r_{23} are large, a region in which the three hydrogen atoms are independent of one another. There are two valleys parallel to the r_{12} and r_{23} axes, and the floors of these valleys rise to a *maximum* as the interatomic distances become small, at a point that is a *minimum* with respect to position along line C, a *saddle point*. The most complete, two-dimensional representation of a potential energy surface is one in which contour curves of constant energy are plotted as a function of r_{12} and r_{23}. Such a representation is shown in Fig. 12-10, based on an early valence bond calculation by Polanyi

Figure 12-10. An early contour diagram for the $H + H_2 = H_2 + H$ reaction, calculated from a valence bond model. [By permission, from H. Eyring, *J. Am. Chem. Soc.*, *53*, 2540 (1931).] The numbers on each contour curve represent relative energy in units of 4.184×10^3 J-mol^{-1}; reaction coordinate (— · — · —).

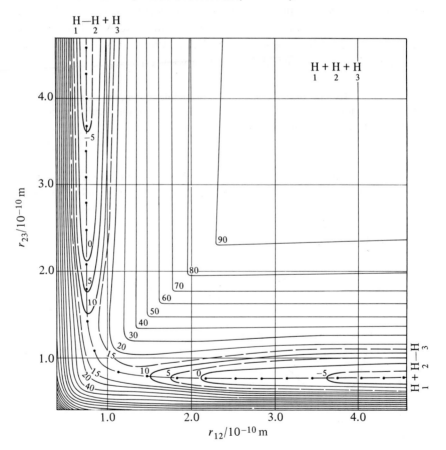

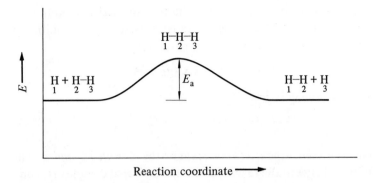

Figure 12-11. Scheme of the height of the potential energy surface for $H + H_2 = H_2 + H$ as a function of position along the reaction coordinate of Fig. 12-10.

and Eyring.[8] The numbers adjacent to each contour represent the energy of the system on an arbitrary scale. The dashed curve from the lower right to the upper left represents the path of minimum energy from the initial state,

$$\underset{1}{H} + \underset{2}{H-H}_{3},$$

to the final state,

$$\underset{1}{H-H}_{2} + \underset{3}{H}.$$

When the energy of the system is plotted against position along this curve, the *reaction coordinate*, the curve in Fig. 12-11 is obtained. Near the maximum, motion along the reaction coordinate corresponds to the asymmetrical stretching vibration of the

$$\underset{1}{H} - \underset{2}{H} - \underset{3}{H}$$

molecule. The height of the maximum in this curve above the energy of the initial state corresponds to the activation energy of the reaction. For reactions that are not as symmetrical as hydrogen exchange, the saddle point does not occur along the line $r_{12} = r_{23}$, and the activation energy of the forward and reverse reactions are not equal.

8. H. Eyring, *J. Am. Chem. Soc.*, *53*, 2540 (1931); H. Eyring and M. Polanyi, *Z. physik. Chem.* B, *12*, 279 (1931).

Statistical Thermodynamic Treatment.[9]   The fundamental assumption of the transition state model is that the activated complex is in equilibrium with the reactants. Thus, for a bimolecular reaction

$$A(g) + B(g) = C(g) \qquad (12\text{-}60)$$

the transition state model is described by the equation

$$A + B \rightleftharpoons AB^{\ddagger} \longrightarrow C \qquad (12\text{-}61)$$

in which $AB^{\ddagger}$ represents the *transition state* or the *activated complex*. The quasi-equilibrium between the reactants and the activated complex is represented by the equilibrium constant (see Eq. 11-48)

$$K_c = \frac{[AB^{\ddagger}]}{[A]\,[B]}\, c^{\ominus} \qquad (12\text{-}62)$$

As we saw in Sec. 11-7, the equilibrium constant can be expressed in terms of the partition functions of the reactants and products. The appropriate expression for Eq. 12-61 (see Eq. 11-70) is

$$K_c = \frac{z^{\ddagger\,\ominus}}{z_A^{\ominus}\, z_B^{\ominus}}\, L \exp\left(\frac{-\Delta\mathscr{E}_0^{\ddagger}}{RT}\right) \qquad (12\text{-}63)$$

where $\Delta\mathscr{E}_0^{\ddagger} = \mathscr{E}_0^{\ddagger} - \mathscr{E}_{0,A}^{\ominus} - \mathscr{E}_{0,B}^{\ominus}$ and $z^{\ominus}$ is the standard partition function of each species, referred to its ground electronic state. The quantity $\Delta\mathscr{E}_0^{\ddagger}$ can also be considered to be the hypothetical activation energy at 0 K.

The partition function $z^{\ddagger}$ differs from that of a normal molecule in that the vibrational motion along the reaction coordinate (the asymmetrical stretch) occurs at the maximum of a potential energy barrier rather than at the minimum of a potential energy curve. Since there is no restraining force, each vibration results in a passage across the barrier, and the motion can be viewed as a translation. If δ is defined as the *distance along the reaction coordinate* within which the transition state is defined, the partition function for this motion is, from Eq. 8-85,

$$(2\pi m^{\ddagger} kT)^{\frac{1}{2}}\left(\frac{\delta}{h}\right)$$

Thus,

$$K_c = \frac{(2\pi m^{\ddagger} kT)^{\frac{1}{2}}\,(\delta/h)\, z_{\ddagger}^{\ominus}}{z_A^{\ominus}\, z_B^{\ominus}}\, L \exp\left(\frac{-\Delta\mathscr{E}_0^{\ddagger}}{RT}\right)$$

9. S. Glasstone, K. J. Laidler, and H. Eyring, *The Theory of Rate Processes*, McGraw-Hill Book Co., New York, 1941, pp. 184–91.

$$= (2\pi m^{\ddagger} kT)^{\frac{1}{2}} \left(\frac{\delta}{h}\right) K^{\ddagger} \qquad (12\text{-}64)$$

where $z_{\ddagger}$ is the partition function for the remaining motions of the transition state, and $K^{\ddagger}$ is the equilibrium constant expression based on the incomplete partition function $z_{\ddagger}$.

In the transition state model, the rate of production of product is equal to the concentration of transition state species divided by their residence time τ in the transition state.

The residence time τ can be expressed as the ratio of the spatial interval δ that defines the transition state to the speed with which the system passes through this interval. If we assume that this speed is given by the Maxwell distribution expression for the average speed of a molecule in one direction, $\bar{u}_x$ is obtained from $g(u_x^2)$ (Eq. 2-67) as

$$
\begin{aligned}
\bar{u}_x &= \int_0^\infty u_x \, g(u_x^2) \, du_x \\
&= \int_0^\infty u_x \left(\frac{m^{\ddagger}}{2kT}\right)^{\frac{1}{2}} \exp\left(\frac{-m^{\ddagger} u_x^2}{2kT}\right) du_x \\
&= \left(\frac{kT}{2\pi m}\right)^{\frac{1}{2}} \qquad (12\text{-}65)
\end{aligned}
$$

Exercise 12-14. Verify Eq. 12-65.

Thus,

$$\tau = \frac{\delta}{(kT/2\pi m^{\ddagger})^{\frac{1}{2}}} \qquad (12\text{-}66)$$

From Eqs. 12-62 and 12-64,

$$
\begin{aligned}
[AB^{\ddagger}] &= K_c \, [A] \, [B]/c^{\ominus} \\
&= (2\pi m^{\ddagger} kT)^{\frac{1}{2}} \left(\frac{\delta}{h}\right) K^{\ddagger} \, [A] \, [B]/c^{\ominus}
\end{aligned}
$$

so that the reaction rate is

$$
\begin{aligned}
r &= \frac{[AB^{\ddagger}]}{\tau} \\
&= \frac{kT}{h} \left(\frac{K^{\ddagger} \, [A] \, [B]}{c^{\ominus}}\right) \qquad (12\text{-}67)
\end{aligned}
$$

Exercise 12-15. Verify Eq. 12-67.

We learned in Sec. 12-2 that the rate of a bimolecular reaction is

$$r = k_r \, [A] \, [B]$$

so that the transition state model expression for the rate constant is

$$k_r = \left(\frac{kT}{h}\right)\left(\frac{K^{\ddagger}}{c^{\ominus}}\right) \tag{12-68}$$

$$= \frac{1}{c^{\ominus}}\left(\frac{kT}{h}\right)\frac{z_{\ddagger}^{\ominus}}{z_A^{\ominus} \, z_B^{\ominus}} \, L \exp\left(\frac{-\Delta \mathscr{E}_0^{\ddagger}}{RT}\right) \tag{12-69}[10]$$

Exercise 12-16. What are the units of k_r obtained from Eq. 12-69, if $c^{\ominus}$ has the units of mol-dm^{-3}? Compare with the results of Exercise 12-6.

Since k_r is expressed in terms of $K^{\ddagger}$ in Eq. 12-68, we can also express it in terms of the *free energy of activation* $\Delta \mathscr{G}^{\ddagger}$, the *enthalpy of activation* $\Delta \mathscr{H}^{\ddagger}$, and the *entropy of activation* $\Delta \mathscr{S}^{\ddagger}$. That is, from Eq. 11-45 and Eq. 10-95,

$$K^{\ddagger} = \exp\left(\frac{-\Delta \mathscr{G}^{\ddagger}}{RT}\right) \tag{12-70}$$

$$= \exp\left(\frac{-\Delta \mathscr{H}^{\ddagger}}{RT}\right)\exp\left(\frac{\Delta \mathscr{S}^{\ddagger}}{R}\right) \tag{12-71}$$

so that,

$$k_r = \frac{1}{c^{\ominus}}\left(\frac{kT}{h}\right)\exp\left(\frac{-\Delta \mathscr{H}^{\ddagger}}{RT}\right)\exp\left(\frac{\Delta \mathscr{S}^{\ddagger}}{R}\right) \tag{12-72}$$

It is convenient, for comparison to the collision model (Eq. 12-59), to express Eq. 12-72 in terms of the experimental activation energy E_a. From Eq. 12-52,

$$E_a = RT^2 \frac{d \ln k_r}{dT}$$

Thus, from Eq. 12-68, with $c^{\ominus} = p^{\ominus}/RT$,

$$E_a = RT^2 \left(\frac{2}{T} + \frac{\Delta \mathscr{H}^{\ddagger}}{RT^2}\right)$$

$$= \Delta \mathscr{H}^{\ddagger} + 2RT \tag{12-73}$$

---

10. This expression for k_r differs from that usually presented, as, for example, in S. Glasstone, K. J. Laidler, and H. Eyring, *The Theory of Rate Processes*, McGraw-Hill Book Co., 1941, pp. 184–91. The form given here has the advantage that $K^{\ddagger}$ is a dimensionless quantity, as an equilibrium constant should be, and the factor of $1/c^{\ominus}$ provides the units of concentration that are appropriate for k_r.

Exercise 12-17. Verify Eq. 12-73. Note that the differentiation is carried out at constant P, since we are using a value of $c^\ominus$ that is determined by $p^\ominus$.

Substituting from Eq. 12-73 for $\Delta\mathscr{H}^\ddagger$ in Eq. 12-72, we obtain

$$k_r = \frac{e^2\, RT}{p^\ominus}\left(\frac{kT}{h}\right)\exp\left(\frac{\Delta\mathscr{S}^\ddagger}{R}\right)\exp\left(\frac{-E_a}{RT}\right) \qquad (12\text{-}74)$$

By comparison with Eq. 12-59, we see that the factor in the transition-state model comparable to the factor PZ' is

$$\frac{e^2\, RT}{p^\ominus}\left(\frac{kT}{h}\right)\exp\left(\frac{\Delta\mathscr{S}^\ddagger}{R}\right) \qquad (12\text{-}75)$$

This expression differs from the collision model formally in that it has a T^2 dependence rather than a $T^{\frac{1}{2}}$ dependence, but kinetic data are not sufficiently precise to distinguish between the two models on this basis. The factor $\exp(\Delta\mathscr{S}^\ddagger/R)$ is more sensitive to structural differences among reactants and to the nature of the activated complex than the collision frequency Z', and it includes the orientation factor P.

The transition-state model permits a quantitative calculation of k_r from the partition functions. Data on the energy levels of the reactants and a reasonable estimate of the structure of the transition state are required. Such calculations have been made for reactions 3, 4, 5, and 6 of Table 12-1. Table 12-2 compares the transition-state calculation of $(e^2 RT/p^\ominus)\,(kT/h)\,\exp(\Delta\mathscr{S}^\ddagger/R)$ with the collision model calculation of Z' and the experimental value of the quantity B of Eq. 12-53; all together, they are called the "frequency factor". Table 12-2 indicates clearly the superior sensitivity of the transition-state model to differences among reactions. Transition-state frequency factors are generally lower than the experimental values and closer to experiment than collision frequency factors. The latter are generally higher than the experimental values.

Table 12-2. Frequency factor/(1-mol^{-1}-s^{-1})

Reaction[a]	Experimental	Collision	Transition state
3. $NO + O_3 = NO_2 + O_2$	8.0×10^8	4.7×10^{10}	4.4×10^8
4. $NO_2 + O_3 = NO_3 + O_2$	5.9×10^9	6.3×10^{10}	1.4×10^8
5. $NOCl + Cl = NO + Cl_2$	1.1×10^{10}	5.7×10^{10}	4.4×10^9
6. $COCl + Cl = CO + Cl_2$	4.0×10^{11}	6.5×10^{10}	1.8×10^9

D. R. Herschbach, H. S. Johnston, K. S. Pitzer, and R. F. Powell, *J. Chem. Phys.*, 25, 736 (1956).
[a] Numbering as in Table 12-1.

Detailed calculations of the partition functions are difficult for complex reactions, such as reactions 1 and 2 of Table 12-1. Estimates of the partition-function factor have been made by taking into account only the conversion of rotational degrees of freedom in the reactants to vibrational degrees of freedom in the transition state. For reactions 1 and 2, the resulting factor, 1.5×10^{-6} dm^3-mol^{-1}-s^{-1}, accounts well for the probability factor, 10^{-5}, required for the collision model.[11]

The partition-function formulation of the rate constant permits an order-of-magnitude estimate of the probability factor even when a detailed calculation cannot be done. Consider the case of two nonlinear reactants A and B, with n_A and n_B atoms, respectively, that form a nonlinear transition state with $n_A + n_B$ atoms.

The expression

$$\frac{1}{c^\ominus} \frac{z_\ddagger^\ominus}{z_A^\ominus z_B^\ominus}$$

is equal to

$$\frac{(z_\ddagger^\ominus / \mathscr{V}^\ominus)}{(z_A^\ominus / \mathscr{V}^\ominus)(z_B^\ominus / \mathscr{V}^\ominus)}$$

since

$$c^\ominus = \frac{1}{\mathscr{V}^\ominus}$$

If we pass over the differences in partition functions between molecules and concentrate on the distinction between kinds of partition functions and if we define $f_{tran}, f_{rot}, f_{vib}$ as the translational, rotational, and vibrational partition functions per degree of freedom, then we can express the partition functions as[12]

$$\frac{z_A^\ominus}{\mathscr{V}^\ominus} = f_{tran}^3 \, f_{rot}^3 \, f_{vib}^{3n_A - 6} \tag{12-76}$$

$$\frac{z_B^\ominus}{\mathscr{V}^\ominus} = f_{tran}^3 \, f_{rot}^3 \, f_{vib}^{3n_B - 6} \tag{12-77}$$

and

$$\frac{z_\ddagger^\ominus}{\mathscr{V}^\ominus} = f_{tran}^3 \, f_{rot}^3 \, f_{vib}^{3(n_A + n_B) - 7} \tag{12-78}$$

11. G. B. Kistiakowsky and J. R. Lacher, *J. Am. Chem. Soc.*, *58*, 132 (1936).
12. The f_{tran}^3 used here is $z_{tran}^\ominus / \mathscr{V}$.

where $AB^{\ddagger}$ has one fewer vibrational degrees of freedom than a normal molecule. Thus,

$$\frac{(z_{\ddagger}^{\ominus}/\mathscr{V}^{\ominus})}{(z_A^{\ominus}/\mathscr{V}^{\ominus})(z_B^{\ominus}/\mathscr{V}^{\ominus})} = \frac{f_{vib}^5}{f_{tran}^3 f_{rot}^3} \tag{12-79}$$

To estimate the probability factor, we can use an analogous expression for a hard-sphere collision model in which two molecules having only translational degrees of freedom form as transition state a rigid rotor having two rotational degrees of freedom. For this case

$$\frac{(z_{\ddagger}^{\ominus}/\mathscr{V}^{\ominus})}{(z_A^{\ominus}/\mathscr{V}^{\ominus})(z_B^{\ominus}/\mathscr{V}^{\ominus})} = \frac{f_{rot}^2}{f_{tran}^3} \tag{12-80}$$

Exercise 12-18. Verify Eq. 12-80. Show that when the high temperature expressions f_{rot}^2 and f_{tran}^3 are substituted in Eq. 12-69, the result is equivalent to the collision model.

The probability factor is just the ratio of the right-hand sides of Eq. 12-79 and Eq. 12-80, or

$$P \approx \frac{f_{vib}^5}{f_{rot}^5} \tag{12-81}$$

Since the orders of magnitudes of the partition functions are $f_{tran} \approx 10^8 - 10^9$, $f_{rot} \approx 10^1 - 10^2$, and $f_{vib} \approx 10^0 - 10^1$, the ratio in Eq. 12-81 is of the order of magnitude of 10^{-5}.

Exercise 12-19. Calculate the probability factor for a reaction in which two linear molecules form a nonlinear transition state.

Exercise 12-20. Verify the orders of magnitude given above for the partition functions, using the results of Exercises 8-20, 8-21, and 8-22.

Unusually high frequency factors for some unimolecular reactions at high pressure are more easily explained using a transition-state model than a collision model. Though values predicted by the exact models mentioned in Sec. 12-2 are about 10^{13} s^{-1}, which is the order of magnitude of a bond vibration frequency, many unimolecular decompositions have experimental frequency factors of the order of 10^{14} to 10^{16}, as shown in Table 12-3. The transition-state model attributes the large values of the frequency factor to a positive value of $\Delta S^{\ddagger}$, in contrast to a negative value of $\Delta S^{\ddagger}$ for most bimolecular reactions. The positive entropy of activation is attributed to the existence of a transition state in which the two prospective fragments undergo free rotation rather than the hindered rotation in the reactant molecule. Also, the force constants of the transition state are much smaller than in the reactant molecule as a result of the additional freedom.

Table 12-3. Frequency factor of some unimolecular reactions

Decomposition of:	Frequency factor/s^{-1}
$(CH_3)_3-C-O-O-C-(CH_3)_3$	5×10^{16}
$CH_3N = NCH_3$	5×10^{15}
$C_2H_5N = NC_2H_5$	5×10^{15}
$Hg(CH_3)_2$	2×10^{13}
$Hg(C_2H_5)_2$	1×10^{14}
$Hg(n-C_3H_7)_2$	2×10^{15}

K. J. Laidler, *Chemical Kinetics*, 2nd ed., McGraw-Hill Book Co., New York, 1965, p. 167.

A striking example of the importance of the entropy of activation is the denaturation of proteins. The conversion of a protein molecule in solution from its compact, native, biologically active state to a random, disordered, denatured state is very rapid, despite a large activation energy. This suggests that $\Delta S^{\ddagger}$ is large and positive because the transition state resembles the disordered denatured state.

Molecular beam methods

Both the collision model and the transition-state model are based on average values over many microscopic states of reacting systems. The averaging process results in loss of information about the microscopic states involved. Reactions between molecules having narrowly defined ranges of energy can be observed using molecular beams. Such observations have made it clear that the probability of a reaction is a continuous function of the kinetic energy of the reactants. This is in contrast to the assumption of the simple collision model that reaction occurs only if the energy is greater than E_a and that energy greater than E_a does not change the probability of reaction.

From the point of view of quantum mechanics, it would be useful to observe the reactions of molecules in specific quantum states and to be able to observe the resulting quantum states of the products, in particular, the distribution of the energy of the products among translational, vibrational, and rotational levels. Molecular beam studies have resulted in some progress in this direction; details can be found in the literature.[13]

13. W. L. Fitz and S. Datz, *Ann. Rev. Phys. Chem.*, *14*, 61 (1963); H. S. Johnston, *Gas Phase Reaction Rate Theory*, Ronald Press, New York, 1966, Chap. 1. I. Amdur and G. G. Hammes, *Chemical Kinetics*, McGraw-Hill Book Co., New York, 1966, Chap. 9.

12-4. Reactions in solution

Reaction rates for reactions in solution are described similarly to reaction rates for species in gases even though their environments differ. The rates of many reactions in solution are proportional to an integral power of the concentration of a reactant, and the magnitudes of the activation energy and frequency factor are comparable to the values for reactions in the gas phase. The *law of mass action*, that the rate of an elementary reaction is proportional to the concentration of each reactant raised to a power equal to the coefficient of the reactant in the chemical equation, was formulated by Guldberg and Waage from observations on reactions in solution. We shall discuss reaction order, mechanisms, activation energy, and entropy of activation for reactions in solution in much the same way as we have for gases.

12-5. Photochemistry

The energy of photons in the visible and ultraviolet regions of the spectrum is great enough to produce electronic excitation. The reactions that occur as a result of absorption of photons are called *photochemical* reactions.

Excitation

When a photon is absorbed by a molecule, the molecule is raised to an excited energy state. That excited state may be a *singlet* state, in which there are equal numbers of electrons of opposite spin, or a *triplet* state, in which the total spin quantum number S is equal to 1, with three quantized values of M_s, 1, 0, and -1, corresponding to three different quantized orientations in an external magnetic field. (Sec. 5-2). Figure 6-12 is a schematic representation of an excitation to an excited singlet state in terms of the potential energy curves of the ground and excited states. A transition due to a photon with energy great enough to excite the molecule beyond line A results in dissociation of the excited molecule, and absorption is continuous beyond that energy. If the excited state has no minimum in its potential energy curve, as in curve D in Fig. 5-16, then excitation to that state always leads to dissociation.

Einstein developed the fundamental quantum-mechanical model for the interaction of radiation and matter in his classic paper, "The Quantum Theory of Radiation".[14] If ψ_m and ψ_n are two quantized energy levels of an atom or molecule such that $\varepsilon_m > \varepsilon_n$, then in a closed system in the presence

14. A. Einstein, *Physik. Z.*, *18*, 121–28 (1917); Boorse and Motz, *The World of the Atom*, Vol. II, Basic Books, New York, 1966, pp. 884–902.

of radiation of frequency $v = (\varepsilon_m - \varepsilon_n)/h$ the rate of absorption of radiation of frequency v is

$$-\frac{d\rho_v}{dt} = B_{nm}\,\rho_v\,N_n \tag{12-82}$$

where ρ_v is the energy density of radiation of frequency v, N_n is the number of molecules in quantum level n, and B_{nm} is called the *Einstein coefficient of stimulated absorption*.

De-excitation

An excited molecule can undergo one of several processes. It can re-radiate the absorbed photon before it loses any energy in collisions with other molecules. This process is called *resonance emission* and is most likely to occur at very low pressure. In the presence of radiation of frequency $v = (\varepsilon_m - \varepsilon_n)/h$ the rate of *stimulated emission* is

$$\frac{d\rho_v}{dt} = B_{mn}\,\rho_v\,N_m \tag{12-83}$$

Figure 12-12. Scheme of the transitions involved in fluorescence. (By permission, from P. W. Atkins, *Molecular Quantum Mechanics*, Clarendon Press, Oxford, 1970, p. 379.)

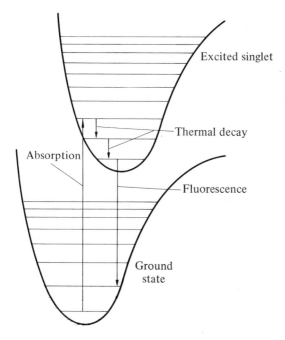

where B_{mn} is the *Einstein coefficient of stimulated emission*. In the presence or absence of radiation, spontaneous emission occurs at a rate

$$\frac{d\rho_v}{dt} = A_{mn} N_m \qquad (12\text{-}84)$$

where A_{mn} is the *Einstein coefficient of spontaneous emission*.

Alternatively, an excited molecule can collide with other molecules and reach vibrational thermal equilibrium before emission. Then a transition from the lowest vibrational level of the excited state to an excited vibrational level of the ground state occurs, as illustrated schematically in Fig. 12-12. This process is called *fluorescence* and takes place from 10^{-9} to 10^{-6} s after excitation. Figures 6-12 and 12-12 indicate that photons emitted in fluorescence generally have less energy than the absorbed photons. It is characteristic for fluorescence emission spectra to occur on the long wavelength, low frequency side of absorption spectra.

In general, excitation from a singlet ground state to a triplet excited state does not occur because the selection rules formally forbid transitions between states of different values of S. If, however, the potential energy curve of an excited singlet state crosses the potential energy curve of an excited triplet state, a transition from one to the other can occur, a non-radiative process called *intersystem crossing*, as illustrated schematically in Fig. 12-13. The subsequent return of the system to the singlet ground state by emission of radiation is very slow, since it is a forbidden transition, and the lifetime of an excited triplet state varies from 10^{-4} s to minutes and hours. Such radiation is called *phosphorescence*. Molecules in such an excited triplet state are likely to undergo photochemical reaction, since their lifetime is relatively long. The frequency of phosphorescence emission is usually lower than that of fluorescence.

Chemical reaction

An important parameter of a photochemical reaction is the *quantum yield*, Φ, which is defined as the number of molecules reacted per photon of light absorbed. The amount of reaction can be determined by a variety of chemical and physical techniques. Photons usually are counted either with a *thermopile* or a *chemical actinometer*. Instruments that detect individual photons in low intensity beams of light have also recently become available; they are called photon counters.

A thermopile converts absorbed light into thermal energy. The increase in temperature is a measure of light intensity, and the instrument is calibrated with standard lamps. If an energy E is recorded for a light beam of wave-

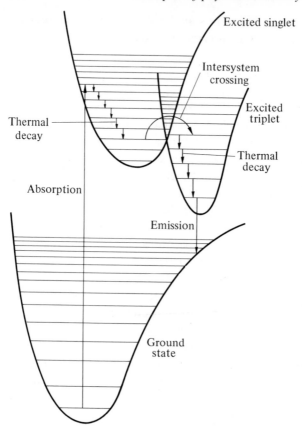

Figure 12-13. Scheme of the transitions involved in phosphorescence. (By permission, from P. W. Atkins, *Molecular Quantum Mechanics*, Clarendon Press, Oxford, 1970, p. 381.)

length λ impinging on a thermopile for a time t, then the number of photons per unit time is $E\lambda/(thc)$. For comparison with chemical quantities measured in moles, it is common to express the amount of light absorbed in *einsteins*, where an einstein is one mole of photons.

Exercise 12-21. Calculate the energy of one einstein of radiation of the mercury line at 2537×10^{-10} m in joules.

A chemical actinometer depends on the use of a photochemical reaction of known quantum yield to measure light intensity. A common reaction used in actinometers is the decomposition of oxalic acid, sensitized by uranyl ion. The uranyl ion UO_2^{2+} absorbs light in the wavelength region from 250 to 440 nm. The excited UO_2^{2+} transfers the absorbed energy to the oxalic acid, probably while forming a complex with the $(COOH)_2$, and the oxalic acid decomposes. The reaction can be described as

$$UO_2^{2+} + hv \longrightarrow (UO_2^{2+})^*$$

$$(UO_2^{2+})^* + (COOH)_2 \longrightarrow UO_2^{2+} + CO_2 + CO + H_2O$$

The number of moles of $(COOH)_2$ decomposed can be determined by a titration with $KMnO_4$. Since the quantum yield[15] is 0.57, the number of einsteins absorbed is approximately twice the number of moles of $(COOH)_2$ decomposed.

The value of Φ varies from zero, when no reaction occurs on absorption of light, to very large numbers. Quantum yields less than one indicate that deactivation occurs by other processes in addition to chemical reactions. Quantum yields greater than one indicate a chain reaction. In the photochemical decomposition of HI, the reaction steps are:

$$HI + hv = H + I$$

$$H + HI = H_2 + I$$

$$I + I + M = I_2 + M$$

where M is any third body. Since two moles of HI are decomposed for each einstein of radiation, the quantum yield is two, in the absence of any other process of de-excitation. The hydrogen-chlorine reaction has a quantum yield of 10^4 to 10^6. The photochemical step is

$$Cl_2 + hv = 2\,Cl$$

This is followed by two chain-propagating steps,

$$Cl + H_2 = HCl + H$$

$$H + Cl_2 = HCl + Cl$$

and a chain-terminating step,

$$Cl + Cl + M = Cl_2 + M$$

Since the chain-propagating steps may occur many times before a pair of Cl atoms happen to collide with a third body or on the wall, the quantum yield is very high.

15. W. G. Leighton and G. S. Forbes, *J. Am. Soc.*, *52*, 3139 (1930).

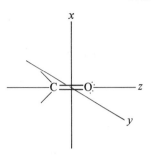

Figure 12-14. A coordinate system for the description of the orbitals of the carbonyl group.

Exercise 12-22. In the photolysis of acetone the reaction is

$$CH_3COCH_3 + h\nu \longrightarrow CH_3 + COCH_3$$

$$CH_3CO \longrightarrow CO + CH_3$$

$$2\,CH_3 \longrightarrow C_2H_6$$

Under the conditions of the experiment, the quantum yield of CO is 1.0. Calculate the mass of CO produced by irradiation with an 8-W Hg lamp (with 85% of its output at the 253.7 nm line) for a period of 1 h.

The carbonyl group is one of the most important photochemically active functional groups in organic chemistry. In terms of the geometry indicated in Fig. 12-14, it is convenient to consider the bonding of the carbonyl group as the interaction of an sp^2 orbital of C and the p_z orbital of O to form a σ bond and the interaction of the p_x orbitals of the C and O to form a π bond. The O 2s orbital is at a low-lying energy level, and the O $2p_y$ is occupied by an unshared pair of electrons and designated n. A reasonable set of localized molecular-orbital energy levels is shown in Fig. 12-15, in which the bonding σ and π orbitals and the non-bonding orbital on the O atom are occupied.

The absorption wavelengths for the lower energy transitions that have been observed for formaldehyde, H_2CO, are indicated in Table 12-4.[16] The absorption bands at shorter wavelength are attributed by Herzberg to transitions from the non-bonding O orbital to higher energy non-bonding orbitals.

16. From *Molecular Spectra and Molecular Structure*, Vol. 3, by G. Herzberg, p. 612. Copyright © 1966 by Litton Educational Publishing, Inc. Reprinted by permission of Van Nostrand Reinhold Co.

Table 12-4. Some electronic transitions and absorption wavelengths for formaldehyde

Transition	Wavelength/nm
Triplet n—π^*	360.0–396.7
Singlet n—π^*	230.0–353.0
Singlet n—σ^*	165.0–175.0

The photons absorbed by the carbonyl group are not energetic enough to cause dissociation of the strong $C=O$ bond, but the excitation seems to be transferred to a neighboring $C-C$ bond, since the products of irradiation of acetone at 280 nm are the methyl and acetyl radicals, CH_3 and CH_3CO, and further dissociation occurs to CH_3 and CO. The combination of the methyl radicals leads to C_2H_6 as a stable product.

Flash photolysis
It is difficult to obtain direct evidence for the intermediates postulated to be involved in photochemical reactions, since they are short-lived and present in low concentrations. In 1950, Norrish and Porter reported the technique of flash photolysis, in which a very high intensity flash of duration about 10^{-6} s is used to illuminate the reaction mixture. Following the initial illumination, which is sufficiently intense to produce a relatively high concentration of intermediates, a rapid series of precisely timed flashes of lower intensity are used to record the absorption spectrum of the intermediates, either photographically or photoelectrically.

The initial work in flash photolysis was carried out using gas discharge

Figure 12-15. Scheme of the energy levels of the carbonyl group, showing the electron configuration for (a) the ground state, (b) the triplet $n-\pi^*$ excited state, and (c) the singlet $n-\pi^*$ excited state.

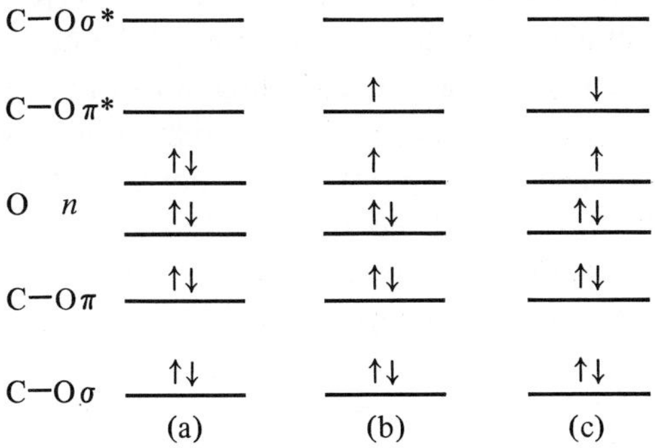

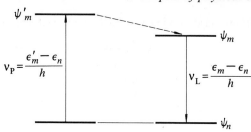

Figure 12-16. Scheme of the energy levels in an "optically pumped" laser; radiationless transition (– – –).

lamps for the initial exciting pulse. Thereafter the *laser* became the source of choice for these experiments. The term laser is an acronym for *light amplification* by *stimulated emission* of *radiation*. The constants B_{mn} and B_{nm} of Eq. 12-83 and Eq. 12-82 are equal,[17] and the net rate of change of energy density of radiation through spontaneous emission and stimulated absorption and emission is

$$\frac{d\rho_v}{dt} = A_{mn} N_m + B_{mn} \rho_v (N_m - N_n) \qquad (12\text{-}85)$$

If N_m is greater than N_n, an inverted population, then $d\rho_v/dt$ is positive, and the light intensity is amplified. Amplification, and the inverted population distribution, which is contrary to the Boltzmann distribution, cannot occur if ψ_m and ψ_n are the only energy levels. If there is another excited level ψ_m' at a higher energy than ψ_m, and there can be a radiationless transition from ψ_m' to ψ_m, as shown schematically in Fig. 12-16, then irradiation of the system with an intense beam of frequency v_p equal to $(\varepsilon_m' - \varepsilon_n)/h$, the "pumping" radiation, can produce a population at ψ_m that is greater than that at ψ_n. Spontaneous emission from ψ_m to ψ_n will then stimulate emission of the same frequency. Stimulated emission is of the same phase and travels in the same direction as the stimulating radiation, so laser radiation is coherent. If the laser is a long cylinder with a fully reflecting mirror at one end and a partially reflecting mirror at the other, then the radiation moves back and forth between the mirrors, stimulating further emission as it does so. Only radiation moving in the direction of the cylindrical axis emerges from the exit mirror instead of leaking through the walls, so that a laser beam is highly directional. The stimulated emission is also highly monochromatic.

17. P. W. Atkins, *Molecular Quantum Mechanics*, Clarendon Press, Oxford, 1970, pp. 223–24, 387–90.

These properties make lasers an extremely useful light source for scientific and practical purposes,[18] including highly selective photochemical activation and measuring the distance to the moon.

Photosynthesis and vision

Two of the most important photochemical reactions are photosynthesis and the visual process. Photosynthesis encompasses a complex series of steps by which solar energy is trapped as the chemical energy of carbohydrates in green plants. Oxygen is given off as a by-product. The initial light absorption results in electronic excitation of two pigments found in the chloroplasts of all green plants, chlorophyll a and chlorophyll b.

It has been suggested[19] that one result of excitation in chlorophyll a is a cyclic transport of the excited electron through a series of oxidation-reduction reactions and back to chlorophyll a. In the process, the energy released is used to synthesize adenosine triphosphate (ATP), which is used as a source of chemical free energy in that part of the photosynthesis reaction that can proceed in the dark. It has been suggested that some excited chlorophyll a molecules reduce nicotinamide adenine dinucleotide phosphate (NADP), which then participates in the reduction of CO_2 to carbohydrate. The electron given up by chlorophyll a in this reaction is thought to be replaced by an electron from excited chlorophyll b that is carried through the series of oxidation-reduction reactions to chlorophyll a. The resultant oxidized form of chlorophyll b is thought to oxidize OH^- from water to O_2 and to recover its electron in this way. Both the physical and the chemical details of the photosynthetic reaction and their relation to the highly organized structure of the chloroplast remain to be determined, but a framework for further research has been established. Recent experiments with a tunable laser spectrometer that uses picosecond (10^{-12} s) pulses indicate that the initial events in photosynthesis occur in less than 40 ps after absorption of a photon.[20]

The mechanism of the reception of visual signals by the mammalian eye is another important photochemical problem. It is known that the receptor molecule in the retina is rhodopsin, a complex of the protein opsin with vitamin A aldehyde, a carotenoid. Very little is known of the mechanism by

18. A. L. Schawlow, *Lasers and Light*, W. H. Freeman, San Francisco, 1969.
19. L. P. Vernon and M. Avron, "Photosynthesis", *Ann. Rev. Biochem.*, *34*, 269 (1965); R. P. Levine, "The mechanism of photosynthesis", *Scientific American*, December, 1969, p. 58.
20. K. J. Kaufmann, P. L. Dutton, T. L. Netzel, J. S. Leigh, and P. M. Rentzepis, *Science*, *188*, 1301–4 (1975).

which the energy received as a photon of light is converted into an electrical signal to the brain.[21]

12-6. Catalysis

A *catalyst* can be defined as a substance that decreases the activation energy of a reaction by providing an alternative mechanism. The catalyst must participate in the reaction but in general is regenerated at the completion of the reaction. To take into account the catalytic activity of reactants and products the concentration of which does change as a result of the reaction, Bell proposed as an operational definition of a catalyst "*a substance which appears in the rate expression to a power higher than that to which it appears in the stoichiometric equation*".[22] Also, a catalyst that is regenerated at the completion of a reaction cannot change the equilibrium yield of a reaction (see Chapters 10 and 11), but must catalyze the forward and reverse reactions to the same extent. We shall be concerned in this section with catalysis in homogeneous solutions, whereas catalysis of reactions in gases and solutions by solid surfaces will be discussed in Chapter 18.

Enzyme catalysis

One of the most important examples of homogeneous catalysis is the catalysis of reactions in biological systems by enzymes. Enzymes are complex protein molecules, and recent research has been devoted to the elucidation of the relation of the detailed structure of enzymes to their catalytic activity. Information in this area, which we shall not discuss, can be found in the book *Mechanisms of Homogeneous Catalysis*.[23]

Consider a model, first proposed by Michaelis, Menten, and Haldane, in which the enzyme E combines with a substrate S to form an intermediate enzyme-substrate complex ES, which then reacts to regenerate the enzyme and produce the product P.

$$E + S \underset{k_2}{\overset{k_1}{\rightleftarrows}} ES$$
$$ES \xrightarrow{k_3} E + P \tag{12-86}$$

The rate of reaction is

$$r = \frac{d[P]}{dt} = k_3 [ES] \tag{12-87}$$

21. G. Wald, *Science*, *162*, 230 (1968); W. E. Abramson and S. E. Ostroy, *Progr. Biophys. Mol. Biol.*, *17*, 170 (1967).
22. R. P. Bell, *Acid-Base Catalysis*, Clarendon Press, Oxford, 1941.
23. M. L. Bender, *Mechanisms of Homogeneous Catalysis from Protons to Proteins*, Wiley-Interscience, New York, 1971.

and the rate of change of concentration of the enzyme-substrate complex is

$$\frac{d[ES]}{dt} = k_1 [E] [S] - k_2 [ES] - k_3 [ES] \qquad (12\text{-}88)$$

The model is used with data on the *initial* rate of reaction and under conditions in which $[S]_0 \gg [E]_0$, so that $[S] \approx [S]_0$. The steady-state assumption (Sec. 12-2) is usually applied to simplify the mathematical treatment, so we assume

$$\frac{d[ES]}{dt} = 0$$

and

$$k_1 [E] [S]_0 = (k_2 + k_3) [ES] \qquad (12\text{-}89)$$

Since

$$[E] = [E]_0 - [ES]$$

Eq. 12-89 can be written as

$$k_1 \{[E]_0 - [ES]\} [S]_0 = (k_2 + k_3) [ES]$$

$$[ES] = \frac{k_1 [E]_0 [S]_0}{k_1 [S]_0 + k_2 + k_3}$$

$$= \frac{[E]_0 [S]_0}{[S]_0 + \{(k_2 + k_3)/k_1\}} \qquad (12\text{-}90)$$

Substituting from Eq. 12-90 in Eq. 12-87, we obtain

$$r_0 = \frac{k_3 [E]_0 [S]_0}{[S]_0 + \{(k_2 + k_3)/k_1\}} \qquad (12\text{-}91)$$

The quotient of rate constants in the denominator of Eq. 12-91 is frequently called the *Michaelis constant* K_M, and k_3 is sometimes called the *catalytic constant*. When $k_2 \gg k_3$, K_M is the *dissociation constant* of the enzyme-substrate complex. A typical graph of initial rate of reaction against initial substrate concentration is shown in Fig. 12-17. At sufficiently high substrate concentrations $[S]_0 \gg (k_2 + k_3)/k_1$, and

$$r_0 = k_3 [E]_0 = v_{max} \qquad (12\text{-}92)$$

accounting for the plateau at high substrate concentrations. At low substrate concentrations, Eq. 12-91 reduces to

$$r_0 = \frac{k_3}{K_M} [E]_0 [S]_0 = \frac{v_{max}}{K_M} [S]_0 \qquad (12\text{-}93)$$

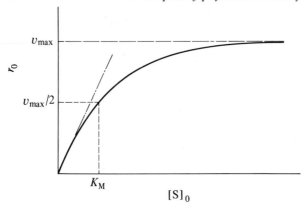

Figure 12-17. An idealized plot of the initial rate of an enzyme-catalyzed reaction as a function of the initial substrate concentration. The limiting rate at high substrate concentration is v_{max}; the substrate concentration when $r_0 = v_{max}/2$ is K_M; and the slope at low substrate concentrations is v_{max}/K_M ($— \cdot — \cdot —$).

Thus from data at low and high substrate concentrations, both v_{max} and v_{max}/K_M can be determined. The quantities v_{max} and v_{max}/K_M are the two independent kinetic parameters that characterize a simple enzyme mechanism, such as Eq. 12-86. The *zero-order rate constant* at high substrate concentration is v_{max}, and the *first-order rate constant* at low substrate concentrations is v_{max}/K_M. The Michaelis constant K_M is not an independent kinetic parameter, but simply the ratio of the two independent parameters.

For many reactions it is not possible to obtain data close enough to the plateau region to determine v_{max}. When Eq. 12-91 is rearranged by taking the inverse of both sides, the result[24] is

$$\frac{1}{r_0} = \frac{1}{v_{max}} + \frac{K_M}{v_{max}} \frac{1}{[S]_0} \qquad (12\text{-}94)$$

A plot of $1/r_0$ against $1/[S]_0$ should be linear, as illustrated in Fig. 12-18, with intercept equal to $1/v_{max}$ and slope equal to K_M/v_{max}. Since such a plot involves a considerable extrapolation to the intercept, and since a small absolute uncertainty in $1/v_{max}$ produces a large relative uncertainty in the value of v_{max}, an alternative graphical treatment has been suggested. When Eq. 12-91 is divided by $[S]_0$ and rearranged, the result[25] is

$$\frac{r_0}{[S]_0} = \frac{v_{max}}{K_M} - \frac{1}{K_M} r_0 \qquad (12\text{-}95)$$

24. H. Lineweaver and D. Burk, *J. Am. Chem. Soc.*, **56**, 658 (1934).
25. G. S. Eadie, *J. Biol. Chem.*, **146**, 85 (1942).

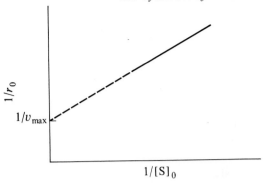

Figure 12-18. An idealized double reciprocal (Lineweaver-Burk) plot of enzyme kinetic data used to evaluate v_{max} and v_{max}/K_M; slope $= K_M/v_{max}$.

Exercise 12-23. Verify Eq. 12-95.

Thus, a plot of $r_0/[S]_0$ against r_0 should be linear, as illustrated in Fig. 12-19, with vertical intercept equal to v_{max}/K_M, horizontal intercept equal to v_{max} and slope equal to $-1/K_M$.

12-7. Rapid reaction kinetics[26]

Reaction rates are usually studied by mixing two reagents at some chosen temperature and measuring the change in concentration of a reactant or product with time. Even if the time-dependent concentration is measured

Figure 12-19. An idealized (Eadie) plot of $r_0/[S]_0$ against r_0 for an enzymatic reaction, used to evaluate v_{max} and v_{max}/K_M; slope $= -1/K_M$.

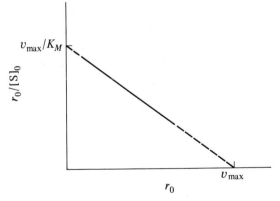

26. I. Amdur and G. G. Hammes, *Chemical Kinetics*, McGraw-Hill Book Co., New York, 1966, Chap. 6.

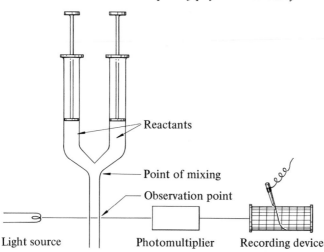

Reactants

Point of mixing

Observation point

Light source   Photomultiplier   Recording device

Figure 12-20.   Scheme of a Chance rapid-mixing apparatus. Syringes force the reactants into the mixing chamber at a fixed distance from the observation point.

by a rapid, automatically recorded physical method, rates of reactions with half-times less than about 10 s cannot be measured because of the limitations of mixing time. In 1923, Hartridge and Roughton developed a rapid-flow technique, with mixing times as short as 10^{-3} s. In the 1940's and 1950's, Britton Chance added electronic recording to further improved mixing techniques. A schematic diagram of his mixing apparatus is shown in Fig. 12-20. If the syringes are pushed down at a constant rate, the rate of flow and the distance between the point of mixing and the observation point determine the time interval between mixing and observation. The absorbance of the solution at a suitable wavelength can be recorded. To obtain a series of points, only the rate of flow need be varied. In the *stopped flow* method, the syringes are pushed rapidly down to bring the mixed reagents to the observation point and then stopped. The absorbance of the solution at some appropriate wavelength can then be recorded as a function of time. A complete kinetic curve can also be obtained by means of the *accelerated-flow* method. The syringes are pushed down at a steadily increasing speed, and the speed and the absorbance of the solution at the observation point are recorded simultaneously.

By means of these techniques Chance was able to detect intermediate enzyme-substrate complexes in a number of enzyme reactions and to follow the kinetics of the approach to the steady state. He was thereby able to determine the rate constants k_2 and k_3 of Eq. 12-86.

Beyond the range of 10^{-3} s, however, it is necessary to produce a nonequilibrium state by some other means than mixing two reagents. We have already mentioned flash photolysis, in which the energy of light is used to produce excited species. Reactions with half-times of the order of 10^{-6} s can be observed using this technique.

The *relaxation* methods introduced by Eigen in the 1950's are among the most versatile ways to study rapid reactions. The term relaxation is used to describe a return to equilibrium when a system previously at equilibrium is perturbed. The rate of return to equilibrium is observed by measurement of some physical property of the system that can be recorded rapidly and automatically, such as light absorption or electrical conductivity. The perturbation must be produced in a time short with respect to the *relaxation time* of the reaction, the reciprocal of the apparent first-order rate constant that describes the return to equilibrium after a perturbation.

By the use of spark discharges through a liquid sample, a temperature rise of about 6 K can be obtained in 10^{-7} s in a volume of 1 cm^3. This method is called the *temperature-jump* method. The temperature remains essentially constant during the time required for the system to reach an equilibrium state at the new temperature. Very rapid application or release of the pressure on a system is used in the *pressure-jump* method. Because of the time required for the propagation of the pressure change, reaction times longer than 10^{-6} s are required. The absorption of acoustical waves at frequencies varying from several hundred Hertz to 500 MHz provides a flexible method of studying chemical relaxation phenomena. The pressure and temperature changes accompanying the acoustic waves constitute the perturbation of the chemical system, and the frequency dependence of the absorption provide information about the relaxation times characteristic of the system. Relaxation times between 10^{-3} and 10^{-10} s can be studied by this technique. Among the reactions studied by relaxation methods are those in Table 12-5.

Table 12-5. Some rate constants determined by relaxation methods

Reaction	$k_1/(\text{dm}^3\text{-mol}^{-1}\text{-s}^{-1})$	k_2/s^{-1}
$H^+ + OH^- \underset{k_2}{\overset{k_1}{\rightleftharpoons}} H_2O$	1.4×10^{11}	2.5×10^{-5}
$H^+ + HCO_3^- \underset{k_2}{\overset{k_1}{\rightleftharpoons}} H_2CO_3$	4.7×10^{10}	$\sim 8 \times 10^6$
$OH^- + NH_4^+ \underset{k_2}{\overset{k_1}{\rightleftharpoons}} NH_3 + H_2O$	3.4×10^{10}	6×10^5

M. Eigen and L. De Maeyer, "Relaxation methods", in *Technique of Organic Chemistry*, Vol. VIII, Part II (S. L. Friess, E. S. Lewis, and A. Weissberger, eds.), Interscience Publishers, New York, 1963, pp. 1034–35.

As pointed out in Sec. 6-5, information about rates of rapid reactions can also be obtained from the line widths and peak separations in nuclear mag-

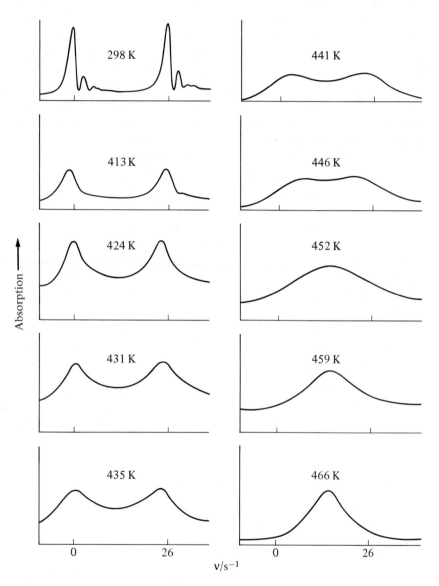

Figure 12-21. Nuclear magnetic resonance spectrum of N, N-dimethylnitrosamine as a function of temperature, at 40 MHz. [Reprinted with permission, from C. E. Looney, W. D. Phillips, and E. L. Reilly, *J. Am. Chem. Soc.*, *79*, 6136(1957) Copyright by the American Chemical Society.]

netic resonance and electron paramagnetic resonance spectra. Relaxation times between 10^{-1} s and 10^{-6} s are measurable by nuclear magnetic resonance methods and relaxation times between 10^{-4} s and 10^{-9} s are measurable by electron paramagnetic resonance methods.

Looney and co-workers used nuclear magnetic resonance to measure the effect of temperature on the hindered rotation in N, N-dimethylnitrosamine.[27] At room temperature, as shown in Fig. 12-21, there are two peaks,

N, N-dimethylnitrosamine

one for the *cis* methyl protons and the other for the *trans* methyl protons, because rotation is hindered and the frequency of rotation is small compared to the separation of the peaks, 26 s^{-1}. As the temperature increases and the thermal energy becomes larger with respect to the barrier to internal rotation, the peaks broaden and coalesce, and finally become a single peak at the average position between the two initial peaks. For temperatures at or below the coalescing temperature, the frequency of exchange v_e is given by the expression[28]

$$v_e = \{2\pi^2[(v_A - v_B)^2 - (\Delta v)^2]\}^{\frac{1}{2}} \qquad (12\text{-}96)$$

where v_A and v_B are the frequencies of the two peaks at very low temperature, when exchange is very slow, and Δv is the separation between the peaks. Thus, v_e can be calculated at any temperature from the experimental spectra, such as those shown in Fig. 12-21. A plot of log v_e against $1/T$ was found to be linear, and the value of E_a, equal to the height of the potential energy barrier to internal rotation, was found to be 9.6×10^4 J-mol^{-1}. At the coalescence temperature, $\Delta v = 0$,

$$v_e = 2\pi^2(v_A - v_B)^2$$
$$= 2.7 \times 10^4 \text{ s}^{-1}$$

and the time of exchange is 3.8×10^{-5} s, one-half the residence time of either of the states represented by the two peaks.

27. C. E. Looney, W. D. Phillips, and E. L. Reilly, *J. Am. Chem. Soc.*, 79, 6136 (1957).
28. J. A. Pople, W. G. Schneider, and H. J. Bernstein, *High-Resolution Nuclear Magnetic Resonance*, McGraw-Hill Book Co., New York, 1959, p. 224.

Problems

12-1. Measurement of the change in light absorption as a function of time is one of the most common physical methods used to follow the rate of a chemical reaction. The quantity measured is the absorbance $A = -\log(I/I_0)$, where I is the intensity of the transmitted light and I_0 is the intensity of the incident light. For the reaction $A = B$, the absorbance at any time t is $A = \varepsilon_A[A]l + \varepsilon_B[B]l$, where ε_A and ε_B are the molar extinction coefficients of A and B and l is the optical path length. Derive an expression for the rate of reaction in terms of dA/dt, and derive a method for obtaining the rate constant for a first-order reaction. Hint: What are the initial and final conditions?

12-2. Plot the data of Exercise 12-2 to test whether they fit the relationships for a second-order reaction. Over how many multiples of $t_{\frac{1}{2}}$ should measurements be made to be reasonably certain to distinguish between first-order and second-order kinetics?

12-3. The following data were obtained for the pressure as a function of time in reaction 1 of Table 12-1 at 564.3 K [G. B. Kistiakowsky and J. R. Lacher, *J. Am. Chem. Soc.*, 58, 123 (1936)]. Derive the appropriate functions of the pressure to plot to test these data for second-order and first-order kinetics and calculate the rate constant for the order that better fits the data.

t/s	$P_{acrolein}/133$ Pa	$P_{butadiene}/133$ Pa
0	418.2	240.0
63	412.3	233.7
181	401.9	222.7
384	385.1	204.9
542	373.5	192.7
745	359.7	178.3
925	349.0	167.0
1145	337.1	154.5
1374	326.6	143.4
1627	315.5	131.9
1988	302.4	118.2

12-4. For the reaction $A \rightarrow$ products, derive an expression for the half-time as a function of initial concentration of A, assuming that the reaction is nth order in A ($n \neq 1$).

12-5. Derive the exact relationship between the empirical E_a obtained from Eq. 12-52 and the quantity E_a in Eq. 12-58 for the collision theory model.

12-6. C. N. Hinshelwood and T. E. Green, [*J. Chem. Soc.*, 129, 730 (1926)] measured the rate of the reaction

$$2\,NO(g) + 2\,H_2(g) = N_2(g) + H_2O(g)$$

at 1099 K. Some of their data is given below. (a) Deduce the order of the reaction with respect to NO and H_2. (b) Calculate the rate constant at 1099 K, indicating clearly the units of k_r. (c) Calculate the Arnhenius activation energy for the reaction.

Initial pressures:		Initial rate:
$p_0(H_2)/133$ Pa	$p_0(NO)/133$ Pa	$r_0/(133$ Pa-s$^{-1})$
400	359	0.150
400	300	0.103
400	152	0.025
300	400	0.174
300	310	0.092
300	232	0.045
289	400	0.160
205	400	0.110
147	400	0.079
404	300	0.103
302	300	0.035
244	300	0.072
147	300	0.059
117	300	0.052
104	300	0.045

T/K	k_r (relative)
1099	476
1061	275
1024	130
984	59
956	25.3
904	5.3

12-7. The following mechanism has been suggested for the thermal decomposition of ozone ($2 O_3 = 3 O_2$) in the pressure of O_2. The notation M indicates any gas molecule and is a way of incorporating a possible dependence on total pressure.

$$M + O_3 \underset{k_2}{\overset{k_1}{\rightleftharpoons}} O_2 + O + M$$
$$O + O_3 \xrightarrow{k_3} 2 O_2$$

(a) Use the steady-state approximation for the concentration of O to derive an expression for the rate of disappearance of O_3 as a function of k_1, k_2, k_3, [O_2], [O_3], and [M]. (b) The experimental rate law has been found to be

$$\frac{-d[O_3]}{dt} = k_{obs} \frac{[O_3]^2}{[O_2]}$$

What are the relationships among magnitudes that must hold for the equation derived in (a) to reduce to this form?

12-8. It has been found that eight photons of wavelength 700 nm are required to produce one molecule of O_2 and to incorporate one atom of C in carbohydrate from CO_2

and H_2O in photosynthesis. The value of $\Delta \mathscr{G}^{\ominus}$ for the reaction is 4.77×10^5 J-mol^{-1}. Calculate the "efficiency" of this reaction, defined as the ratio of the biosynthetic work ($\Delta \mathscr{G}^{\ominus}$) obtained to the light energy absorbed. What would the temperature of the high temperature reservoir in a reversible heat engine have to be to obtain a conversion factor equal to this "efficiency," if the low temperature reservoir is at 298 K?

13 · *The solid state*

In Sec. 7-3 we discussed the use of X-ray and neutron diffraction to determine the arrangements of atoms in crystals. Now we shall consider some physical and chemical models that have been developed to account for the arrangements thus observed as well as for the physical properties of crystalline substances.

13-1. Atomic arrangements in solids

Representation of atoms and ions in crystals as spheres
Though, as we emphasized in Chapters 4 and 5, atoms and molecules do not have rigid boundaries, a rigid-sphere model is useful. It is convenient to consider the geometric requirements for the closest packing of spheres of fixed radius as determining the arrangement of atoms in simple crystals.

Diffraction experiments with crystals provide information on the distance between nuclei, and many methods have been proposed by which the experimental internuclear distance is partitioned into a sum of the radii of the atoms or ions. In many crystals in which the negative ion is much larger than the positive ion, a relatively constant anion-anion distance is observed. This result indicates that the anions are closest packed. As a result, the radii of the anions can be calculated, and this information can in turn be used to calculate cation radii in other crystals. Pauling pointed out that the ionic radii thus calculated are not strictly constant but are functions of the relative sizes of the oppositely charged ions in the crystal and of the number of nearest neighbors.[1] Nevertheless, tables of ionic radii provide reasonable approximations to internuclear distances in crystals whose structure has not been determined. The comparisons for several systems are shown in Table 13-1.

Table 13-1. Ionic radii and radius sums for several alkali halide crystals together with the observed internuclear distances

Ions	Ionic radii[a]/10^{-10} m	Radius sum/10^{-10} m	Observed distance[b]/10^{-10} m
Na^+, Cl^-	0.95, 1.81	2.76	2.82
Li^+, F^-	0.60, 1.36	1.96	2.01
Li^+, I^-	0.60, 2.16	2.76	3.00
Cs^+, F^-	1.69, 1.36	3.05	3.00

[a] L. Pauling, *Nature of the Chemical Bond*, 3rd ed., Cornell University Press, Ithaca, 1960, p. 514.
[b] R. W. G. Wyckoff, *Crystal Structures*, Interscience Publishers, New York, 1948.

Similarly, covalent radii can be assigned such that a sum of covalent radii provides an approximate value for bond lengths in covalent compounds. Known bond lengths obtained from diffraction data for crystals and from spectroscopic data for gaseous molecules are used to calculate covalent radii. Covalent radii are listed by Pauling.[2] The most comprehensive source of data on bond lengths is *Tables of Interatomic Distances and Configurations in Molecules and Ions*, published by the Chemical Society of London. Covalent radii of the halogens, quite different from the ionic radii in Table 13-1 are 0.099 nm for Cl, 0.064 nm for F, and 0.133 nm for I.

Slater[3] proposed that a single set of *atomic radii* be used to obtain bond lengths both for ionic and covalent bonds. He proposed the use of covalent radii for this purpose, pointing out that the ionic radius of a cation is approximately 0.085 nm smaller than the covalent radius of the same atom and that the ionic radius of an anion is approximately 0.085 nm larger than the covalent radius of the same atom, so that the sum of either pair of radii leads to the same value.

Self-consistent field calculations[4] have confirmed the suggestion Huggins made in 1926 that the atomic radius could be correlated with the radius of maximum ψ^2 of the outermost electron shell. The atomic radius and the

1. L. Pauling, *Nature of the Chemical Bond*, 3rd ed., Cornell University Press, Ithaca, 1960, pp. 511–19.
2. L. Pauling, op. cit., p. 224.
3. J. C. Slater, *Quantum Theory of Molecules and Solids*, Vol. 2, McGraw-Hill Book Co., New York, 1965, pp. 95–108.
4. J. C. Slater, *op. cit.*, p. 103.

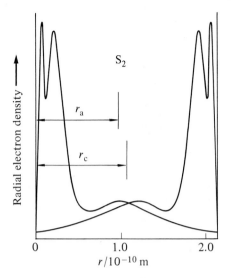

Figure 13-1. A plot of radial atomic electron density as a function of distance from the nucleus for two sulfur atoms at a distance equal to the internuclear distance in S_2. The atomic radius (r_a) is indicated at the maximum radial density of the outermost shell. The covalent radius (r_c) is indicated at the intersection of the density curves of the two atoms. (By permission, from W. L. Bragg, *The Crystalline State*, G. Bell and Sons, London, 1949.)

covalent radius coincide closely because the outermost shells of the two atoms forming a covalent bond overlap, as shown for S_2 in Fig. 13-1. The radial electron density of one S atom is plotted against the distance r from its nucleus. The radial electron density of a second S atom is plotted against distance from its nucleus in the direction of the first S atom on the same graph, with the two nuclei separated by the equilibrium internuclear distance in S_2.

A corresponding diagram showing radial electron density curves for KCl at the internuclear distance in the crystal is shown in Fig. 13-2. The solid curves represent the radial density distribution for K^+ and Cl^-, and the dashed curve shows the position of the 4s shell in the K atom. The curve for the Cl^- ion differs very little from that of the Cl atom. The radii of the free K^+ ion and the free Cl^- ion are much smaller than the corresponding ionic radii in the crystal. The latter are the radii read at the intersection of the distribution curves. There is closed-shell repulsion between the ions when the outer shells overlap only slightly. As Slater points out, however, the sum of the atomic radii of K and Cl does not differ much from the sum of the ionic radii of K^+ and Cl^-.

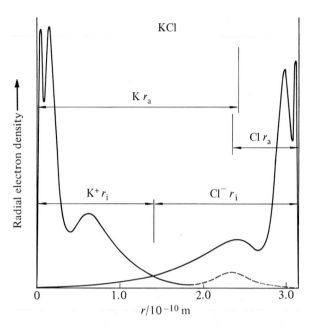

Figure 13-2. A plot of radial electron density as a function of distance from the nucleus for a K atom and a Cl atom at a distance equal to the internuclear distance in a KCl crystal. The atomic radii (r_a) are indicated as before and the ionic radii (r_i) are indicated at the intersection of the density curves. (By permission, from W. L. Bragg, *The Crystalline State*, G. Bell and Sons, London, 1949.)

Closest packing of uniform spheres

The closest packed arrangement of uniform spheres in a plane is that in which each sphere is surrounded by six nearest neighbors, as shown in Fig. 13-3. A second layer of the same spheres can rest in the hollows formed by the first layer, in the positions indicated by the shaded circles in Fig. 13-4a. When a third layer of spheres is placed on the second, there are two possible arrangements; they are positioned either directly above the spheres in the first layer or in a third position. Figure 13-5a represents a view of the layers from the top. The solid circles represent the centers of the spheres in the first layer, labeled A, the shaded circles represent the centers of the spheres in the second layer, labeled B, and the open circles represent the alternate positions of the centers of the spheres in the third layer, labeled C. Figures 13-4a and 13-5b are structural models of the alternative modes of closest packing.

The arrangement ABAB . . . is called *hexagonal closest packing* and the arrangement ABCABC . . . is called *cubic closest packing*. The later arrangement constitutes a face-centered cubic lattice with the 111 planes as the

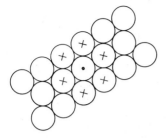

Figure 13-3. Closest packing of spheres in a plane, showing six nearest neighbors about each sphere in a hexagonal array.

closest packed planes. In both arrangements, each sphere has twelve nearest neighbors, six in the plane, three below, and three above. In hexagonal closest packing, the triangles formed by the spheres above and below have the same orientation; in cubic closest packing, the two triangles are at an angle of 60° with respect to each other. The noble gases and many metals crystallize in either cubic or hexagonal closest packing, as do many substances in which the molecules are polyatomic but spherically symmetrical, such as methane.[5]

Figure 13-4. Hexagonal closest packing (a) in which the first layer of spheres is labeled A, the second layer of spheres is labeled B, and the third layer of spheres is over the first. The repeating pattern from layer to layer is ABAB.... . (b) A structural model of hexagonal closest packing (by permission, from A. F. Wells, *The Third Dimension in Chemistry*, Clarendon Press, Oxford, 1968.).

(a)   (b)

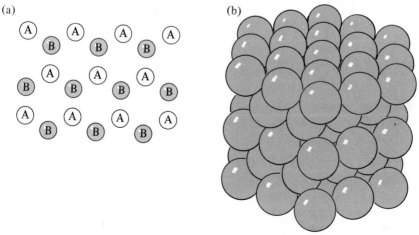

5. Both cubic and hexagonal closest packing were suggested as important arrangements of atoms in crystals by W. Barlow, *Nature*, *29*, 186, 205, 404 (1883), 30 years before these structures were determined by X-ray diffraction.

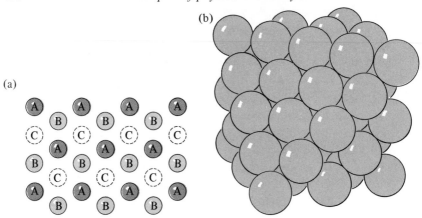

Figure 13-5. Cubic closest packing (a) in which the first layer of spheres is labeled A, the second layer of spheres is labeled B, and the third layer of spheres is labeled C. The repeating pattern from layer to layer is ABC ABC.... (b) A structural model of cubic closest packing (by permission, from A. F. Wells, *The Third Dimension in Chemistry*, Clarendon Press, Oxford, 1968.).

The effect of radius ratio

It might be expected that in a crystal lattice composed of equal numbers of two different atoms of the same size there should be twelve nearest neighbors of one kind about each atom of the other kind. A brief examination of a model is enough to demonstrate that such an arrangement is not possible.

Exercise 13-1. Verify the preceding sentence from a consideration of a plane containing equal numbers of closest packed spheres of two different kinds.

When there are two different sizes of atoms, it is useful to consider that the smaller atoms occupy holes in a closest packed lattice of the larger atoms. A close packed lattice contains triangular holes (coordination number 3), tetrahedral holes (coordination number 4), and octahedral holes (coordination number 6), as illustrated in Fig. 13-6. Consideration of the geometry of the holes leads to the observation that, given large spheres of unit radius, spheres that can fill the holes have the radii shown in Table 13-2.

Table 13-2. Radius ratio for filled holes of several different coordinations

Triangular	$(1 - \cos 30°)/\cos 30° = 0.155$
Tetrahedral	$(1 - \cos 35°16')/\cos 35°16' = 0.225$
Octahedral	$2 \cos 45° - 1 = 0.414$

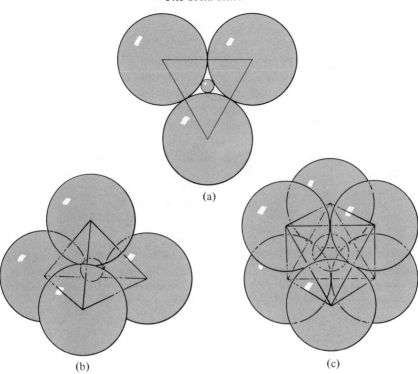

Figure 13-6. Holes of (a) triangular, (b) tetrahedral, and (c) octahedral coordination in a lattice of closest packed spheres.

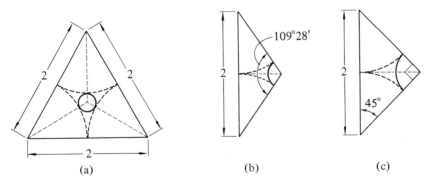

Figure 13-7. Geometric construction to illustrate minimum radius ratio for (a) triangular, (b) tetrahedral, and (c) octahedral coordination, showing the inscribed sphere tangent to closest packed larger sphere. Lines ($l = 2$) represent distance between centers of larger spheres; inscribed sphere (——); large spheres (– – –).

If the smaller sphere in a crystal is larger than that listed, the coordination number is maintained, but the larger spheres are not in contact. When the sphere becomes large enough, the arrangement corresponds to the next higher coordination number. When the radius ratio is greater than 0.645 the lattice expands to eight nearest neighbors in a square anti-prism arrangement, which is no longer closest packed. Above 0.732, a body-centered cubic arrangement is stable.

Exercise 13-2. Verify the minimum radius ratio values given above. Hint: See the geometric construction in Fig. 13-7.

Table 13-3 shows the correlation between the coordination number predicted by the radius-ratio rule and that found experimentally for oxides of a number of metals. The agreement is seen to be very good.

Table 13-3. Observed and predicted coordination for oxides of several elements

Ion	Radius ratio	Predicted coordination number	Observed coordination number
B^{3+}	0.20	3 or 4	3, 4
Be^{2+}	0.25	4	4
Li^+	0.34	4	4
Si^{4+}	0.37	4	4, 6
Al^{3+}	0.41	4 or 6	4, 5, 6
Ge^{4+}	0.43	4 or 6	4, 6
Mg^{2+}	0.47	6	6
Na^+	0.54	6	6, 8
Ti^{4+}	0.55	6	6
Sc^{3+}	0.60	6	6
Zr^{4+}	0.62	6 or 8	6, 8
Ca^{2+}	0.67	8	7, 8, 9

Reprinted from L. Pauling, *The Nature of the Chemical Bond*, p. 546. 3rd ed. Copyright 1939 and 1940, 3rd ed. © 1960 by Cornell University Press. Used by permission of Cornell University Press.

Though the correlation between the radius-ratios shown in Table 13-2 and the coordination found in crystals is quite good, the alkali halide crystals show several exceptions that are not easily explained. The radius ratios of these crystals are shown in Table 13-4, and all exhibit octahedral coordination except CsCl, CsBr, and CsI, which are body-centered cubic. Pauling, who contributed greatly to the development of a set of consistent crystal radii for the monatomic ions, has discussed in some detail the problem of explaining the alkali halide structures.[6]

6. L. Pauling, *The Nature of the Chemical Bond*, 3rd ed., Cornell University Press, Ithaca, 1960, pp. 519–32.

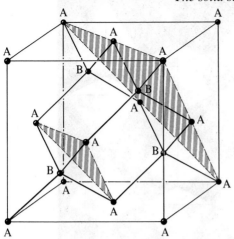

Figure 13-8. A tetrahedral diamond lattice inscribed in a cubic cell. The 111 planes are shaded. (By permission, from W. L. Bragg, *The Crystalline State*, G. Bell and Sons, London, 1949.)

Table 13-4. Ionic radius ratios of alkali halide crystals

	Li	Na	K	Rb	Cs
F	0.44	0.70	0.98	1.09	1.24
Cl	0.33	0.52	0.73	0.82	0.93
Br	0.31	0.49	0.68	0.76	0.87
I	0.28	0.44	0.62	0.69	0.78

A. F. Wells, *Structural Inorganic Chemistry*, Clarendon Press, Oxford, 1945, p. 98.

Structural classification of crystals

Before we consider in detail the types of bonding that occur in crystals, it is convenient to classify crystals on a purely structural basis, following Wells.[7] The simplest type, the *three-dimensional infinite complex*, comprises those crystals in which no neutral polyatomic unit smaller than the entire crystal can be detected. Examples include: NaCl, in which Na^+ ions fill the octahedral holes of a cubic closest-packed lattice of Cl^- ions (see Fig. 7-13); diamond, in which every C atom is covalently bonded to four tetrahedrally arranged C atoms (see Fig. 13-8); the noble gases, in which only London forces hold the atoms in the crystal; metallic crystals; silica, in which each SiO_4 tetrahedron shares vertices with four other tetrahedra (see Fig. 13-9a); and the framework silicates, in which small positive ions occupy holes in a three-dimensional lattice of SiO_4 and AlO_4 tetrahedra (see Fig. 13-9b).

7. A. F. Wells, *Structural Inorganic Chemistry*, Clarendon Press, Oxford, 1945, pp. 140–53.

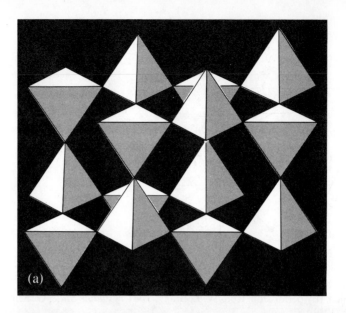

Figure 13-9. (a) A lattice of linked SiO_4 tetrahedra in cristobalite, a form of SiO_2. (b) Alternating SiO_4 and AlO_4 tetrahedra in a lattice of a framework silicate, paracelsian ($Ba\,Al_2\,Si_2\,O_8$). (By permission, from A. F. Wells, *The Third Dimension in Chemistry*, Clarendon Press, Oxford, 1968.)

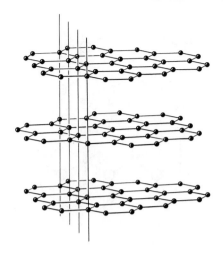

Figure 13-10. The graphite lattice. (By permission, from W. L. Bragg, *The Crystalline State*, G. Bell and Sons, London, 1949.)

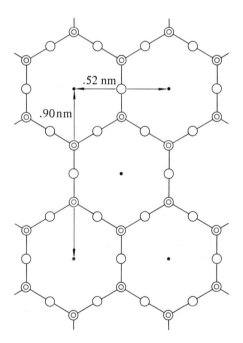

Figure 13-11. Lattice of a platelike silicate, with each SiO_4 or AlO_4 tetrahedron linked to three other tetrahedra in a plane. [By permission, from W. L. Bragg, *Z. Krystall.*, 74, 237–305 (1930).]

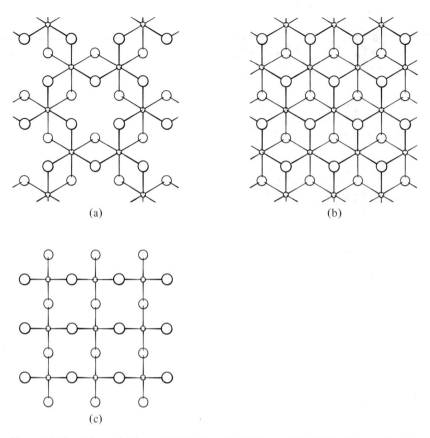

Figure 13-12. Planar lattices of (a) $CrCl_3$, (b) $CdCl_2$, and (c) HgI_2. The small circles represent metal atoms in a plane, and the large circles represent halogen atoms above or below the plane of the metal atoms. The halogen atoms are projected normally on the plane of the metal atoms. (By permission, from A. F. Wells, *Structural Inorganic Chemistry*, 3rd Ed., Clarendon Press, Oxford, 1963, p. 147.)

A number of crystals contain *infinite two-dimensional complexes*. Familiar examples are graphite, in which infinite planes of aromatic rings are bonded to each other by weak intermolecular forces (see Fig. 13-10), and the platelike silicates, based on infinite planar networks in which each SiO_4 or AlO_4 tetrahedron is linked with three other tetrahedra, as in Fig. 13-11. The latter class includes talc, the micas, and the clay minerals. Several metal halides, such as $CrCl_3$, $CdCl_2$ and HgI_2 (see Fig. 13-12), are also in this class.

One-dimensional infinite complexes are reflected in the macroscopic properties of the substance in the fibrous silicate, asbestos. The fundamental

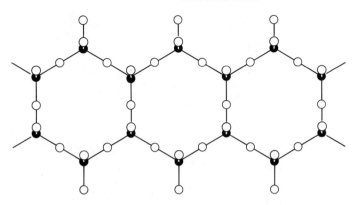

Figure 13-13. The lattice structure of asbestos, showing SiO_4 tetrahedra linked in hexagonal rings in a linear array. (By permission, from W. L. Bragg, *Z. fur Krystall.*, *74*, 237–305, 1930.)

unit of the crystal is a chain of hexagonal rings sharing parallel sides, with each ring formed of six SiO_4 tetrahedra, as in Fig. 13-13. A simpler chain silicate structure is found in the pyroxenes, in which each SiO_4 tetrahedron is linked to only two other tetrahedra in a linear fashion, as in Fig. 13-14. Other examples of one-dimensional infinite complexes include SiS_2, and $PdCl_2$, with the structures shown in Fig. 13-15.

All other types of crystals have as their fundamental units finite complexes, either molecules or complex ions. If a molecule or complex ion is approxi-

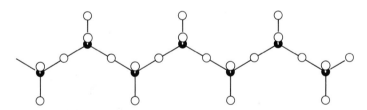

Figure 13-14. The lattice structure of the pyroxenes, showing SiO_4 tetrahedra linked in a chain. (By permission, from W. L. Bragg, *Z. fur Krystall.*, *74*, 237–305, 1930.)

$$\overset{\diagdown}{\underset{\diagup}{Si}}\overset{S}{\underset{S}{\diagup\diagdown}}\overset{\diagdown}{\underset{\diagup}{Si}}\overset{S}{\underset{S}{\diagup\diagdown}}\overset{\diagdown}{\underset{\diagup}{Si}}\overset{\diagdown}{\diagup}$$

$$\overset{\diagdown}{\underset{\diagup}{Pd}}\overset{Cl}{\underset{Cl}{\diagup\diagdown}}\overset{\diagdown}{\underset{\diagup}{Pd}}\overset{Cl}{\underset{Cl}{\diagup\diagdown}}\overset{\diagdown}{\underset{\diagup}{Pd}}\overset{\diagdown}{\diagup}$$

(a) (b)

Figure 13-15. Atomic arrangements in linear complexes of (a) SiS_2 and (b) $PdCl_2$.

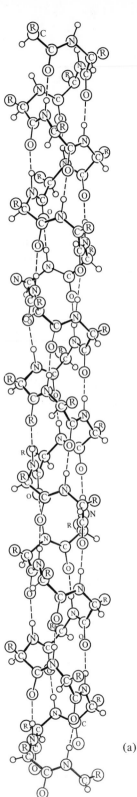

Figure 13-16. (a) Model of an α-helix. [By permission, from L. Pauling, R. B. Corey, and H. R. Branson, *Proc. Natl. Acad. Sci., U.S.A.*, *37*, 205–11 (1951).] (b) Models of coiled α-helices that represent the structure of α-keratin. [By permission, from L. Pauling and R. B. Corey, *Nature*, *171*, 59–61 (1953).] (c) Models of polyethyleneterephthalate (polyester): configuration of the molecule and arrangement in the crystal. [By permission, from R. deP. Daubeney, C. W. Bunn, and C. J. Brown, *Proc. Roy. Soc.* (*London*) *A226*, 531–42(1954).]

(a)

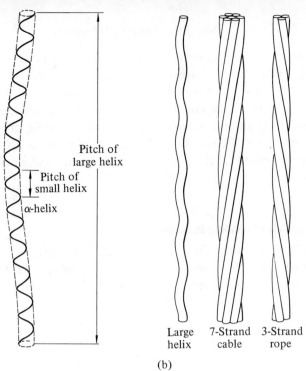

(b)

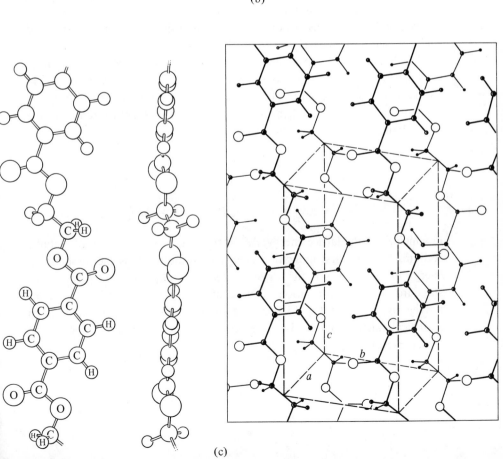

(c)

mately spherical, the packing in the crystal can be described in the same way as simple infinite three-dimensional complexes. The more anisotropic the finite complex, the more the crystal structure resembles that of a one- or two-dimensional infinite complex. For example, the structure of protein fibers and polyethylene terephthalate closely resemble one-dimensional infinite complexes, though the molecules are of finite length. Models that have been proposed for these substances of high molecular mass are shown in Fig. 13-16.

13-2.   Models of bonding in solids

A complete description of crystals must include consideration of the forces between units as well as the geometric arrangement of units. In this section we shall present some models used to describe ideal crystals.

Ionic crystals

The simplest crystal model is that used to describe an ideal ionic lattice, one in which the lattice points are occupied by monatomic ions. A stable crystalline lattice is formed as the result of the opposing effects of coulombic interaction between ions of opposite charge and the repulsion between filled shells of electrons as described by the Pauli principle.

The *Born-Haber cycle* provides a thermodynamic basis for calculating the *lattice energy* of an ionic solid, the energy required to convert the crystal to gaseous ions at 0 K. Consider the formation of crystalline NaCl from the elements. Since reactions 2 through 6 of Table 13-5 together represent the same change of state as reaction 1,

$$\Delta\mathscr{E}_1 = \Delta\mathscr{E}_2 + \Delta\mathscr{E}_3 + \Delta\mathscr{E}_4 + \Delta\mathscr{E}_5 + \Delta\mathscr{E}_6$$

and

$$\mathscr{E}_L = \Delta\mathscr{E}_v + I_1 + \tfrac{1}{2} D_0 - EA - \Delta\mathscr{E}f_0^{\ominus} \tag{13-1}$$

in which $\mathscr{E}_L$ is the lattice energy, $\Delta\mathscr{E}_v$ is the heat of vaporization of metallic sodium, I_1 is the first ionization energy of sodium, D_0 is the dissociation energy of Cl_2, $-EA$ is the *electron affinity* of chlorine, and $\Delta\mathscr{E}f_0^{\ominus}$ is the standard energy of formation of NaCl at 0 K. The values of all the terms on the right side of Eq. 13-1 can be obtained from experiment or from statistical calculations, so the cycle provides a *thermodynamic* value for the lattice energy.

Table 13-5.   Reactions of the Born-Haber cycle for NaCl

1. $Na(c) + \frac{1}{2} Cl_2(g) = NaCl(c)$	$\Delta\mathscr{E}_1 = \Delta\mathscr{E}f_0^{\ominus}$
2. $Na(c) = Na(g)$	$\Delta\mathscr{E}_2 = \Delta\mathscr{E}_v$
3. $Na(g) = Na^+(g) + e^-$	$\Delta\mathscr{E}_3 = I_1$
4. $\frac{1}{2} Cl_2(g) = Cl(g)$	$\Delta\mathscr{E}_4 = \frac{1}{2} D_0$
5. $Cl(g) + e^- = Cl^-(g)$	$\Delta\mathscr{E}_5 = -EA$
6. $Na^+(g) + Cl^-(g) = NaCl(c)$	$\Delta\mathscr{E}_6 = -\mathscr{E}_L$

Born also suggested a model that could be used to calculate the lattice energy for a crystal of known structure. Consider a Na^+ ion in a crystal of NaCl. There are six Cl^- ions at a distance a, twelve Na^+ ions at a distance $\sqrt{2}\,a$, eight Cl^- ions at a distance $\sqrt{3}\,a$, and so on. The electrostatic energy of a Na^+ ion in this lattice, first calculated by Madelung, is

$$\varepsilon_e = -\frac{6\,z_+z_-\,e^2}{4\pi\varepsilon_0 a} + \frac{12\,z_+^2 e^2}{4\pi\varepsilon_0\,(\sqrt{2}\,a)} - \frac{8\,z_+z_-\,e^2}{4\pi\varepsilon_0\,(\sqrt{3}\,a)} + \cdots$$

$$= -\frac{z^2 e^2}{4\pi\varepsilon_0 a}\left[6 - \frac{12}{\sqrt{2}} + \frac{8}{\sqrt{3}}\cdots\right] \tag{13-2}$$

in which $z = z_+ = z_-$. The series in brackets converges conditionally and therefore has a sum that depends on the arrangement of the terms. The order that is chosen, with successive terms for ions at successively greater distances, is the one for which the infinite sum is a suitable approximation for the finite sum that describes a finite crystal.[8] The sum of the series is called the Madelung constant M_L and it has the value 1.746 for the NaCl structure and 1.763 for the CsCl structure (body-centered cubic). The electrostatic energy for one mole of NaCl is

$$\mathscr{E}_e = -\frac{Lz^2 e^2}{4\pi\varepsilon_0 a} M_L \tag{13-3}$$

Since ions in a crystal reach an equilibrium position, there must be a repulsive force between ions, which is attributed to the Pauli principle requirement that electrons of like spin remain apart. Born and Landé expressed the potential energy due to repulsion as B/a^n, where B and n are parameters to be determined for each crystal. If we abbreviate the right side

8. E. L. Burrows and S. F. A. Kettle, *J. Chem. Educ.*, *52*, 58–59 (1975); R. Courant, *Differential and Integral Calculus*, Vol. I, Blackie, London, 1937, pp. 369–75.

of Eq. 13-3 as $-A/a$, we can express the energy of a crystal as a function of the variable interionic distance r as

$$\mathscr{E} = -\frac{A}{r} + \frac{B}{r^n} \tag{13-4}$$

The energy is a minimum at the equilibrium internuclear distance, so that

$$\left(\frac{d\mathscr{E}}{dr}\right)_{r=a} = 0 \tag{13-5}$$

The application of Eq. 13-5 to Eq. 13-4 leads to a value for B of

$$B = \frac{Aa^{n-1}}{n} \tag{13-6}$$

Exercise 13-3.　Verify Eq. 13-6.

Thus, we can rewrite Eq. 13-4 as

$$\mathscr{E} = -\frac{A}{r} + \frac{Aa^{n-1}}{nr^n}$$

$$= -\frac{A}{a}\left[\frac{a}{r} - \frac{1}{n}\left(\frac{a}{r}\right)^n\right] \tag{13-7}$$

At $r = a$, $\mathscr{E}$ is the lattice energy,

$$\mathscr{E}_L = -\frac{A}{a}\left[1 - \frac{1}{n}\right] \tag{13-8}$$

The parameter n can be calculated from the compressibility of the crystal at 0 K. The higher the value of n, the more rapidly the repulsive force varies with r, and the more *incompressible* the crystal. The compressibility κ is defined as

$$\kappa = -\frac{1}{\mathscr{V}}\left(\frac{\partial \mathscr{V}}{\partial P}\right)_T \tag{13-9}$$

From Eq. 10-46,

$$\left(\frac{\partial \mathscr{E}}{\partial \mathscr{V}}\right)_T = T\left(\frac{\partial P}{\partial T}\right)_V - P$$

so that, at 0 K,

$$\left(\frac{\partial \mathscr{E}}{\partial \mathscr{V}}\right)_T = -P$$

and

$$\left(\frac{\partial^2 \mathscr{E}}{\partial \mathscr{V}^2}\right)_T = -\left(\frac{\partial P}{\partial \mathscr{V}}\right)_T$$

$$= \frac{1}{\kappa \mathscr{V}} \tag{13-10}$$

Also,

$$\left(\frac{\partial^2 \mathscr{E}}{\partial \mathscr{V}^2}\right)_T = \frac{\partial}{\partial \mathscr{V}}\left(\frac{\partial \mathscr{E}}{\partial r} \cdot \frac{\partial r}{\partial \mathscr{V}}\right)$$

$$= \frac{\partial \mathscr{E}}{\partial r} \cdot \frac{\partial^2 r}{\partial \mathscr{V}^2} + \frac{\partial^2 \mathscr{E}}{\partial r^2}\left(\frac{\partial r}{\partial \mathscr{V}}\right)^2 \tag{13-11}$$

From Eq. 13-7,

$$\left(\frac{\partial^2 \mathscr{E}}{\partial r^2}\right) = -\frac{2A}{r^3} + \frac{(n+1)Aa^{n+1}}{r^{n+2}} \tag{13-12}$$

The volume occupied by one mole of ions is

$$\mathscr{V} = 2Lr^3$$

$$\frac{\partial \mathscr{V}}{\partial r} = 6Lr^2$$

and

$$\frac{\partial r}{\partial \mathscr{V}} = \frac{1}{6Lr^2} \tag{13-13}$$

At $r = a$, $(\partial \mathscr{E}/\partial r) = 0$,

$$\left(\frac{\partial^2 \mathscr{E}}{\partial \mathscr{V}^2}\right) = \frac{(n-1)A}{a^3} \cdot \frac{1}{36L^2 a^4}$$

and

$$\kappa = \frac{18La^4}{(n-1)A} \tag{13-14}$$

Exercise 13-4. Verify Eq. 13-14.

Born and Landé found values of n near 9 for the alkali halides. With a value of n determined, one can compare calculated values of $\mathscr{E}_L$ from Eq. 13-8 with thermodynamic values of $\mathscr{E}_L$ from Eq. 13-1. Table 13-6 shows the value of $\mathscr{E}_L$ and the quantities from which it is calculated by means of Eq. 13-1 for the alkali halides. Also shown are the values obtained by Fajans and Schwarz for $\mathscr{E}_L$, using Eq. 13-8, and the values of $\mathscr{E}_L$ obtained by Born and Mayer, using a more rigorous exponential form for the repulsive potential energy and taking into account the attraction between ions due to the London force (see below) as well as the Coulomb force. The results of the thermodynamic calculation and those of the model calculations agree within $2-3\%$, and there is relatively little difference between the two model calculations.

Table 13-6. Calculated lattice energies at 0 K for the alkali halides ($kJ\text{-}mol^{-1}$)

Halide	$\Delta\mathscr{E}_v{}^a$	$I_1{}^a$	$\frac{1}{2}D_0{}^a$	$-EA^b$	$-\Delta\mathscr{E}f_0^{\ominus a}$	$\mathscr{E}_L$ (Eq. 13-1)	$\mathscr{E}_L$ (Eq. 13-8)^c	$\mathscr{E}_L{}^d$
LiF	152.7	507.7	76.90	-332.7	614.1	1018.7	1067	1005
LiCl	152.7	507.7	120.0	-348.7	410.8	842.5	837	833
LiBr	152.7	507.7	117.9	-324.5	352.3	806.1	782	788
LiI	152.7	507.7	107.2	-295.6	273.1	745.1	711	728
NaF	109.0	495.7	76.90	-332.7	571.0	919.9	929	893
NaCl	109.0	495.7	120.0	-348.7	409.0	785.0	761	766
NaBr	109.0	495.7	117.9	-324.5	361.9	760.0	720	731
NaI	109.0	495.7	107.2	-295.6	290.0	706.3	665	686
KF	90.92	418.7	76.90	-332.7	564.6	818.4	805	794
KCl	90.92	418.7	120.0	-348.7	440.3	721.2	682	692
KBr	90.92	418.7	117.9	-324.5	394.2	697.2	653	667
KI	90.92	418.7	107.2	-295.6	329.6	650.8	609	631
RbF	*85.81	409.1	76.90	-332.7	551.3	790.4	761	760
RbCl	*85.81	409.1	120.0	-348.7	432.6	698.8	653	672
RbBr	*85.81	409.1	117.9	-324.5	391.2	679.5	623	642
RbI	*85.81	409.1	107.2	-295.6	330.4	636.9	586	608
CsF	*78.78	381.8	76.90	-332.7	532.9	737.7	715	727
CsCl	*78.78	381.8	120.0	-348.7	435.0	666.9	607	637
CsBr	*78.78	381.8	117.9	-324.5	396.6	650.9	582	612
CsI	*78.78	381.8	107.2	-295.6	338.8	611.0	548	582

[a] Values calculated from data in NBS Circular 500, "Selected Values of Chemical Thermodynamic Properties", 1952. Values of $\Delta\mathscr{E}_v$ with an asterisk are those at 298 K, which differ insignificantly from those at 0 K. Values for $-\Delta\mathscr{E}f_0^{\ominus}$ for all salts except NaCl are values at 298 K + 2 $kJ\text{-}mol^{-1}$.
[b] From R. S. Berry and C. W. Reimann, *J. Chem. Phys.*, **38**, 1540 (1963).
[c] K. Fajans and E. Schwartz, *Z. phys. Chem.*, 1931, Bodenstein Festb., 717.
[d] M. Born and J. Mayer, *Z. Physik*, **75**, 1 (1932).

Molecular crystals

Molecular crystals contain a finite unit in which interatomic distances are less than distances between units. Frequently, these units correspond to the molecules of the substance that exist in the gas phase at higher temperatures. The simplest molecular crystals are those of the noble gas atoms. Since the units of these crystals are spherical and electrically symmetrical, it is surprising that they exert intermolecular forces at all. In 1930, London developed a quantum-mechanical explanation for these forces.

Though the noble gas atoms are electrically symmetrical over a time average, at any instant of time the positions of the electrons with respect to the nucleus define an instantaneous dipole, in which the center of positive charge and the center of negative charge do not correspond. This instantaneous dipole induces a dipole in neighboring atoms that results in a net attractive force. The magnitude of the induced dipole is dependent on the polarizability α of the atom. London was able to derive an expression for the attractive force between two like atoms as

$$F = \frac{9\,I\alpha^2}{2\,r^7} \tag{13-15}$$

where I is the first ionization energy of the electrons, α is the polarizability, and r is the intermolecular distance. The corresponding potential energy of the interaction is

$$U = -\frac{3\,I\alpha^2}{4\,r^6} \tag{13-16}$$

Since the polarizability of a molecule increases with the number of electrons, the London forces increase with the mass of the molecule.

The noble gases crystallize in a cubic closest-packed arrangement, and the net attractive potential energy is obtained by summing pair by pair the interactions over the crystal structure as was done in calculating the Madelung constant for ionic crystals. The equilibrium internuclear distance is determined by the balance between the London attractive force and the closed-shell repulsion. Since the London forces are relatively weak, the overlap of the radial electron density distribution is very small, as shown by the curve for argon in Fig. 13-17.

Molecules with permanent dipoles also attract one another through dipole-dipole and dipole-induced–dipole forces. The potential energy functions for these two forces are, respectively,

$$U = -\frac{2}{3kT}\frac{\mu^4}{r^6} \tag{13-17}$$

and

$$U = -\frac{\mu^2\alpha}{r^6} \tag{13-18}$$

where μ is a dipole moment and α is a polarizability. These two potential energy functions are spherically averaged over the angular dependence of the dipole interaction; the temperature dependence of the dipole-dipole energy is a result of the Boltzmann distribution of relative orientations with different mutual potential energies.

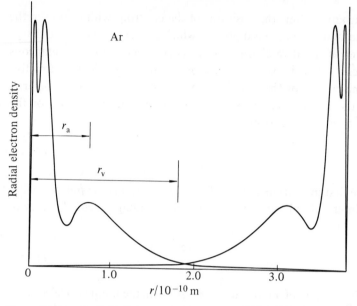

Figure 13-17. Radial electron density curves as a function of distance from the nucleus for two argon atoms at a distance equal to the internuclear distance in the crystal, showing the van der Waals radius (r_v) greater than the atomic radius (r_a) due to closed shell repulsion. (By permission, from W. L. Bragg, *The Crystalline State*, G. Bell and Sons, London, 1949.)

The relative magnitudes of the coefficient of the $1/r^6$ term for dipole, induction, and London forces are given in Table 13-7 for several molecules.

Table 13-7. $r^6 U/(10^{-79}$ J-m$^6)$[a]

Molecule	Dipole	Induction	London
He	0	0	1.23
Ne	0	0	4.67
A	0	0	55.4
Kr	0	0	107.
Xe	0	0	233.
Cl_2	0	0	321.
CH_4	0	0	112.
CO	0.0034	0.057	67.5
HI	0.35	1.68	382.
HBr	6.2	4.05	176.
HCl	18.6	5.4	105.
NH_3	84.	10.	93.
H_2O	190.	10.	47.

[a] J. Hirschfelder, C. F. Curtis, and R. B. Bird, *Molecular Theory of Gases and Liquids*, John Wiley & Sons, Inc., New York, 1954, pp. 966, 988. H. Margenau, *J. Chem., Phys.*, 6, 897 (1938); F. London, *Trans. Faraday Soc.*, 33, 8 (1937).

It can be seen that the London forces are of major importance even between molecules with permanent dipole moments. The values given for the dipole and induction energies of polar molecules, being spherically averaged, are appropriate for large intermolecular distances and are smaller than the corresponding values in a crystal, in which the dipoles would be oriented in positions of minimum potential energy.

The unusually large dipole-dipole potential energies for NH_3 and H_2O reflect the presence of "hydrogen bonds", strong dipole interactions between a H atom attached to F, O, or N and an unshared pair of electrons of a different F, O, or N atom. Hydrogen bonds are responsible for the special features of a number of crystal structures, the most important of which is

Figure 13-18. Arrangement of molecules in the ice crystal, showing the tetrahedral geometry about each oxygen. The large spheres represent oxygen atoms and the small spheres represent one possible set of positions of the hydrogen atoms. (Reprinted from L. Pauling, *The Nature of the Chemical Bond*, 3rd. Ed. Copyright 1939 and 1940, third edition © 1960 by Cornell University. Used by permission of Cornell University Press.)

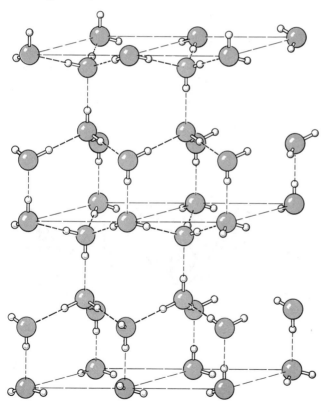

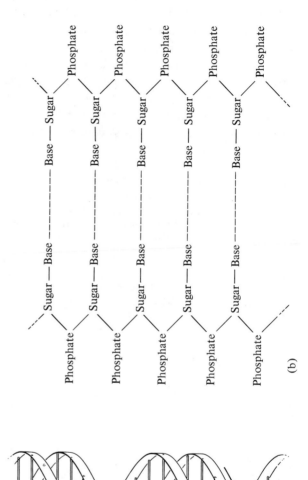

(b)

(a)

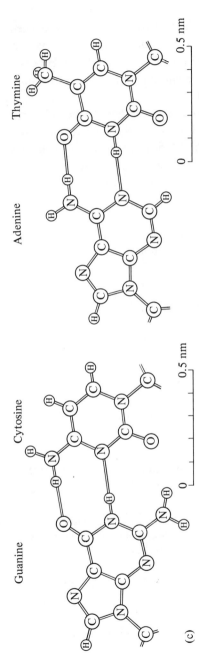

Figure 13-19. The Watson-Crick, double helix model of nucleic acids, showing the hydrogen bonding between complementary chains. (a) Schematic diagram; (b) scheme of chemical structure of a pair of nucleic acid chains; (c) pairing of bases by hydrogen bonding. [By permission, from J. D. Watson and F. H. Crick, *Nature, 171,* 965 (1953).]

ice. Each O atom in a crystal of ice is surrounded by four other O atoms, each at a distance of 0.276 nm with a H atom in between (Fig. 13-18). The open lattice structure has a density less than that of liquid water. Neutron diffraction (Fig. 7-17) shows that two of the hydrogens about each oxygen are at a distance of 0.101 nm, very close to the bond distance in the vapor, and the other two are at a distance of 0.175 nm. That two of the O–H bonds are normal is also supported by an O–H stretching absorption in the infrared that differs very little from that in the vapor.

Other substances in which hydrogen bonds are important include proteins, as in the structure shown in Fig. 13-16, and the nucleic acids that carry genetic information. The Watson and Crick model of DNA utilizes hydrogen bonds formed between complementary purine and pyrimidine bases on the two strands of the double helix (Fig. 13-19).

Covalent crystals

The diamond lattice (Fig. 13-8) is the classic example of a crystal in which every atom is joined to its neighbors by covalent bonds. Both the high bond energy and the strongly directional nature of covalent bonds contribute to the mechanical strength and the high melting point of diamond. The enthalpy of vaporization of diamond[9] is 714.8 kJ-mol^{-1}; the bond energy of a C-C single bond in aliphatic compounds is 346 kJ-mol^{-1} (Table 9-4).

Exercise 13-5. Why is the enthalpy of vaporization of diamond approximately twice the C–C bond energy? Hint: How many bonds must be broken per carbon atom in vaporizing a diamond crystal? Count each bond only once.

Each C atom is surrounded by four nearest neighbors arranged tetrahedrally at a distance of 0.154 nm. The resulting lattice fills only 34% of the available space. A body-centered cubic lattice with eight nearest neighbors fills 68% of the available space and a face-centered cubic lattice with twelve nearest neighbors fills 74% of the available space.

Exercise 13-6. Verify the percentage of space filled for a tetrahedral lattice, a body-centered cubic lattice, and a face-centered cubic lattice, assuming the atoms to be spheres of a diameter equal to the nearest neighbor internuclear distance. (See Fig. 13-8.)

The group IV elements Si and Ge also form primarily covalent crystals. Like C, both have an sp^3 valence state outer electron configuration. Other

9. "Selected Values of Chemical Thermodynamic Properties", U.S. NBS Technical Note 270–3, 1968.

covalent crystals include SiC, and the III·V compounds borazon, BN, and germanium arsenide, GeAs. Like diamond, their electronic structure can be described in terms of localized electron pair bonds between nearest neighbor atoms; the number of bonds and the geometry are determined by the combination of s and p orbitals and the presence of eight valence electrons about each atom (Sec. 5-4).

Metallic crystals

Metallic crystals are easily distinguished from all other crystals by their opacity, their reflectivity, and their electrical and thermal conductivity. The range of values of hardness and melting point is wide, from liquid mercury, which melts at 234 K, to tungsten, which melts at 3643 K.

By the first decade of this century, Drude and Lorentz had developed a classical electron theory of the properties of metals that explained the conductivity of metals very well. Since the metallic elements have relatively low ionization energies, they proposed that the valence electrons of metallic atoms constituted a collection of "free" electrons moving randomly in the field of a lattice of positive ions. When an electrical field is applied to a sample of metal, the electrons are accelerated in a direction opposite to that of the field (since they are negatively charged). The acceleration is

$$a = -\frac{eE}{m} \tag{13-19}$$

where e is the proton charge, E is the magnitude of the electrical field, and m is the mass of the electron.

The acceleration is effectively reduced to zero at each collision of an electron with the lattice as a result of the random motion. Thus, the average velocity attained between collisions, the *drift velocity*, is

$$v_d = \tfrac{1}{2} a t_0 \tag{13-20}$$

where t_0 is the average time between collisions. Since the random motion results in no net displacement, it is the drift velocity that is observed in conductivity measurements. If the velocity imposed by the electrical field is small compared to the average speed of random motion, we can assume that the time between collisions is determined by the latter speed; that is

$$t_0 = \frac{\lambda}{\bar{v}} \tag{13-21}$$

where λ is the mean free path between collisions, and $\bar{v}$ is the average speed of random motion.

Thus,

$$v_{\mathrm{d}} = \frac{1}{2} \frac{a\lambda}{\bar{v}}$$

$$= -\frac{1}{2} \frac{e}{m\bar{v}} E \tag{13-22}$$

If there are n_A free electrons per unit volume in the metal, the magnitude j of the current density is

$$j = -n_A e v_{\mathrm{d}}$$

$$= \frac{1}{2} \frac{n_A e^2 \lambda}{m\bar{v}} E \tag{13-23}$$

$$= \sigma E \tag{13-24}$$

where σ is the electrical conductivity. Since σ is a characteristic of the metal that depends only on the temperature and is independent of the applied field, Eq. 13-30 is an expression of *Ohm's law*.

A simple qualitative description of the bonding of atoms in metallic crystals is provided by a molecular orbital model developed by Bloch in 1928. When two atoms form a bond, corresponding atomic orbitals of the two atoms form two linear combinations, one bonding and the other antibonding, each of which is a molecular orbital. The energy of each of these orbitals as a function of internuclear distance is shown as curves B and D of Fig. 5-16 for H_2. For aggregates involving many atoms, as in a crystal, there are a number of molecular orbitals equal to the product of the number of atoms and the number of atomic orbitals in a constituent atom. These molecular orbitals then form a *band* of closely spaced energy levels.

The results of calculations for sodium crystals are shown schematically in Fig. 13-20. It can be seen that an inner orbital, such as the 2p, is not affected by the interaction and forms part of the ion core of the crystal. The 3s band, with one electron per atom in the crystal, is half-filled. Most of the electrons in this band are at a lower energy than in the isolated atom, reflecting their role in the lattice energy of the crystal.

Since the 3s band is half-filled and since it overlaps with the vacant 3p band, there are many vacant levels of only slightly higher energy than the occupied levels to which the valence electrons can be excited, either thermally or by an external electrical field. This model is consistent, then, with the high thermal and electrical conductance of metallic electrons as well as with conventional bonding theory. Also, since the orbitals extend over all the atoms in the crystal, the model is consistent with the free electron model.

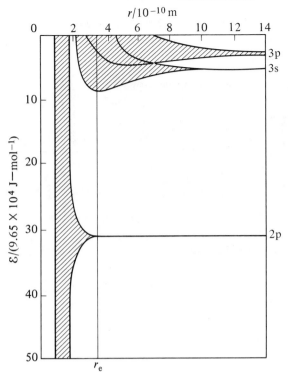

$r/10^{-10}$ m

Figure 13-20. Energy bands of a sodium crystal as a function of internuclear distance; equilibrium internuclear distance (r_e). (By permission, from J. C. Slater, *Quantum Theory of Molecules and Solids*, Vol. 2, McGraw-Hill Book Co., New York, 1965, p. 208.)

Though the Drude-Lorentz free electron model led to an equation of the correct form for the electrical conductivity, it could not be used to explain the heat capacity data for metallic crystals. According to classical equipartition theory (Sec. 2-3), an alkali metal crystal, with one valence electron per atom in the free electron gas, should have a molar heat capacity of $\frac{9}{2} R$, $3 R$ for the vibrational motion of the positive ions that form the lattice, and $\frac{3}{2} R$ for the translational motion of the electron gas.

The older empirical results summarized in the law of Dulong and Petit indicated a high temperature equipartition limit of $3 R$, consistent with contributions to the heat capacity only from the lattice vibrations. The temperature dependence of the heat capacity exhibited in the data of Nernst and others was explained on the basis of quantum theory by Einstein, Debye, and Born and von Kármán by taking into account only the lattice vibrations

(Sec. 8-3). The discrepancy was resolved when Fermi applied the Pauli principle to the statistics of an ideal gas,[10] and Sommerfeld treated the electron gas of a metal in terms of the resulting Fermi-Dirac statistics.

Fermi-Dirac statistics. The valence electrons in a metallic crystal are considered to be quantum-mechanical particles in a three-dimensional infinite potential well (neglecting the variation in potential due to the lattice ions). The quantized energy levels are those given in Eq. 8-49 for an ideal gas. That is,

$$\varepsilon - \varepsilon_0 = \frac{(n_x^2 + n_y^2 + n_z^2)\, h^2}{8ma^2}$$

where n_x, n_y, and n_z are the three translational quantum numbers, m is now the mass of the electron, and a is the length of a cube edge of the crystal. Since the Pauli principle specifies that no two electrons can have all their quantum numbers in common, only two electrons of opposite spin can have the three translational quantum numbers in common. It is convenient to define a quantum state for an electron as including the spin quantum number; with this definition there are two quantum states for each set of values of n_x, n_y, and n_z, and only one electron can occupy each quantum state. Thus, in contrast to a collection of Bose-Einstein particles, which can all occupy the lowest quantum state of the system at 0 K, a collection of electrons in a crystal must occupy states of relatively high energy even at 0 K due to the limitation imposed by the Pauli principle.

The form of the Fermi-Dirac distribution can be derived as follows. Consider a group of g_j electron quantum states sufficiently dense to include a large number of states but including a sufficiently narrow range of energies that they can be said to share the same energy, $\varepsilon_j - \varepsilon_0$. Let N_j be the number of electrons in this group of states. Since electrons are described by the Pauli principle, each quantum state is either vacant or contains only one electron. The number Ω_j of ways in which N_j *indistinguishable* particles can be distributed among g_j *distinguishable* states, $N_j < g_j$, is the same as the number of combinations of N_j-filled and $(g_j - N_j)$-unfilled numbered containers. That is,

$$\Omega_j = \frac{g_j!}{N_j!\,(g_j - N_j)!} \tag{13-25}$$

Exercise 13-7. Verify Eq. 13-25 by enumerating the possible arrangements of three indistinguishable objects among five numbered containers.

10. E. Fermi, *Z. Physik*, **36**, 902–12 (1926), as translated in H. A. Boorse and L. Motz, *The World of the Atom*, Vol. II, Basic Books, New York, 1966, pp. 1321–31.

Since the number of arrangements within any group of states is independent of the number in any other group, the total number of arrangements Ω of electrons among electron quantum states is the product of the number of arrangements in each group. That is,

$$\Omega = \prod_{j=0}^{\infty} \Omega_j$$

$$= \prod_{j=0}^{\infty} \frac{g_j!}{N_j! \, (g_j - N_j)!} \tag{13-26}$$

As before, it will be convenient to take the natural logarithm of Ω.

$$\ln \Omega = \sum_{j=0}^{\infty} [\ln (g_j!) - \ln (N_j!) - \ln (g_j - N_j)!] \tag{13-27}$$

By the application of Stirling's approximation,[11] Eq. 13-27 reduces to

$$\ln \Omega = \sum_{j=0}^{\infty} [g_j \ln g_j - N_j \ln N_j - g_j \ln (g_j - N_j) + N_j \ln (g_j - N_j)] \tag{13-28}$$

$$= \sum_{j=0}^{\infty} \left[g_j \ln \frac{g_j}{g_j - N_j} - N_j \ln \frac{N_j}{g_j - N_j} \right] \tag{13-29}$$

As in Sec. 8-2, we find the most probable distribution by minimizing $\delta \ln \Omega$ with respect to changes in the values of the N_j subject to the conditions of constant total energy and constant number of electrons. The method of undetermined multipliers gives the condition of maximum probability as

$$\sum_{j=0}^{\infty} \left[\ln \frac{N_j}{g_j - N_j} + \mu + \beta \, (\varepsilon_j - \varepsilon_0) \right] \delta n_j = 0 \tag{13-30}$$

Since the constraints of constant energy and total number of electrons are included in the undetermined multipliers μ and β, all the variations δn_j are independent, and the condition of Eq. 13-30 can be met only if the coefficient of each δn_j in the sum is equal to zero. Thus, for all j,

$$\ln \frac{N_j}{g_j - N_j} = - \mu - \beta \, (\varepsilon_j - \varepsilon_0)$$

and

$$\frac{g_j}{N_j} - 1 = \exp \mu \exp [\beta \, (\varepsilon_j - \varepsilon_0)] \tag{13-31}$$

At temperatures sufficiently high that $g_j/N_j \gg 1$, Eq. 13-31 reduces to the Boltzmann approximation to the Fermi-Dirac distribution, identical to

---

11. Stirling's approximation is valid only for very large numbers. Many groups of electron states will be filled or nearly filled, so that $g_j - N_j$ may not be large. This mathematical difficulty can be circumvented by finding the most probable distribution for a sufficiently large ensemble of systems of a given kind rather than for a single system.

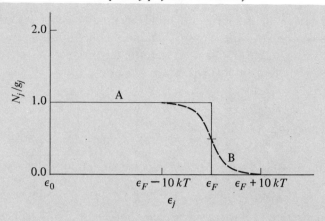

Figure 13-21. Fractional occupancy N_j/g_j of the energy levels of a metallic crystal as a function of ε_j for the Fermi-Dirac distribution. Curve A, at a temperature of 0 K; curve B, at temperature T higher than 0 K.

Eq. 8-60 for the Boltzmann approximation to the Bose-Einstein distribution. Rearranging Eq. 13-31, we obtain

$$N_j = \frac{g_j}{\exp \mu \exp [\beta (\varepsilon_j - \varepsilon_0)] + 1} \tag{13-32}$$

As before, β is set equal to $1/kT$, and Eq. 13-32 becomes

$$N_j = \frac{g_j}{\exp \mu \exp [(\varepsilon_j - \varepsilon_0)/kT] + 1} \tag{13-33}$$

The identification of μ is more difficult than for the Boltzmann approximation, but it can be shown that μ is approximately equal to $\varepsilon_0 - \varepsilon_F$, where ε_F is the energy of the *Fermi level, the highest filled level at* 0 K.[12] Then Eq. 13-33 becomes

$$N_j = \frac{g_j}{\exp [(\varepsilon_j - \varepsilon_F)/kT] + 1} \tag{13-34}$$

At 0 K, according to Eq. 13-34, $N_j = g_j$ (that is, all states are occupied) if $\varepsilon_j < \varepsilon_F$ and $N_j = 0$ if $\varepsilon_j > \varepsilon_F$. This result is shown graphically as curve A in Fig. 13-21. As the temperature is raised, some electrons are excited from states with energy $\varepsilon_j < \varepsilon_F$ to states with energy $\varepsilon_j > \varepsilon_F$, and the distribution is that shown as curve B in Fig. 13-21.

A numerical example illustrates how Fermi-Dirac statistics explains the

12. R. W. Gurney, *Introduction to Statistical Mechanics*, McGraw-Hill Book Co., New York, 1949, p. 251.

lack of contribution from electronic motion to the heat capacity of a metal. Consider one mole of metallic sodium, a crystal in the shape of a cube 2.8×10^{-2} m on a side. The right side of Eq. 8-51, when multiplied by two to take into account electron spin, gives the number of electron quantum states with energy less than $\varepsilon - \varepsilon_0$ for the model we are using. If we choose ε equal to ε_F, then the number of quantum states is equal to 6.02×10^{23}, the number of valence electrons in the crystal. Solving for the value of $\varepsilon_F - \varepsilon_0$, we obtain

$$\varepsilon_F - \varepsilon_0 = 1.2 \times 10^{-19} \text{ J} \qquad (13\text{-}35)$$

Exercise 13-8. Verify Eq. 13-35.

At 298 K, kT is equal to 4.1×10^{-21} J, so that kT is small compared to the energy of the Fermi level. We can use Eq. 8-52 (with a factor of two) to calculate the number of electron quantum states between ε_F and $\varepsilon_F + kT$, the number to which electrons might be thermally excited at room temperature. The result is 4.3×10^{21}, less than 1% of the number of states below ε_F.

Exercise 13-9. Verify the value of 4.3×10^{21} states within kT of ε for sodium at 298 K.

At temperatures close to 0 K, the electronic heat capacity can be shown to be directly proportional to T. As pointed out in Sec. 8-3, Debye's model for the heat capacity of lattice vibrations leads to a T^3 dependence at low temperatures. Therefore, at a sufficiently low temperature, the electronic heat capacity will be larger than the lattice heat capacity. Measurements between 0.005 K and 1 K confirm this prediction.[13]

The Electrical Conductivity of Metals. Equation 13-24 has the form of Ohm's law, but the results are puzzling if we use it to obtain numerical values of the mean free path. For example, the conductivity σ of sodium is $2.1 \times 10^7 \ \Omega^{-1}$-m^{-1}. From the data given above $n_A = 2.5 \times 10^{28}$ m^{-3}. On the assumption that $\varepsilon_F - \varepsilon_0 = \frac{1}{2} mv^2$, the value of v is 5.1×10^5 m-s^{-1}, so λ is 1.5×10^{-8} m.

Exercise 13-10. Verify the value of 1.5×10^{-8} m for λ, starting with Eqs. 13-23 and 13-24 and the data given above.

13. C. Kittel, *Introduction to Solid State Physics*, 4th ed., John Wiley & Sons, Inc., New York, 1971, pp. 249–55.

It seems strange that an electron can travel many lattice spacings through a metallic crystal before it collides with a lattice ion. If the electron is viewed as a wave rather than a particle, however, the difficulty disappears. Like X-rays and light, electron waves traverse periodic structures freely except for the special cases in which reflection occurs. This behavior also forms the basis for the theory of X-ray and electron diffraction. If a metallic crystal were a perfect lattice, and if the lattice ions were perfectly motionless, the crystal would have zero electrical resistance. Resistance is a result of the scattering of the electron wave due to the thermal oscillations of the lattice and to the presence of impurities and irregularities in the lattice. When expressed quantitatively, this point of view accounts well for the decrease of electrical resistance with decreasing temperatures.[14, 15] Also, since the scattering of the electron waves by the oscillating lattice ions involves an exchange of energy, this model can account for the Joule heating of electrical conductors.

The Band model. If the periodic variation of the potential of lattice ions is considered, the free electron model can account for the bands of molecular orbitals discussed above. For this purpose, the trigonometric functions that were given as solutions to the particle-in-a-well problem in Eq. 5-9 are less suitable than the equivalent complex exponential solutions. The former represent standing waves and the latter traveling waves.

For a free electron model, the traveling wave solution to Eq. 5-5 is (substituting ψ for x)

$$\psi = A \exp (ikx) \qquad (13\text{-}36)$$

and

$$k = \left(\frac{8 \pi^2 m \varepsilon}{h^2} \right)^{\frac{1}{2}} \qquad (13\text{-}37)$$

Exercise 13-11. Verify that Eq. 13-36 is a solution of Eq. 5-5.

Since a metallic crystal is very large compared to the interatomic spacing, the band structure should be independent of the size of the crystal. Instead

---

14. J. C. Slater, *Quantum Theory of Molecules and Solids*, Vol. 3 McGraw-Hill Book Co., New York, 1967, pp. 12–18.
15. The sharp transition to a superconducting state near 0 K in some metals and alloys requires a quite different explanation.

of setting boundary conditions at the edges of the crystal, it is convenient to use *periodic boundary conditions* such that

$$\psi(x + L) = \psi(x) \tag{13-38}$$

where L is some finite distance smaller than the crystal but large with respect to the interatomic spacing. Applying the condition of Eq. 13-38 to the solution in Eq. 13-36, we obtain

$$A \exp[ik(x + L)] = A \exp(ikx)$$

or

$$\exp(ikL) = 1 \tag{13-39}$$

By the nature of the complex exponential, Eq. 13-39 requires that

$$kL = \pm n(2\pi)$$

or

$$k = \pm n\frac{2\pi}{L} \tag{13-40}$$

where $n = 0, 1, 2, \dots$.

When the value of k from Eq. 13-40 is substituted in Eq. 13-37, the resulting expression for the energy is

$$\varepsilon = \frac{k^2 h^2}{8\pi^2 m} \tag{13-41}$$

$$= \frac{n^2 h^2}{2mL^2} \tag{13-42}$$

Since

$$\varepsilon = \frac{p^2}{2m}$$

where p is the momentum,

$$p = \frac{kh}{2\pi} \tag{13-43}$$

$$= \pm \frac{nh}{L}$$

The De Broglie wavelength of the electron is $\lambda = h/p$, so that the *wavenumber* k is

$$k = \frac{2\pi}{\lambda} \tag{13-44}$$

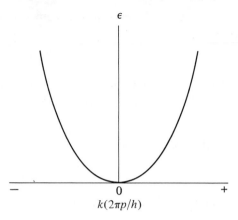

Figure 13-22. Parabolic dependence of the energy of an electron in a metal on the wavenumber $k = (2\pi p/h)$, according to the free-electron model.

The quadratic dependence of ε on k is shown in Fig. 13-22 for a free electron in one dimension. Positive values of k correspond to waves moving in the positive x direction, and negative values of k correspond to waves moving in the negative x direction. When the periodic potential of the lattice ions is taken into account, the wave function in Eq. 13-36 must be multiplied by a periodic function whose period is a, the interatomic spacing in the crystal.[16] For most values of k, the perturbing potential has little effect and the plane waves are good approximations to the behavior of the electrons in the crystal. When $k = \pm n\pi/a$, however, that is, when $\lambda = 2\pi/k = 2a/n$ or $n\lambda = 2a$, the Bragg relation for reflection at an angle of 90° (see Eq. 7-16) is met. For these values of k neither of the traveling waves $\exp(i\pi x/a)$ or $\exp(-i\pi x/a)$ are good solutions. The appropriate solutions are the two linear combinations:

$$\psi_+ = \exp\left(\frac{i\pi x}{a}\right) + \exp\left(-\frac{i\pi x}{a}\right) = 2\cos\left(\frac{\pi x}{a}\right)$$

and

$$\psi_- = \exp\left(\frac{i\pi x}{a}\right) - \exp\left(-\frac{i\pi x}{a}\right) = 2i\sin\left(\frac{\pi x}{a}\right)$$

If the origin is taken at a lattice ion position, ψ_+ has its maxima at lattice ion positions and therefore has a low energy, whereas ψ_- has its minima at

16. C. Kittel, *Introduction to Solid State Physics*, 4th ed., John Wiley & Sons, Inc., New York, 1971, pp. 296–306.

lattice ion positions and therefore has a high energy. *There are no permitted solutions between these two energies, and this range is the gap between permitted bands of solutions or between energy bands.* The dependence of ε on k for this model is shown in Fig. 13-23. The solid line represents the energy of electrons in a periodic potential, and the dashed line shows the corresponding curve for the free electron.

The electron states of three-dimensional crystals are described by specifying the values of k_x, k_y, and k_z. It is not possible to draw diagrams of ε as a function of the three components of k, as was done for the one-dimensional case in Fig. 13-22 and Fig. 13-23. But, contour surfaces of constant energy in k space for various crystal structures have been constructed. The contour surface of constant energy within which all levels are filled and outside of which all levels are vacant at 0 K is called the *Fermi surface.*

Consideration of the arrangement of energy bands in k space (momentum space) and of their population enables us to understand the conductance properties of metals and of insulators and semiconductors as well. Figure 13-24 is a schematic representation of a one-dimensional analogue of the 3s band in an alkali metal such as sodium. Only states in the lower half of

Figure 13-23. The dependence of the energy of an electron in a metal on the wavenumber k, taking into account the wave nature of the electron and the periodic potential due to the positive lattice. Bands of allowed energies and finite energy gaps result.

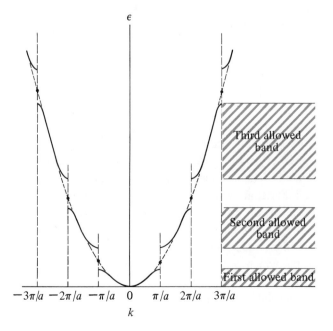

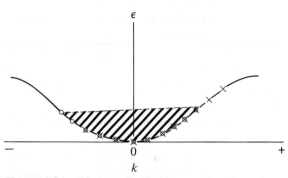

Figure 13-24. Electron distribution as a function of wavenumber $k = (2\pi p/h)$ in the absence (o) and presence (x) of an electric field for electrons in the highest partly filled band in sodium metal, showing a net momentum and, hence, conductance in the presence of the field.

the band are filled. In the absence of an electrical field, the occupied states are symmetrically arranged about $k = 0$. The effect of an applied electric field is to increase the momentum of each electron in the direction opposite to the field. This is possible for sodium because there are vacant *momentum* states to which the electrons can be shifted. The result is a net momentum of the electrons, which is observed as a current.

It might appear that a divalent metal like magnesium, which has the 3s level filled, would not be a conductor. As indicated in Fig. 13-20, the 3s and 3p bands overlap, so that there are many available momentum states in the 3p band for the electrons in magnesium to occupy under the influence of an electric field.

Figure 13-25 shows the energy bands of diamond as a function of internuclear distance. At the equilibrium internuclear distance, the 2s and 2p bands have combined to form a band with enough states to hold four electrons per carbon atom (closely related to the sp³ hybrid orbitals of the carbon atom). The corresponding antibonding band has a much higher energy. Since the diamond crystal has four valence electrons per atom, the lower *valence band* is filled. When an electric field is applied, there are no momentum states available that would permit a net momentum of the electrons, so that diamond is an insulator. If the field is strong enough to excite the electrons to the much higher energy vacant band, the crystal will conduct, and this phenomenon is known as *dielectric breakdown*.

Silicon and germanium, which crystallize in the diamond structure, have an energy band pattern much like diamond, as shown in Fig. 13-25. The crucial difference is that the equilibrium internuclear distance is much farther to the right on the diagram. Thus, the gap between the filled band and the empty band is much smaller, and the thermal energy at normal temperatures is great enough to excite some electrons to the vacant band. Substances such as these are called *intrinsic semiconductors*, because the conduction properties are intrinsic to the structure. In contrast to the metals, in which resistance increases with increasing temperature, the resistance of intrinsic semiconductors decreases with increasing temperature.

When an electron in a semiconductor is excited to the conduction band, it leaves a hole in the valence band. As a result, electrical conductivity occurs in both bands, with the hole and the electron moving in opposite directions under the influence of the electric field.

The conducting property of Si and Ge can be increased by addition of suitable impurities. When a trivalent atom, such as Al or B, occurs in a silicon lattice, it has only three valence electrons to contribute instead of four. It is said to be *electron-deficient*, and there is a hole in the valence band for each such atom in the lattice. When As or Sb atoms are impurities, the extra electron goes into the conduction band. The former is called a *p-type semiconductor*, for the positive holes that are added. The latter is called an *n-type semiconductor*, for the negative electrons that are added.

The insulating property of ionic crystals also can be explained by calculation of their energy bands. Figure 13-26 shows the results of calculations

Figure 13-25. Energy bands as a function of internuclear distance for C (diamond), Si, and Ge. [By permission, from A. Holden, *The Nature of Solids*, Columbia University Press, 1965, p. 202 and G. Kimball, *J. Chem. Phys.*, *3*, 560 (1935).]

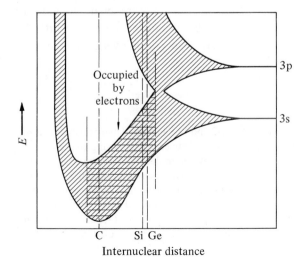

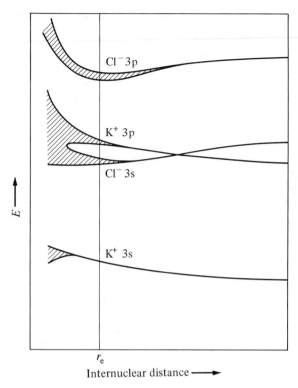

Figure 13-26. Energy bands as a function of internuclear distance for KCl crystals. The equilibrium internuclear distance is r_e. [By permission, from L. P. Howland, *Phys. Rev.*, *109*, 1927 (1958).]

for KCl. All the filled bands characteristic of the ions are separated and very far below the lowest empty band. The bands are affected very little by changes in internuclear distance, indicating very little overlap of the filled orbitals of oppositely charged ions.

Problems

13-1. The structure of ice described in Secs. 7-2 and 13-2 indicated two possible positions of equal probability for a hydrogen atom between neighboring oxygen atoms. Thus for the two hydrogen atoms in a water molecule there are four possible arrangements (2^2). If one considers the four hydrogen atoms about each oxygen in the crystal, there are sixteen possible arrangements, but only six of these, or three-eighths of the total are compatible with two covalently bonded and two hydrogen-bonded hydrogens to each oxygen. Thus there are $(4)(\frac{3}{8}) = \frac{3}{2}$ possible arrangements per water molecule or $(\frac{3}{2})^L$ arrangements of a mole of water molecules in a crystal. If these arrangements are frozen into the crystal as it is cooled to 0 K, what is the configurational entropy at 0 K? Giauque and Ashley and Giauque and Stout found a value [*Phys. Rev.*, *43*, 81 (1933); *J. Am. Chem. Soc.*, *58*, 1144 (1936)] of 3.43 J-mol^{-1}-K^{-1} for the entropy at 0 K by comparison of the entropy of the gas calculated from the partition function and that calculated from heat capacity measurements. Pauling [*J. Am. Chem. Soc.*, *57*, 2680 (1935)] developed the theoretical model for the residual entropy at 0 K.

13-2. How many C atoms are contained in the unit cell shown in Fig. 13-8? Remember that some of the atoms shown are shared with other unit cells.

13-3. The energy gaps between the highest filled band and the lowest vacant band for Ge, Si, and C (diamond) are 6.94 × 10^4 J-mol^{-1}, 1.06 × 10^5 J-mol^{-1}, and 5.40 × 10^5 J-mol^{-1}, respectively. At what temperature does kT become equal to the energy gap for each of these crystals? At what wavelengths of light would photons have sufficient energy to excite electrons to the conduction band, producing *photo-conductivity*?

14 · *Liquids*

At low temperatures, all substances can exist at equilibrium as crystalline solids. As we saw in Chapter 13, these solids are highly ordered structures in which the molecules are at a low potential energy due to attractive forces and at a low entropy because only the lowest of widely spaced energy levels are occupied. Because of the strong forces between lattice units in the crystal, there is substantial resistance to deformation, and the shape of the crystal is independent of the shape of its container. At sufficiently high temperatures and low pressures, all substances exist at equilibrium as a dilute gas. The molecules are at a high potential energy because they are too far apart for attractive forces to be significant and at a high entropy because there are many more accessible, closely spaced, energy levels than there are molecules. A sample of gas has no fixed shape, and it fills any container in which it is placed. For a limited range of intermediate temperatures and pressures, the equilibrium state is the liquid state. Intermolecular distances are comparable to those in crystals, but there is no long-range order. Liquids, like gases, are fluids, and there is no resistance to deformation. They adopt the shape of a container, but the liquid has a volume independent of the container.

At any pressure, the transition between crystalline and liquid states occurs at a sharply defined temperature with a finite change of enthalpy, entropy, and volume. At temperatures below the critical point (Sec. 1-3), the transition between liquid and gas is similarly sharp, but above the critical point it is not possible to obtain a boundary between liquid and gaseous phases. As pointed out in Fig. 1-8, it is possible to go from a state that is clearly gas to one that is clearly liquid without having more than a single phase at any time and thus without observing a boundary between phases.

Any satisfactory model of the liquid state must afford an explanation of these properties of liquids and the relationship between the liquid state and the crystalline and gaseous states.

14-1. Theories of the liquid state

Liquids have neither the regular arrangement of solids nor the large inter-molecular distances of gases. Thus, a satisfactory model of the liquid state can neither consider the molecules independent, as in a gas, nor as having a constant number of nearest neighbors, as in a crystal.

The van der Waals model

The earliest attempts to develop a model of the liquid state grew out of observations of the continuity of the liquid and gaseous states and the existence of a critical point. An early effort was that of van der Waals in his Ph.D. dissertation in 1873. The equation he derived (Eq. 14-1) took into account the volume b from which a gas molecule is excluded due to the finite size of the molecules and the decrease in pressure $n^2 a/V^2$ from that of an ideal gas due to intermolecular forces.

$$\left(P + \frac{n^2 a}{V^2} \right)(V - nb) = nRT \tag{14-1}$$

Since the excluded volume is considered to be a constant independent of the fluid density, the van der Waals equation assumes a "hard-sphere" repulsive force. It can be shown[1] that the form of the correction to the pressure is consistent with an attractive force with potential energy proportional to r^{-6}. The potential energy curve is shown in Fig. 14-1.

A plot of the van der Waals equation is an oscillatory curve for the isotherms that represent temperatures below the critical temperature. The real behavior of two-phase fluid systems shows a line of constant pressure with a discontinuous change of slope at each end of the line where it enters one-phase regions. Thus, the van der Waals model, or one derived by a more rigorous statistical-mechanical approach, predicts a phase transition in the sense that it shows the volume as a three-valued function of the pressure but is inadequate in that it cannot reproduce the discontinuity of slope.

The position of the constant-pressure line on the oscillatory curve can be

1. T. L. Hill, *Statistical Thermodynamics*, Addison-Wesley, Reading, 1960, pp. 286–89.

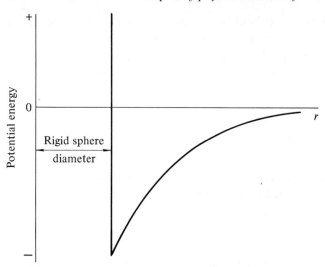

Figure 14-1.   The potential energy of interaction of two rigid spheres with an attractive potential energy proportional to r^{-6}.

determined by a thermodynamic criterion, however.[2] A two-phase isotherm is plotted in Fig. 14-2. The condition of equilibrium between the liquid at D and the gas at A, at constant pressure and temperature, is (from Eq. 10-71)

$$\Delta \mathscr{G} = \mathscr{G}_D - \mathscr{G}_A = 0$$

where $\Delta \mathscr{G}$ is the free energy of transfer for one mole of substance from gas to liquid. Since (from Eq. 10-87),

$$\left(\frac{\partial \mathscr{G}}{\partial P} \right)_T = \mathscr{V}$$

$$\Delta \mathscr{G} = \int_1^2 \left(\frac{\partial \mathscr{G}}{\partial P} \right)_T dP$$

$$= \int_1^2 \mathscr{V} \, dP$$

From Fig. 14-2,

$$\mathscr{G}_B - \mathscr{G}_A = \int_A^B \mathscr{V} \, dP = \text{Area I} + \text{Area II} + \text{Area III}$$

$$\mathscr{G}_C - \mathscr{G}_B = \int_B^C \mathscr{V} \, dP = - \text{Area II} - \text{Area III} - \text{Area IV} - \text{Area V}$$

2. *The Scientific Papers of James Clerk Maxwell*, Vol. II, Cambridge University Press, Cambridge, 1890, p. 425.

and

$$\mathscr{G}_\mathrm{D} - \mathscr{G}_\mathrm{C} = \int_\mathrm{C}^\mathrm{D} \mathscr{V}\, dP = \text{Area V}$$

Therefore,

$$\mathscr{G}_\mathrm{D} - \mathscr{G}_\mathrm{A} = \text{Area I} - \text{Area IV} \tag{14-2}$$

and Area I must equal Area IV.

Exercise 14-1. Verify Eq. 14-2.

Lattice models

An alternative approach to the theory of liquids starts with their resemblance to crystals. Debye and Menke[3] were among the first to study the scattering of X-rays by liquids. They studied the scattering by liquid mercury, both because the mercury atoms acted as strong scatterers and because the scattering pattern of a monatomic liquid should provide information about molecular distribution without complications due to intramolecular interference. The resultant intensity of scattered radiation as a function of scatter-

Figure 14-2. The Maxwell construction for finding the constant pressure line (AD) for an oscillatory isotherm in a 2-phase region; temperature = T, a constant.

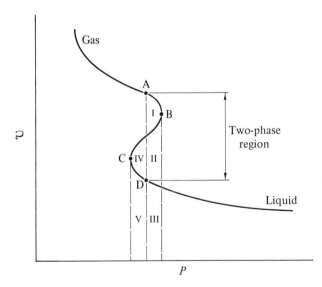

3. P. Debye and H. Menke, *Physik. Z.*, *31*, 797 (1930). *Collected Papers of Peter J. W. Debye*, New York, Interscience Publishers, 1954, p. 133.

ing angle was transformed by Fourier integration to a radial distribution function that represents the probability of finding a molecule in a volume at a distance r from any given molecule (Sec. 7-2).

Their plot of the radial distribution function is shown as the solid curve in Fig. 14-3. The quantity $g(r)$ represents the ratio of the number density of molecules in any volume element to the average number density in the sample. Thus, a constant value of $g(r)$ equal to one represents a random distribution. The dotted curve shows the behavior of a gas at a low pressure with an attractive interaction between molecules. The heights of the vertical lines represent the relative values of $g(r)$ at the discrete distances characteristic of a face-centered cubic crystal. The distribution function should be equal to zero at distances less than the molecular diameter, since no molecules can be closer than this value. The curve for $r < 2.2 \times 10^{-10}$ m shows non-zero values due to the low precision of the measurement of scattered intensity at large angles. The broad peaks of the curve for the liquid, in contrast to the sharp peaks for the solid, show that the order present in the liquid is short range and statistical. The results also show that the distance of closest approach is the same in the liquid as in the crystal, but that there are fewer nearest neighbors in the liquid.

Henry Eyring and his students developed a model of the liquid state based on the resemblance between liquids and crystals that they called the

Figure 14-3. The radial distribution function $g(r)$ for liquid mercury (——) obtained from the X-ray scattering results of Debye and Menke. The corresponding curve for a gas ($\cdots$) and the vertical lines corresponding to the discrete distribution of a face-centered cubic crystal are also shown. [By permission, adapted from P. Debye and H. Menke, *Physik. Z.*, *31*, 797 (1930). *The Collected Papers of Peter J. W. Debye*, Interscience Publishers, Inc., New York, 1954.]

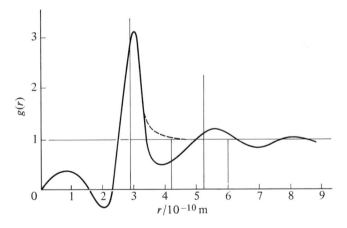

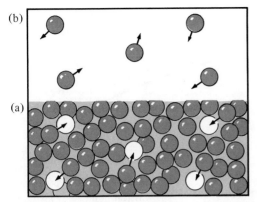

Figure 14-4. Scheme of the relation between holes in the liquid (a) and molecules in the equilibrium vapor (b) in the Eyring significant-structure theory of liquids. (After International Science and Technology, March, 1963.).

"significant structure" theory of liquids.[4] They attribute the decrease in density on melting a normal solid to the addition of mobile holes of molecular size to the crystalline lattice. The molecules adjacent to a hole are free to translate into it, thus leaving a new hole, and these molecules are considered to have gas-like properties. The picture that emerges is that of a random fine-grained mixture of solid- and gas-like regions with neither the solid- nor the gas-like regions large enough to constitute nuclei for the formation of a solid or vapor phase. This picture is consistent with the observation that liquids can be supercooled below the normal freezing point and superheated above the normal boiling point.

As the temperature of a liquid is increased from the triple point to the critical point along the vapor-liquid equilibrium curve (see Fig. 15-3), the increased number of molecules in the vapor is reflected by an increased number of vacancies in the liquid, as indicated in Fig. 14-4. This model provides a simple explanation of the *law of rectilinear diameters*, enunciated by Cailletet and Mathias in 1886. They suggested that the mean density of liquid and vapor in equilibrium should be a constant, independent of temperature, and equal to the critical density. Actually, the density of the liquid decreases more rapidly with increasing temperature than the density of the vapor increases, due to thermal expansion of the lattice. The mean density, however, is a linear function of the temperature; the critical density

4. H. Eyring, *J. Chem. Phys.*, *4*, 283 (1936); M. S. Jhon and H. Eyring, in *Physical Chemistry, An Advanced Treatise*, Vol. VIII A (D. Henderson, ed.), Academic Press, New York, 1971, pp. 335–75.

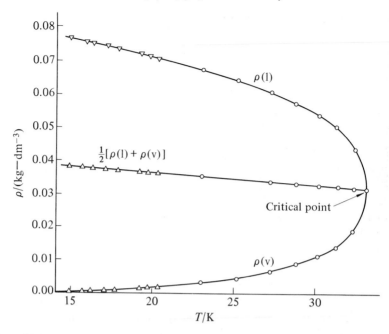

Figure 14-5. Data on the density of liquid and equilibrium vapor for H_2 as a function of temperature up to the critical point, illustrating the law of rectilinear diameters of Cailletet and Mathias. [By permission, from E. Mathias, C. A. Crommelin, and H. Kamerlingh-Onnes, *Comm. Phys. Lab. Leiden,* No. 154b (1921).]

can be obtained by extrapolation of this function to the critical temperature, as shown for H_2 in Fig. 14-5.

The partition function for the liquid is an appropriate combination of the partition function for the solid-like molecules, usually based on the Einstein model, and the partition function for the gas-like molecules, usually that of the ideal gas. Thus, the molecular properties of the solid and the gas are built in to the model, and there are no adjustable parameters. We shall compare the results of this model with those of several others and with experimental values in the next section.

Shortly after the publication of Eyring's first paper on significant structure theory, Lennard-Jones and Devonshire published a series of papers on a comprehensive theory of fluids.[5] This theory is based on the conception that molecules of liquid are confined to "cells" as a result of the repulsive forces

5. J. E. Lennard-Jones and A. F. Devonshire, *Proc. Roy. Soc., London, A163,* 53 (1937); *A165,* 1 (1938).

of nearest neighbor molecules. In contrast to similar cells in crystals, the cells in liquids can move relatively freely throughout the volume of the liquid.

In this model, the interaction between a molecule in one cell and the molecules in the nearest neighbor cells, assumed fixed at the centers of their cells, is described by the Lennard-Jones potential energy function, first used to describe interactions in imperfect gases. That function is

$$U(r) = -\frac{a}{r^6} + \frac{b}{r^{12}} \tag{14-3}$$

The form of the potential energy function is illustrated in Fig. 14-6, in which σ is the approximate *collision diameter*, r_e is the intermolecular distance at the minimum in the potential energy curve, and ε is the depth of the potential energy minimum. The attractive term, $-a/r^6$, is of the same form as the intermolecular forces described in Eqs. 13-15 through 13-18. The exponent in the repulsive term is somewhat arbitrary, and a value of 12 is convenient

Figure 14-6. The Lennard-Jones potential energy function [$u(r)$]. The parameters are the negative of the energy at the minimum (ε), the equilibrium intermolecular distance (r_e), and the collision diameter at zero kinetic energy (σ).

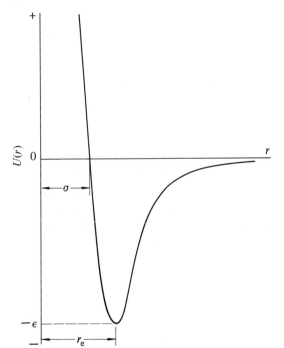

mathematically. In terms of the parameters σ, r_e, and ε, Eq. 14-3 can also be expressed as

$$U(r) = -2\varepsilon\left(\frac{r_e}{r}\right)^6 + \varepsilon\left(\frac{r_e}{r}\right)^{12} \tag{14-4}$$

$$= 4\varepsilon\left[-\left(\frac{\sigma}{r}\right)^6 + \left(\frac{\sigma}{r}\right)^{12}\right] \tag{14-5}$$

Exercise 14-2. Show that Eqs. 14-4 and 14-5 are consistent with Eq. 14-3 and with the definitions of r_e, σ, and ε.

Like the van der Waals equation, the Lennard-Jones and Devonshire model is a two-parameter model (σ and ε or r_e and ε), and is thus consistent with the law of corresponding states (Sec. 1-3). The reduced variables, in terms of the parameters of this model, are, for a face-centered cubic lattice,[6]

$$T_r = \frac{kT}{1.30\,\varepsilon} \tag{14-6}$$

$$\mathscr{V}_r = \frac{\mathscr{V}}{1.25\,Lr_e^3} \tag{14-7}$$

and

$$P_r = \frac{Pr_e^3}{0.614\,\varepsilon} \tag{14-8}$$

The crucial problem in any theory of the liquid state based on a lattice model is that of introducing disorder into the model. In Eyring's treatment, disorder is introduced with the random arrangement of holes in the lattice. In the model of Lennard-Jones and Devonshire, disorder is introduced by the somewhat arbitrary addition of a "communal entropy" to the liquid compared to that of a solid.[7] We can see how such an entropy change is calculated if we apply Eq. 8-93 for the entropy of a crystal, with its distinguishable units, and the partition function of Eq. 8-85 for a particle free to move in a box of volume V/N; and if we apply Eq. 8-94 for the entropy of a liquid, with its *indistinguishable* units, and the partition function of Eq. 8-85 for a particle free to move in a box of volume V. Thus,

$$S_{\text{crys}} = kN\ln z + \frac{E - E_0}{T}$$

6. T. L. Hill, *Introduction to Statistical Thermodynamics*, Addison-Wesley, Reading, 1960, p. 295.
7. J. Hirschfelder, D. Stevenson, and H. Eyring, *J. Chem. Phys.*, 5, 896 (1937).

$$= kN \ln \left\{ \left[\frac{2\pi mkT}{h^2} \right]^{\frac{3}{2}} \frac{V}{N} \right\} + \frac{E - E_0}{T}$$

$$= kN \ln \left\{ \left[\frac{2\pi mkT}{h^2} \right]^{\frac{3}{2}} V \right\} + \frac{E - E_0}{T} - kN \ln N \qquad (14\text{-}9)$$

$$S_{\text{liq}} = kN \ln z + \frac{E - E_0}{T} - kN \ln N + kN$$

$$= kN \ln \left\{ \left[\frac{2\pi mkT}{h^2} \right]^{\frac{3}{2}} V \right\} + \frac{E - E_0}{T} - kN \ln N + kN \qquad (14\text{-}10)$$

and the communal entropy of the liquid is

$$S_{\text{liq}} - S_{\text{crys}} = kN \qquad (14\text{-}11)$$

This model predicts that the molar entropy of fusion is equal to R, or $8.3 \text{ J-mol}^{-1}\text{-K}^{-1}$. The experimental values given in Table 14-1 indicate that this result is correct as to order of magnitude for monatomic solids.

Table 14-1. Enthalpy and entropy of fusion for several monatomic liquids

Substance	$\Delta \mathscr{H}_f/(\text{kJ-mol}^{-1})$	T_f/K	$\Delta \mathscr{S}_f/(\text{J-mol}^{-1}\text{-K}^{-1})$
Ne	0.324	24.6	13.2
Ar	1.21	83.0	14.6
Kr	1.37	116.6	11.7
Xe	3.10	161.6	19.2
Na	2.64	371.0	7.1
K	2.40	336.6	7.1
Ag	11.3	1234	9.2
Hg	2.33	234	10.0

Since the Lennard-Jones and Devonshire model only describes systems with each cell occupied by one molecule, it agrees relatively well with experiments in the dense liquid range, where each molecule is confined to its cell by its neighbors, but it agrees poorly with experiments near the critical region.

In a refinement of their earlier work, Lennard-Jones and Devonshire developed a model with which they were able to describe a phase transition between the crystalline and the liquid states.[8] They picture a solid as consisting of two interpenetrating lattices. In a perfectly ordered state, all the sites

8. J. E. Lennard-Jones and A. F. Devonshire, *Proc. Roy. Soc., London*, *A169*, 317 (1939).

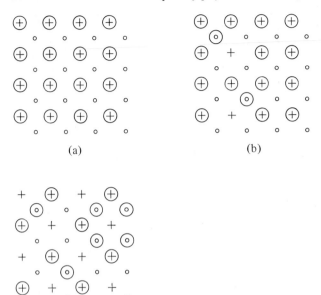

Figure 14-7.   Scheme of the Lennard-Jones and Devonshire model. (a) Complete order, with all atoms ($\bigcirc$) on the α sites ($+$) and all the β sites ($\circ$) vacant. (b) Partial disorder, with several atoms on the β sites and several α sites vacant. (c) Complete disorder, with a random distribution of atoms between the α and β sites. [By permission, from J. E. Lennard-Jones and A. F. Devonshire, *Proc. Roy. Soc.*, *A169*, 317 (1939).]

on one lattice, the α sites, are occupied by atoms and all the sites on the other lattice, the β sites, are occupied by holes, as shown in Fig. 14-7, a. The long-range order Σ is defined as $2q - 1$, where q is the fraction of atoms on α sites. As the temperature is increased, some atoms migrate to β sites, which are at a higher potential energy due to repulsive forces that are predominant at distances less than the equilibrium distance between α sites. An example with a few such displacements that still retains long-range order is shown in Fig. 14-7, b. At sufficiently high temperature, there is a random distribution of atoms among α and β sites, as shown in Fig. 14-7, c, and long-range order has disappeared.

At sufficiently high pressure, over a wide range of temperatures, the lattice is in a perfectly ordered state, with all the atoms on α sites. For a lattice that could be expanded without losing its order, the pressure of the lattice due to the kinetic energy of the molecules and the repulsive forces between them decreases monotonically with increasing volume at constant temperature. In addition to this effect of increasing volume, the Lennard-Jones and Devonshire lattice exhibits increased disorder with increasing volume because the movement of molecules to β sites is facilitated by the increased spacing. The initial result of the increase in disorder is an increase in pressure, due to increased repulsive forces, followed by a decrease in pressure when the volume becomes large enough to permit easy movement of molecules to β sites. A detailed consideration of the model leads to an oscillatory isotherm like those below the critical temperature in Fig. 1-8. As in the case of liquid-gas equilibrium, this behavior suggests a phase transition at constant pressure. The model also leads to a critical temperature above which a solid-liquid transition does not occur, a phenomenon that has not yet been observed even at the highest pressures attainable. Predictions of the volume change on melting and the entropy of fusion of argon and nitrogen agree well with experiment, however, as shown in Table 14-2. The inclusion of holes in this model, which leads to an entropy of mixing among lattice sites, accounts for an entropy of fusion greater than the communal entropy. Good agreement was also found between predicted and observed values for the boiling points and entropies of vaporization of argon and nitrogen.

Table 14-2. Some properties of argon and nitrogen, compared to values calculated from the Lennard-Jones and Devonshire model

	T_M/K	$\Delta \mathscr{V}(\%)$		$\Delta \mathscr{S}_f/(\text{J-mol}^{-1}\text{-K}^{-1})$	
		calc.	obs.	calc.	obs.
Ar	83.8	13	12	14.4	14.0
N_2	63.2	8	9	10.8	11.4

	T_B/K			$\Delta \mathscr{S}_v/(\text{J-mol}^{-1}\text{-K}^{-1})$	
	calc.	obs.		calc.	obs.
Ar	87.3	87.4		73.2	72.0
N_2	75.0	77.2		69.9	72.4

J. E. Lennard-Jones and A. F. Devonshire, *Proc. Roy. Soc., London, A169*, 317 (1939).

Distribution function theories

A rigorous theory of the liquid state would utilize a *distribution function*, which describes the probability of finding other molecules as a function of distance from some central molecule, and a potential energy function to obtain the partition function. Kirkwood and Born and Green developed theories of this kind, but they were forced by mathematical difficulties to use approximate potential energy functions and simplified distribution functions.[9] As a result of this effort, Kirkwood stated:

> While these methods (*lattice models*) may be regarded as useful alternatives to our own, we believe that liquid structure cannot be adequately described in terms of a lattice blurred by thermal motion, but that the local order in liquids manifested in the radial distribution function is of an essentially different nature from the long range order in crystals.

Geometric model

J. D. Bernal proposed a model of the liquid state that took Kirkwood's reservations into account.[10] He found that he could describe *essentially disordered* but closest packed arrays as the packing of irregular polyhedra with a large fraction of pentagonal faces. A pentagonal figure is rare in crystalline structures. Also, he found that a ball and stick model constructed with intermolecular distances appropriate to a liquid distribution function could not be compressed to a density between that of a liquid and that of a crystal without producing a transition to a crystalline state. He suggests, therefore, "that there is an absolute impossibility of forming a homogeneous assembly of points, of volume intermediate between those of long-range order and closest packed disorder."

Perturbation theory

All the theories we have discussed thus far lead to the conclusion that liquids are packed quite closely and that the repulsive forces are the predominant factors in determining the structure of liquids. Recently Barker and Henderson[11] have developed a *perturbation theory of liquids* based on this idea. They begin with an exact solution for a hypothetical fluid composed of hard spheres and then apply deviations from the hard sphere model due to the repulsive and the attractive portions of the potential energy curve as separate perturbations. Though their model is not applicable to any real liquid, they obtained good agreement with computer simulation results using the same potential energy function. The major advantages of this approach are the direct relationship to the physical factors of importance to liquid structure and the relatively modest requirements in computer time.

Computer simulation

The advent of large, high-speed digital computers has stimulated attempts

9. J. G. Kirkwood and E. M. Boggs, *J. Chem. Phys.*, *10*, 394 (1942).
10. J. D. Bernal, *Nature*, *183*, 141 (1959); *185*, 68 (1960); *Scientific American*, *203*, 124 (1960).
11. J. A. Barker and D. Henderson, *Accounts of Chem. Res.*, *4*, 303–7 (1971).

to calculate the properties of liquids by means of computer simulation. In one such attempt, called the *Monte Carlo method*,[12] a finite number of particles with an assumed potential energy of interaction are placed in a rectangular cell. By the method of periodic boundary conditions (Eq. 13-38), the cell is assumed to be repeated periodically in all directions to fill space. The molecules are placed in some arbitrary initial configuration and then moved one at a time to obtain other configurations, which can be averaged to obtain the statistical-mechanical equilibrium properties of the liquid. The Monte-Carlo designation refers to a technique that ensures the choice of configurations with sufficiently high probability that a reasonable average can be obtained with a practical number of configurations.

The *molecular dynamics* method of Alder and Wainwright[13] also begins with some assumed configuration of a finite number of molecules in a rectangular cell. Successive new configurations are obtained by numerical solution of the classical differential equations of motion of a many-body system. From an averaging over the configurations, we obtain information on the equilibrium properties, whereas the time dependence can provide information on dynamic properties. Figure 14-8, a shows the calculated trajectories of 32 rigid sphere particles during 3000 collisions under conditions corresponding to the liquid state. Figure 14-8, b shows similar trajectories for conditions corresponding to the crystalline state. It is a tribute to the method that the solid-liquid transition can be simulated in this way.

The comparison of Monte-Carlo and molecular dynamics calculations with experimental results will provide an experimental test of the fundamental assumptions of statistical mechanics and kinetic theory, respectively. They do require very large amounts of computer time, however, and they do not provide information on the physical factors involved in the stability of the liquid state. In this connection, a quotation from Wigner and Seitz in another context is of interest.[14]

> If one had a great calculating machine, one might apply it to the problem of solving the Schrödinger equation for each metal and obtain thereby the interesting physical quantities. . . . It is not clear, however, that a great deal would be gained by this. Presumably, the results would agree with the experimentally determined quantities and nothing vastly new would be learned from the calculation. It would

12. N. Metropolis, A. W. Rosenbluth, M. N. Rosenbluth, A. H. Teller, and E. Teller, *J. Chem. Phys.*, *21*, 1087 (1953).
13. B. J. Alder and T. E. Wainwright, *J. Chem. Phys.*, *27*, 1208 (1957); *Scientific American*, *201*, 113 (1950).
14. E. P. Wigner and F. Seitz, *Solid State Physics*, *1*, 97 (1955); as quoted in C. Kittel, *Introduction to Solid State Physics*, 4th ed., John Wiley & Sons, New York, 1971, p. 125.

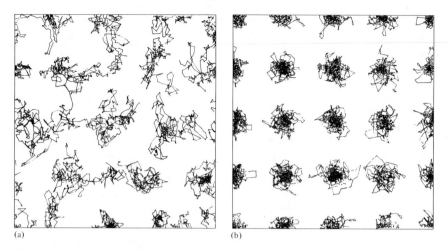

Figure 14-8. Molecular dynamics computer simulation model of paths of molecular motion in (a) liquid and (b) crystal. [Adapted by permission, from B. J. Alder and T. E. Wainwright, *Scientific American, 201,* 113–126, October (1959).]

be preferable instead to have a vivid picture of the behavior of the wave functions, a simple description of the essence of the factors which determine cohesion and an understanding of the origins of variation in properties from metal to metal.

14-2. Equilibrium properties of liquids

P-V-T data

The equation of state of a fluid provides a primary source of data to test any theory of liquids. Since most theories are two-parameter theories, consistent with the law of corresponding states, they predict that the compressibility factor $z = P\mathscr{V}/RT$ (Sec. 1-3) is a universal function of the reduced pressure and reduced temperature or the reduced density and the reduced temperature. The compressibility factor for different substances at their critical points should be the same. Thus, both the theoretical value of the critical compressibility factor and the temperature and pressure dependence of the compressibility factor are commonly used to test theories.

Table 14-3 shows values of the critical compressibility factor $P_c\mathscr{V}_c/RT_c$ for several groups of substances.

Clearly the law of corresponding states is not universally valid, but it is a useful approximation to the behavior of groups of similar molecules.

Table 14-3. The critical compressibility factor

Essentially spherical nonpolar molecules		Hydrocarbons		Polar molecules	
He	0.300	Ethane	0.267	CH_3CN	0.181
H_2	0.304	Propane	0.270	H_2O	0.224
Ne	0.296	Butane	0.257	NH_3	0.238
Ar	0.291	*i*-Pentane	0.268	CH_3OH	0.220
Xe	0.293	Benzene	0.265	CH_3Cl	0.258
N_2	0.292	Propylene	0.273	C_2H_5Cl	0.269
O_2	0.292				
CH_4	0.290				
CO_2	0.287				

J. O. Hirschfelder, C. F. Curtiss, and R. B. Bird, *Molecular Theory of Gases and Liquids*, J. Wiley & Sons, New York, 1954, p. 237, by permission.

Table 14-4 shows values for the reduced critical constants calculated from several models compared with the average of the experimental values for several spherical nonpolar molecules, for which the models should provide the best description. Though the models of Eyring and Lennard-Jones and Devonshire show better agreement with experiment than the van der Waals equation, it is clear that none provides a satisfactory description of the *P-V-T* behavior of real fluids.

Table 14-4. Reduced critical constants calculated from several models compared to the average experimental value for spherical nonpolar molecules[a]

	$\mathscr{V}_c/(r_e^3 L)$	kT_c/ε	$P_c r_e^3/\varepsilon$	$P_c \mathscr{V}_c/RT_c$
Experimental	2.12	1.29	0.179	0.292
Lennard-Jones and Devonshire	1.25	1.30	0.614	0.590
Cernuschi and Eyring[b]	1.41	2.74	0.663	0.342
van der Waals	6.28	0.296	0.0177	0.375

[a]T. L. Hill, *Statistical Thermodynamics*, Addison-Wesley, Reading, 1960, p. 297.
[b]F. Cernuschi and H. Eyring, *J. Chem. Phys.*, 7, 547 (1939).

Trouton's rule
Trouton[15] proposed as an empirical generalization in 1884 the rule that the ratio of the molar enthalpy of vaporization to the normal boiling point for

15. F. Trouton, *Phil. Mag.* (5), *18*, 54 (1884).

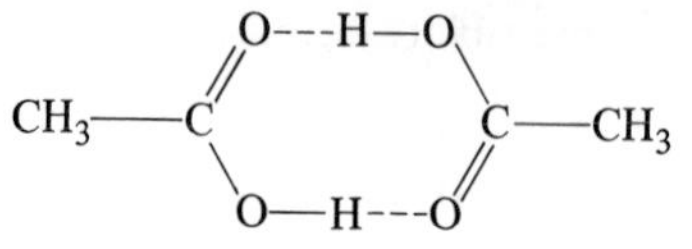

Figure 14-9. Dimeric acetic acid, which persists in the vapor phase, accounting for the small value of $\Delta \mathscr{S}_v$.

all normal liquids is approximately constant and equal to 92 J-mol^{-1}-K^{-1}. Since (Eq. 10-95 and Eq. 10-71)

$$\Delta \mathscr{G} = \Delta \mathscr{H} - T\Delta \mathscr{S} = 0$$

for a system at equilibrium at constant temperature and pressure, Trouton's rule states that

$$\Delta \mathscr{S}_{vap} = \Delta \mathscr{H}_{vap}/T = \text{constant} \tag{14-12}$$

This empirical rule is consistent with lattice theories of the liquid state, which consider a molecule in the liquid confined to a small volume by the repulsive forces of its neighbors. Gas molecules, on the other hand, are free to move in the entire volume of the container. If we use Eq. 8-94 for the entropy of both liquid and gas, and the partition function of Eq. 8-85, with the only difference between the two phases the volume available to the molecule, for one mole

$$\Delta \mathscr{S}_{vap} = R \ln \frac{\mathscr{V}_{gas}}{\mathscr{V}_{liq}} \tag{14-13}$$

Exercise 14-3. Verify Eq. 14-13.

Substitution of Trouton's constant in Eq. 14-13 leads to a ratio of V_{gas}/V_{liq} of 6×10^4. Since the volume of one mole of gas is approximately 20×10^{-3} m^3, this gives a value for $\mathscr{V}_{liq}$ of approximately 0.3 cm^3. This result is not unreasonable, if it is considered to be the free volume available to the molecule in its "cell", rather than the total volume.

Table 14-5 shows values of $\Delta \mathscr{S}_{vap}$ at 1.01×10^5 Pa pressure for a number of substances.

Hildebrand noticed that $\Delta \mathscr{S}_{vap}$ showed less variation if it were calculated at constant volume of vapor rather than at constant pressure. His results are shown in the last column of Table 14-5, and are consistent with the lattice model calculation. The most striking deviations are shown by liquids such as water, with $\Delta \mathscr{S}_{vap}$ equal to 108 J-mol^{-1}-K^{-1} at 373 K and 1.01×10^5 Pa.

It is not surprising that the entropy of vaporization is greater than normal for a liquid in which highly directional molecular interactions, such as the hydrogen bond, are important. The low value of 57.7 J-mol^{-1}-K^{-1} for $\Delta \mathscr{S}_{vap}$ of acetic acid is explained by the persistence of hydrogen-bonded dimers (as in Fig. 14-9) in the vapor phase.

Table 14-5. Entropy of vaporization at constant pressure and at constant volume for some nonpolar liquids[a]

Substance	T_B/K	$\Delta \mathscr{S}_{vap}/(J\text{-mol}^{-1}\text{-K}^{-1})$; $(P = 1.33 \times 10^4 \text{ Pa})$	$\Delta \mathscr{S}_{vap}/(J\text{-mol}^{-1}\text{-K}^{-1})$; $(\mathscr{V} = 19.7 \text{ dm}^3\text{-mol}^{-1})$
N_2	79	91.5	114.7
O_2	97	94.8	114.7
Cl_2	235	112.2	115.6
Pentane	307	109.7	112.2
Hexane	341	110.6	113.1
CCl_4	347	111.4	112.2
Benzene	351	113.9	113.9
$SnCl_4$	385	113.1	113.1
Hg	627	112.2	108.9
Cd	935	123.1	109.7
Zn	1188	125.5	109.7

[a] J. H. Hildebrand, *J. Am. Chem. Soc.*, *37*, 970–78 (1915).

Surface tension
It is well known that liquids have a surface that is independent of the size of the container and that droplets of liquid tend to acquire a spherical shape. This behavior is explained by ascribing a *tension* to the surface that minimizes the surface area. A simple demonstration of surface tension can be made by enclosing a film of liquid in a rectangular frame with three fixed sides and one sliding side, as shown schematically in Fig. 14-10. For a given value of w, the force **F** required to stretch the film is found to be independent of l. The work done in stretching (extending) the film is

$$\delta W = \mathbf{F}_{ex} \cdot \delta \mathbf{l}$$

$$= F_{ex} \, \delta l \tag{14-14}$$

If we define the *surface tension* γ as the force per unit length of line normal to the direction of the force, then

$$\gamma = \frac{F_{ex}}{2w} \tag{14-15}$$

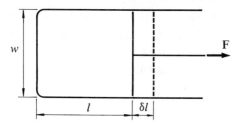

Figure 14-10. A model illustrating the surface tension of a liquid film.

since the film has two sides, and

$$\delta W = \gamma(2w\,\delta l)$$
$$= \gamma\delta A \qquad (14\text{-}16)$$

where δA is the infinitesimal change in area on stretching the film. From Eq. 10-78, for a reversible process,

$$\delta G = \delta W$$

so that

$$\delta G = \gamma\delta A$$

and

$$\gamma = \left(\frac{\partial G}{\partial A}\right)_{T,P} \qquad (14\text{-}17)$$

Thus, the surface tension is the rate of increase of free energy with increasing area, and the equilibrium state of minimum free energy is the state of minimum area. The units of surface tension are newtons per meter or joules per square meter. From a molecular point of view, it is reasonable that molecules on the surface, attracted to other molecules in only one direction, should be in a higher energy state than molecules in the interior of the liquid attracted to neighbors on all sides.

As a result of surface tension there is a difference of pressure on the two sides of a curved liquid surface.[16] We shall consider only the simple case of a spherical surface. A hemispherical surface of radius R, with center at the origin and equator in the x-y plane, is shown in Fig. 14-11. Let P' be the pressure outside the surface and P'' be the pressure inside the surface. The net pressure on the surface is $P'' - P'$, and the magnitude of the net force acting normal to the element of area δA is

$$\delta F = (P'' - P')\,\delta A$$

16. P. Laplace, *Oeuvres*, 4, 364 (1880); F. G. Donnan, *Phil. Mag.*, 3, 305 (1902).

The z component of the force is

$$\delta F_z = (P'' - P') \cos \theta \, \delta A \qquad (14\text{-}18)$$

The total force in the z direction is obtained by integrating Eq. 14-18 over the angle ϕ from 0 to 2π and over the angle θ from 0 to $\pi/2$, noting that δA is equal to $R^2 \sin \theta \, d\theta \, d\phi$ (see Exercise 2-9).

Thus,

$$F_z = \int_0^{\pi/2} \int_0^{2\pi} (P'' - P') \cos \theta \, R^2 \sin \theta \, d\theta \, d\phi \qquad (14\text{-}19)$$

$$= \pi R^2 \, (P'' - P') \qquad (14\text{-}20)$$

Exercise 14-4. Verify Eq. 14-20.

From the symmetry of the surface about the z axis, the components in the x-y plane of the force due to the pressure difference must add to zero. Thus, at equilibrium, F_z must be equal to the force of the surface tension acting along the circumference of the hemisphere, $2\pi R \gamma$.

Therefore,

$$\pi R^2 \, (P'' - P') = 2\pi R \gamma$$

Figure 14-11. Geometry of a hemispherical surface of radius R, with the equator in the x-y plane and the center at the origin. The element of area δA is equal to $R^2 \sin \theta \, \delta \theta \, \delta \phi$.

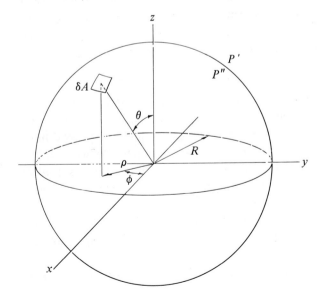

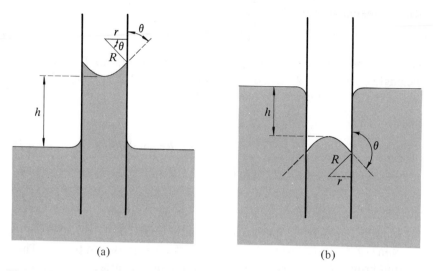

Figure 14-12. (a) Capillary rise due to surface tension when $\theta < \pi/2$. (b) Capillary depression due to surface tension when $\theta > \pi/2$.

and

$$(P'' - P') = \frac{2\gamma}{R} \tag{14-21}$$

When a capillary is inserted into a sample of liquid, most liquids rise in the capillary (mercury is a prominent exception; the level of mercury in a capillary is depressed). Whether liquids rise or fall in a capillary is determined by the contact angle θ between the gas-liquid interface and the solid-liquid interface.

The contact angle of the liquid in Fig. 14-12, a, which wets the glass surface, is less than 90°, the liquid rises in the capillary, and the meniscus is concave upward. The contact angle of the liquid in Fig. 14-12, b, which does not wet the glass surface, is greater than 90°, the liquid is depressed in the capillary, and the meniscus is concave downward. This behavior can be related to the pressure difference across a curved surface given in Eq. 14-21, and the relationship can be used to calculate the surface tension of a liquid from the measured value of capillary rise.[17]

If the pressure at the surface of the bulk liquid in Fig. 14-12, a is P_0, then the pressure in the air just above the liquid meniscus is $P_0 - h\rho'g$, where ρ'

17. G. E. Kimball, in H. S. Taylor and S. Glasstone, *A Treatise on Physical Chemistry*, 3rd ed., Vol. II, D. Van Nostrand Co., New York, 1951, pp. 415–16.

is the density of the air and g is the gravitational acceleration. Similarly, the pressure in the liquid just below the meniscus is $P_0 - h\rho g$, where ρ is the density of the liquid. Thus, the difference in pressure between the inside and outside of the meniscus is

$$P'' - P' = (\rho - \rho')hg \qquad (14\text{-}22)$$

When we equate the right sides of Eq. 14-21 and Eq. 14-22 and rearrange, we obtain

$$\gamma = \frac{(\rho - \rho')hgR}{2} \qquad (14\text{-}23)$$

Since it is easier to measure the radius of the capillary than the radius of curvature of the meniscus, and since $r = R \cos \theta$,

$$\gamma = \frac{(\rho - \rho')hgr}{2 \cos \theta} \qquad (14\text{-}24)$$

When θ is less than $90°$, then, h is positive, and, when θ is greater than $90°$, h is negative. For most liquids that wet glass, such as water, θ is equal to $0°$, and Eq. 14-24 can be written as

$$\gamma = \frac{(\rho - \rho')hgr}{2} \qquad (14\text{-}25)$$

Similarly, θ is equal to $180°$ for mercury, and here

$$\gamma = \frac{-(\rho - \rho')hgr}{2} \qquad (14\text{-}26)$$

Table 14-6 shows the surface tensions of a variety of liquids. It can be seen that liquids with weak intermolecular forces have a relatively low surface tension, whereas those liquids with strong intermolecular forces have a high surface tension.

Table 14-6. Surface tension of some liquids

Substance	T/K	$\gamma/(\text{N-m}^{-1})$
H_2O	273	0.076
C_2H_5OH	273	0.023
Benzene	273	0.032
Hexane	273	0.021
Hg	273	0.478
KCl	1073	0.096
AgCl	723	0.125
Pb	598	0.510

J. R. Partington, *An Advanced Treatise on Physical Chemistry*, Vol. II, Longman Group, Ltd., London, 1951, pp. 192–94.

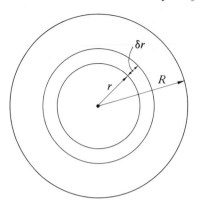

Figure 14-13. Geometry for the derivation of Poiseuille's law for the flow of a liquid through a cylindrical tube.

The theoretical description of the surface tension of liquids has been attempted by Kirkwood and Buff, Lennard-Jones and Corner, and Eyring and Jhon, but these treatments are beyond the scope of this text.[18]

14-3. The Viscosity of liquids

Though liquids, like gases, are fluids and show no resistance to deformation, the incompressibility of liquids leads to quite different flow properties from those of gases. Equation 2-39, Newton's law of viscous flow, does apply to liquids as well as gases. The most common method of measuring the viscosity of liquids is by timing the flow of a liquid through a capillary. The theory of this method was developed in 1840 by Poiseuille, a French physician interested in the flow of blood in the circulatory system.

Consider a liquid flowing through a cylindrical tube of radius R, whose geometry is shown schematically in Fig. 14-13. The liquid at the center is moving at some maximum velocity under the influence of a pressure drop ΔP, while the layer of liquid adjacent to the wall of the cylinder is at rest. The force required to maintain the cylindrical shell of liquid between r and $r + \delta r$ at a velocity v over its whole length l is, from Eq. 2-39,

$$f = \eta(2\pi rl) \frac{dv}{dr} \tag{14-27}$$

18. J. G. Kirkwood and F. P. Buff, *J. Chem. Phys.*, *17*, 338 (1949); J. E. Lennard-Jones and J. Corner, *Trans. Faraday Soc.*, *36*, 1156 (1940); M. S. Jhon and H. Eyring, in *Physical Chemistry, an Advanced Treatise*, Vol. VIII A, (D. Henderson, ed.), Academic Press, 1971, pp. 365–68.

This force is equal and opposite in sign to the pressure drop across the tube multiplied by the cross-sectional area;

$$f = - \Delta P(\pi r^2) \tag{14-28}$$

Equating the two expressions for the force, and rearranging, we obtain

$$\delta v = -\frac{r \Delta P}{2l\eta} \delta r \tag{14-29}$$

On integration, the result is

$$v = -\frac{(\Delta P) r^2}{4l\eta} + \text{constant} \tag{14-30}$$

Since $v = 0$ when $r = R$, the constant of integration is $\Delta P R^2 / 4l\eta$, and Eq. 14-30 can be written as

$$v = \frac{\Delta P}{4l\eta} (R^2 - r^2) \tag{14-31}$$

The volume of liquid flowing through the tube per unit time dV/dt is obtained by integrating the flow through the cylindrical shell, $2\pi r\, \delta r\, v$, from $r = 0$ to $r = R$. The result is

$$\frac{dV}{dt} = \int_0^R \frac{2\pi \Delta P}{4l\eta} (R^2 - r^2)\, r\, dr$$

$$= \frac{\pi \Delta P R^4}{8l\eta} \tag{14-32}$$

which is *Poiseuille's law*.

Exercise 14-5. Verify Eq. 14-32.

Though absolute values of viscosity can be obtained by measuring the rate of flow of liquids through capillaries of known dimensions, it is more common to make relative measurements, with a liquid of known viscosity as a standard. Figure 14-14 shows a Cannon-Fenske viscometer, commonly used for this purpose. A fixed volume of liquid is inserted into the left tube of the viscometer and drawn into the right tube above mark a by suction or pressure. The liquid is then allowed to flow down, and the time required for the meniscus to pass from a to b is recorded. Since the volumes of different liquids used is kept constant, the height of the liquid head during flow from a to b is the same for each liquid, and the value of ΔP is proportional to the

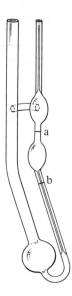

Figure 14-14. A Cannon-Fenske viscometer.

density ρ. The flow time for each liquid is inversely proportional to the velocity of flow. The values of l and R are constant for a given viscometer. Given the flow times and densities of two liquids, their viscosities are related as

$$\frac{\eta_2}{\eta_1} = \frac{t_2\rho_2}{t_1\rho_1}$$

(14-33)

The Couette viscometer measures directly the force required to maintain a velocity gradient between concentric cylinders. This method also has the advantage that the velocity gradient can be varied in order to test the applicability of Eq. 2-39.

A third method, which is most useful for liquids of high viscosity, makes use of Stoke's law: the frictional force on a sphere of radius r falling through a liquid of viscosity η with a velocity v is

$$f = 6\pi\eta vr$$

(14-34)

When a steady velocity has been reached, the frictional force is equal to the gravitational force on the sphere

$$f_g = \frac{4}{3}\pi r^3 g \left(\rho - \rho_{\text{liq}}\right)$$

(14-35)

where ρ is the density of the sphere and ρ_{liq} is the density of the liquid. Thus,

$$\eta = \frac{2}{9} \frac{gr^2 (\rho - \rho_{liq})}{v} \tag{14-36}$$

The mechanism of viscous resistance in liquids is quite different from that in gases. As we pointed out in Sec. 2-4, viscosity in gases is the result of transfer of momentum across a velocity gradient by random motion of the molecules. In liquids, resistance to flow is due to the difficulty of moving molecules past close-packed neighbors. Eyring and his co-workers have applied the transition state model of reaction rates to the process of viscous flow. They visualize the motion of a molecule in the liquid from one equilibrium site of low energy through an intermediate transition state of higher energy and then to a new equilibrium state. Thus the process involves an activation energy, and they predict that the temperature dependence of the viscosity is given by a relation of the form

$$\eta = A \exp\left(\frac{\Delta\mathscr{E}_{vis}}{RT}\right) \tag{14-37}$$

It has since been found that the logarithm of the viscosity is to a good approximation a linear function of $1/T$ and that the values of ΔE_{vis}, the activation energy of viscous flow, vary from one-half to one-third the energy of vaporization for many substances. These results indicate that the holes that must be formed in the process of viscous flow are smaller than molecular dimensions, since energy equal to the energy of vaporization is required to form a hole of molecular size.

14-4. Liquid water [19]

The most thoroughly studied liquid is water. It is abundant and has interesting properties from a theoretical point of view. Everyone knows the human body is about 65% water; fewer are aware that the mass of water in the earth's crust present as water of hydration in crystals is one-half as great as the mass of water in the oceans.

Water vapor molecules are mutually attracted by all the forces discussed in Sec. 13-2 in connection with molecular crystals. In ice, each molecule is hydrogen-bonded to four other water molecules. When ice sublimes, two hydrogen bonds are broken per molecule. Thus the energy of a hydrogen

19. D. Eisenberg and W. Kauzmann, *The Structure and Properties of Water*, Clarendon Press, Oxford, 1969.

bond in ice is approximately equal to $\frac{1}{2}$ ($\Delta\mathscr{H}_s$). Other interactions, dispersion and repulsion, are also involved when ice sublimes, and there is disagreement about their magnitude, but the hydrogen bond energy seems close to the 21 kJ-mol^{-1} noted in Table 14-7. It is easy to see from the enthalpy change on fusion $\Delta\mathscr{H}_f = 6$ kJ-mol^{-1} that about 30% of the hydrogen bonds in ice can be broken when ice melts. A comparison of the electronic, vibrational, and rotational energies in Table 14-7 with the value 2.48×10^3 J-mol^{-1} for RT at 298 K indicates clearly that water molecules at this temperature are largely in the lowest vibrational state, but that a number of rotational levels must be significantly populated.

Table 14-7. Some properties of the water molecule

Energy of formation from atoms at 0 K	-9.1771×10^5 J-mol^{-1}
$\Delta\mathscr{E}f_0^{\ominus} = \Delta\mathscr{H}f_0^{\ominus}$	-2.3891×10^5 J-mol^{-1}
Bond length, $O-H$	$0.9572 \pm 0.0003 \times 10^{-10}$ m
Bond angle, $H-O-H$	$104.52° \pm 0.05°$
Normal modes of vibration	
Symmetrical stretching	$\nu = 1.0970 \times 10^{14}$ s^{-1}
Asymmetrical stretching	$\nu = 1.1267 \times 10^{14}$ s^{-1}
Bending	$\nu = 4.7844 \times 10^{13}$ s^{-1}
Hydrogen bond energy	~ 21 kJ-mol^{-1}
$\Delta\mathscr{H}_f$	6 kJ-mol^{-1}
Potential energy of interaction at 5×10^{-10} m	
Dipole-dipole	-7.1×10^2 J-mol^{-1}
Dipole-induced–dipole	-40 J-mol^{-1}
London forces	-3×10^2 J-mol^{-1}
Excitation energy	
First excited electronic state	9.65×10^5 J-mol^{-1}
Lowest vibrational excitation	1.91×10^4 J-mol^{-1}
Rotational excitation	$\sim 5 \times 10^2$ J-mol^{-1}

The density increase of about 10% that is observed when ice melts is assumed to be due to the collapse of the open tetrahedral lattice of the hydrogen-bonded network. The resulting liquid water is more closely packed than ice, but it is at a higher potential energy, presumably as a result of the breaking of orientation-dependent hydrogen bonds. X-ray diffraction measurements indicate that the minimum distance between liquid water molecules is 0.29 nm and that no order exists beyond 0.8 nm. At a distance of 0.3 nm, there are 4.4 nearest neighbors at 274 K and 4.9 at 356 K. Thus, a tetrahedral geometry about water molecules is retained in the liquid state, but with some collapse of the open structure. The density continually increases with temperature from 273 to 277 K, indicating that collapse of the ice structure contributes more than thermal vibration over this temperature

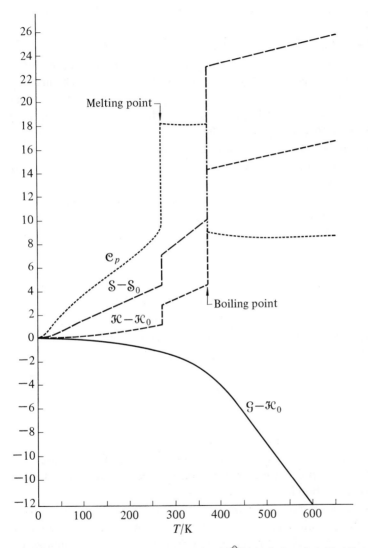

Figure 14-15. The molar heat capacity $[\mathscr{C}_p^{\ominus}/(4.184)\ \text{J-mol}^{-1}\text{-K}^{-1})]$ ($\cdots$), enthalpy $[(\mathscr{H}^{\ominus} - \mathscr{H}_0^{\ominus})/(4.184\ \text{J-mol}^{-1})]$ (---), entropy $[\mathscr{S}^{\ominus} - \mathscr{S}_0^{\ominus})/(8.368\ \text{J-mol}^{-1}\text{-K}^{-1})]$ (---), and Gibbs free energy ($\mathscr{G}^{\ominus} - \mathscr{H}_0^{\ominus})/(4.184\ \text{J-mol}^{-1})$ (——), of solid, liquid, and gaseous water as a function of temperature. (By permission, from D. Eisenberg and W. Kauzmann, *The Structure and Properties of Water*, Clarendon Press, Oxford, 1969.)

513

range to the structural characteristics of water. The instantaneous positions of the molecules cannot be obtained from X-ray diffraction data, since the time required for measurements is long compared to the speed of molecular motion in the liquid. The radial distribution function gives a time-average estimate of molecular positions and no information about the position of hydrogen atoms.

Figure 14-15 shows the molar thermodynamic properties $\mathscr{C}_{p}^{\ominus}$, $\mathscr{S}^{\ominus} - \mathscr{S}_{0}^{\ominus}$, $\mathscr{H}^{\ominus} - \mathscr{H}_{0}^{\ominus}$, and $\mathscr{G}^{\ominus} - \mathscr{H}_{0}^{\ominus}$ as a function of temperature for ice, liquid water, and steam at 1.01×10^{5} Pa pressure. Between 0 K and 80 K, the increase in the heat capacity of ice is due almost entirely to a hindered translational motion. Between 150 K and the melting point, hindered rotation (usually called *libration*) also contributes significantly.

Though the heat capacity of ice and steam can be accounted for in terms of thermal excitation of mechanical degrees of freedom, such as the hindered translation and rotation in ice or the translational, rotational, and vibrational motions in steam, the heat capacity of liquid water is too large to be accounted for in this way. The large heat capacity of liquid water is explained by assuming that there is a change in the structure of the liquid with temperature in addition to a change in the excitation of energy levels corresponding to mechanical degree of freedom.

Of the many models that have been proposed for the structure of water, two different classes are still considered. One class treats the structure change that accounts for the heat capacity as a transition among discrete species, such as a two-state transition between hydrogen-bonded and free water molecules, between bulky species and dense species, or between lattice and interstitial species. The most complex of the discrete state models is that of Nemethy and Scheraga; they divide water molecules into classes with 0, 1, 2, 3, or 4 hydrogen bonds. All these models adjust a number of parameters to produce a best fit to the thermodynamic data; they are not consistent with spectroscopic data, however, which indicate a broad continuum of environments for water molecules in the liquid rather than discrete states. In particular, there is no experimental criterion for deciding what constitutes a broken hydrogen bond.

The other class of models assumes that all water molecules have four nearest neighbors at all temperatures but that the hydrogen bonds are distorted to various degrees that result in apparent increases in the number of nearest neighbors and in the increase in density between 273 and 277 K. They are consistent with the spectroscopic data that indicate a broad continuum of environments in liquid water. The distorted hydrogen bonds they postulate have been observed in the forms of ice that are stable at high pressure (Fig. 15-9).

Problems

14-1. Start with the data in Column 3 of Table 14-5 and calculate the entropy of vaporization at a constant volume of 19.7 dm³, assuming that the vapors can be described as ideal gases.

14-2. Simple models frequently are adequate for the calculation of several properties to at least the correct order of magnitude. Consider the example of CCl_4: $\rho = 1.6$ kg-dm^{-3}; $M = 0.154$ kg-mol^{-1}; $\Delta \mathscr{H}_v = 30$ kJ-mol^{-1} at its boiling point, 350 K. (a) Assume a model of closest-packed spherical molecules, with 26% of the volume "empty" space. Compute the radius r of a CCl_4 molecule from the density and molar mass. (b) A molecule in closest-packed structure has a co-ordination number of 12. Therefore, the number of nearest neighbor pairs in a mole of crystal is 6 L. Estimate ε in the Lennard-Jones potential from $\Delta \mathscr{H}_{vap}$. (c) Calculate the constant a in the Lennard-Jones potential for CCl_4. (d) Plot the three terms U, $-a/r^6$, and b/r^{12}, on the same graph. (e) Estimate the collision diameter from the Lennard-Jones parameters and compare with the result of a.

14-3. From the following data on the viscosity of liquids as a function of temperature, calculate $\Delta \mathscr{E}_{vis}$ for each liquid (at several different temperatures if appropriate) and compare to values of $\Delta \mathscr{E}_{vap}$ for each liquid (at the same temperature if $\Delta \mathscr{E}_{vis}$ varies with temperature).

(a) H_2O

T/K	$\eta/(10^{-3}$ Pa-s)	T/K	$\eta/(10^{-3}$ Pa-s)
264	2.55	283	1.307
265	2.46	293	1.002
266	2.43	303	0.7975
267	2.25	313	0.6529
268	2.12	323	0.5468
271	1.93	333	0.4665
273	1.787	343	0.4042

R. C. Weast (ed.), Handbook of Chemistry and Physics, 54th ed. CRC Press, Cleveland, 1973, pp. F43 to F51. Copyright © CRC Press, 1973. Used by permission of CRC Press.

(b) C_6H_6

T/K	$\eta/(10^{-3}$ Pa-s)	$P/1.01 \times 10^5$ Pa
363	0.286	2.38
373	0.261	2.72
398	0.210	4.08
423	0.171	6.8
448	0.144	10.2
473	0.121	15.3
498	0.100	21.4
523	0.082	30.3
548	0.065	43.2
553	0.062	44.2
558	0.059	47.6
561.5 (T_c)	0.056	49.0

Landolt-Börnstein, *Zahlenwerte und Functionen*, 6th ed., Vol. II, Part 5a, Springer-Verlag, Berlin, 1969, p. 158; J. R. Heiks and E. Orban, *J. Phys. Chem.*, 60, 1025 (1956).

(c) $NaNO_3$

T/K	$\eta/(10^{-3}$ Pa-s$)$	T/K	$\eta/(10^{-3}$ Pa-s$)$
588	2.92	648	2.13
589	2.90	654	2.05
593	2.82	660	2.00
599	2.72	666	1.95
603	2.64	679	1.84
608	2.56	685	1.79
613	2.52	703	1.66
623	2.38	721	1.56
631	2.28	732	1.51
640	2.20	743	1.44
643	2.17		

Landolt-Bornstein, *Zahlenwerk und Functionen*, 6th ed., Vol. II, Part 5a, Springer-Verlag, Berlin, 1969, p. 142; P. I. Procenko and O. N. Razumowskaya, *Zurn., prikl. khim.*, 38, 2355 (1965); 40, 2576 (1967).

(d) Hg

T/K	$\eta/(10^{-3}$ Pa-s$)$	T/K	$\eta/(10^{-3}$ Pa-s$)$
273	1.661	371	1.263
290	1.572	411	1.168
292	1.558	452	1.106
293	1.547	466	1.079
308	1.476	501	1.035
336	1.360	535	1.005
353	1.299	572	0.975

Landolt-Bornstein, *Zahlenwerk und Functionen*, 6th ed., Vol. II, Part 5a, Springer-Verlag, Berlin, 1969, p. 125; M. Pluss, *Z. anorg. Chem.*, 93, 1 (1915).

(e) C_6H_6

T/K	$\eta/(10^{-3}$ Pa-s$)$	$P/1.01 \times 10^5$ Pa	$\mathcal{V}/(dm^3 \cdot mol^{-1})$
303	6.15	120	*0.0890
323	5.97	420	0.0886
	5.75	370	0.0890
	5.30	240	0.0900
	4.80	120	*0.0912
343	4.97	435	0.0900
	4.62	335	*0.0912
	4.16	200	0.0923
363	3.87	320	*0.0923
	3.31	97	0.0948

A. Jobling and A. S. C. Lawrence, *Proc. Roy. Soc., London*, A206, 257 (1951).

Obtain a value for $\Delta\mathcal{E}_{vis}$ from the data marked with asterisks (approximately at constant volume), and compare with the lov ɔr temperature data in (b). Other data in this problem are at constant pressure. Which condition should yield a more meaningful value of $\Delta \mathscr{E}_{vis}$?

14-4. The following data were obtained for the density of liquid benzene and its saturated vapor as a function of temperature. [J. Timmermans, *Physico-chemical Constants of Pure Organic Compounds*, Elsevier Publishing Co., Amsterdam, 1950, p. 141; S. Young, *J. Chem. Soc.* (*London*), 77, 1145 (1900); *81*, 707, 777 (1902); *Sci. Proc. Roy. Soc. Dublin*, N. S. XII, 374 (1909–1910).] Plot ρ_1, ρ_v, and $(1/2)(\rho_1 + \rho_v)$ as a function of temperature on the same graph and calculate the critical density, at 561.7 K.

	$\rho/(\text{kg-dm}^{-3})$				$\rho/(\text{kg-dm}^{-3})$	
T/K	Liquid	Vapor	T/K		Liquid	Vapor
273	0.90006	0.00012	433		0.7185	0.01734
283	0.8895	0.0002	443		0.7043	0.02087
293	0.8790	0.0004	453		0.6906	0.02487
303	0.8685	0.0006	463		0.6758	0.02977
313	0.8576	0.0008	473		0.6605	0.03546
323	0.8466	0.0011	483		0.6432	0.04207
333	0.8357	0.0015	493		0.6255	0.05015
343	0.8248	0.002040	503		0.6065	0.05977
353	0.8145	0.002732	513		0.5851	0.07138
363	0.8041	0.003610	523		0.5609	0.08554
373	0.7927	0.004704	533		0.5328	0.1038
383	0.7809	0.006042	543		0.4984	0.1287
393	0.7692	0.007675	553		0.4514	0.1660
403	0.7568	0.009851	557.5		0.4213	—
413	0.7440	0.01176	559.3		0.4078	—
423	0.7310	0.01437	561.2		0.3856	—

14-5. The following data [P. W. Bridgman, *Proc. Natl. Acad. Sci., U.S.*, *11*, 603 (1925).] are for the viscosity of H_2O as a function of pressure at two different temperatures. Explain the difference between the two sets of data on the basis of some reasonable model of the structure of water.

$P/10^6$ Pa	$\eta_{rel}(H_2O)$	
	273 K	348 K
0.098	1.000	0.222
49	0.938	0.230
98	0.921	0.239
147	0.932	0.247
196	0.957	0.258
294	1.024	0.278
392	1.111	0.302
490	1.218	0.333
588	1.347	0.367
686	—	0.404

15 · *Phase equilibria and the phase rule*

This chapter describes the thermodynamics of equilibria among solid, liquid, and gaseous phases. The treatment is primarily macroscopic and phenomenological.

15-1. The phase rule

The *phase rule* was first derived by J. Willard Gibbs in the paper. "On the Equilibrium of Heterogeneous Substances", in which he laid the basis for classical thermodynamics.[1] It is an orderly scheme for the description of multiphase, multicomponent systems at equilibrium. Emphasis here is on graphical description and the use of *phase diagrams*.

Definitions

A *phase* is a homogeneous part of a system with properties distinct from the properties of other parts. The *number of phases*, P, is independent of the number of separate *regions* with identical properties; for example, an ice-water system has two phases independent of the number of pieces of ice in the system.

When no chemical reactions occur, the *number of components*, C is the number of chemical species in the system.

The *number of degrees of freedom*, F, is the minimum number of inde-

1. J. W. Gibbs, *Trans. Conn. Acad. Sci.*, III, 108–248, October, 1875–May, 1876, and 345–524, May 1877–July 1878; *Collected Works*, Vol. I, Yale University Press, New Haven, 1928, pp. 55–353.

pendent intensive variables the value of which must be specified to fix the values of all the other intensive variables. The value of F is also the *maximum number of intensive variables* that can be varied, within limits, without the appearance or disappearance of a phase. Even a one-phase, one-component system can be described by a large number of intensive variables, such as density, refractive index, viscosity, temperature, pressure, and molar heat capacity. Yet we know that fixing the values of any two of these variables will determine the values of all the others. The phase rule provides a method for computing the value of F for any multicomponent, multiphase system and is expressed in the equation

$$F = C - P + 2 \qquad (15\text{-}1)$$

Conditions of equilibrium
Mechanical equilibrium. Consider a system of fixed total volume and uniform temperature throughout. According to Eq. 10-64, the condition of equilibrium for such a system is

$$\delta A = 0$$

If phase I of the system changes its volume, with a concurrent change in the volume of phase II,

$$\delta A = \delta A_{\mathrm{I}} + \delta A_{\mathrm{II}}$$

Since the temperature is uniform (Eq. 10-89),

$$\delta A_{\mathrm{I}} = -P_{\mathrm{I}} \delta V_{\mathrm{I}}$$

and

$$\delta A_{\mathrm{II}} = -P_{\mathrm{II}} \delta V_{\mathrm{II}}$$

Thus,

$$\delta A = -P_{\mathrm{I}} \delta V_{\mathrm{I}} - P_{\mathrm{II}} \delta V_{\mathrm{II}} = 0 \qquad (15\text{-}2)$$

Since the total volume is fixed,

$$\delta V_{\mathrm{II}} = -\delta V_{\mathrm{I}}$$

and

$$P_{\mathrm{II}} \delta V_{\mathrm{I}} - P_{\mathrm{I}} \delta V_{\mathrm{I}} = 0 \qquad (15\text{-}3)$$

The condition of Eq. 15-3 can be met only if $P_{\mathrm{II}} = P_{\mathrm{I}}$, a condition of *mechanical equilibrium*. As pointed out in Sec. 14-2, this condition does not apply if the phase boundary is curved and if the radius of curvature is sufficiently small. In that case, the surface properties must be considered to be additional variables of the system. Also, if a phase boundary is constrained to remain fixed ($\delta V_\mathrm{I} = \delta V_\mathrm{II} = 0$) as in an osmotic pressure experiment (see Sec. 16-5), the pressure on two phases in equilibrium can differ.

Thermal equilibrium. Consider a system at equilibrium at fixed energy and volume in which an infinitesimal amount of heat δQ is transferred from phase I to phase II. For a system at equilibrium, such a change must be reversible, and (Eq. 10-38 and Eq. 10-27)

$$\delta S = \delta S_\mathrm{I} + \delta S_\mathrm{II}$$

$$= -\frac{\delta Q}{T_\mathrm{I}} + \frac{\delta Q}{T_\mathrm{II}} = 0 \tag{15-4}$$

The condition of Eq. 15-4 can be met only if $T_\mathrm{I} = T_\mathrm{II}$, a condition of *thermal equilibrium*.

Equilibrium for transfer of matter between phases. Consider a system at equilibrium at constant temperature and pressure in which δn moles of a substance are transferred from phase I to phase II. The change in the Gibbs free energy, which must be zero at equilibrium (Eq. 10-70), is given by (Eq. 11-4),

$$\delta G = -\mu_\mathrm{I}\, \delta n + \mu_\mathrm{II}\, \delta n = 0 \tag{15-5}$$

where μ_I and μ_II are the chemical potentials of the substance in phase I and phase II, respectively. The condition of Eq. 15-5 can be met only if $\mu_\mathrm{I} = \mu_\mathrm{II}$, a condition of *transfer equilibrium*.

Derivation of the phase rule

The minimum number of intensive variables needed to specify the state of a phase containing C components is $C + 1$, the $C - 1$ mole fractions required to specify the composition of the phase plus the temperature and pressure. Therefore, for P phases there are $P(C + 1)$ such variables. From the conditions for mechanical, thermal, and transfer equilibrium there are $(C + 2)(P - 1)$ independent relationships that limit the variation of these variables.

Exercise 15-1. Verify the statements above for a system of two phases, I and II and two components, α and β, by listing the variables and the relationships.

The number of degrees of freedom of the system is the difference between the number of intensive variables and the number of relationships that limit their variation. Thus,

$$F = P(C + 1) - (C + 2)(P - 1)$$

$$= C - P + 2 \tag{15-6}$$

When a component exists in some phases but not others, there is one composition variable fewer in the number of intensive variables and one transfer equilibrium relationship fewer in the number of relationships for every phase from which it is absent. The value of F is thus independent of such a circumstance.

Chemically reacting systems. If a system contains N chemical species, and they can undergo R *independent* chemical reactions, the number of intensive variables is $P(N + 1)$ and the number of relationships is $(N + 2)(P - 1) + R$, since there is an equilibrium constant expression for each chemical reaction. Thus, for this system

$$F = P(N + 1) - [(N + 2)(P - 1) + R]$$

$$= N - R - P + 2 \tag{15-7}$$

Equation 15-7 is consistent with Eq. 15-6, if $C = N - R$. This expression is clearly correct if no reactions are possible, since the number of components in a non-reacting system is defined as the number of chemical species. The general definition of the number of components is the *minimum number of chemical species* from which all phases in the system of arbitrary composition can be prepared. For example, there are three chemical species in the system $CaCO_3(s)$, $CaO(s)$, $CO_2(g)$, but the chemical reaction

$$CaCO_3(s) = CaO(s) + CO_2(g)$$

can take place. The minimum number of species required to prepare all phases in the system is two, which is equal to $N - R$.

Exercise 15-2. Show that two species are required in order to prepare all three phases: $CaCO_3(s)$, $CaO(s)$, and $CO_2(g)$, in arbitrary proportions.

15-2.   One-component systems

The phase diagram
According to Eq. 15-6, the phase rule for a one-component system is

$$F = 3 - P \tag{15-8}$$

Since the minimum number of phases is one, the maximum number of degrees of freedom is two. Similarly, since the minimum value of F is zero, the maximum number of coexistent phases at equilibrium is three. When two phases are present, $F = 1$.

Since $F_{max} = 2$, the value of any other intensive variable can be expressed as a function of two intensive variables. Temperature and pressure are usually chosen as two of the intensive variables, and the molar volume is commonly

Figure 15-1. A P-$\mathscr{V}$-T surface for a substance that contracts on freezing. Here T_1 represents an isotherm below the triple-point temperature, T_2 represents an isotherm between the triple point and the critical point, T_c is the critical temperature, and T_4 represents an isotherm above the critical temperature. Points g, h, and i represent the molar volumes of solid, liquid, and vapor, respectively, in equilibrium at the triple point. Points e and d represent the molar volumes of solid and liquid, respectively, in equilibrium at temperature T_2 and the corresponding equilibrium pressure. Points c and b represent the molar volumes of liquid and vapor, respectively, in equilibrium at temperature T_2 and the corresponding equilibrium pressure. (By permission, from Sears-Salinger, *Thermodynamics, Kinetic Theory, and Statistical Thermodynamics*, Addison-Wesley, Reading, Mass., 1975.)

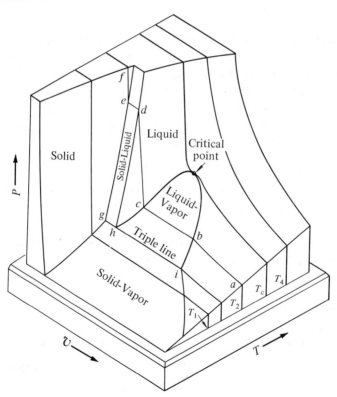

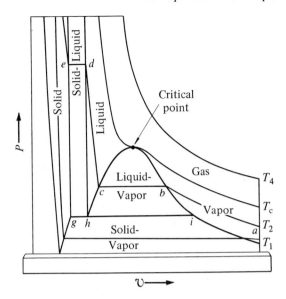

Figure 15-2. A projection of the surface in Fig. 15-1 on the *P-𝒱* plane. Isotherms and points are designated as in Fig. 15-1. (By permission, from Sears-Salinger, *Thermodynamics, Kinetic Theory, and Statistical Thermodynamics,* Addison-Wesley, Reading, Mass., 1975.)

taken as the third. The choice of which two are independent is arbitrary and depends on the experimental situation being described. Any variable that is a function of two other variables can be represented as a surface in three-dimensional space, and the equation of state of a pure substance is frequently represented in this way. A schematic diagram of such a surface is shown in Fig. 15-1. One-phase regions are shown as areas, since two variables are independent. Two-phase regions are represented by their boundary curves; once the value of one variable is fixed, the values of the other two are also determined. Two points, one on each boundary curve, represent the intensive variables for the two phases in equilibrium and each pair of points is joined by a *tie-line.* When three phases coexist, the system is invariant; three points on a line at fixed temperature and pressure represent the intensive variables for the three phases in equilibrium.

It is also convenient to show the projections of the *P-V-T* surface on the three coordinate planes. Figure 15-2 shows the projection of the surface on the *P-V* plane, together with a series of *isotherms* that represent the behavior of the system at several fixed temperatures. One-phase regions are represented by areas and have two degrees of freedom. Along a given isotherm,

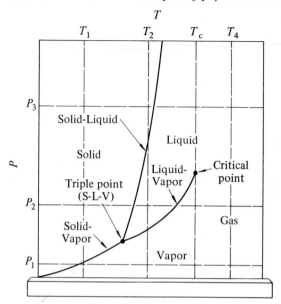

Figure 15-3. A projection of the surface in Fig. 15-1 on the *P-T* plane. (By permission, from Sears-Salinger, *Thermodynamics, Kinetic Theory, and Statistical Thermodynamics* Addison-Wesley, Reading, Mass., 1975.)

either the pressure or the molar volume can be varied freely. The coexistence of two phases is also represented by areas, but such systems have only one degree of freedom. Along any two-phase isotherm, the pressure and molar volume of the two phases are determined by the intersection of the isotherm with the boundary curves of the region. A three-phase system is invariant, with the properties of all three phases determined by the intersection of the *triple-point isotherm* with the solid-liquid, solid-vapor, and liquid-vapor boundary curves. Labeled points correspond to those on Fig. 15-1.

Figures 15-3 and 15-4 show the projections of the *P-V-T* surface on the *P-T* and *V-T* planes. Isotherms are labeled to correspond to the isotherms in Fig. 15-2. The former is the conventional *phase diagram* for a one-component system. Two-phase systems are represented by curves and three-phase regions by a single point, since the pressure is the same for all phases. Though the liquid-vapor equilibrium curve is limited to the span between the *triple point* and the *critical point*, the solid-vapor equilibrium curve extends as close to the origin as such systems have been investigated, and there seems to be no critical point beyond which there is no solid-liquid phase transition. Below the triple point pressure, the liquid phase does not exist at equilibrium, and the solid phase is converted directly to gas by *sublimation*.

Exercise 15-3. Sketch graphs of the volume of the system as a function of pressure along the dotted lines T_1, T_2, T_c, and T_4 in Fig. 15-3.

Exercise 15-4. Sketch graphs of the volume of the system as a function of the temperature along the lines P_1, P_2, and P_3 in Fig. 15-3.

Exercise 15-5. Sketch graphs of the pressure of the system as a function of volume along the lines T_1, T_2, T_c, and T_4 in Fig. 15-4.

Quantitative relationships

At a fixed pressure and temperature, the equilibrium state of a one-component system is determined by the relative values of the chemical potential of the solid, liquid, and gaseous phases. At low temperature the solid phase is stable, so it must have the lowest chemical potential. At constant pressure, the variation of the chemical potential with temperature is given by Eq. 11-36 and Eq. 11-38 as

$$\left(\frac{\partial \mu}{\partial T}\right)_P = -\mathscr{S}$$

where $\mathscr{S}$ is the molar entropy for a pure substance. Since the molar entropy is always positive, the chemical potential of each phase decreases with increasing temperature. From Sec. 8-7 we know that $\mathscr{S}_{gas} > \mathscr{S}_{liq} > \mathscr{S}_{sol}$, so that

$$-\left(\frac{\partial \mu}{\partial T}\right)_{gas} > -\left(\frac{\partial \mu}{\partial T}\right)_{liq} > -\left(\frac{\partial \mu}{\partial T}\right)_{sol} \qquad (15\text{-}9)$$

Figure 15-4. A projection of the surface in Fig. 15-1 on the T-$\mathscr{V}$ plane.

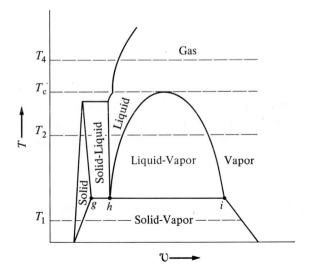

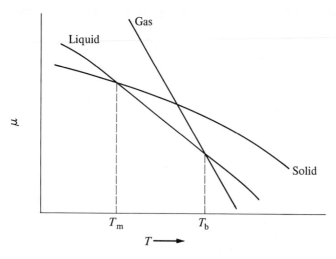

Figure 15-5. Scheme of the variation of the chemical potentials of the solid, liquid, and gaseous phases of a pure substance as a function of temperature above the triple point pressure, showing phase transitions (T_m, melting point; T_b, boiling point) at equilibrium at the intersections between curves.

Figure 15-5 shows schematically the variation of the chemical potential of the solid, liquid, and gaseous phases of a pure substance with temperature at constant pressure. At T_m, the melting point, the chemical potentials of solid and liquid are equal and the two phases are in equilibrium. Between T_m and T_b the liquid phase is the stable phase, and at T_b, the boiling point, liquid and gaseous phases are in equilibrium. Above T_b the gaseous phase is the stable state of the substance.

Exercise 15-6. Sketch curves similar to those in Fig. 15-5 for a pressure below the triple-point pressure.

The Clausius-Clapeyron equation. The liquid vapor curve in Fig. 15-3 represents both the variation with pressure of the *boiling point*—the temperature at which liquid and vapor are in equilibrium, and the variation with temperature of the *vapor pressure* of the liquid—the pressure of the vapor in equilibrium with the liquid. At equilibrium,

$$\mu_{liq} = \mu_{gas}$$

If the temperature is changed by an infinitesimal amount δT, and the system adjusts to a new equilibrium by a change of pressure δP, then for the new equilibrium state

$$\mu_{liq} + \delta\mu_{liq} = \mu_{gas} + \delta\mu_{gas}$$

and

$$\delta\mu_{liq} = \delta\mu_{gas} \tag{15-10}$$

Since each chemical potential is a function of T and P, we may write

$$\left(\frac{\partial\mu_{liq}}{\partial T}\right)_P \delta T + \left(\frac{\partial\mu_{liq}}{\partial P}\right)_T \delta P = \left(\frac{\partial\mu_{gas}}{\partial T}\right)_P \delta T + \left(\frac{\partial\mu_{gas}}{\partial P}\right)_T \delta P$$

But (Eqs. 11-36, 11-38),

$$\left(\frac{\partial\mu}{\partial T}\right)_P = -\mathscr{S}$$

and (Eqs. 11-37, 11-39)

$$\left(\frac{\partial\mu}{\partial P}\right)_T = \mathscr{V}$$

Thus,

$$-\mathscr{S}_{liq}\,\delta T + \mathscr{V}_{liq}\,\delta P = -\mathscr{S}_{gas}\,\delta T + \mathscr{V}_{gas}\,\delta P$$

and

$$\frac{\delta P}{\delta T} = \frac{\mathscr{S}_{gas} - \mathscr{S}_{liq}}{\mathscr{V}_{gas} - \mathscr{V}_{liq}}$$

$$= \frac{\Delta\mathscr{S}_{vap}}{\Delta\mathscr{V}_{vap}}$$

In the limit as $\delta T \to 0$,

$$\frac{dP}{dT} = \frac{\Delta\mathscr{S}_{vap}}{\Delta\mathscr{V}_{vap}} \tag{15-11}$$

An equation of the same form is applicable to solid-gas and solid-liquid equilibria. The general form is called the *Clapeyron-Clausius equation*.

In most phase transitions at constant pressure, there is a zero change in the Gibbs free energy and a finite change in entropy and volume. Since the latter quantities are first derivatives of the free energy, such transitions are called *first-order transitions*, transitions in which a discontinuity appears in the first derivative.

The entropy and volume change of transition have the same sign for both solid to gas and liquid to gas transitions, and the slope of the equilibrium curve is always positive. The entropy change is positive for the solid to liquid

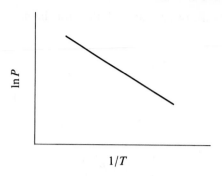

Figure 15-6. Scheme of a plot of the natural logarithm of the vapor pressure of a pure liquid as a function of $1/T$; slope $= -\Delta \mathscr{H}_{vap}/R$.

transition, but the volume change may be positive or negative. The slope of the equilibrium curve depends on the sign of $\Delta \mathscr{V}$, which is positive for most substances but negative for water.

If we substitute for $\Delta \mathscr{S}_{vap}$ from Eq. 10-57 we can write Eq. 15-11 as

$$\frac{dP}{dT} = \frac{\Delta \mathscr{H}_{vap}}{T \Delta \mathscr{V}_{vap}} \tag{15-12}$$

Since the volume of vapor is much greater than the volume of liquid, $\Delta \mathscr{V}_{vap} \approx \mathscr{V}_{gas}$, and

$$\frac{dP}{dT} = \frac{\Delta \mathscr{H}_{vap}}{T \mathscr{V}_{gas}} \tag{15-13}$$

If the vapor is assumed to be ideal, $\mathscr{V}_{gas} = RT/P$, and Eq. 15-13 becomes

$$\frac{dP}{P} = \frac{\Delta \mathscr{H}_{vap}}{RT^2} dT \tag{15-14}$$

Equation 15-14 can be integrated if $\Delta \mathscr{H}_{vap}$ is known as a function of T. To a good approximation, $\Delta \mathscr{H}_{vap}$ is constant over a limited temperature range. With this assumption, Eq. 15-14 can be integrated to yield

$$\ln P = -\frac{\Delta \mathscr{H}_{vap}}{RT} + \text{constant} \tag{15-15}$$

$$\ln \frac{P_2}{P_1} = -\frac{\Delta \mathscr{H}_{vap}}{R} \left[\frac{1}{T_2} - \frac{1}{T_1} \right] \tag{15-16}$$

If $\Delta \mathscr{H}_{vap}$ is constant in the temperature range in which measurements are made, $\ln P$ is a linear function of $1/T$, as shown schematically in Fig. 15-6.

The slope of the line is $-\Delta \mathcal{H}_{vap}/R$. If log P is plotted, the slope is $-\Delta \mathcal{H}_{vap}/(2.303\ R)$.

A more exact treatment can be obtained by introducing the compressibility factor $z = P\mathcal{V}_{gas}/RT$ (Sec. 1-3). Equation 15-13 can be rearranged to

$$\frac{d\ln P}{d(1/T)} = -\frac{T^2}{P}\frac{dP}{dT} \qquad (15\text{-}17)$$

$$= -\frac{T\Delta \mathcal{H}_{vap}}{P\mathcal{V}_{gas}}$$

Exercise 15-7.   Verify Eq. 15-17.

Substituting from the definition of z for $P\mathcal{V}_{gas}/T$, we obtain

$$\frac{d\ln P}{d(1/T)} = -\frac{\Delta \mathcal{H}_{vap}}{Rz} \qquad (15\text{-}18)$$

If ln P is a linear function of $1/T$, then $\Delta \mathcal{H}_{vap}/z$ is a constant. The value of $\Delta \mathcal{H}_{vap}$ at any point in the experimental range can then be determined from an experimental value of z or from a value of z estimated as a function of the reduced pressure and reduced temperature from Fig. 1-9. If a plot of ln P against $1/T$ is curved, the slope at any point is given by Eq. 15-18 and $\Delta \mathcal{H}_{vap}$ can then be calculated as above.

Experimental determination of vapor pressure.   The experimental data to be fitted to Eq. 15-15 or Eq. 15-18 are obtained either by measuring the boiling point as a function of applied pressure or by measuring the vapor pressure as a function of temperature.[2] In measurements of the boiling point, the thermometer bulb is wrapped with absorbent material so that the temperature measured is that of the vapor in equilibrium with refluxing liquid or liquid dripped on to the material from a dropping funnel.

One procedure for measuring the vapor pressure uses the isoteniscope, sketched in Fig. 15-7. The upper end of the apparatus is attached to a vacuum system with a controlled variable pressure. The isoteniscope is placed in a thermostat. After the air has been removed from the space between bulb a

---

2. G. W. Thomson, in A. Weissberger (ed.), *Technique of Organic Chemistry*, Vol. 1, of *Physical Methods of Organic Chemistry*, 3rd ed., Part 1, Chapt. 9, Interscience Publishers, Inc., New York, 1959; F. Daniels, J. W. Williams, P. Bender, R. A. Alberty, C. D. Cornwell, and J. E. Harriman, *Experimental Physical Chemistry*, 7th ed., McGraw-Hill Book Co., New York, 1970, pp. 47–59; D. P. Shoemaker and C. W. Garland, *Experiments in Physical Chemistry*, McGraw-Hill Book Co., New York, 1962, pp. 159–66.

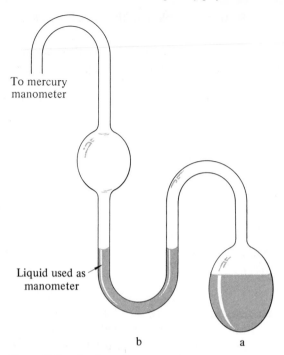

Figure 15-7. An isoteniscope, used to measure the vapor pressure of a liquid as a function of the temperature.

and U-tube b by boiling out the liquid at reduced pressure, the vapor pressure of the liquid in bulb a is measured by finding the pressure of the vacuum system at which the levels of the liquid in the two arms of the U-tube are equal. The measurements can be repeated at other temperatures. This method is most useful for liquids with vapor pressures from about 200 Pa to 1×10^5 Pa.

The *Knudsen method* is used to measure the vapor pressure of relatively nonvolatile solids at high temperatures. A weighed, finely powdered sample of solid is placed in a Knudsen cell, a vessel with a very fine orifice. If the orifice is small compared with the mean free path of the vapor, the number of molecules passing through the orifice into a vacuum is proportional to the vapor pressure. It can be shown from kinetic theory that

$$P = m\left(\frac{2\pi RT}{M}\right)^{\frac{1}{2}} \tag{15-19}$$

where m is the mass of vapor escaping per second. The Knudsen cell is placed in a thermostat and the change of mass of the cell and contents during a measured time interval is determined.

Exercise 15-8. If m is expressed in kg-s^{-1} and R is expressed as J-K^{-1}-mol^{-1}, show that P is in N-m^{-2} (Pa).

Dependence of vapor pressure on pressure. Most measurements of the boiling point of a liquid and, therefore, indirectly of the vapor pressure are made in the presence of additional components, the gases of the atmosphere. It is interesting to determine whether the presence of these components has an appreciable effect on the results of the measurement.

Let p_{gas} be the partial pressure of the vapor in the gas phase above the liquid at equilibrium. At equilibrium at constant temperature and pressure in the absence of other gases,

$$\mu_{liq} = \mu_{gas}$$

If the temperature is held constant but the pressure is increased by adding an additional gas not soluble in the liquid, the changes in chemical potential of liquid and gas are equal after a new equilibrium is reached.

$$\delta\mu_{liq} = \delta\mu_{gas}$$

The only change affecting the liquid is the total pressure, so

$$\delta\mu_{liq} = \left(\frac{\partial\mu_{liq}}{\partial P}\right)_T \delta P$$

$$= \mathscr{V}_{liq}\,\delta P \tag{15-20}$$

The vapor is now part of a mixture of gases, and, from Eq. 11-34,

$$\delta\mu_{gas} = RT\,\delta(\ln p_{gas}) \tag{15-21}$$

where p_{gas} is the partial pressure of the vapor in the gas mixture. Equating the right sides of Eq. 15-20 and Eq. 15-21, we obtain

$$\mathscr{V}_{liq}\,\delta P = RT\,\delta(\ln p_{gas}) \tag{15-22}$$

If we integrate Eq. 15-22 from $P = P_0$, the vapor pressure of the liquid in the absence of other gases, to a total pressure P, and from $p_{gas} = P_0$ to p_{gas}, the result is

$$\mathscr{V}_{liq}\,(P - P_0) = RT\ln\frac{p_{gas}}{P_0}$$

or

$$\ln\frac{p_{gas}}{P_0} = \frac{\mathscr{V}_{liq}\,(P - P_0)}{RT} \tag{15-23}$$

If P is large with respect to P_0, the right side of Eq. 15-23 reduces to $P\mathscr{V}_{liq}/RT$. The quantity RT/P is the volume of gas divided by the total number of moles

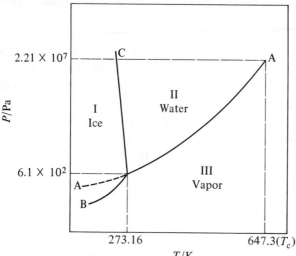

Figure 15-8.   Phase diagram for water at low pressure.

in a mixture (Sec. 1-2), and is also equal to the partial molar volume of each
gas (Eq. 11-39), so that Eq. 15-23 can be written

$$\ln \frac{p_{gas}}{P_0} = \frac{\mathscr{V}_{liq}}{\mathscr{V}_{gas}} \tag{15-24}$$

Since the volume of one mole of gas is very large compared to one mole of
liquid, the ratio on the right is very small. Thus the logarithm is close to
zero and the ratio is close to one.

Exercise 15-9.   Use Eq. 15-23 to calculate the vapor pressure of H_2O at 373 K, the
normal boiling point, under the pressure of 1.01×10^8 Pa of gas insoluble in the liquid.
The density of $H_2O(l)$ at 373 K is 958 kg-m^{-3}.

The system water
The phase behavior of water is of great importance because of the role of
water in biological and geological systems and because an understanding
of the behavior of pure water forms the basis of the treatment of solutions
in which water is a solvent.

Figure 15-8 shows the phase diagram for water at moderate pressures. The
triple point of water, because of its high degree of reproducibility, forms the
basis of the definition of the Kelvin temperature scale and is assigned the

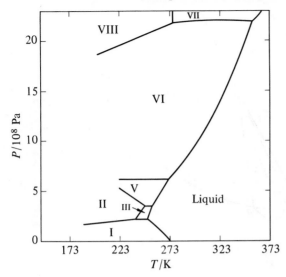

Figure 15-9. Phase diagram for water at high pressure. [By permission, from D. Eisenberg and W. Kauzmann, *The Structure and Properties of Water*, Clarendon Press, Oxford, 1969, and B. Kamb, *Science, 150,* 205 (1965).]

value 273.16 K. The triple-point pressure is 6.1×10^2 N-m^{-2} (4.6 mm Hg).[3] The normal freezing point, at 1.01×10^5 Pa pressure, is at 273.15 K. The negative slope of the ice-water equilibrium line is due to the negative volume change on fusion. The critical constants for water are[4]: $T_c = 647.3$ K, $P_c = 2.21 \times 10^7$ N-m^{-2} (218.3 atm), and $\mathscr{V}_c = 59.1 \times 10^{-6}$ m^3-mol^{-1}.

Tamman discovered in 1900 that ice can exist in solid forms other than ordinary ice at high pressure, and his work has been extended by Bridgman. Figure 15-9 shows the phase diagram of ice at high pressures. The normal ice I-liquid-vapor triple point and the liquid-vapor equilibrium curve are not discernible on this scale.

The system sulfur[5]
Sulfur exists in two crystalline forms, rhombic and monoclinic, both of which have S_8 molecules as units of the crystal. Pure rhombic sulfur is the

3. D. Eisenberg and W. Kauzmann, *The Structure and Properties of Water*, Clarendon Press, Oxford, 1969, p. 93.
4. E. S. Nowak and P. E. Liley, *J. Heat Transfer, 83C,* 1 (1961).
5. J. Zernike, *Chemical Phase Theory*, N. V. Uitgevers-Maatschappij AE. E. Kluwer, Antwerp, 1955, pp. 22, 26, 440–52.

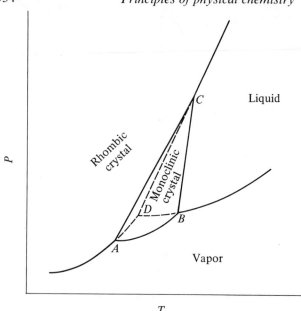

Figure 15-10. A schematic phase diagram for sulfur. Point *A*, 368.5 K, 0.9 Pa. Point *B*, 392.4 K, 3.7 Pa. Point *C*, 424 K, 1.3 × 10⁸ Pa. Point *D*, 386 K, 2.7 Pa.

stable form at low temperatures. At 368.5 K and 0.9 N-m^{-2}, there is a triple point at which rhombic sulfur, monoclinic sulfur, and sulfur vapor (also S_8) are at equilibrium. The monoclinic crystals have a triple point with liquid and vapor at 392.4 K and 3.7 N-m^{-2}. At high pressure, there is a rhombic-monoclinic-liquid triple point at 424 K and 1.3 × 10⁸ N-m^{-2}, which represents the highest pressure at which monoclinic crystals can exist at equilibrium. The relationships among the phases obtained from pure crystals are shown in Fig. 15-10, which is distorted to permit the inclusion of all the triple points. If rhombic sulfur is heated rapidly past the triple point at *A*, a *metastable* rhombic-liquid-vapor triple point occurs at 386 K and 2.7 N-m^{-2}. The metastable solid-liquid curve for rhombic sulfur is continuous with the curve above the triple point at *C*.

If the liquid sulfur is maintained above 395 K for some time and then cooled, the triple points *A*, *B*, and *D* are found at 368.7, 387.7, and 283 K, respectively. This result has been attributed to the transformation of some S_8 molecules to S_4 and polymers of S_4. A detailed discussion can be found in Zernike.

The system helium
Helium has a critical temperature just above 5 K so that only the part of the

phase diagram at very low temperature is of interest. This phase diagram, shown in Fig. 15-11, exhibits several unusual features: (a) There is no triple point at which solid, liquid, and gas are at equilibrium. (b) There is a transition at 2.2 K in which liquid He^I, the form of liquid present at higher temperatures, is transformed to liquid He^{II}, which has rather remarkable properties. (c) The solid-liquid equilibrium curve has a horizontal slope at low temperatures, so that liquid He^{II} is stable to as low a temperature as has so far been reached.

If the Clausius-Clapeyron equation, in the form of Eq. 15-11, applies to the solid-liquid equilibrium, then the solid and the liquid He^{II} have the same entropy at 0 K, which should be zero according to the third law of thermodynamics (Sec. 8-4). Helium is the only substance that remains liquid down to 0 K and the only liquid with zero entropy. The entropy of liquid helium above and below the transition at 2.2 K is shown in Fig. 15-12.

Since there is no discontinuity in the entropy or the volume in the transition between He^I and He^{II}, the transition is not first order. It has been suggested that it is a second-order transition, since the heat capacity behavior shown in Fig. 15-13 suggests a discontinuity in the heat capacity, a second derivative of the free energy. The way in which the heat capacity changes as

Figure 15-11. A phase diagram for ^4He at temperatures below 6 K, showing the λ-transition to liquid He^{II}. [By permission, from W. H. Keesom and K. Clusius, *Proc. Kon. Akad. Weten. Amsterdam*, **34**, 605 (1931).]

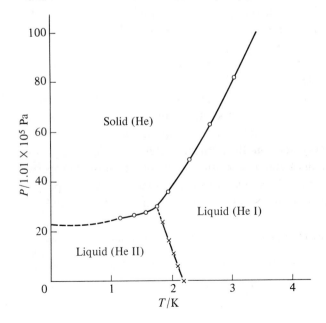

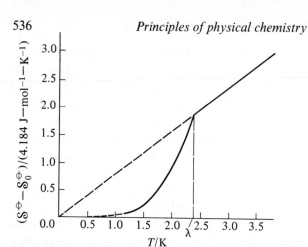

Figure 15-12. The standard entropy of liquid He^{II} and liquid He^{I} as a function of temperature; note that there is no discontinuity in the entropy at the λ-transition (λ). [By permission, from K. Mendelssohn, *Science*, *127*, 215 (1958).]

the transition is approached differentiates the transition from the expected behavior of a second-order transition, and it is designated a λ-transition because of the shape of the heat capacity curve. The density goes through a maximum at the same temperature as the maximum in the heat capacity.

Liquid He^{II} exhibits the property of *superfluidity*. If a beaker is immersed in a sample of He^{II}, the fluid will fill the beaker to the level of the liquid outside by film flow over the edge of the beaker. Similarly, if the beaker containing liquid is lifted out of the He^{II}, the liquid will flow out of the beaker and return to the large sample. The viscosity of He^{II} is 10^{-6} that of He^{I}.

Helium molecules of the normal isotope ^{4}He, since they are composed of an even number of elementary particles, are *bosons* (Sec. 4-3), and obey Bose-Einstein statistics (Sec. 8-5). Einstein suggested that bosons should exhibit unusual properties at low temperatures due to their "condensation" into the lowest energy state, since there is no limitation in the number of molecules in an energy state in Bose-Einstein statistics. This suggestion, modified by London and Landau, explained many of the unusual properties of He^{II} and was confirmed by a study of the properties of the isotope ^{3}He, a Fermi liquid, which does not exhibit a λ-transition or superfluidity.[6]

6. An elementary discussion of the properties of liquid helium and the history of their discovery can be found in K. Mendelssohn, *The Quest for Absolute Zero*, McGraw-Hill Book Co., New York, 1966, Chapt. 10, and in J. Wilks, *An Introduction to Liquid Helium*, Clarendon Press, Oxford, 1970.

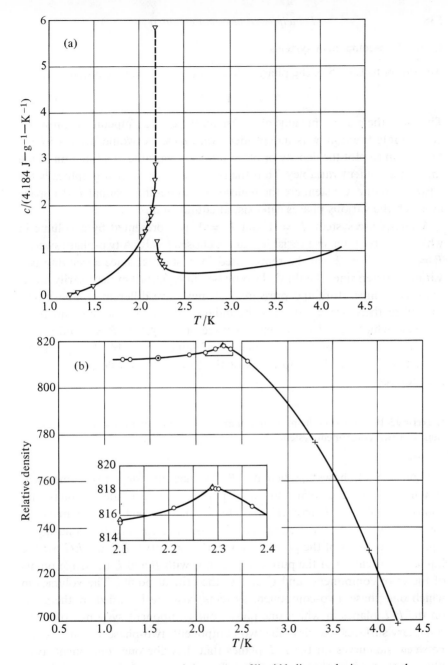

Figure 15-13. (a) The heat capacity per gram of liquid helium under its saturated vapor pressure near the λ-transition. [By permission, from W. H. Keesom and A. P. Keesom, *Physica*, 2, 557 (1935).] (b) The density of liquid helium in the vicinity of the λ-transition, showing that there is no discontinuity at the transition temperature. [By permission, from H. Kamerlingh-Onnes and J. D. A. Boks, *Comm. Phys. Lab. Leiden*, No. 170b (1924).] Maximum density occurs at the temperature of the λ-transition.

537

15-3. Two-component systems

According to Eq. 15-6, the phase rule for a two-component system is

$$F = 4 - P \qquad (15\text{-}25)$$

Therefore the maximum value of F is three. If we take temperature, pressure, and a mole fraction x as independent variables, we would need a fourth dimension to plot the density or other intensive variable as a function of the three independent variables. Thus the most we can represent graphically for a two component system are the regions in a three-dimensional T-P-x space in which the various phases can exist at equilibrium.

A one-phase system, $P = 1$ and $F = 3$, is represented by a volume in which all three intensive variables can be chosen freely. A two-phase system, $P = 2$ and $F = 2$, can be represented by a surface, since specifying two variables determines the third. Two phases in equilibrium necessarily are at the same temperature and pressure, but may have different compositions. Therefore there are usually two boundary surfaces in the diagram for a region in which two phases coexist. A three-phase system, $P = 3$ and $F = 1$, can be represented by a curve, since specifying one variable determines the other two. Since the compositions of the three phases differ, three curves are required.

Exercise 15-10. What is the maximum number of phases that can coexist at equilibrium in a two-component system?

Figure 15-14 shows a schematic P-T-x diagram for a two-component system in which the solid and liquid phases as well as the gas phase are completely miscible. Curves AB and DE are the solid-gas equilibrium curves for components A and B, respectively, BH and EF are the solid-liquid equilibrium curves of the pure components, and curves BC and EG are the liquid-vapor curves of the pure components, with B and E the triple points of the pure components and C and G the critical points. The volumes in which one-phase, two-component systems exist are bounded by the areas on the P-T planes in which one-phase, one-component systems exist. The boundary surfaces that describe two-component, two-phase systems extend between the curves on the P-T planes that describe one-component, two-phase systems. The three curves that describe three phases in equilibrium meet at the triple points of the pure components.

It is customary to plot pressure against mole fraction at constant temperature, or temperature against mole fraction at constant pressure in order

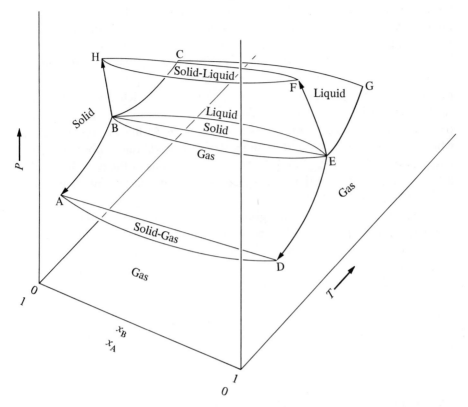

Figure 15-14. Scheme of a P, T, x phase diagram of a two-component system in which solid and liquid phases are completely miscible, with triple points (B, E) and the critical points (C, G) of the pure components. (By permission, from A. Reisman, *Phase Equilibria*, Academic Press, New York, 1970.)

to present phase data in two dimensions. The remainder of our discussion of two-component systems will be represented in this way.

Completely miscible systems

A constant-temperature section through the three-dimensional phase diagram in Fig. 15-14 at a temperature below the two triple points is shown schematically in Fig. 15-15. The pure solids have vapor pressures $P_{A,s}^*$ and $P_{B,s}^*$ at this temperature. Since the temperature is fixed, a one-phase system has two degrees of freedom and is represented by an area, as indicated for the solid and gaseous phases, the only ones that can exist at equilibrium below the triple points. A two-phase system has one degree of freedom, and the composition of the two phases in equilibrium is given by the curves that bound the two-phase region. If we have a system of composition $x_{B,1}$ at low pressure and slowly increase the pressure, we observe a decrease in volume, but no change of phase until P_1 is reached. At that pressure, solid of composition given by point N precipitates. As the pressure is raised further, more solid precipitates and both phases become richer in component A. When point R is reached, the solid has the composition $x_{B,1}$, and the last trace of gas has the composition given by point O. At higher pressures the whole system is solid, and only a slight decrease in volume is observed. At any pressure between $P_{A,s}^*$ and $P_{B,s}^*$ the compositions of the two phases in equilibrium are given by the intersections of the constant pressure line with the solid and gas curves.

At temperatures between the higher of the two triple points and the lower of the two critical points in Fig. 15-14, one can observe both liquid-gas and solid-liquid equilibria, as indicated in Fig. 15-16. The vapor pressures of the pure liquids at this temperature are $P_{A,1}^*$ and $P_{B,1}^*$, and the fusion pressures of the pure solids at this temperature are $P_{A,f}^*$ and $P_{B,f}^*$.

Exercise 15-11. Label the one-phase and two-phase regions in Fig. 15-16 and indicate the composition of pairs of phases at equilibrium in the liquid-gas and liquid-solid regions.

Figure 15-15. A P-x section through Fig. 15-14 at a constant temperature below the triple points of the pure components. The liquid phase cannot exist at equilibrium in this temperature range.

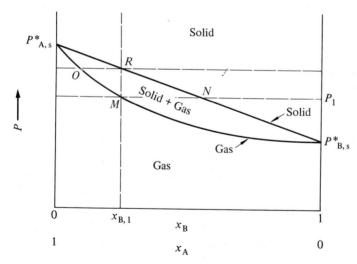

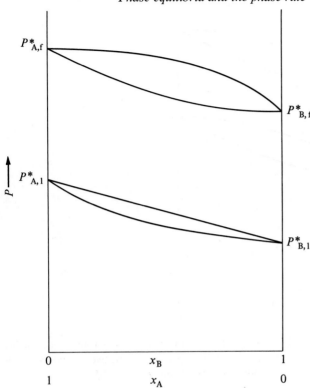

Figure 15-16. A *P-x* section through Fig. 15-14 at a constant temperature between the higher triple point and the lower critical point of the two components.

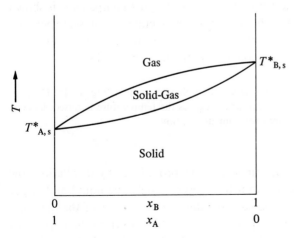

Figure 15-17. A *T-x* section through Fig. 15-14 at a constant pressure below the triple point pressures of the two components.

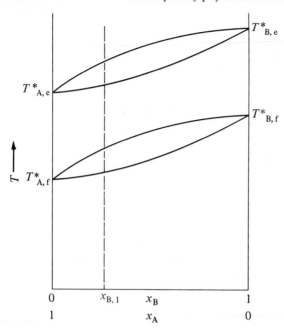

Figure 15-18. A T-x section through Fig. 15-14 at a pressure between the higher triple point pressure and the lower critical point pressure of the two components. The evaporation temperatures of the pure components are $T_{A,e}^*$ and $T_{B,e}^*$.

A constant pressure section through the phase diagram of Fig. 15-14 at a pressure below the triple point pressures of the pure components is shown schematically in Fig. 15-17. The corresponding section at a pressure between the higher of the triple points of the pure components and the lower of the critical points of the pure components is shown in Fig. 15-18.

Exercise 15-12. Label the one-phase and two-phase regions in Fig. 15-18. Describe the behavior of the system of composition $x_{B,1}$ as the temperature is raised, labeling any points on the graph required for your description.

Liquid-vapor equilibria

Because of their importance in separation procedures by distillation, the portions of a T-x phase diagram that describe only the liquid and vapor phases are frequently considered separately from the rest of the diagram. The simplest kind of liquid-vapor diagram is that shown in the upper part of Fig. 15-18 and reproduced in Fig. 15-19. When a liquid phase of composition $x_{B,1}$ is heated to a temperature corresponding to line NM, the vapor produced has a composition corresponding to point N and the liquid has a composition corresponding to point M.

The relative amounts of the two phases at equilibrium is given by the relationship

$$\frac{\text{Moles vapor}}{\text{Moles liquid}} = \frac{\text{Length } WM}{\text{Length } NW} \qquad (15\text{-}26)$$

Equation 15-26 is known as the *lever rule*, and can be applied to any two-phase system at equilibrium. When the distillate N is condensed and redistilled, vapor and liquid fractions L and K are obtained. When the residue M is redistilled, vapor and liquid fractions Q and O are obtained. In each step the vapor becomes richer in the lower boiling component and the liquid becomes richer in the higher boiling component. With enough redistillations relatively pure A and B can be obtained.

Exercise 15-13. Verify the lever rule. Hint: Consider a conservation equation relating the total number of moles of B in the system to the number of moles of B in the liquid and vapor phases.

Figure 15-19. A representation of liquid vapor equilibrium on a T-x diagram, with two completely miscible liquids, showing the effect of successive condensations and distillations on the compositions of the more volatile and the less volatile fractions.

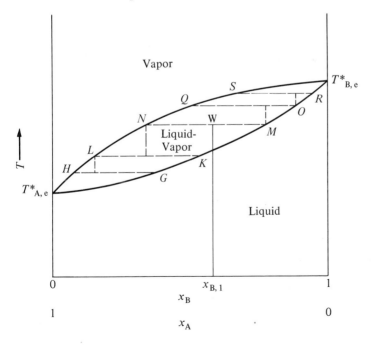

This procedure is tedious and time-consuming. All the steps in the process can be carried out continuously and simultaneously in a distillation column, such as that shown schematically in Fig. 15-20. The boiler at the bottom of the column is heated, and the column is either insulated or carefully heated to maintain a steady temperature gradient, with the lowest temperature at the top of the column. At each level, vapor rising from a lower level comes to equilibrium with liquid condensing from a higher level. In a continuous distillation, the liquid to be distilled is fed in at the middle of the column at the same rate at which distillate is removed at the top and condensate is removed from the boiler. In a steady state, the liquid in the boiler is richer

Figure 15-20. Diagram of a bubble-cap fractional distillation apparatus. (By permission, from A. Findlay, *The Phase Rule*, rev. by A. N. Campbell and N. O. Smith, Dover Publications, Inc., New York, 1951.)

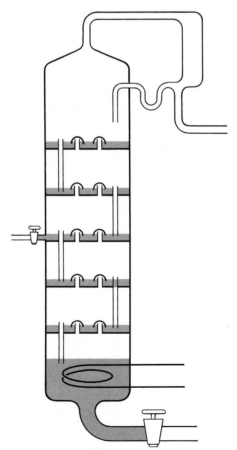

in the higher boiling component, and the vapor condensed at the top is richer in the lower boiling component. If the number of levels in the column is great enough for a given liquid pair, essentially pure components can be obtained.

In practice, a distillation column is not used at equilibrium, since the purpose of the process is to remove purified material from the system. The effectiveness of the separation decreases as the rate of removal of distillate is increased. A practical compromise between rate and separation is reached in each case. It is more common to use a column packed with glass or metal helices to provide a surface on which liquid and vapor can come to equilibrium rather than bubble-cap plates.

Azeotropic mixtures. In the examples we have discussed thus far, the boiling point of a mixture is always between the boiling points of the pure components. For many pairs of liquids, the boiling point-composition curve shows a maximum or a minimum. An example of commercial importance is the ethanol-water system. When ethanol is produced by fermentation of sugar, the highest concentration of ethanol that can be obtained is determined by the level at which the solution inhibits further action by the yeast, usually less than 20% alcohol by volume. Relatively dilute solutions are also obtained in the chemical synthesis of ethanol by hydration of ethylene. More concentrated solutions are obtained by distillation, but the product of distillation is 96% ethanol, 4% water by mass rather than pure alcohol. This result can be understood in terms of the boiling-point diagram in Fig. 15-21. At concentrations of ethanol between 0 and x_A, fractional distillation produces pure water in the boiler and 96% ethanol in the distillate. Once this concentration of ethanol has been reached, it boils without a further change in composition, and no higher concentration can be reached by distillation. Such a constant-boiling mixture is called an *azeotrope*.

Pure ethanol can be obtained by treatment of the minimum-boiling mixture with a dehydrating agent, such as quicklime. Commercial "absolute" alcohol is obtained by adding benzene to 96% ethanol. Upon distillation, a ternary azeotrope of benzene, ethanol, and water is distilled off, leaving ethanol with only a trace of benzene.

If distillation were carried out at only one pressure it might be concluded that an azeotropic mixture is a compound of fixed composition. Variation of the pressure, however, changes the composition of the azeotrope.

Liquid pairs with limited miscibility. As we shall see from our discussion of solutions in Chapter 16, minimum-boiling mixtures are characteristic of liquid pairs in which interaction between unlike molecules is less strong than interaction between like molecules. If this difference in interaction is sufficiently great, the system exhibits limited miscibility or complete immiscibility. One example of the former situation is the water-isobutyl alcohol system, the boiling-point diagram of which is shown in Fig. 15-22. A sample of over-all composition given by point H has two phases of composition A and D at the lowest temperature in the diagram. As the system temperature is raised the compositions of the two phases follow lines AB and DC and are joined by a horizontal line at any temperature. This two-phase region has one degree of freedom, in agreement with Eq. 15-25, since the diagram represents conditions at a fixed pressure. At the temperature of line BFC, a vapor phase of composition N appears in equilibrium with liquid 1 of composition B and liquid 2 of composition C. This three-phase system is invariant, and the temperature cannot increase until liquid 2 disappears. (If the initial composition of the system were between F and D, liquid 1 would disappear at the invariant point.)

Above the invariant temperature, the system consists of a liquid phase in equilibrium with a vapor phase. These phase compositions change with

Figure 15-21. A *T-x* diagram of the liquid vapor equilibrium in an ethanol-water system, showing the minimum boiling point at 96 mass percent ethanol (89 mole percent) at atmospheric pressure. [W. A. Noyes and R. R. Warfel, *J. Am. Chem. Soc.*, **23**, 463–68 (1901).] x_A is at 96 mass percent ethanol.

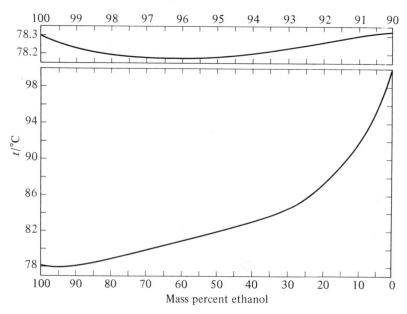

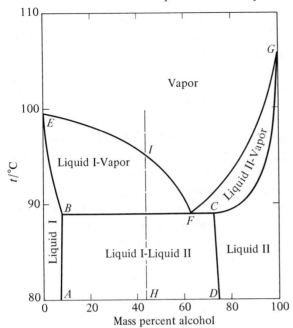

Figure 15-22. A boiling point diagram for water-isobutyl alcohol mixtures at 1×10^5 Pa, showing the effect of partial miscibility in the liquid phase. (By permission, from L. P. Hammett, *Introduction to the Study of Physical Chemistry*, McGraw-Hill Book Co., New York, 1952.)

temperature along the curves *BE* and *FG*, respectively. This two-phase region also has one degree of freedom. When the temperature reaches point *I*, the last of the liquid phase disappears and only vapor is present. In the one-phase area representing vapor, the system has two degrees of freedom.

Some regions in Fig. 15-22 are one-phase regions, whereas other regions are two-phase. The one-phase regions are truly areas with two degrees of freedom, and each point in the area represents the composition and temperature of the system. The compositions of the two phases of a two-phase system are given by two curves joined by a constant temperature line. These boundary curves would be expected to be continuous and have a continuous first derivative. The one-phase areas shown in Fig. 15-22 do not fulfill this criterion.

If two liquids are completely immiscible, the boiling-point diagram takes the form shown schematically in Fig. 15-23. At temperatures below the line *DCF*, the two pure liquid phases coexist. At the temperature corresponding to *DCF*, a vapor phase appears. This is the temperature at which the sum of the vapor pressures is equal to the pressure on the system. Since three phases are present, the system is invariant. The composition of the vapor phase is fixed; the mole fraction of each component is proportional to its vapor pressure at that temperature.

As the system is heated at this temperature, one of the liquid phases decreases in size until it disappears. If the over-all composition of the system is to the left of C, B disappears first; if the over-all system composition is to the right of C, A disappears first. Once there are only two phases present the system has one degree of freedom and the composition of the vapor phase moves up either CG or CH until the gas composition is the same as that of the system and the last of the liquid has disappeared.

A phase diagram, such as Fig. 15-23, explains the technique of *steam distillation*, in which a large excess of water is heated with a water-immiscible compound to distill the compound at a temperature well below the boiling point.

When the limited miscibility of a pair of liquids occurs only at temperatures below the liquid-vapor equilibrium curves, it is common to indicate only the liquid-phase solubility behavior in the phase diagram. Three major types of behavior are observed: (a) systems in which the mutual solubility increases

Figure 15-23. Scheme of a boiling point diagram for two completely immiscible liquids.

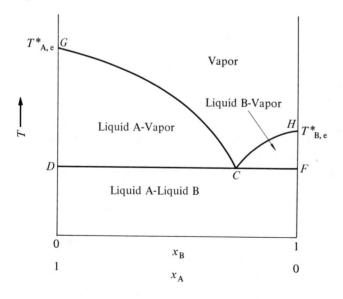

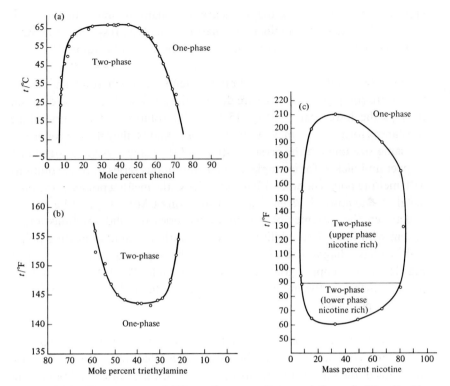

Figure 15-24. Temperature-solubility diagrams for partially miscible liquids. (a) Phenol-water, with an upper critical solution temperature. [By permission, from A. N. Campbell and A.J.R. Campbell, *J. Am. Chem. Soc.*, *59*, 2481 (1937).] (b) Triethyl-amine-water, with a lower critical solution temperature. [R. T. Lattey, *Phil. Mag.* [6], *10*, 397 (1905).] (c) Nicotine-water, with both an upper and a lower critical solution temperature. [C. S. Hudson, *Z. physik. Chem.*, *47*, 113 (1904).]

with *increasing* temperature and in which there is an *upper critical solution temperature* (Fig. 15-24, a); (b) systems in which the mutual solubility increases with *decreasing* temperature and in which there is a *lower critical solution temperature* (Fig. 15-24, b); and (c) systems in which there are both a lower and an upper critical solution temperature (Fig. 15-24, c).

Solid-liquid equilibria
In many systems the range of temperatures and pressures of most interest are those in which the vapor phase is not present at equilibrium and in which *condensed* phase diagrams are used to describe these systems. The major

applications of such phase diagrams are the separation and purification of solid species, the study of mineral formation in geological systems, the study of alloys and intermetallic compounds, and the study of ceramics.[7]

Completely miscible systems. Simple systems in which solid and liquid phases are completely miscible are described by a condensed phase diagram such as the lower part of Fig. 15-18. Solid solutions of two crystalline substances form only if they are *isomorphous*; that is, they must have nearly the same size units in the crystal lattice and the same crystalline structure. Copper and nickel, for example, form a continuous series of solid solutions with melting points between 1356 and 1728 K, the melting points of the pure metals.[8] The mineral *olivine* is a solid solution of Mg_2SiO_4 and Fe_2SiO_4.[9]

Between the extremes of a continuous series of solid solutions and a system with limited miscibility is a system with a minimum melting point, with a phase diagram similar to that of the vapor-liquid equilibrium of Fig. 15-21. The copper-gold system is an example. When the system composition is that at the minimum, the melting solid and the liquid have the same composition.[10]

Systems of limited miscibility. Solid-liquid systems in which the two pure solids are completely immiscible are represented by condensed phase diagrams of the same geometry as Fig. 15-23. Silver chloride and potassium chloride form a system of this type; its phase diagram is shown in Fig. 15-25. The point B is called the *eutectic* point, and the composition of the liquid in equilibrium with the two solid phases at that point is called the *eutectic composition*. Since three phases are present at equilibrium, the system is invariant. Such systems can provide a constant temperature environment at temperatures for which one-component solid-liquid equilibria cannot be found. The curves DB and BE can be viewed as describing the effect of temperature on the solubility of one component in the pure liquid phase of the other component or as describing the effect of one dissolved component on the melting point of the other component. The two are mathematically equivalent, as we shall see in Chapter 16. Other systems that exhibit a phase

7. H. C. Yeh, in *Phase Diagrams, Materials Science and Technology*, Vol. I (A. M. Alper, ed.), Academic Press, New York, 1970, Chapt. IV.
8. American Society for Metals, *Metals Handbook*, Metals Park, Novelty, Ohio, 1948, pp. 1146–1268.
9. R. Kern and A. Weisbrod, *Thermodynamics for Geologists*, Freeman, Cooper and Co., San Francisco, 1967, p. 197; N. L. Bowen and J. R. Schairer, *Am. J. Sci.*, 29, 151–217 (1935).
10. M. Hansen, *Composition of Binary Alloys*, McGraw-Hill Book Co., New York, 1958.

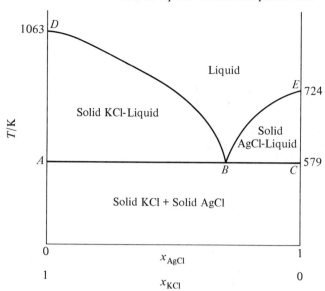

Figure 15-25. *T-x* phase diagram for solid and liquid phases in the system AgCl-KCl, showing the *eutectic* point (*B*) and complete immiscibility of the pure solids. (By permission, from A. Findlay, *The Phase Rule*, rev. by A. N. Campbell and N. O. Smith, Dover Publications, Inc., New York, 1951.)

diagram of this kind include alloys of bismuth and cadmium and the system diopside-anorthite.[11] That a eutectic mixture is not a compound can be deduced from the observation that composition varies with pressure and from microscopic and X-ray diffraction examination of the solid.

In some systems, an intermediate solid compound is formed that melts to liquid of the same composition. The phase diagram for the system formic acid-formamide is shown in Fig. 15-26. The diagram can be seen as two eutectic diagrams side by side, with the eutectic points *M* and *N*. The melting point of the compound in this diagram is called a congruent melting point.

When an intermediate solid compound melts to form a solid phase of a different composition, the melting point is called an *incongruent melting point* or a *peritectic point*. The system sodium-potassium exhibits such behavior, as shown in Fig. 15-27. If the solid compound Na_2K is heated, it melts at 280 K to form solid Na and liquid of composition *C*. This is an invariant point as long as any solid remains. Point *A* is a eutectic of solid K

11. A. Findlay, *The Phase Rule*, revised ed. (A. N. Campbell and N. O. Smith, eds.), Dover Publications, Inc., New York, 1951, p. 139.

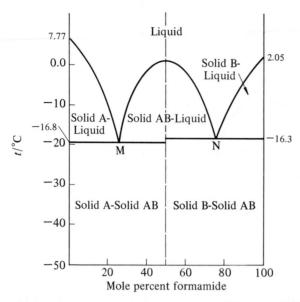

Figure 15-26. *T-x* phase diagram for the system formic acid-formamide, showing formation of a solid compound at 50 mole percent, with a congruent melting point. [By permission, from S. English and W. E. S. Turner, *J. Chem. Soc.*, *107*, 774 (1915).]

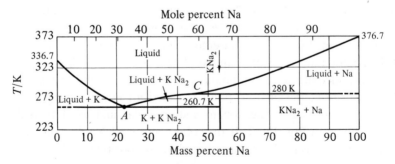

Figure 15-27. *T-x* phase diagram for the system Na-K, showing formation of a compound at 66.7 mole percent Na with an incongruent melting point (*peritectic* point *C*), as well as a *eutectic* point *A* at approximately 33 mole percent Na. (By permission, from C. J. Smithells, *Metals Reference Book*, 4th Ed., Vol. II, Butterworths, London, 1967.)

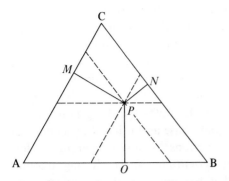

Figure 15-28. The triangular diagram that represents the composition of a three-component system at a fixed temperature and pressure.

and solid Na_2K. This system is used when a liquid alkali metal is required at temperatures below the melting points of Na or K.

For systems of solid solutions having limited miscibility, a phase diagram is of the form shown in Fig. 15-22. Some examples are silver-copper, mercurous iodide-silver iodide, lead-antimony, and cadmium-zinc.

15-4. Three-component systems

According to Eq. 15-6, the phase rule for a three-component system is

$$F = 5 - P \qquad (15\text{-}27)$$

Since the maximum value of F is four, it is possible to represent such a system graphically only for fixed values of several variables. The method used most commonly is for a system at constant pressure. The geometry used is that of a prism in which the vertical axis represents the temperature and the cross section is an equilateral triangle. As indicated in Fig. 15-28, the vertices represent pure components A, B, and C, and each side of the triangle represents a two-component system. If we take the altitude of the triangle as the unit of length and use mole fraction as our measure of composition, any point P within the triangle represents a system the over-all composition of which is given by the relations

$$\text{Mole fraction A} = \text{Length } PN$$
$$\text{Mole fraction B} = \text{Length } PM \qquad (15\text{-}28)$$
$$\text{Mole fraction C} = \text{Length } PO$$

since the sum of the perpendiculars from a point in the interior of an equilateral triangle to each side is equal to the length of the altitude.

Consider a condensed phase diagram in which each two-component system forms a simple eutectic (Fig. 15-25) and in which the addition of a third component to any of the two-component eutectics further lowers the melting point. The three-dimensional phase diagram is shown schematically in Fig. 15-29. At any fixed temperature, the behavior of the system as a function of composition is given by a triangular section of the prism parallel to the base. At any temperature above the highest melting point of the components, the system is a single liquid phase and it has three degrees of freedom, temperature and two mole fractions. At any temperature below the ternary eutectic temperature, the system consists of varying amounts of the three pure solid phases, and it has only one degree of freedom, temperature. At the ternary eutectic temperature, three solid phases and a liquid phase are present and the system is invariant.

A section through the solid at a temperature between the ternary eutectic and the lowest temperature binary eutectic is shown in Fig. 15-30. The area within *JKH* represents a one-phase liquid system with two degrees of freedom at constant temperature. The areas *AJH*, *CJK*, and *BHK* represent two-phase systems containing a pure solid phase in equilibrium with a liquid phase the composition of which can vary along the curves *JH*, *JK*, or *KH*. These systems have one degree of freedom at constant temperature. If, for example, the over-all system composition is given by *M*, solid *A* is in equilibrium with liquid of composition *N*, and the relative amounts of the

Figure 15-29. The triangular prism that describes the phase behavior of a three-component system at constant pressure. The lines normal to the plane of the triangle *ABC* at the vertices represent one-component systems as a function of temperature. T_A, T_B, and T_C represent the melting points of the pure components; points *D*, *E*, and *F* represent the AB, AC, and BC binary eutectics, respectively; *G*, a ternary eutectic.

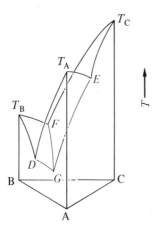

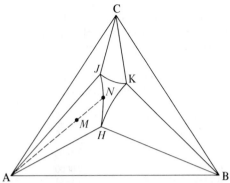

Figure 15-30. The triangular phase diagram that represents a cross section of the prism (Fig. 15-29) between the ternary eutectic and the lowest temperature binary eutectic. Area *JKH* represents a one-phase liquid system. Areas *AJH*, *BHK*, and *CKJ* represent two-phase systems in which solid *A* is in equilibrium with liquid along *JH*, solid *B* is in equilibrium with liquid along *HK*, and solid *C* is in equilibrium with liquid along *KJ*, respectively. Areas *ABH*, *BCK*, and *ACJ* represent three-phase systems in which solid *A* and *B* are in equilibrium with liquid *H*, solid *B* and *C* are in equilibrium with liquid *K*, and solid *A* and *C* are in equilibrium with liquid *J*, respectively.

two phases is given by the lever rule (Eq. 15-26). The areas *AJC*, *BCK*, and *AHB* represent three-phase areas, for example phases *A*, *B*, and *H*, and these systems are invariant. In all these cases, however, the temperature can be varied without the appearance or disappearance of a phase until the ternary eutectic temperature is reached.

Problems

15-1. (a) How many components are there in a system in which water vapor has been heated to 2000 K, so that H_2O, H_2, O_2, OH, O, and H are present at equilibrium? (b) How many components are there in a system in which an arbitrary mixture of H_2 and O_2 has been heated to 2000 K, so that H_2O, H_2, O_2, OH, H, and O are present at equilibrium?

15-2. Scatchard, Wood, and Mochel [*J. Phys. Chem.*, **43**, 119 (1939)] measured the vapor pressure of benzene and cyclohexane over a range of temperatures, and their smoothed data are shown below.

	$P/10^4$ Pa:	
T/K	Benzene	Cyclohexane
303.00	1.5829	1.6185
313.00	2.4341	2.4601
323.00	3.6178	3.6234
333.00	5.2213	5.1900
343.00	7.3443	7.2513
353.00	10.099	9.9067

Plot log P against $1/T$ for both liquids. If the plots are linear, calculate $\Delta \mathscr{H}_{vap}$ for each liquid over the temperature range of the data. If the plots are not linear, measure the slopes of the curves at the four intermediate temperatures and calculate $\Delta \mathscr{H}_{vap}$ at each temperature.

15-3. Scatchard, Wood, and Mochel also measured the vapor pressure, liquid composition, and vapor composition for a series of benzene-cyclohexane mixtures. Their data at 313.15 K are given below.

	Mole fraction benzene:	
$P/10^4$ Pa	Liquid	Vapor
2.4595	0.0000	0.0000
2.5986	0.1282	0.1657
2.6747	0.2354	0.2766
2.7293	0.3685	0.3912
2.7476	0.4932	0.4950
2.7350	0.6143	0.5909
2.6891	0.7428	0.6979
2.5999	0.8656	0.8205
2.4327	1.0000	1.0000

Plot a P-x diagram for the system benzene-cyclohexane at 313.15 K from these data, showing the liquid and vapor composition as a function of pressure. Calculate the over-all composition of the system when the number of moles of liquid and vapor are equal at 2.5986×10^4 Pa.

15-4. The value of $\Delta \mathscr{H}_f$ for bismuth is 1.048×10^4 J-mol^{-1} at the normal melting point, 544 K. Bridgman [*Phys. Rev., 6,* 28 (1915)] has found that the melting

Figure 15-31. A condensed phase diagram for CaF_2-$CaCl_2$. [Plato, *Z. physik. Chem.,* *58,* 363 (1907).]

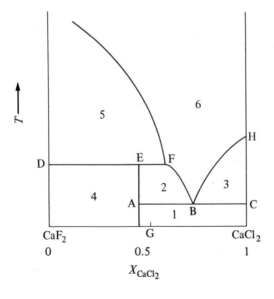

point decreases 0.00342 K for a pressure increase of 9.8×10^4 Pa. Calculate ΔV for the melting process. If the density of liquid bismuth is 10.04 kg-dm^{-3} at the melting point, what is the density of the solid?

15-5. Describe the changes that take place in a sodium-potassium system of 80 mole % Na when the system is slowly heated from an initial temperature of 220 K, including the appearance and disappearance of phases and the presence of invariant points. Relate your description to the number of degrees of freedom given by the phase rule. (See Fig. 15-27.)

15-6. The phase diagram in Fig. 15-31 is for the system calcium fluoride—calcium chloride [Plato, *Z. phys. Chem.*, *58*, 363 (1907)]. Describe the two-phase and one-phase areas and the three-phase lines. Describe the changes that take place in the system if it is heated slowly, starting at point *G*.

15-7. Draw the triangular diagram corresponding to the phase diagram in Fig. 15-29 at a temperature between binary eutectic points *D* and *F*. Describe the one-phase, two-phase, and three-phase areas, labeling any points required for the description.

15-8. The freezing points of acetamide-acetic acid mixtures were determined by Sisler, Davidson, Stoenner, and Lyon [*J. Am. Chem. Soc.*, *66*, 1888 (1944)]. Reprinted with permission, Copyright by the American chemical society.

x_{CH_3COOH}	T_f/K
0.000	353.4
0.115	345.7
0.209	336.7
0.300	325.1
0.402	308.8
0.515	281.2
0.519	273.0
0.536	272.7
0.558	272.6
0.579	271.4
0.603	270.6
0.625	269.3
0.682	264.0
0.699	261.5
0.700	262.1
0.789	273.2
0.890	282.1
1.000	289.8

Plot T_f/K against x_{CH_3COOH} and draw the phase diagram, indicating the significance of the curves and areas in the diagram.

16 · *Solutions of nonelectrolytes*

Many phases of variable composition appeared in the phase diagrams of two-component and three-component systems in Chapter 15. Such a phase, a homogeneous portion of a system of a composition that can vary within limits, is called a *solution*. Since the oceans, the interior of living cells, and the rocks and minerals of the earth either are solutions or are formed from solutions, the broad applicability of an understanding of solutions is clear. Emphasis in this chapter will be on the analytical relationships among intensive variables, the relationships that are represented graphically by equilibrium curves and surfaces in phase diagrams. Solutions of ionic species will be considered in Chapter 17.

16-1. Basic definitions

Solutions can be gaseous, liquid, or solid, but we will be concerned in this chapter primarily with liquid and solid solutions, since gaseous solutions have been discussed in Secs. 11-2, 11-3, and 11-7. Though there is no fundamental distinction among components of a solution, it is sometimes convenient to designate as *solvent* that component that is present in highest concentration or that component the pure form of which is of the same phase as the solution. For example, water is called the solvent in a sucrose-water solution even when the mass of sucrose present is much greater than the mass of water. The remaining components are then called *solutes*.

Composition measures
We have used mole fraction most often as a composition measure. When it

is convenient to designate one component as a solvent, it may also be convenient to express the composition with respect to each *solute* as a ratio relative to the amount of *solvent*. The *molality* m_B of a solute B is defined as the number of moles of solute per unit mass of solvent. Thus

$$m_B = n_B/m_A' \tag{16-1}$$

where m_A' is the mass of solvent, usually in kilograms. We shall commonly designate the solvent as component A. Like the mole fraction the molality is a measure of composition that is independent of temperature.

The relationship between the mole fraction and the molality can be obtained from the definition of mole fraction.

$$x_B = \frac{n_B}{n_A + n_B} \tag{16-2}$$

$$= \frac{n_B}{m_A'/M_A + n_B}$$

$$= \frac{n_B/m_A'}{1/M_A + n_B/m_A'}$$

$$= \frac{m_B}{1/M_A + m_B}$$

$$= \frac{m_B M_A}{1 + m_B M_A} \tag{16-3}$$

In sufficiently dilute solution, $n_B \ll n_A$, $m_B M_A \ll 1$, and Eq. 16-3 becomes

$$x_B \approx m_B M_A \tag{16-4}$$

Exercise 16-1. If M_A is 0.050 kg-mol^{-1}, how small must m_B be so Eq. 16-4 will be correct within an error of 1%?

A third expression for the composition of a solution defines the *concentration*, c_B or [B] as

$$c_B = \frac{n_B}{V} \tag{16-5}$$

where V is the volume of the solution, usually in cubic decimeters. The concentration and molality are related by the expression

$$c_B = \frac{m_B \rho}{1 + m_B M_B} \tag{16-6}$$

since

$$n_B = m_B \, m_A{}' \qquad (16\text{-}7)$$

and

$$V = \frac{m_A{}' + m_B{}'}{\rho}$$

$$= \frac{m_A{}' (1 + m_B \, M_B)}{\rho} \qquad (16\text{-}8)$$

where $m_B{}'$ is the mass of solute B and ρ is the density of the solution in kilograms per cubic meter. In sufficiently dilute solution, ρ approaches ρ_A, the density of the solvent, $m_B M_B \ll 1$, and Eq. 16-6 becomes

$$c_B \approx m_B \, \rho_A \qquad (16\text{-}9)$$

Exercise 16-2. If M_B is equal to 10 kg-mol^{-1} (a small macromolecule), what must the value of m_B be to have $m_B M_B$ less than 0.01? What is the corresponding concentration of the solution in mass percent?

We can solve Eq. 16-3 and Eq. 16-6 for m_B, equate the two expressions and obtain a relationship between x_B and c_B,

$$x_B = \frac{c_B \, M_A}{\rho + c_B \, (M_A - M_B)} \qquad (16\text{-}10)$$

In sufficiently dilute solution, ρ approaches ρ_A and $c_B(M_A - M_B) \ll \rho_A$, so that

$$x_B \approx \frac{c_B \, M_A}{\rho_A} \qquad (16\text{-}11)$$

Exercise 16-3. Verify Eq. 16-10.

16-2. Ideal solutions. Raoult's law

The simplest kind of solution is one in which the two kinds of molecules in the solution have the same size and in which the interaction between like and unlike molecules is the same. Then there is no volume change on mixing, so $\Delta V = 0$; and no enthalpy change on mixing, so $\Delta H = 0$. Such a solution is called an *ideal solution*. If solute and solvent differ only by a single isotopic substitution, their solution is nearly ideal. Similarly, benzene-toluene, $CHCl_3 - CHBr_3$, and $Mg_2SiO_4 - Fe_2SiO_4$ solutions are nearly ideal.

Ideal solid solution

The calculation of the entropy of mixing in a binary crystal (Sec. 8-4) corresponds to this model and leads to an entropy of mixing (from Eq. 8-48) of

$$\Delta S_{\text{mix}} = -Rn_A \ln x_A - Rn_B \ln x_B \tag{16-12}$$

Since

$$\Delta H_{\text{mix}} = 0$$

$$\Delta G_{\text{mix}} = -T\Delta S_{\text{mix}}$$

$$= RT n_A \ln x_A + RT n_B \ln x_B \tag{16-13}$$

From Eq. 11-30,

$$G = n_A \mu_A + n_B \mu_B$$

The free energy of the pure components is

$$G^* = n_A \mu_A{}^* + n_B \mu_B{}^* \tag{16-14}$$

and

$$\Delta G_{\text{mix}} = G - G^*$$

$$= n_A(\mu_A - \mu_A{}^*) + n_B(\mu_B - \mu_B{}^*) \tag{16-15}$$

Equating the right sides of Eq. 16-13 and Eq. 16-15 and rearranging, we have

$$n_A[\mu_A - \mu_A{}^* - RT \ln x_A] + n_B[\mu_B - \mu_B{}^* - RT \ln x_B] = 0 \tag{16-16}$$

Since Eq. 16-16 is valid for all values of n_A and n_B, the quantities in brackets must be identically equal to zero, and

$$\mu_A = \mu_A{}^* + RT \ln x_A \tag{16-17}$$

Thus, Eq. 16-17 relates the chemical potential of a component of an ideal solid solution to the mole fraction of that component in the solution and to the chemical potential of the pure solid component at the same temperature and pressure.

Solid-vapor equilibrium

If an ideal solid solution is in equilibrium with a vapor phase, as in the phase diagram of Fig. 15-15, the conditions of equilibrium are

$$\mu_{A,\,s} = \mu_{A,\,\text{vap}}$$

and a corresponding equation for component B. If we use Eq. 16-17 for the chemical potential in the solid phase and Eq. 11-33 for the chemical potential in the vapor phase (assuming the vapor to behave as a mixture of ideal gases), the condition of equilibrium becomes

$$\mu_{A,s}{}^* + RT \ln x_{A,s} = \mu_{A,gas}^{\ominus} + RT \ln\left(\frac{p_A}{p^{\ominus}}\right) \qquad (16\text{-}18)$$

For pure solid A, the condition of equilibrium with its vapor is

$$\mu_{A,s}{}^* = \mu_{A,gas}^{\ominus} + RT \ln\left(\frac{p_A{}^*}{p^{\ominus}}\right) \qquad (16\text{-}19)$$

Subtracting Eq. 16-19 from Eq. 16-18, we obtain

$$RT \ln x_{A,s} = RT \ln\left(\frac{p_A}{p_A{}^*}\right)$$

or

$$p_A = p_A{}^* \, x_{A,s} \qquad (16\text{-}20)$$

which is *Raoult's law*. This equation was obtained empirically by Raoult in 1887[1]. A similar equation applies to component B, and the total pressure above the solid is

$$P = p_A + p_B$$
$$= p_A{}^* \, x_{A,s} + p_B{}^* \, x_{B,s}$$
$$= p_A{}^* - x_{B,s}\,(p_A{}^* - p_B{}^*) \qquad (16\text{-}21)$$

where $x_{A,s} = 1 - x_{B,s}$ for a two-component system. The pressure-mole fraction diagram for this system at constant temperature is shown in Fig. 16-1. The curve for the total pressure is the same as the upper curve in Fig. 15-15. The composition of the vapor phase at each pressure is calculated from the partial pressures of the components in the vapor.

Exercise 16-4. Calculate $x_{A,vap}$ and $x_{B,vap}$ for an ideal solution in which the mole fractions in the solid are $x_{A,s}$ and $x_{B,s}$ and the vapor pressures of the pure solids are $p_A{}^*$ and $p_B{}^*$.

Ideal liquid solutions
Equation 16-17 applies to a component of an ideal liquid solution as well as to a component of an ideal solid solution. This result can best be understood in terms of a lattice model of liquids (Sec. 14-1). If we assume that the volume occupied by the molecules of the two components is the same and that the interactions between unlike and like molecules are the same, the

1. F. M. Raoult, *Compt. Rend.*, *104*, 1430 (1887); *Z. physik. Chem.*, 2, 353 (1888).

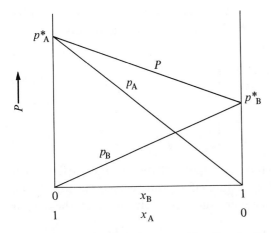

Figure 16-1. The pressure-composition diagram for an ideal binary solid mixture, one that is described by Raoult's law.

only difference between the solution and the pure liquids is the increased configurational entropy provided by the increased number of arrangements on the lattice. This increased number of arrangements is the same on the disordered liquid lattice as on the ordered crystalline lattice. Raoult's law then follows from Eq. 16-17 just as it does for solid solutions.

The equation of the boiling point curve for an ideal liquid pair can be obtained from the temperature dependence of the equilibrium expression. For a component of a liquid solution in equilibrium with a vapor mixture, where both are ideal, the condition of equilibrium is

$$\mu_{A,\,liq}{}^* + RT \ln x_{A,\,liq} = \mu_{A,\,gas}^{\ominus} + RT \ln\left(\frac{p_A}{p^{\ominus}}\right) \qquad (16\text{-}22)$$

From the definition of partial pressure

$$p_A = x_{A,\,gas}\, P$$

so that Eq. 16-22 becomes

$$\mu_{A,\,liq}{}^* + RT \ln x_{A,\,liq} = \mu_{A,\,gas}^{\ominus} + RT \ln x_{A,\,gas} + RT \ln\left(\frac{P}{p^{\ominus}}\right)$$

Dividing by RT, we obtain

$$\frac{\mu_{A,\,liq}{}^*}{RT} + \ln x_{A,\,liq} = \frac{\mu_{A,\,gas}^{\ominus}}{RT} + \ln x_{A,\,gas} + \ln\left(\frac{P}{p^{\ominus}}\right) \qquad (16\text{-}23)$$

The last term on the right is a constant at a fixed pressure, so its derivative with respect to temperature is zero. From Eq. 10-107 and the discussion of partial molar quantities in Sec. 11-2,

$$\left[\frac{\partial(G/T)}{\partial T}\right]_P = -\frac{H}{T^2} \tag{16-24}$$

so that

$$\left[\frac{\partial(\mu/T)}{\partial T}\right]_P = -\frac{\mathscr{H}}{T^2} \tag{16-25}$$

where $\mathscr{H}$ is the molar enthalpy for a pure phase and the partial molar enthalpy of a component of a mixture. If we differentiate Eq. 16-23 with respect to temperature at constant pressure and apply Eq. 16-25, the result is

$$-\frac{\mathscr{H}_{A, \text{liq}}^*}{RT^2} + \left(\frac{\partial \ln x_{A, \text{liq}}}{\partial T}\right)_P = -\frac{\mathscr{H}_{A, \text{gas}}^{\ominus}}{RT^2} + \left(\frac{\partial \ln x_{A, \text{gas}}}{\partial T}\right)$$

Rearranging, we have

$$\left(\frac{\partial \ln x_{A, \text{gas}}}{\partial T}\right)_P - \left(\frac{\partial \ln x_{A, \text{liq}}}{\partial T}\right)_P = \frac{\mathscr{H}_{A, \text{gas}}^{\ominus} - \mathscr{H}_{A, \text{liq}}^*}{RT^2}$$

or

$$\left[\frac{\partial \ln (x_{A, \text{gas}}/x_{A, \text{liq}})}{\partial T}\right]_P = \frac{\Delta \mathscr{H}_{\text{vap}, A}}{RT^2} \tag{16-26}$$

The value of $\Delta \mathscr{H}_{\text{vap}, A}$ is the standard enthalpy of vaporization of pure A if the pressure is equal to 1.01×10^5 Pa. (See Table 9-2.)

For a sufficiently narrow range of temperature, $\Delta \mathscr{H}_{\text{vap}}$ can be taken to be constant over the boiling range of the solutions. With this assumption Eq. 16-26 can be integrated to obtain

$$\left[\ln\left(\frac{x_{A, \text{gas}}}{x_{A, \text{liq}}}\right)\right]_{T_2} - \left[\ln\left(\frac{x_{A, \text{gas}}}{x_{A, \text{liq}}}\right)\right]_{T_1} = \frac{\Delta \mathscr{H}_{\text{vap}, A}}{R}\left(\frac{1}{T_1} - \frac{1}{T_2}\right) \tag{16-27}$$

A similar equation can be derived for component B. Thus, if the liquid and vapor composition at one temperature are known, the ratios $x_{A, \text{gas}}/x_{A, \text{liq}}$ and $x_{B, \text{gas}}/x_{B, \text{liq}}$ can be calculated at any other temperature in the boiling range. These values, together with the relationship $x_A = 1 - x_B$, for liquid and vapor phases permit the calculation of the composition of liquid and vapor at each temperature. An equivalent procedure can be used to describe solid-liquid equilibria in which both solid and liquid phases are ideal solutions.[2]

2. I. M. Klotz and R. M. Rosenberg, *Chemical Thermodynamics*, 3rd ed., W. A. Benjamin, Menlo Park, 1972, pp. 303–5.

Equilibrium between a solid and an ideal liquid solution

Consider an ideal solution, saturated and in equilibrium with pure solid solute. Assume that the solid phases of solute and solvent are completely immiscible, as in the two-phase regions shown in Fig. 15-25.

The condition of equilibrium is

$$\mu_{B,s}^* = \mu_{B,liq}$$

Since the liquid phase is an ideal solution, we may use Eq. 16-17 to substitute for $\mu_{B,liq}$ and write

$$\mu_{B,s}^* = \mu_{B,liq}^* + RT \ln x_{B,liq} \tag{16-28}$$

Pure component B is a solid at the temperature of the experiment, so $\mu_{B,liq}^*$ must refer to *supercooled* liquid B at the same temperature. Solving for $x_{B,liq}$, the mole fraction of B in the saturated solution, we obtain

$$\ln x_{B,liq} = \frac{\mu_{B,s}^* - \mu_{B,liq}^*}{RT} \tag{16-29}$$

Since the chemical potential of pure solid B and the chemical potential of pure supercooled liquid B are constant at a fixed temperature and pressure, the solubility is constant at a fixed temperature and pressure. The solubility is not only independent of the amount of solid solute, as would be true for any two-component system according to the phase rule, but the solubility is also independent of the *solvent*, as long as the solution is ideal.

The effect of temperature on the solubility can be determined by rearranging Eq. 16-29 to the form

$$\ln x_{B,liq} = \frac{1}{R} \left(\frac{\mu_{B,s}^*}{T} - \frac{\mu_{B,liq}^*}{T} \right)$$

and differentiating with respect to T at constant P. The result is

$$\left(\frac{\partial \ln x_{B,liq}}{\partial T} \right)_P = \frac{1}{R} \left(-\frac{\mathscr{H}_{B,s}^*}{T^2} + \frac{\mathscr{H}_{B,liq}^*}{T^2} \right)$$

$$= \frac{\Delta \mathscr{H}_{f,B}}{RT^2} \tag{16-30}$$

in which $\Delta \mathscr{H}_{f,B}$ is the enthalpy change on fusion of pure B. Written in this way, Eq. 16-30 describes the effect of temperature on the solubility of component B in solvent A. The equilibrium temperature can also be viewed as the freezing point of a solution of component A in solvent B. If we invert Eq. 16-30, we obtain

$$\left(\frac{\partial T_f}{\partial \ln x_{B,liq}} \right) = \frac{RT^2}{\Delta \mathscr{H}_{f,B}} \tag{16-31}$$

which describes the effect of the mole fraction of *solvent* B on the freezing point of the solvent. We see that the change in the freezing point depends only on the mole fraction of solvent and is independent of the nature of the solute as long as the solution is ideal. Integration of either Eq. 16-30 or Eq. 16-31 gives the equation of one of the liquid composition curves in the two-phase regions of Fig. 15-25. The intersection of the two liquid composition curves gives the eutectic point.

16-3. The dilute solution. Henry's law

Very few solutions follow Raoult's law over the entire range of compositions.

Figure 16-2. The pressure composition diagram for the system acetone-carbon disulfide, a system that exhibits positive deviation from ideality. [J. von Zawidsky, *Z. physik. Chem.*, *35*, 129 (1900).]. Experimental (——), ideal solution (– – –).

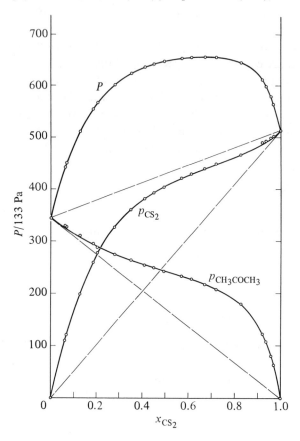

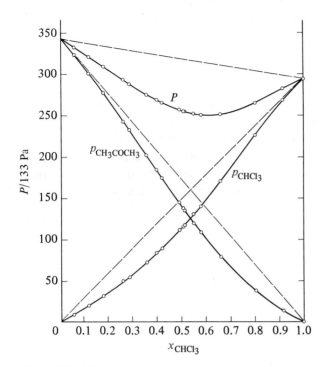

Figure 16-3. The pressure-composition diagram for the system acetone-chloroform, a system that exhibits negative deviation from ideality. [J. von Zawidski, *Z. physik. Chem.*, **35**, 129 (1900).] Experimental (——), ideal solution (– – –).

Some show *positive* deviations from ideality, as in Fig. 16-2 for acetone and carbon disulfide; others show negative deviations from ideality, as in Fig. 16-3 for acetone and chloroform. Positive deviations occur when the interaction between unlike molecules is weaker than interactions between like molecules. They are characterized by a positive ΔH of mixing and a positive ΔV of mixing. Negative deviations occur when interaction between unlike molecules is stronger than interaction between like molecules. They are characterized by a negative ΔH of mixing and a negative ΔV of mixing. In the case of acetone and chloroform, a hydrogen bond is formed between the hydrogen in chloroform and the carbonyl oxygen of acetone.

Though the solutions described in Fig. 16-2 and Fig. 16-3 do not follow Raoult's law over the whole range of composition, it can be seen that the partial pressure curve of each component approaches the Raoult's law line asymptotically as the mole fraction of the component approaches one. We can say that Raoult's law is a *limiting law* for the solvent in very dilute solutions, or

$$\lim_{x_{A, liq} \to 1} \left(\frac{p_A}{x_{A, liq}} \right) = p_A{}^* \tag{16-32}$$

The vapor pressure of the solute in very dilute solution approaches linear behavior asymptotically, but the line is not the Raoult's law line. In the case of positive deviation from ideality, the slope of the limiting line is greater than the slope of the Raoult's law line; in the case of negative deviation from ideality, the slope of the limiting line is less than the slope of the Raoult's law line. The equation for the limiting line of the solute is

$$p_B = k_{B, A} x_{B, A} \tag{16-33}$$

or, more accurately,

$$\lim_{x_{B, A} \to 0} \left(\frac{p_B}{x_{B, A}} \right) = k_{B, A} \tag{16-34}$$

where $k_{B, A}$ is a constant for a given solute in a particular solvent and is not equal to $p_B{}^*$. Equation 16-33 was first used by Henry[3] to describe the solubility of gases in liquids, but Henry's law is generally applicable to all solutions when they are sufficiently dilute.

The expression for the chemical potential of a solute that follows Henry's law can be obtained from the condition of equilibrium between liquid and vapor phases.

$$\mu_{B, A} = \mu_{B, gas}$$

If the vapor phase is considered an ideal mixture of ideal gases (Eq. 11-33), the equilibrium condition becomes

$$\mu_{B, A} = \mu_{B, gas}^{\ominus} + RT \ln \left(\frac{p_B}{p^{\ominus}} \right) \tag{16-35}$$

Substituting for p_B in Eq. 16-35 from Eq. 16-33, we have

$$\mu_{B, A} = \mu_{B, gas}^{\ominus} + RT \ln \left[\frac{(k_{B, A})(x_{B, A})}{p^{\ominus}} \right]$$

$$= \mu_{B, gas}^{\ominus} + RT \ln \left(\frac{k_{B, A}}{p^{\ominus}} \right) + RT \ln x_{B, A} \tag{16-36}$$

3. W. Henry, *Phil. Mag.*, *16*, 90 (1803).

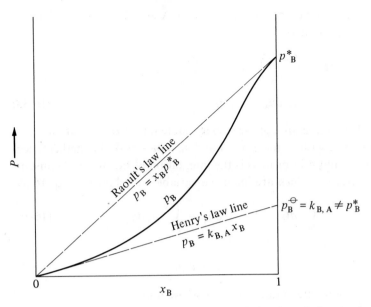

Figure 16-4. A representation of the hypothetical unit mole fraction standard state for the solute in a binary solution.

Since the first two terms on the right in Eq. 16-36 are constant at constant temperature and pressure for a given solvent-solute pair, it is convenient to choose their sum as the standard chemical potential of component B on the mole-fraction scale. Thus Eq. 16-36 becomes

$$\mu_{B,A} = \mu_{B,A}^{\ominus} + RT \ln x_{B,A} \qquad (16\text{-}37)$$

It can be seen that $\mu_{B,A}$ is equal to $\mu_{B,A}^{\ominus}$ when $x_{B,A}$ is equal to one. But, the standard state is not pure B, it is the hypothetical state of unit mole fraction found by extrapolating along the Henry's law line. The relationship is indicated in Fig. 16-4. The standard state has properties like those of a very dilute solution of B in A. The standard chemical potential, like the Henry's law constant, depends on the solvent as well as the solute.

Henry's law and Raoult's law
For solutions sufficiently dilute so that the solute follows Henry's law, it can be shown that the solvent follows Raoult's law. In any two-component phase at constant temperature and pressure, the Gibbs free energy is a

function of the mole numbers of components A and B and the total differential can be expressed as

$$\delta G = \left(\frac{\partial G}{\partial n_A}\right)_{T,P} \delta n_A + \left(\frac{\partial G}{\partial n_B}\right)_{T,P} \delta n_B$$

$$= \mu_A \, \delta n_A + \mu_B \, \delta n_B \tag{16-38}$$

If we think of building up a phase of fixed mole ratio from an infinitesimal size to a finite size with n_A moles of A and n_B moles of B, the mathematical correlate of this physical process is the integration of Eq. 16-38 at constant values of μ_A and μ_B, which are intensive variables, as shown in Eq. 16-39.

$$\int_0^G dG = \int_0^{n_A} \mu_A \, dn_A + \int_0^{n_B} \mu_B \, dn_B \tag{16-39}$$

or

$$G = \mu_A n_A + \mu_B n_B \tag{16-40}$$

If we take the total differential of Eq. 16-40, we find

$$\delta G = \mu_A \, \delta n_A + n_A \, \delta \mu_A + \mu_B \, \delta n_B + n_B \, \delta \mu_B \tag{16-41}$$

A comparison of Eq. 16-41 and Eq. 16-38 shows that

$$n_A \, \delta \mu_A + n_B \, \delta \mu_B = 0 \tag{16-42}$$

an expression derived by Gibbs and Duhem and known as the *Gibbs-Duhem equation*. If we divide Eq. 16-42 by the quantity $(n_A + n_B)$, we obtain the Gibbs-Duhem equation expressed only in terms of intensive variables

$$x_A \, \delta \mu_A + x_B \, \delta \mu_B = 0 \tag{16-43}$$

If the dependence of the chemical potential of one component on composition is known, the Gibbs-Duhem equation can be used to find the composition dependence of the chemical potential of the other component.

For a solution in which the solute follows Henry's law

$$\mu_B = \mu_{B,A}^{\ominus} + RT \ln x_B$$

$$\delta \mu_B = RT \, \delta \ln x_B$$

$$= RT \left(\frac{\delta x_B}{x_B}\right) \tag{16-44}$$

Substituting from Eq. 16-44 in Eq. 16-43 and solving for $\delta \mu_A$, we have

$$\delta \mu_A = -RT \left(\frac{\delta x_B}{x_A}\right) \tag{16-45}$$

Since $x_B = 1 - x_B$, $\delta x_B = -\delta x_A$ and Eq. 16-45 becomes

$$\delta \mu_A = RT\left(\frac{\delta x_A}{x_A}\right)$$

$$= RT\,\delta \ln x_A \tag{16-46}$$

If we integrate Eq. 16-46 from a state of pure solvent, $x_A = 1$, to some value of $x_A < 1$ we obtain

$$\int_{\mu_A*}^{\mu_A} d\mu_A = RT \int_{\ln x_A = 0}^{\ln x_A} d \ln x_A$$

or

$$\mu_A - \mu_A* = RT \ln x_A$$

which is identical to Eq. 16-17, from which we derived Raoult's law.

Equilibrium between a solid and a dilute solution
If the solubility of B in A is sufficiently small that B follows Henry's law in the saturated solution, we can derive an expression for the mole fraction of B in the saturated solution. The condition of equilibrium is

$$\mu_{B, s}* = \mu_{B, A}$$

$$= \mu_{B, A}^{\ominus} + RT \ln x_{B, A, satd}$$

Thus,

$$\ln x_{B, A, satd} = \frac{\mu_{B, s}* - \mu_{B, A}^{\ominus}}{RT} \tag{16-47}$$

which differs from Eq. 16-29 in that the solubility depends on the properties of the solvent as well as the solute.

Distribution equilibrium
If solute B dissolves in two solvents, A and C, which are immiscible, and if B follows Henry's law in each solvent, we can derive an expression for the equilibrium distribution of B between A and C when the two solutions are in contact. The condition of equilibrium is

$$\mu_{B, A} = \mu_{B, C}$$

or

$$\mu_{B, A}^{\ominus} + RT \ln x_{B, A} = \mu_{B, C}^{\ominus} + RT \ln x_{B, C}$$

Therefore

$$\ln \frac{x_{B, A}}{x_{B, C}} = \frac{\mu_{B, C}^{\ominus} - \mu_{B, A}^{\ominus}}{RT} \tag{16-48}$$

or

$$\frac{x_{B,A}}{x_{B,C}} = \exp\left[\frac{(\mu_{B,C}^{\ominus} - \mu_{B,A}^{\ominus})}{RT}\right] \qquad (16\text{-}49)$$

$$= K_D \qquad (16\text{-}50)$$

since all the quantities on the right side of Eq. 16-49 are constants at a fixed temperature and pressure. Thus, within the concentration range in which Henry's law is followed, the ratio of mole fractions of the solute in the two solvents is independent of the total amount dissolved. Eq. 16-50 is called the *Nernst distribution law*, and K_D is called the *distribution constant*.

Exercise 16-5. Show that the ratio of mole fractions in Eq. 16-49 is the same as the ratio of the solubilities of B in the two solvents, as given by Eq. 16-47.

16-4.   The general case. The activity functions and excess functions

The solute and the solvent in dilute solution follow Henry's law and Raoult's law, respectively, both of which are consistent with a logarithmic dependence of the chemical potential on the mole fraction, as in Eq. 16-17 and Eq. 16-37. To describe the relationship between chemical potential and mole fraction over the entire range of composition we can either use a different function for each solvent-solute pair or we can *define* a new function of the mole fraction such that the chemical potential depends logarithmically on this function. In the latter case, the unique characteristics of a solute-solvent pair are incorporated in the way this function depends on the mole fraction.

The activity function
G. N. Lewis[4] chose the latter alternative when he defined the *activity* a_j of component j by the relationship

$$\mu_j = \mu_j^{\ominus} + RT \ln a_j \qquad (16\text{-}51)$$

for any component j. It is clear from Eq. 16-51 that the numerical value of a_j depends on the choice of $\mu_j^{\ominus}$, that the activity in the standard state is equal to one, and that the activity is a dimensionless quantity.

For a component of an ideal solution we can see from Eq. 16-17 and Eq. 16-51 that the activity is equal to the mole fraction if the standard state is taken to be the *pure component at the same temperature and pressure as the solution*. A quantity called the *activity coefficient f* is defined as

$$f_j = \frac{a_j}{x_j} \qquad (16\text{-}52)$$

4. G. N. Lewis, *Proc. Am. Acad.*, **43**, 259 (1907).

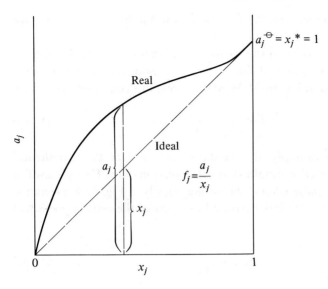

Figure 16-5. A representation of the activity and activity coefficient of a component when the standard state is the pure component.

and is a measure of the deviation from ideality in a real solution.[5] For a binary liquid (or solid) solution, ideal behavior is always approached in the limit as $x_j \to 1$. Thus, we can complete the definition of the activity for components of such a solution with the statement

$$\lim_{x_j \to 1} \left(\frac{a_j}{x_j} \right) = 1 \tag{16-53}$$

The relationships among the activity, the mole fraction, and this choice of standard state are indicated schematically in Fig. 16-5.

When it is convenient to distinguish between solvent and solute, the limiting Raoult's law behavior of the solvent and the limiting Henry's law behavior of the solute provide guides for the choice of standard states. For the solvent in the limit of dilute solutions Eq. 16-17 applies. Therefore, this equation, Eq. 16-51, Eq. 16-52, and Eq. 16-53, together form a basis for the definition of a and f, just as for the components of a binary solution. The graphical relationships shown in Fig. 16-5 also apply. The activity coefficient

5. An alternative choice of standard state that is commonly used is that of the pure component at 1.01×10^5 Pa pressure. This choice has the unfortunate result of including the effect of pressure on the chemical potential of the pure component in the activity of the component in solution. It would seem to be better to consider the effect of pressure separately and explicitly.

in this case is a measure of the deviation from the limiting Raoult's law behavior.

Equation 16-37 applies to the solute in the limit of infinite dilution. In this limiting case, the activity defined in Eq. 16-51 should become equivalent to the mole fraction in Eq. 16-37. We can express this condition in the form

$$\lim_{x_{B,A} \to 0} \left(\frac{a_{B,A}}{x_{B,A}} \right) = 1 \tag{16-54}$$

For this condition to apply, the standard state must be the hypothetical state of unit mole fraction obtained by extrapolation of the Henry's law line to $x_{B,A} = 1$. This relationship is shown graphically in Fig. 16-6. Deviation from Henry's law behavior is measured in terms of the activity coefficient defined as

$$f_{B,A} = \frac{a_{B,A}}{x_{B,A}} \tag{16-55}$$

Figure 16-6. A representation of the activity and the activity coefficient for the solute when the standard state is the hypothetical unit mole fraction standard state.

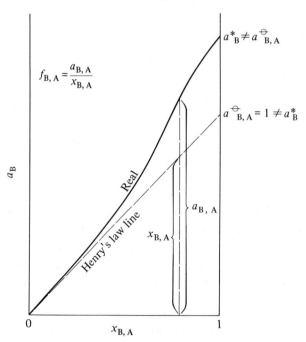

It is common to use the molality to describe the composition of a dilute solution. If the solution is dilute enough so that Henry's law applies, Eq. 16-4 is valid. When x_B from Eq. 16-4 is substituted in Eq. 16-33, it leads to a form of Henry's law appropriate to the molality scale.

$$p_B = k_{B, A} M_A m_B$$

$$= k_{B, A, m} m_B \qquad (16\text{-}56)$$

Expressed as a limiting law, Eq. 16-56 becomes

$$\lim_{m_B \to 0} \left(\frac{p_B}{m_B} \right) = k_{B, A, m} \qquad (16\text{-}57)$$

When x_B from Eq. 16-4 is substituted in Eq. 16-36, the result is

$$\mu_{B, A} = \mu_{B, gas}^{\ominus} + RT \ln \left(\frac{k_{B, A}}{p^{\ominus}} \right) + RT \ln (m_B M_A) \qquad (16\text{-}58)$$

For the standard state of the solute in the solution

$$\mu_{B, A, m}^{\ominus} = \mu_{B, gas}^{\ominus} + RT \ln \left(\frac{k_{B, A}}{p^{\ominus}} \right) + RT \ln (m_B^{\ominus} M_A) \qquad (16\text{-}59)$$

Subtraction of Eq. 16-59 from Eq. 16-58 gives

$$\mu_{B, A} = \mu_{B, A, m}^{\ominus} + RT \ln \left(\frac{m_B}{m_B^{\ominus}} \right) \qquad (16\text{-}60)$$

for the chemical potential of the solute in dilute solutions. The chemical potential is a property of the solute that is independent of the composition measure used, but the standard chemical potential has a value that depends on the composition measure and the arbitrary choice of standard state. This distinction is emphasized in the subscripts for the two quantities.

Since the mole fraction scale has limits of 0 and 1, the choice of $x = 1$ as a standard state, either real or hypothetical, is a natural one. The molality scale has no upper limit, so the choice of $m_B^{\ominus}$ is completely arbitrary. It has been conventional to choose $m_B^{\ominus} = 1$ mol-kg^{-1}.

For a dilute solution, Eq. 16-51 should be equivalent to Eq. 16-60 so that

$$\lim_{m_B \to 0} \left(\frac{a_{B, m}}{m_B / m_B^{\ominus}} \right) = 1 \qquad (16\text{-}61)$$

The standard state is then the hypothetical one molal standard state obtained when the Henry's law line is extrapolated to $m_B = 1$. This relationship is shown graphically in Fig. 16-7. Neither the real state at $m_B = 1$ (point D)

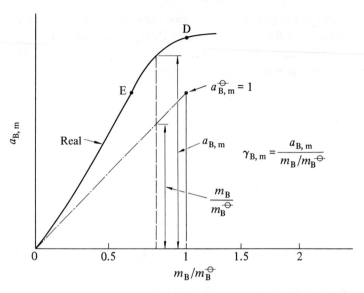

Figure 16-7. A representation of the activity and the activity coefficient for the solute when the standard state is the hypothetical unit molality ratio standard state.

or the real state with $a_B = 1$ (point E) has all the properties of the standard state. The measure of deviation from Henry's law behavior is the activity coefficient $\gamma_{B, A}$, defined as

$$\gamma_{B, A} = \frac{a_{B, m}}{m_B/m_B^{\ominus}} \tag{16-62}$$

For the activity on a concentration scale, the standard state is conventionally chosen as the hypothetical state obtained when the Henry's law line is extrapolated to $c_B = 1 \text{ mol-dm}^{-3}$.

Reasoning like that used for the molality scale leads to the following definitions for activity on the concentration scale:

$$\lim_{c_B \to 0} \left(\frac{p_B}{c_B} \right) = k_{B, A, c} = \frac{k_{B, A} M_A}{\rho_A} \tag{16-63}$$

$$\mu_{B, A} = \mu_{B, A, c}^{\ominus} + RT \ln \left(\frac{c_B}{c_B^{\ominus}} \right) \tag{16-64}$$

$$\lim_{c_B \to 0} \left(\frac{a_{B, c}}{c_B/c_B^{\ominus}} \right) = 1 \tag{16-65}$$

$$y_B = \frac{a_{B, c}}{c_B/c_B^{\ominus}} \tag{16-66}$$

where y_B is the activity coefficient of B on the concentration scale.

Table 16-1 summarizes the relationships for activity and composition measures.

Table 16-1. Relationships for activity and activity coefficient, and standard state, for several composition measures

Component	Composition measure	Limiting law for chemical potential	Activity coefficient	Standard state[a]
In binary solution	x_j	$\mu_j = \mu_j{}^* + RT \ln x_j$	$f_j = \dfrac{a_j}{x_j}$	Pure component
Solvent in solution	x_A	$\mu_A = \mu_A{}^* + RT \ln x_A$	$f_A = \dfrac{a_A}{x_A}$	Pure solvent
Solute in solution	x_B	$\mu_{B,A} = \mu_{B,A}^{\ominus} + RT \ln x_B$	$f_B = \dfrac{a_B}{x_B}$	Hypothetical unit mole fraction
Solute in solution	m_B	$\mu_{B,A} = \mu_{B,A,m}^{\ominus} + RT \ln\left(\dfrac{m_B}{m_B^{\ominus}}\right)$	$\gamma_B = \dfrac{a_{B,m}}{m_B/m_B^{\ominus}}$	Hypothetical one molal
Solute in solution	c_B	$\mu_{B,A} = \mu_{B,A,c}^{\ominus} + RT \ln\left(\dfrac{c_B}{c_B^{\ominus}}\right)$	$\gamma_B = \dfrac{a_{B,c}}{c_B/c_B^{\ominus}}$	Hypothetical one mole per cubic decimeter

[a] At same temperature and pressure as the solution

Pressure dependence of the activity. Differentiation of Eq. 16-51 with respect to pressure at constant temperature and composition leads to the result

$$\left(\frac{\partial \mu_j}{\partial P}\right)_{T,x} = \left(\frac{\partial \mu_j^{\ominus}}{\partial P}\right)_{T,x^{\ominus}} + RT\left(\frac{\partial \ln a_j}{\partial P}\right)_{T,x} \qquad (16\text{-}67)$$

or

$$\left(\frac{\partial \ln a_j}{\partial P}\right)_{T,x} = \frac{\mathscr{V}_j - \mathscr{V}_j^{\ominus}}{RT} \qquad (16\text{-}68)$$

For the solvent in solution, $\mathscr{V}_A^{\ominus} = \mathscr{V}_A^*$, since the standard state is the pure solvent. For the solute in solution, from Eq. 16-36 and Eq. 16-37,

$$\mu_{B,A}^{\ominus} = \mu_{B,gas}^{\ominus} + RT \ln\left(\frac{k_{B,A}}{p^{\ominus}}\right) \qquad (16\text{-}69)$$

Differentiation yields

$$\left(\frac{\partial \mu_{B,A}^{\ominus}}{\partial P}\right)_{T,x} = \mathscr{V}_B^{\ominus} = RT\left(\frac{\partial \ln k_{B,A}}{\partial P}\right)_{T,x} \qquad (16\text{-}70)$$

since the standard state of the gas is by definition at a fixed value of $p^\ominus$ equal to 1.01×10^5 Pa. Differentiation of Eq. 16-36 with respect to pressure at constant T and x gives a constant value of $\mathcal{V}_B$ equal to $\mathcal{V}_B^\ominus$ from Eq. 16-70 along the Henry's law line. Since the only real solution on the Henry's law line is that at infinite dilution, $\mathcal{V}_B^\ominus$ is the partial molar volume at infinite dilution, or

$$\mathcal{V}_B^\ominus = \mathcal{V}_B^\infty \tag{16-71}$$

Exercise 16-6. Show that $\mathcal{V}_B^\ominus = \mathcal{V}_B^\infty$ for the solute on the molality and concentration scales.

Temperature dependence of the activity. Equation 16-51 can be rearranged to read

$$\frac{\mu_j}{T} = \frac{\mu_j^\ominus}{T} + R \ln a_j \tag{16-72}$$

Differentiation of Eq. 16-72 with respect to temperature at constant pressure and mole fraction leads to the result

$$\left[\frac{\partial(\mu_j/T)}{\partial T}\right]_{P,x} = \left[\frac{\partial(\mu_j^\ominus/T)}{\partial T}\right]_{P,x^\ominus} + R\left(\frac{\partial \ln a_j}{\partial T}\right)_{P,x} \tag{16-73}$$

Substitution from Eq. 16-25 and rearrangement yields

$$\left(\frac{\partial \ln a_j}{\partial T}\right)_{P,x} = \frac{\mathcal{H}_j^\ominus - \mathcal{H}_j}{RT^2} \tag{16-74}$$

For the solvent in solution, $\mathcal{H}_A^\ominus = \mathcal{H}_A^*$, since the standard state is the pure solvent. The quantity $\mathcal{H}_A^\ominus - \mathcal{H}_A$ is just the negative of the *partial molar enthalpy of solution* of the solvent. For the solute in solution, from Eq. 16-36 and Eq. 16-37,

$$\frac{\mu_{B,A}^\ominus}{T} = \frac{\mu_{B,gas}^\ominus}{T} + R \ln\left(\frac{k_{B,A}}{p^\ominus}\right)$$

Differentiation yields

$$-\frac{\mathcal{H}_{B,A}^\ominus}{T^2} = -\frac{\mathcal{H}_{B,gas}^\ominus}{T^2} + R\left(\frac{\partial \ln k_{B,A}}{\partial T}\right)_{P,x} \tag{16-75}$$

Rearrangement of Eq. 16-36 and differentiation yields a constant value of $\mathcal{H}_{B,A}$ equal to $\mathcal{H}_{B,A}^\ominus$ from Eq. 16-75 along the Henry's law line. Since the only real solution on the Henry's law line is at infinite dilution

$$\mathcal{H}_{B,A}^\ominus = \mathcal{H}_{B,A}^\infty \tag{16-76}$$

The quantity $\mathscr{H}_{B, A}^{\ominus} - \mathscr{H}_{B, A}$ to be used in Eq. 16-74 for the solute is just the *partial molar enthalpy of dilution* of the solute to *infinite dilution*.

The activity and vapor pressure. The measurement of the vapor pressure of the components of a binary liquid mixture or of the solvent component of a solution is the most direct method for determination of the activity. At equilibrium the chemical potential of a component is the same in the liquid and vapor phases. Thus, if the vapor phase is a mixture of ideal gases,

$$\mu_A = \mu_{A, \text{gas}}^{\ominus} + RT \ln\left(\frac{p_A}{p^{\ominus}}\right)$$

For the standard state, which is pure A at the same temperature and pressure,

$$\mu_A^{\ominus} = \mu_{A, \text{gas}}^{\ominus} + RT \ln\left(\frac{p_A^*}{p^{\ominus}}\right)$$

Thus,

$$\mu_A - \mu_A^{\ominus} = RT \ln\left(\frac{p_A}{p_A^*}\right) \tag{16-77}$$

Equating $\mu_A - \mu_A^{\ominus}$ from Eq. 16-77 and Eq. 16-51, we obtain

$$a_A = \frac{p_A}{p_A^*} \tag{16-78}$$

If the vapor is not an ideal gas, the fugacity can be substituted for the pressure (Eq. 11-73) and

$$a_A = \frac{f_A}{f_A^*} \tag{16-79}$$

Vapor pressure measurements are also useful in determining the activity of a volatile solute. Since the chemical potential is the same for B in the liquid and vapor phases at equilibrium

$$\mu_{B, A} = \mu_{B, \text{gas}}^{\ominus} + RT \ln\left(\frac{p_B}{p^{\ominus}}\right) \tag{16-80}$$

If we choose the hypothetical state of unit mole fraction as the standard state, from Eq. 16-33 the vapor pressure of B in this state is

$$p_B^{\ominus} = k_{B, A}$$

so that

$$\mu_{B, A}^{\ominus} = \mu_{B, \text{gas}}^{\ominus} + RT \ln\left(\frac{k_{B, A}}{p^{\ominus}}\right) \tag{16-81}$$

Subtracting Eq. 16-81 from Eq. 16-80, we have

$$\mu_{B,A} - \mu_{B,A}^{\ominus} = RT \ln\left(\frac{p_B}{k_{B,A}}\right) \qquad (16\text{-}82)$$

Equating $\mu_{B,A} - \mu_{B,A}^{\ominus}$ from Eq. 16-82 and Eq. 16-51, we obtain

$$a_B = \frac{p_B}{k_{B,A}} \qquad (16\text{-}83)$$

Thus, measurement of the vapor pressure and the Henry's law constant is sufficient to calculate the activity of a volatile solute if the vapor can be considered an ideal gas.

Consider the data of Scatchard, Wood, and Mochel for mixtures of benzene and cyclohexane at 313.15 K given in Problem 15-3. The data are

Figure 16-8. The pressure-composition diagram for the system benzene (C_6H_6)-cyclohexane (C_6H_{12}) at 313.15 K: total pressure (○), benzene (□), cyclohexane (△). [Data from G. Scatchard, S. E. Wood, and J. M. Mochel, *J. Phys. Chem.*, **43**, 119 (1939).] Experimental (——), ideal (– – –).

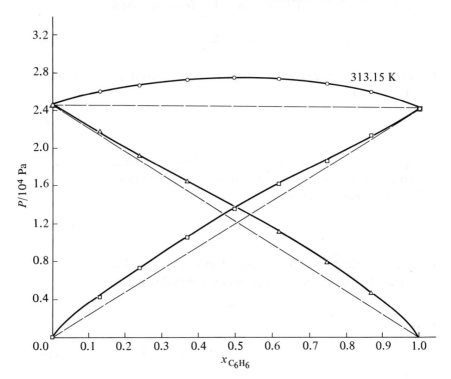

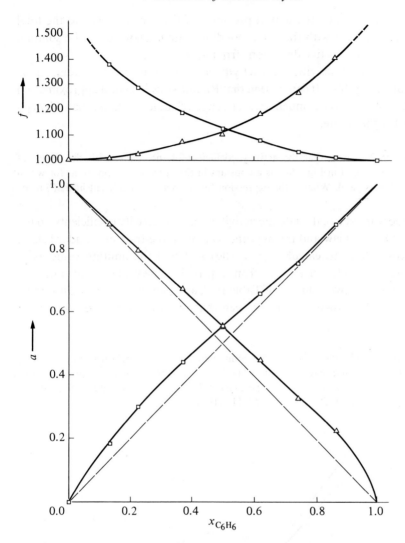

Figure 16-9. Activities and activity coefficients for the system benzene-cyclohexane at 313.15 K, pure component standard state; benzene (□), cyclohexane (△). [From data of G. Scatchard, S. E. Wood, and J. M. Mochel, *J. Phys. Chem.*, *43*, 119 (1939).]

reproduced in the first three columns of Table 16-2. The partial pressures in the vapor are calculated from the vapor composition and the total pressure. The activity of each component of the liquid is calculated from Eq. 16-78, and thus the *pure component* is chosen as the standard state. The activity coefficients are calculated from Eq. 16-55.

Figure 16-8 shows the partial pressures of the components and the total pressure, together with the corresponding Raoult's law lines. The solution clearly exhibits positive deviations from ideality.

The calculated activities and activity coefficients are plotted against mole fraction in Fig. 16-9. It can be seen that Raoult's law is a good approximation for the solvent over a much greater range of mole fractions than Henry's law is for the solute.

Exercise 16-7. Show that the activity coefficient of the *solute*, with the choice of standard state used in Fig. 16-9, is a constant in the range of compositions for which Henry's law is valid. What is the expression for the activity coefficient in that range?

If one wanted to calculate the activities and the activity coefficients on the basis of Henry's law and the hypothetical unit mole-fraction standard state, one would have to calculate p_j/x_j (liq) and find the limiting value as x_j approaches zero to calculate k_j from Eq. 16-34. The values of this ratio are given in the last two columns of Table 16-2, and they are plotted against mole fraction in Fig. 16-10. It can be seen that the ratio is not approaching a

Figure 16-10. The ratio p_j/x_j for the components of the system benzene-cyclohexane at 313.15 K, showing that the Henry's law line has not been reached, even in the most dilute solutions; benzene ($\times$), cyclohexane ($\circ$). [Data from G. Scatchard, S. E. Wood and J. M. Mochel, *J. Phys. Chem.*, *43*, 119 (1939).]

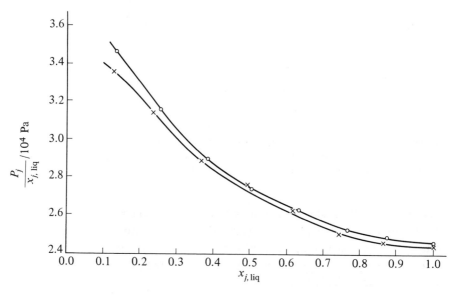

constant value in the most dilute solutions studied, and that many more data in very dilute solutions would be required in order to calculate activities on the basis of Henry's law.

Table 16-2. Activity calculations for benzene-cyclohexane at 313.15 K

Mole fraction benzene		$P/10^4$ Pa	$p/10^4$ Pa		$a_j = p_j/p_j{}^*$		$f_j = a_j/x_j$ (liq)		p_j/x_j (liq)	
Liquid	Vapor		Ben-zene	Cyclo-hexane	Ben-zene	Cyclo-hexane	Ben-zene	Cyclo-hexane	Ben-zene	Cyclo-hexane
0.0000	0.0000	2.4595	0.0000	2.4595	0.0000	1.0000	—	1.0000	—	2.4595
0.1282	0.1657	2.5986	0.4306	2.1680	0.1769	0.8815	1.380	1.011	3.3588	2.4868
0.2354	0.2766	2.6747	0.7398	1.9349	0.3040	0.7867	1.291	1.029	3.1427	2.5306
0.3685	0.3912	2.7293	1.0677	1.6616	0.4387	0.6756	1.190	1.070	2.8974	2.6312
0.4932	0.4950	2.7476	1.3601	1.3875	0.5588	0.5641	1.133	1.113	2.7577	2.7378
0.6143	0.5909	2.7350	1.6161	1.1189	0.6640	0.4549	1.081	1.179	2.6308	2.9010
0.7428	0.6979	2.6891	1.8767	0.8124	0.7711	0.3303	1.038	1.284	2.5265	3.1586
0.8656	0.8205	2.5999	2.1332	0.4667	0.8765	0.1898	1.012	1.412	2.4644	3.4725
1.0000	1.0000	2.4339	2.4339	0.0000	1.0000	0.0000	1.000	—	2.4339	—

Date of G. Scatchard, S. E. Wood, and J. M. Mochel, *J. Phys. Chem.*, 43, 119 (1939).

Excess thermodynamic functions

The activity function is a useful device for describing the behavior of individual *components* of solutions and the deviation of that behavior from the behavior of an ideal or a dilute solution. To describe the properties of a *solution* as they deviate from ideality, Scatchard[6] devised the *excess thermodynamic functions*. The free energy of mixing for an ideal solution is given by Eq. 16-13 as

$$\Delta G^{id}_{mix} = n_A RT \ln x_A + n_B RT \ln x_B$$

The *excess free energy of mixing* ΔG^E_{mix} of a real solution is defined as

$$\Delta G^E_{mix} = \Delta G_{mix} - \Delta G^{id}_{mix} \tag{16-84}$$
$$= \Delta G_{mix} - n_A RT \ln x_A - n_B RT \ln x_B$$

In an analogous fashion, we can define

$$\Delta S^E_{mix} = \Delta S_{mix} + n_A R \ln x_A + n_B R \ln x_B$$

and other excess functions of mixing. All the thermodynamic relations that

6. G. Scatchard and C. L. Raymond, *J. Am. Chem. Soc.*, 60, 1278 (1938).

connect ordinary thermodynamic functions also apply to excess functions. Thus

$$\Delta H_{mix}^E = \Delta G_{mix}^E + T \Delta S_{mix}^E$$

$$= \Delta H_{mix} \qquad (16\text{-}85)$$

Exercise 16-8.   Verify Eq. 16-85.

Guggenheim[7] has defined a class of solutions he calls *simple mixtures*, for which

$$\Delta \mathscr{G}_{mix}^E = x_A x_B w \qquad (16\text{-}86)$$

where w is a quantity with the dimensions of energy that depends on T and P and the nature of the two components but is independent of the composition. A positive value of w corresponds to positive deviations from ideality and negative values of w correspond to negative deviations from ideality. When the positive deviation is so large that $w = 2RT$, it can be shown that phase separation occurs.

16-5.   Colligative properties

The activity of a nonvolatile solute cannot be determined by measuring its vapor pressure. The properties of the solvent can, however, be related to the activity and the mole fraction of the solute. If we substitute $(1 - x_B)$ for x_A in Raoult's law (Eq. 16-20), the result is

$$p_A^* - p_A = p_A^* x_B \qquad (16\text{-}87)$$

Thus, for an ideal solution, or for the limiting case of a real solution that is sufficiently dilute, the decrease in the vapor pressure of the solvent is proportional to the mole fraction of solute and independent of the nature of the solute. A property that depends on the relative number of solute particles and that is independent of their nature is called a *colligative property*. In addition to vapor pressure lowering, boiling point elevation, freezing point depression, and osmotic pressure are colligative properties. At higher concentrations, specific interactions affect the properties of solutions, and the properties called colligative are no longer independent of the nature of the solute and solvent.

7. E. A. Guggenheim, *Thermodynamics*, 4th ed., North Holland Publishing Co., Amsterdam, 1959, p. 250.

Vapor pressure measurements

When the vapor pressure of the solvent is measured as a function of solution composition, the activity of the solvent based on the pure solvent standard state can be calculated (Eq. 16-78) and the activity of the solute can be obtained by use of the Gibbs-Duhem equation (Eq. 16-43). From Eq. 16-51,

$$\delta \mu_A = RT \, \delta \ln a_A$$

and

$$\delta \mu_B = RT \, \delta \ln a_B$$

Substituting in Eq. 16-43, we have

$$x_A RT \, \delta \ln a_A + x_B RT \, \delta \ln a_B = 0$$

or

$$\delta \ln a_A = - \frac{x_B}{x_A} \delta \ln a_B \tag{16-88}$$

From Eq. 16-78, $a_A = p_A/p_A^*$ for the solvent. If the solution is sufficiently dilute that $a_B = x_B$, based on the hypothetical unit mole fraction as standard state, we have

$$\delta \ln \left(\frac{p_A}{p_A^*} \right) = - \frac{x_B}{x_A} \delta \ln x_B$$

$$= - \frac{\delta x_B}{(1 - x_B)} \tag{16-89}$$

Integration of Eq. 16-89 from $p_A = p_A^*$, $x_B = 0$ to p_A, x_B yields

$$\ln \left(\frac{p_A}{p_A^*} \right) = \ln (1 - x_B)$$

or

$$\frac{p_A}{p_A^*} = 1 - x_B$$

which is equivalent to Eq. 16-87. Thus, we can derive Raoult's law from the Gibbs-Duhem equation and the assumption that $a_B = x_B$, the equivalent of assuming Henry's law.

A slight rearrangement of Eq. 16-88 yields

$$\delta \ln a_B = - \frac{x_A}{x_B} \delta \ln a_A \tag{16-90}$$

The left side of Eq. 16-90 cannot be integrated from the lower limit of $a_B = 0$ since the logarithm of zero is an indeterminate quantity. This dilemma can be avoided by transforming Eq. 16-90 into one involving the activity coefficient instead of the activity.[8]

Since

$$x_B = 1 - x_A$$

$$\delta x_B = -\delta x_A$$

and

$$\delta \ln x_B = -\frac{\delta x_A}{x_B}$$

$$= -\frac{x_A}{x_B} \delta \ln x_A \qquad (16\text{-}91)$$

Subtracting Eq. 16-91 from Eq. 16-90, we obtain

$$\delta \ln \left(\frac{a_B}{x_B} \right) = -\frac{x_A}{x_B} \delta \ln \left(\frac{a_A}{x_A} \right)$$

or

$$\delta \ln f_B = -\frac{x_A}{x_B} \delta \ln f_A \qquad (16\text{-}92)$$

The activity coefficients of both solvent and solute approach one as x_B approaches zero, so the integration on the left is no longer a problem.

The integration on the right in Eq. 16-92 cannot be carried out with a lower limit at $x_B = 0$ because the integrand becomes infinite. But, the lower limit of integration can be a small, fixed mole fraction x_B', and

$$\int_{\ln f'_B}^{\ln f_B} d \ln f_B = -\int_{\ln f'_A}^{\ln f_A} \frac{x_A}{x_B} d \ln f_A$$

or

$$\ln \frac{f_B}{f_B'} = -\int_{\ln f'_A}^{\ln f_A} \frac{x_A}{x_B} d \ln f_A \qquad (16\text{-}93)$$

The integral on the right in Eq. 16-93 can be obtained by numerical or graphical integration (Appendix C-2). When the integration has been carried out from x_B' to a series of values of x_B, f_B/f_B' is thereby determined as a

8. I. M. Klotz and R. M. Rosenberg, *Chemical Thermodynamics*, 3rd ed., ©, W. A. Benjamin, Menlo Park, 1972, pp. 352–53.

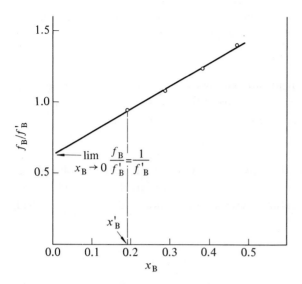

Figure 16-11. Extrapolation of the ratio f_B/f_B' for the solute of a binary solution to infinite dilution to calculate f_B' and to complete calculation of activity of solute from activity of solvent through the Gibbs-Duhem relation. (By permission, from I. M. Klotz and R. M. Rosenberg, *Chemical Thermodynamics*, © W. A. Benjamin, Menlo Park, 1972.)

function of x_B. If f_B/f_B' is plotted against x_B, as in Fig. 16-11, the resulting curve can be extrapolated to $x_B = 0$. Since $\lim_{x_B \to 0} f_B = 1$

$$\lim_{x_B \to 0} \frac{f_B}{f_B'} = \frac{1}{f_B'}$$

Once f_B' is determined by this extrapolation, the values of f_B and thus a_B can be calculated for all the other solutions.

Isopiestic method. If the activity of a solvent in solutions of one non-volatile solute has been determined over a range of composition, the activity of the same solvent in a solution of a different non-volatile solute can be determined by the *isopiestic method.* Two solutions containing different solutes are placed in a container in which the two solutions can come to a liquid-vapor equilibrium. Since both solutions are in equilibrium with the same vapor phase, the chemical potential of the solvent is equal in both solutions. With the same choice of standard states for solvent in the two solutions, the activity of the solvent is also equal in both solutions. An analysis of each solution at equilibrium yields the composition; since the activity of the solvent in one solution, the reference solution, is known, the activity of the solvent in the other solution can be calculated.

Freezing point measurements

Freezing point measurements are among the most precise that can be made on solutions, and they provide the basis for equally precise calculations of solvent activities. For the pure solvent at the freezing point,

$$\mu^*_{A, s} = \mu^*_{A, \text{liq}}$$

When solute is added to the liquid phase (we assume that the solid phases are completely immiscible), the chemical potential of the liquid solvent decreases and the temperature decreases until a new equilibrium is reached, such that

$$\mu^*_{A, s} + \delta\mu^*_{A, s} = \mu^*_{A, \text{liq}} + \delta\mu_{A, \text{liq}}$$

and

$$\delta\mu^*_{A, s} = \delta\mu_{A, \text{liq}} \tag{16-94}$$

Since the chemical potential of the solid solvent depends only on the temperature and since the chemical potential of the solvent in the solution depends on the temperature and the composition of the solution at constant pressure, Eq. 16-94 can be expressed as

$$\left(\frac{\partial \mu^*_{A, s}}{\partial T}\right)_T \delta T = \left(\frac{\partial \mu_{A, \text{liq}}}{\partial T}\right)_{P, x} \delta T + \left(\frac{\partial \mu_{A, \text{liq}}}{\partial x_A}\right)_{T, P} \delta x_A$$

or

$$RT\left(\frac{\partial \ln a_{A, \text{liq}}}{\partial x_A}\right) \delta x_A = (\mathscr{S}_{A, \text{liq}} - \mathscr{S}^*_{A, s}) \delta T$$

$$= \frac{\mathscr{H}_{A, \text{liq}} - \mathscr{H}^*_{A, s}}{T} \delta T \tag{16-95}$$

Exercise 16-9. Verify Eq. 16-95.

The activity of the solvent in solution, like the chemical potential, is a function of temperature and composition at constant pressure, so that (see Eq. 16-74),

$$\delta \ln a_{A, \text{liq}} = \left(\frac{\partial \ln a_{A, \text{liq}}}{\partial T}\right)_{P, x} \delta T + \left(\frac{\partial \ln a_{A, \text{liq}}}{\partial x_A}\right)_{T, P} \delta x_A$$

$$= \frac{\mathscr{H}^{\ominus}_{A, \text{liq}} - \mathscr{H}_{A, \text{liq}}}{RT^2} \delta T + \left(\frac{\partial \ln a_{A, \text{liq}}}{\partial x_A}\right)_{T, P} \delta x_A \tag{16-96}$$

Substituting from Eq. 16-96 for the derivative in Eq. 16-95, we obtain

$$RT\, \delta \ln a_{A,\,liq} - \frac{\mathcal{H}_{A,\,liq}^{\ominus} - \mathcal{H}_{A,\,liq}}{T}\, \delta T = \frac{\mathcal{H}_{A,\,liq} - \mathcal{H}_{A,\,s}^{*}}{T}\, \delta T$$

or

$$\delta \ln a_{A,\,liq} = \frac{\mathcal{H}_{A,\,liq}^{\ominus} - \mathcal{H}_{A,\,s}^{*}}{RT^2}\, \delta T$$

$$= \frac{\Delta\mathcal{H}_{f,\,A}^{*}}{RT^2}\, \delta T \tag{16-97}$$

On the assumption that $\Delta\mathcal{H}_{f,\,A}^{*}$, the enthalpy of fusion of the pure solvent, is constant over the range of temperature involved, integration of Eq. 16-97 from $a_{A,\,liq} = a_{A,\,liq}^{*} = 1$, T_f^{*} to a $a_{A,\,liq}$, T_f yields

$$\ln a_{A,\,liq} = - \frac{\Delta\mathcal{H}_{f,\,A}^{*}}{R} \left[\frac{1}{T_f} - \frac{1}{T_f^{*}} \right]$$

$$= - \frac{\Delta\mathcal{H}_{f,\,A}^{*}}{R} \frac{\Delta T_f}{T_f T_f^{*}} \tag{16-98}$$

Thus, if the enthalpy of fusion of pure solvent is known, a measurement of the freezing point permits the calculation of the solvent activity at temperature T_f. A more rigorous treatment requires information about the temperature dependence of $\Delta\mathcal{C}_{p,\,f,\,A}$ so that Eq. 16-97 can be integrated with $\Delta\mathcal{H}_{f,\,A}^{*}$ as a known function of T. If the pressure is equal to 1.01×10^5 Pa, $\Delta\mathcal{H}_{f,\,A}^{*} = \Delta\mathcal{H}_{f,\,A}^{\ominus}$. It is necessary to know the partial molar enthalpy of solution of the solvent in order to correct the calculated activity to the freezing point of the pure solvent or some other reference temperature (see Eq. 16-74).

The activity of the solvent calculated from the freezing point measurements can be used to calculate the activity of the solute, as in Eq. 16-92. A common use of freezing point data, however, is in the form of a limiting law. If the solution is sufficiently dilute, $a_{A,\,liq} \approx x_{A,\,liq} = (1 - x_{B,\,liq})$ and $T_f \approx T_f^{*}$, so that Eq. 16-98 can be expressed

$$\ln (1 - x_B) = - \frac{\Delta\mathcal{H}_{f,\,A}^{*}(\Delta T_f)}{R(T_f^{*})^2} \tag{16-99}$$

Also, if the solution is sufficiently dilute, $x_B \ll 1$ so that $\ln (1 - x_B)$ can be expanded in a Taylor's series,

$$\ln (1 - x_B) = - x_B - \frac{x_B^2}{2} - \frac{x_B^3}{3} - \cdots$$

and terms involving higher powers can be neglected. Neglecting higher powers and substituting in Eq. 16-99, we obtain

$$\Delta T_f = \frac{R(T_f^*)^2}{\Delta \mathscr{H}_{f,A}^*} x_B$$

Since the solution is dilute, $x_B \approx m_B M_A$ and

$$\Delta T_f = \frac{R(T_f^*)^2 M_A}{\Delta \mathscr{H}_{f,A}^*} m_B \tag{16-100}$$

Thus, in sufficiently dilute solution, freezing point depression is directly proportional to the molality of solute and the proportionality constant depends only on the properties of the solvent. The factor in Eq. 16-100 containing the constants is defined as the *freezing point depression constant* or *cryoscopic constant* of solvent A, $K_{f,A}$, where

$$K_{f,A} = \frac{R(T_f^*)^2 M_A}{\Delta \mathscr{H}_{f,A}^*}$$

A common application of Eq. 16-100 is in the determination of the molar mass of a solute in solution. For this purpose a convenient form of the equation is

$$\frac{\Delta T_f}{m_B'/m_A'} = \frac{K_{f,A}}{M_B} \tag{16-101}$$

Exercise 16-10.   Verify Eq. 16-101.

Since Eq. 16-101 is exact only in the limit of infinite dilution, it is best expressed as

$$\lim_{m_B'/m_A' \to 0} \frac{\Delta T_f}{m_B'/m_A'} = \frac{K_{f,A}}{M_B} \tag{16-102}$$

The freezing point depression is measured as a function of the mass ratio of solute to solvent. The ratio of ΔT_f to the mass ratio is plotted against the mass ratio and extrapolated to m_B'/m_A' equals zero to obtain exact values of molar mass.

Exercise 16-11.   Derive an equation analogous to Eq. 16-102 for boiling point elevation.

Osmotic pressure measurements
Osmosis is the spontaneous transfer of solvent from a dilute solution to a

more concentrated solution through a semipermeable membrane, which is permeable to solvent but not to solute. It is an important process in all biological systems, and thermodynamic considerations make it possible to calculate the maximum work that is attainable from the spontaneous transfer of solvent or the minimum work required to prevent the transfer. The term *osmotic work* is also used to describe the work necessary for the active transport of solutes across a membrane against a concentration gradient in biological systems.

Consider the transfer of one mole of solvent A from a solution of mole fraction x_A to one of mole fraction x_A'.

$$A(x_A) = A(x_A')$$

The molar Gibbs free energy change for the transfer is

$$\Delta \mathscr{G} = \mu_A' - \mu_A$$

$$= \mu_A^{\ominus} + RT \ln a_A' - (\mu_A^{\ominus} + RT \ln a_A)$$

$$= RT \ln \left(\frac{a_A'}{a_A} \right) \tag{16-103}$$

From Eq. 10-79

$$W_{net, \, rev} = \Delta \mathscr{G}$$

$$= RT \ln \left(\frac{a_A'}{a_A} \right) \tag{16-104}$$

If a_A' is greater than a_A, $W_{net, \, rev}$ is positive, and work must be done on the system to carry out a nonspontaneous transfer. If a_A' is less than a_A, $W_{net, \, rev}$ is negative, and the system can do work on the surroundings as a consequence of a spontaneous transfer.

The spontaneous tendency of solvent to pass through a semipermeable membrane into a solution can be measured directly in an *osmotic pressure* experiment. Figure 16-12 is a schematic representation of such an experiment. The bell-shaped compartment above the membrane is filled with solution through the capillary to a height just above or just below the expected equilibrium height. The reference capillary is used to correct for the effect of surface tension (assuming that this is the same for solution and pure solvent). When the pressure on the solution is sufficiently great, the system is at osmotic equilibrium, and there is no further net transfer of solvent into the solution. The difference in height of the solution meniscus and the solvent meniscus in the capillaries is a measure of this pressure.

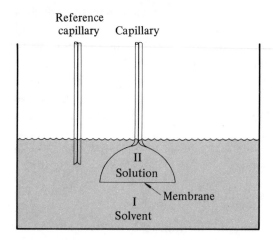

Figure 16-12.   Scheme for an osmotic pressure experiment. Phase I is pure solvent and phase II is solution.

Consider a hypothetical initial state in which pure solvent is on both sides of the membrane at hydrostatic equilibrium. In this situation

$$\mu_{A, I}^{*}(P_{0}) = \mu_{A, II}^{*}(P_{0})$$

When solute is added to phase II, the chemical potential of the solvent in phase II is decreased, and solvent enters phase II until the hydrostatic pressure is great enough to restore the chemical potential to that of the pure solvent. The condition of osmotic equilibrium is

$$\mu_{A, I}^{*}(P_{0}) = \mu_{A, II}(P)$$
$$= \mu_{A, II}^{\ominus}(P) + RT \ln a_{A, II}$$

or

$$RT \ln a_{A, II} = \mu_{A, I}^{*}(P_{0}) - \mu_{A, II}^{\ominus}(P) \qquad (16\text{-}105)$$

Both terms on the right in Eq. 16-105 refer to pure solvent. Thus, at constant temperature and composition the right side of Eq. 16-105 is given by

$$\mu_{A, I}^{*}(P_{0}) - \mu_{A, II}^{\ominus}(P) = \int_{P}^{P_{0}} \frac{\partial \mu_{A}^{*}}{\partial P} \, dP$$

$$= \int_{P}^{P_{0}} \mathscr{V}_{A}^{*} \, dP \qquad (16\text{-}106)$$

Since liquids are relatively incompressible, $\mathcal{V}_A^*$ can be assumed independent of pressure and Eq. 16-106 can be integrated to obtain

$$\mu_{A,1}^*(P_0) - \mu_{A,11}^{\ominus}(P) = \mathcal{V}_A^*(P_0 - P)$$

If we define $P - P_0$ as the osmotic pressure π and substitute in Eq. 16-105, we obtain

$$\ln a_{A,11} = -\frac{\mathcal{V}_A^* \pi}{RT} \tag{16-107}$$

The activity obtained is at pressure P and the small correction to pressure P_0 can be obtained by use of Eq. 16-68.

An important application of Eq. 16-107 is to the determination of the molar mass of macromolecules in dilute solution. In sufficiently dilute solutions, $a_A = x_A$ and Eq. 16-107 becomes

$$\frac{\mathcal{V}_A^* \pi}{RT} = -\ln x_A = -\ln(1 - x_B)$$

As in Eq. 16-99, $\ln(1 - x_B)$ can be expanded in a Taylor's series and terms beyond the first can be neglected. The result is

$$\pi = \frac{RT}{\mathcal{V}_A^*} x_B \tag{16-108}$$

so that osmotic pressure is also a colligative property. In sufficiently dilute solutions, from Eq. 16-11,

$$x_B \approx \frac{c_B M_A}{\rho_A}$$

$$= c_A \mathcal{V}_A^*$$

Substituting in Eq. 16-108, we obtain

$$\pi = RT c_B \tag{16-109}$$

which is *Van't Hoff's law of osmotic pressure*. The molar concentration c_B is related to the mass concentration c_B' by $c_B = c_B'/M_B$, so that Eq. 16-109 can be expressed as

$$\pi = \frac{RT}{M_B} c_B' \tag{16-110}$$

Since Eq. 16-110 is exact only in the limit of infinite dilution, it is best expressed as a limiting law,

$$\lim_{c_B \to 0} \frac{\pi}{c_B'} = \frac{RT}{M_B} \tag{16-111}$$

If π is measured as a function of c'_B, the ratio on the left of Eq. 16-111 can be plotted against c'_B and extrapolated to zero concentration. If the Taylor's series in x_B is retained and converted to a series in c'_B, the result is

$$\pi = \frac{RT}{M_B}(c'_B + D(c'_B)^2 + \ldots) \qquad (16\text{-}112)$$

which is analogous to the virial equation for gases. The constant D is called the *second virial* coefficient, which is given by the slope of a plot of π/c'_B against c'_B at low concentrations.

Osmotic pressure is not a useful technique for the study of solutes of low molar mass, primarily because it is difficult to find membranes that are impermeable to such solutes. It is a commonly used method for the study of macromolecules, however, because osmotic changes are larger than the changes with freezing point depression, boiling point elevation, and vapor pressure lowering.

Exercise 16-12.   Consider a solution of a protein the molar mass of which is 100 kg-mol^{-1}. The mass ratio of the solution is 0.01 kg-kg^{-1} and we can assume that the density of the solution is equal to the density of the pure solvent water, which is 10^6 kg-m^{-3}. The vapor pressure of water is 2670 Pa at 298 K. Calculate the vapor pressure lowering, the boiling point elevation, the freezing point depression, and the osmotic pressure at 298 K.

16-6.   Chemical equilibrium in solution

With the definition of the activity function, it is now possible to describe equilibria in non-ideal systems and in solid and liquid phases as well as the gas phase and still retain the formalism of the equilibrium constant that was described in Chapter 11 for gases. Any deviations from ideality or from dilute solution behavior are then described by the variation of the activity coefficient with composition.

Consider the generalized reaction (Eq. 11-13)

$$sS + uU + \ldots = cC + dD + \ldots$$

for which the condition of equilibrium at constant temperature and pressure is (the equality in Eq. 11-18)

$$\sum_{\text{prod}} (v_j\mu_j) - \sum_{\text{react}} (v_j\mu_j) = 0$$

Equation 16-51 can be applied in Eq. 11-18 to any chemical reaction. The result, when applied to Eq. 11-13, is

$$c(\mu_C^{\ominus} + RT \ln a_C) + d(\mu_D^{\ominus} + RT \ln a_D) + \ldots$$

$$- s(\mu_S^{\ominus} + RT \ln a_S) - u(\mu_U^{\ominus} + RT \ln a_U) = 0 \qquad (16\text{-}113)$$

If the standard chemical potentials are gathered on one side of the equal sign and the logarithmic terms on the other, the result is

$$\ln \frac{a_C^c \, a_D^d}{a_S^s \, a_U^u} = \frac{-[c\mu_C^{\ominus} + d\mu_D^{\ominus} - s\mu_S^{\ominus} - u\mu_U^{\ominus}]}{RT}$$

$$= -\frac{\Delta\mathscr{G}^{\ominus}}{RT} \qquad (16\text{-}114)$$

Since all the terms on the right side are constant at constant *temperature* and *pressure*, the term on the left side is also constant, and the quotient of activities is the *equilibrium constant* in *terms* of *activity*, K_a.

$$K_a = \frac{a_C^c \, a_D^d}{a_S^s \, a_U^u} = \exp \frac{-\Delta\mathscr{G}^{\ominus}}{RT} \qquad (16\text{-}115)$$

or

$$\ln K_a = \frac{-\Delta\mathscr{G}^{\ominus}}{RT} \qquad (16\text{-}116)$$

Pressure dependence of the equilibrium constant
The standard states that we have chosen vary with temperature and pressure, so that $\Delta\mathscr{G}^{\ominus}$ is a function of pressure as well as temperature, and K_a is a function of pressure as well as temperature. From Eq. 10-94,

$$\left(\frac{\partial \Delta\mathscr{G}^{\ominus}}{\partial P} \right)_T = \Delta\mathscr{V}^{\ominus}$$

so that

$$\left(\frac{\partial \ln K_a}{\partial P} \right)_T = -\frac{\Delta\mathscr{V}^{\ominus}}{RT} \qquad (16\text{-}117)$$

Only T and P need be considered as independent variables, since K_a is independent of composition. Equation 16-117 is a quantitative expression of LeChatelier's principle for the effect of a change in pressure. Since $\Delta\mathscr{V}^{\ominus}$ is small for most reactions in solid and liquid phases, K_a is relatively independent of pressure for ambient pressures in the laboratory. Reactions occurring in geological systems or in the depths of the ocean, however, are subject to very high pressure, and predictions about equilibria in such systems based on observations at atmospheric pressure must be corrected for the pressure change with Eq. 16-117.[9]

Temperature dependence of the equilibrium constant
Differentiation of Eq. 16-116 with respect to temperature at constant pressure yields

$$\left(\frac{\partial \ln K_a}{\partial T}\right)_P = -\frac{1}{R}\left[\frac{\partial(\Delta \mathscr{G}^\circ/T)}{\partial T}\right]_P$$

From Eq. 16-24

$$\left[\frac{\partial(\Delta \mathscr{G}^\circ/T)}{\partial T}\right]_P = -\frac{\Delta \mathscr{H}^\circ}{T^2}$$

so that

$$\left(\frac{\partial \ln K_a}{\partial T}\right)_P = \frac{\Delta \mathscr{H}^\circ}{RT^2} \tag{16-118}$$

which is known as the *Van't Hoff equation*. When Eq. 16-118 is integrated over a temperature range narrow enough that $\Delta \mathscr{H}^\circ$ is constant the result is

$$\ln K_a = \text{constant} - \frac{\Delta \mathscr{H}^\circ}{RT} \tag{16-119}$$

or

$$\ln \frac{K_a(T_2)}{K_a(T_1)} = -\frac{\Delta \mathscr{H}^\circ}{R}\left[\frac{1}{T_2} - \frac{1}{T_1}\right] \tag{16-120}$$

If $\Delta \mathscr{H}^\circ$ is constant a plot of $\ln K_a$ against $1/T$ is linear and the slope is $-\Delta \mathscr{H}^\circ/R$. Since

$$\left(\frac{\partial \ln K_a}{\partial T}\right)_P = \left[\frac{\partial \ln K_a}{\partial(1/T)}\right]_P \frac{d(1/T)}{dT}$$

$$= -\frac{1}{T^2}\left[\frac{\partial \ln K_a}{\partial(1/T)}\right]_P$$

and

9. If any of the reactants or products are pure liquids or solids, the molar volume used to calculate $\Delta \mathscr{V}^\circ$ in Eq. 16-117 is the actual molar volume. Since μ_j° is defined at a fixed pressure for pure solids and liquids

$$\mathscr{V}_j^\circ = \left(\frac{\partial \mu_j^\circ}{\partial P}\right)_T \equiv 0$$

and

$$\left(\frac{\partial \ln a_j}{\partial P}\right)_T = \mathscr{V}_j$$

Eq. 16-118 can also be written as

$$\left[\frac{\partial \ln K_a}{\partial(1/T)}\right]_P = -\frac{\Delta\mathscr{H}^{\ominus}}{R} \qquad (16\text{-}121)$$

Even if $\Delta\mathscr{H}^{\circ}$ is not constant, it is clear from Eq. 16-121 that $-\Delta\mathscr{H}^{\ominus}/R$ is the slope at any point of a plot of $\ln K_a$ against $1/T$. The dependence of an equilibrium constant on temperature given in Eq. 11-64 applies to K_a as well as to K_p, but the heat capacities required for reactants and products in solution are the partial molar heat capacities.

The measurement of equilibrium constants and their temperature coefficients is one source of the tabulated thermodynamic data discussed in Sec. 11-4. The tabulated data can, in turn, be used to calculate equilibrium constants for reactions for which direct equilibrium measurements are difficult or impossible.

The values of the activity coefficients are required in order to calculate the equilibrium composition of a reaction mixture from a value of K_a. If the activity is based on a mole fraction scale we can write Eq. 16-115 as

$$K_{a,x} = \frac{x_C^c \, x_D^d}{x_S^s \, x_U^u} \cdot \frac{f_C^c \, f_D^d}{f_S^s \, f_U^u} \qquad (16\text{-}122)$$

$$= K_x \, K_f \qquad (16\text{-}123)$$

If the activity is based on a molality scale we can write Eq. 16-115 as

$$K_{a,m} = \frac{(m_C/m_C^{\ominus})^c \, (m_D/m_D^{\ominus})^d}{(m_S/m_S^{\ominus})^s \, (m_U/m_U^{\ominus})^u} \cdot \frac{\gamma_C^c \, \gamma_D^d}{\gamma_S^s \, \gamma_U^u} \qquad (16\text{-}124)$$

$$= K_m \, K_\gamma \qquad (16\text{-}125)$$

If the activity is based on a concentration scale, we can write Eq. 16-115 as

$$K_{a,c} = \frac{(c_C/c_C^{\ominus})^c \, (c_D/c_D^{\ominus})^d}{(c_S/c_S^{\ominus})^s \, (c_U/c_U^{\ominus})^u} \cdot \frac{y_C^c \, y_D^d}{y_S^s \, y_U^u} \qquad (16\text{-}126)$$

$$= K_c \, K_y \qquad (16\text{-}127)$$

When the solution is ideal or sufficiently dilute, it is possible to take K_f, K_γ, and K_y equal to one and express the equilibrium constant in terms of the composition measure. Even when this simplification is not exact it is frequently a good approximation within the precision of the experimental equilibrium measurements.

The equilibrium constants K_x, K_m, and K_c are all dimensionless quantities, but the numerical value of the equilibrium constant for a particular reaction depends on the composition scale and the standard states on which the constant is based. Similarly, $\Delta\mathscr{G}^{\circ}$ depends on the choice of standard states and composition scale. In contrast, $\Delta \mathscr{G}$ for a given change of state must be independent of the arbitrary choice of standard states and composition measures.

Coupled reactions in biological systems

Photosynthetic organisms capture energy from sunlight and use it to carry out the synthesis of carbohydrates. All organisms oxidize carbohydrates to CO_2 and H_2O and use the energy that is liberated to carry out the synthesis of such macromolecules as proteins and nucleic acids. In thermodynamic terms we can say that a biological system uses the free energy liberated by an *exergonic* reaction to carry out a non-spontaneous, *endergonic*, reaction. The thermodynamic requirement for this process is that the magnitude of $\Delta \mathscr{G}$ for the exergonic reaction is greater than the magnitude for $\Delta \mathscr{G}$ for the endergonic reaction so that the net free energy change for the resulting change of state is negative. Whatever the thermodynamics of the situation, the two reactions *cannot be coupled unless they have a reactant or product in common*.

The oxidation of glucose in a living cell has a standard free energy change[10] of 2.87×10^3 kJ-mol^{-1} at pH 7 and 298 K.

$$C_6H_{12}O_6(aq) + 6 O_2(g) = 6 CO_2(g) + 6 H_2O(l)$$

The process of oxidation takes place in a large number of steps, each of which has a much smaller magnitude of $\Delta \mathscr{G}^{\ominus}$. A number of these steps are coupled to the phosphorylation of adenosine diphosphate (ADP) to form adenosine triphosphate (ATP), a reaction[11] with a standard free energy change at pH 7, $\Delta'\mathscr{G}^{\ominus}$, of $+ 29.3$ kJ-mol^{-1}. The ATP that is produced

$$\text{Adenosine} -O-\underset{\underset{O}{\|}}{\overset{\overset{O^-}{|}}{P}}-O-\underset{\underset{O}{\|}}{\overset{\overset{O^-}{|}}{P}}-O^- + HPO_4^{-2} + H_3O^+$$

$$= \text{Adenosine} -O-\underset{\underset{O}{\|}}{\overset{\overset{O^-}{|}}{P}}-O-\underset{\underset{O}{\|}}{\overset{\overset{O^-}{|}}{P}}-O-\underset{\underset{O}{\|}}{\overset{\overset{O^-}{|}}{P}}-O^- + 2 H_2O \qquad (16\text{-}128)$$

in this way then acts as a common intermediate and carrier of free energy to endergonic reactions in the cell.

10. H. A. Krebs and H. L. Kornberg, *Energy Transformations in Living Matter*, Springer-Verlag, Berlin, 1957.

11. It is more useful to tabulate $\Delta'\mathscr{G}^{\ominus}$ values for biochemical reactions, which usually occur near pH 7 and which frequently involve H_3O^+ as a reactant or product. The value of $\Delta \mathscr{G}^{\ominus}$ refers to unit activity of H_3O^+, which is seven orders of magnitude greater than that found in the cell.

One of the steps in the oxidation of glucose is the oxidation of 3-phosphoglyceraldehyde to 3-phosphoglycerate by pyruvate with a $\Delta'\mathscr{G}^{\ominus}$ of -29.3 kJ-mol^{-1}. If the oxidation occurs with the phosphorylation of 3-phosphoglycerate

$$
\begin{array}{c}
\text{O}^- \quad\quad \text{H}\ \ \text{H}\\
|\quad\quad\quad | \ \ \ |\\
{}^-\text{O}-\text{P}-\text{O}-\text{C}-\text{C}-\text{C}-\text{H} \quad + \quad \text{CH}_3-\text{C}-\text{COO}^- \quad + \quad 2\,\text{H}_2\text{O} =\\
\ \| \quad\quad\quad | \ \ \ | \ \ \|\quad\quad\quad\quad\quad\quad \|\\
\ \text{O} \quad\quad\quad \text{H}\ \ \text{O}\ \ \text{O}\quad\quad\quad\quad\quad\quad \text{O}\\
|\\
\text{H}
\end{array}
$$

3-Phosphoglyceraldehyde   Pyruvate

$$
\begin{array}{c}
\text{O}^- \quad\quad \text{H}\ \ \text{H}\quad\quad\quad\quad\quad \text{H}\\
|\quad\quad\quad | \ \ \ |\quad\quad\quad\quad\quad\ |\\
{}^-\text{O}-\text{P}-\text{O}-\text{C}-\text{C}-\text{C}-\text{O}^- + \text{CH}_3-\text{C}-\text{COO}^- + \text{H}_3\text{O}^+ \quad\quad (16\text{-}129)\\
\ \| \quad\quad\quad | \ \ \ | \ \ \|\quad\quad\quad\quad\quad\ |\\
\ \text{O} \quad\quad\quad \text{H}\ \text{OH}\ \text{O}\quad\quad\quad\quad\ \text{OH}
\end{array}
$$

3-Phosphoglycerate   Lactate

to form 1,3-diphosphoglycerate, followed by the transfer of a phosphate group to ADP, the sum of the coupled reactions, Eq. 16-130 and Eq. 16-131, is the same as the sum of Eq. 16-128 and Eq. 16-129.

3-Phosphoglyceraldehyde $+$ HPO$_4^{-2}$ $+$ Pyruvate $=$

$$
\begin{array}{c}
\text{O}^- \quad\quad \text{H}\ \ \text{H}\quad\quad\quad \text{O}^-\\
|\quad\quad\quad | \ \ \ |\quad\quad\quad\ |\\
{}^-\text{O}-\text{P}-\text{O}-\text{C}-\text{C}-\text{C}-\text{O}-\text{P}-\text{O}^- + \text{Lactate} \quad\quad (16\text{-}130)\\
\ \| \quad\quad\quad | \ \ \ | \ \ \| \quad\quad\ \|\\
\ \text{O} \quad\quad\quad \text{H}\ \text{OH}\ \text{O}\quad\quad\ \text{O}
\end{array}
$$

1,3-Diphosphoglycerate

1,3-Diphosphoglycerate $+$ ADP $=$ 3-Phosphoglycerate $+$ ATP   (16-131)

Exercise 16-13. The value of $\Delta'\mathscr{G}^{\ominus}$ for the reaction

1,3-Diphosphoglycerate $+$ 2 H$_2$O $=$ 3-Phosphoglycerate $+$ HPO$_4^{-2}$ $+$ H$_3$O$^+$

is -49.4 kJ-mol^{-1}. Calculate $\Delta'\mathscr{G}^{\ominus}$ for Eq. 16-130 and Eq. 16-131.

In the course of the oxidation of 1 mole of glucose, 38 moles of ATP are formed. If we consider $\Delta'\mathscr{G}^{\ominus}$ of the hydrolysis of ATP as the free energy available for further biological work, then the efficiency of the process is

$$
\frac{(38)\,(29.3\ \text{kJ-mol}^{-1})}{(2.87 \times 10^3\ \text{kJ-mol}^{-1})} = 0.39
$$

This process makes more efficient use of the chemical free energy of a carbohydrate than the combustion processes used to power electrical generators.

Problems

16-1. Given Eq. 16-15 for the free energy of mixing of an ideal solution and Eq. 16-17 for the chemical potential of the components, show that ΔV of mixing is equal to zero.

16-2. The $\Delta \mathscr{H}_f$ of phenanthrene is 1.86×10^4 J-mol^{-1}, and the melting point is 373 K. Assume that a solution of phenanthrene in benzene is ideal and calculate the solubility of phenanthrene in benzene at 298 K. [J. H. Hildebrand, *J. Am. Chem. Soc.*, *39*, 2297 (1917).]

16-3. Bronsted found that the solubility of monoclinic sulfur is 1.28 ± 0.01 times as great on the average as the solubility of rhombic sulfur at 298.15 K and 1.01×10^5 Pa in a number of organic solvents. The sulfur in solution was present as S_8 molecules. Calculate $\Delta \mathscr{G}$ for the transformation

$$S_8 \text{ (monoclinic)} = S_8 \text{ (rhombic)}$$

at 298.15 K. Which form is stable at 298.15 K? Compare your answer with Fig. 15-10. [J. Bronsted, *Z. physik. Chem.*, *55*, 371–82 (1906).]

16-4. Calculate $\Delta \mathscr{G}_{mix}^E$ for the benzene-cyclohexane solutions described in Table 16-2 [G. Scatchard, S. E. Wood, and J. M. Mochel, *J. Phys. Chem.*, *43*, 119 (1939)], with $n_A + n_B = 1$. Determine w and compare with $2 RT$.

16-5. Harkins and Wampler measured the freezing point depression for solutions of butanol and water of the following mass ratios. Use the data to determine the molar mass of butanol.

$(m_B'/m_A')/10^{-3}$	$\Delta T_f/K$
0.3064	0.007669
1.433	0.035588
1.794	0.04452
2.951	0.07300
3.859	0.9505
5.979	0.14679
7.224	0.17680

By permission, from W. D. Harkins and R. W. Wampler, *J. Am. Chem. Soc.*, *53*, 850 (1931).

16-6. C. Bawn, R. Freeman, and A. Kamaliddin [*Trans. Faraday Soc.*, *46*, 862 (1950)] measured the osmotic pressure of several fractions of the high polymer polystyrene in solution in toluene. Their data are shown below. Calculate the molar mass of each sample of polystyrene.

	$c_B'/(\text{kg-m}^{-3})$	π/Pa
	2.60	9.8
	5.07	31.
I	5.16	32.
	6.54	51.
	9.19	107.
	9.50	115.

$c_B'/(\text{kg-m}^{-3})$	π/Pa
1.55	16.
2.56	27.
2.93	31.
3.80	46.
5.38	75.
7.80	133.
8.68	157.

II is shown to the left of the table, aligned with the 3.80 / 46. row.

16-7. Wall and Rouse [Reprinted with permission from *J. Am. Chem. Soc.*, *63*, 3002 (1941). Copyright by the American Chemical Society.] carried out an isopiestic experiment in which a solution of phenanthrene in benzene and a solution of benzoic acid in benzene were allowed to come to equilibrium in an evacuated space at 329.3 K. After equilibrium was reached, the phenanthrene solution had a mass of 21.68 g and contained 242.6 mg of phenanthrene. The benzoic acid solution had a mass of 24.55 g and contained 323.8 mg of benzoic acid. In benzene solution, benzoic acid forms dimers, but phenanthrene does not. Assume that the solutions are sufficiently dilute so that the solute obeys Henry's law (hypothetical unit mole fraction standard state) and calculate K_x for the reaction

$$2\ C_6H_5COOH\ (\text{benzene}) = (C_6H_5COOH)_2\ (\text{benzene})$$

16-8. Wall and Rouse (see above) carried out similar experiments at 317.1 K. After equilibrium was obtained, a sample of 491.0 mg phenanthrene in benzene had a total mass of 20.11 g and a solution of 665.2 mg benzoic acid in benzene a total mass of 21.78 g. Calculate K_x at this temperature, and calculate $\Delta\mathcal{H}$ for the reaction.

17 · Solutions of electrolytes

Solutions of *electrolytes* contain charged particles, *ions*, which make such solutions conductors of an electrical current. As ionic solids dissolve, a polar solvent separates the ions from their lattice and they enter into solution. Some covalently bonded molecules react with the solvent as they dissolve to form ions in solution. Familiar examples are

$$HCl(g) + H_2O(l) = H_3O^+(aq) + Cl^-(aq) \qquad (17\text{-}1)$$

$$CH_3COOH(l) + H_2O(l) = H_3O^+(aq) + CH_3COO^-(aq) \quad (17\text{-}2)$$

The equilibrium for Eq. 17-1 is far to the right in dilute solution, and HCl, like ionic solids, is called a *strong electrolyte*. The equilibrium constant for Eq. 17-2 is small, and CH_3COOH is called a *weak electrolyte*. In this chapter we shall explore both the electrical and the chemical consequences of the presence of ions in solution.

17-1. Electrolysis and Faraday's laws

In 1834 Michael Faraday obtained the first quantitative evidence that a fundamental unit of electrical charge is associated with atomic units of matter.[1] He passed a direct electrical current through a solution of electrolytes in which two electrodes were immersed, and found that *the increase (or decrease) of the mass of an electrode was proportional to the amount of*

---

1. H. A. Boorse and L. Motz, *The World of the Atom*, Vol. I, Basic Books, New York, 1966, pp. 315–28.

electrical charge passing through the cell. Expressed as Faraday's *first law of electrolysis*, the proportionality can now be deduced from an atomic model in which the charges on ions are an integral multiple of a fundamental charge.

When he compared the amounts of different elements deposited in electrolysis, he found, for example, that 1 mole of zinc atoms was deposited by a charge sufficient to liberate 2 moles of hydrogen atoms. In modern terms, Faraday's *second law of electrolysis* states that *the amounts of two elements deposited by the passage of a given amount of electrical charge are in the ratio of the molar mass of the atom divided by the charge of the ion,* a quantity that has been called the *equivalent mass, M/z.* After Millikan's determination of the charge of the electron, the results of electrolysis experiments could be used to calculate a numerical value for Avogadro's constant.

Exercise 17-1. When a current of 0.0100 A is passed through a solution of $AgNO_3$ for 2.24×10^3 s, 2.50×10^{-5} kg of Ag is deposited at one silver electrode and 2.50×10^{-5} kg of Ag dissolves from the other silver electrode. (a) Calculate the amount of charge required to deposit 1 mole of Ag atoms. (b) What mass of Cu would be deposited in a cell with Cu electrodes immersed in $CuSO_4$ solution in series with the silver cell above?

Exercise 17-2. The charge of the electron is 1.60×10^{-19} coulomb. Calculate Avogadro's constant, using this value and the data in Exercise 17-1.

A *coulometer*, an electrolysis cell in which a known chemical reaction is carried out, is used to measure the electrical charge passing through a circuit precisely. Faraday's laws are used to calculate the charge transported, based on the definition that 9.64867×10^4 coulombs are required to deposit 107.868×10^{-3} kg of silver. This number of coulombs, the charge of 1 mole of electrons, is the *Faraday constant F.*

17-2. The transport of electrical charge in solution

The measurement of transport numbers

Twenty years after Faraday had established the relationship between the amount of chemical change in electrolysis and the amount of electrical charge passed through a cell, Hittorf demonstrated clearly that the current is carried by ions in solution. A Hittorf apparatus is shown schematically in Fig. 17-1. In this example, the apparatus is filled with $AgNO_3$ solution and both positive and negative electrodes are Ag. A source of potential difference and a coulometer are in series with the electrolysis cell. When a charge Q has passed through the cell, $Q/(z_+ F)$ moles of Ag have been deposited at the negative electrode and $Q/(z_+ F)$ moles of Ag have gone into solution at the

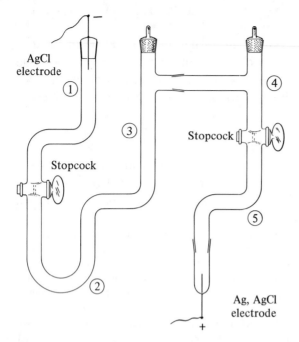

Figure 17-1. A Hittorf apparatus for the determination of transport numbers. [After D. A. MacInnes and M. Dole, *J. Am. Chem. Soc.*, **53**, 1357 (1931).]

positive electrode. The current is carried through the cell by the transport of Ag^+ ions from right to left and the transport of NO_3^- ions from left to right. If t_+ is the fraction of current carried by Ag^+ ions and if t_- is the fraction of current carried by the NO_3^- ions, then $t_+ Q/(z_+ F)$ moles of Ag^+ have been transported from the right compartment to the center compartment and from the center compartment to the left compartment, while $t_- Q/(z_- F)$ moles of NO_3^- have been transported from the left compartment to the center and from the center compartment to the right compartment.

The changes in the right compartment are: (a) the gain of $Q/(z_+ F)$ moles of Ag^+ from the electrode reaction, (b) the loss of $t_+ Q/(z_+ F)$ moles of Ag^+ by transport, and (c) the gain of $t_- Q/(z_- F)$ moles of NO_3^- by transport. Since $t_+ + t_- = 1$ and since $z_+ = z_-$, the net change in the right compartment is the gain of $t_- Q/(z_+ F)$ moles of $AgNO_3$.

The changes in the left compartment are: (a) the loss of $Q/(z_+ F)$ moles of Ag^+ in the electrode reaction, (b) the gain of $t_+ Q/(z_+ F)$ moles of Ag^+ by transport, and (c) the loss of $t_- Q/(z_- F)$ moles of NO_3^- by transport.

The net change in the left compartment is the loss of $t_- Q/(z_+ F)$ moles of $AgNO_3$.

If the changes at the electrodes are small enough so that no diffusion from the electrode compartment has occurred, there should be no change in the center compartment. Transport experiments are carried out under conditions such that changes in the composition of the electrode compartments are small, both to avoid diffusion from one compartment to another and because the value of t is a function of the solution composition.

Thus, the *transport number* t_- can be calculated from the changes in the left and right compartments as

$$t_- = \frac{\text{Decrease in moles } AgNO_3 \text{ on left}}{Q/(z_+ F)}$$

$$= \frac{\text{Increase in moles } AgNO_3 \text{ on right}}{Q/(z_+ F)}$$

and

$$t_+ = 1 - t_-$$

The changes in amounts of dissolved electrolyte in the electrode compartments are usually measured with respect to the solvent, water, which is assumed to be stationary, since it is uncharged. But, ions in aqueous solution are hydrated, so some water is transported with the ions. To correct for this effect, Washburn carried out transport measurements in which such nonelectrolytes as sucrose or urea were also present in the solution; he used these presumably stationary substances as references to determine changes in the amount of electrolyte.

Exercise 17-3. Derive an expression for t_+ in a solution of $CuCl_2$ in which the positive and negative electrodes are Cu.

The moving boundary method. It is difficult to obtain precise results with the Hittorf method because it depends on the measurement of a small difference between two concentrations. More recent measurements of transport numbers have been carried out by the *moving boundary method*, in which the motion of ions in an electrical field is observed directly. Figure 17-2 is a schematic representation of a moving boundary experiment. For example, a solution of KCl is carefully layered over a solution of $CdCl_2$, with the positive electrode at the bottom. At the lower electrode the reaction is

$$Ag(c) + Cl^-(aq) = AgCl(c) + e^-$$

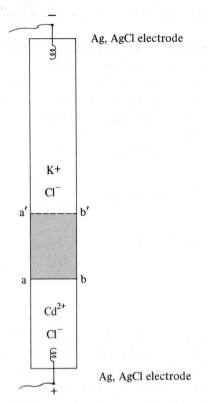

Figure 17-2. Scheme of a moving boundary apparatus for the determination of transport numbers.

and at the upper electrode the reaction is the reverse

$$AgCl(c) + e^- = Ag(c) + Cl^-(aq)$$

In the solution, Cl^- ions move from the top of the apparatus to the bottom and K^+ and Cd^{2+} ions from the bottom to the top. The lower, or indicator, solution must be more dense than the upper solution to prevent mixing at the boundary. As current passes through the solution, the boundary is observed to move from its initial position $a\,b$ to a new position $a'b'$ in a time t. When both solutions are colorless, the boundary can be observed by the difference in index of refraction of the solutions. The number of K^+ ions contained in the volume V between $a\,b$ and $a'b'$ has thus passed the plane $a'b'$ in this time. If a constant-current source is used, the total charge passed through the plane $a'b'$ in time t is $Q = I\,t$, where I is the current. The charge carried by the K^+ ions is $c(K^+)Vz(K^+)F$, where $c(K^+)$ is the concentration of the K^+ ions and $z(K^+)$ is the charge of the K^+ ions. Thus

$$t(K^+) = \frac{c(K^+)Vz(K^+)F}{It} \tag{17-3}$$

The motion of the boundary has to be corrected for volume changes as a result of the electrode reactions, but the measurement is much more direct than the measurements in the Hittorf method.

A sharp initial boundary is produced by one of two methods. In the first method, the two solutions to be joined at the boundary are contained in tubes leading to openings in upper and lower ground glass plates that slide over one another. When the boundary is to be produced, the plates are rotated until the tubes meet and a sharp boundary is formed. In the second method, the lower electrode is made of a metal that produces the indicator ions in the electrode reaction. In this way the boundary is produced *autogenically* as the current is passed through the solution.

The conductivity of solutions of electrolytes

Like the metallic conductors discussed in Sec. 13-2, solutions of electrolytes obey Ohm's law in a form similar to Eq. 13-24,

$$j = \kappa E \tag{17-4}$$

where j is the magnitude of the current density, κ is the conductivity of the electrolyte, and E is the magnitude of the electrical field. If the current density is uniform throughout the cross section of the conductor, $j = I/A$, where I is the current and A is the cross-sectional area of the conductor. Similarly, if κ is uniform throughout the length of the conductor, $E = V/l$, where V is the electrical potential difference between the ends of the conductor and l is the length of the conductor. Substituting these relations in Eq. 17-4, we obtain

$$\frac{I}{A} = \kappa \frac{V}{l} \tag{17-5}$$

The conductor also obeys Ohm's law in its macroscopic form,

$$V = IR \tag{17-6}$$

Combining Eq. 17-5 and Eq. 17-6, we have

$$R = \frac{1}{\kappa} \frac{l}{A} \tag{17-7}$$

The measurement of conductivity. The resistance of an electrolyte solution is measured in a cell such as that shown in Fig. 17-3. The cell is filled with electrolyte solution through tubes b. Electrical contact with the electrodes e is made by sealing the wires connected to the electrode through the wall of the cell into tubes a that can be filled with mercury, into which copper leads can be dipped. The assumptions made in deriving Eq. 17-6 clearly are not fulfilled in the cell, since the current density j would be expected to vary over the cross section of the path through which current flows and the field E would also vary in the cell. Nevertheless, the inverse relationship between R and κ is found to hold and the constant of proportionality is a characteristic constant of the cell, called a cell constant k. Thus, we can write the equivalent of Eq. 17-7 as

$$R = \frac{1}{\kappa} k \tag{17-8}$$

In practice, the cell constant is determined by measuring the resistance of one or more solutions of known conductivity in the cell. The resistance is expressed in ohms, and k has the dimensions of a reciprocal length, so κ is expressed in mho per meter (or ohm^{-1}-m^{-1}).

The resistance of an electrolyte cannot be measured by applying a steady potential difference to the cell, because the resulting electrode reactions change the concentration of electrolyte about the electrode and because the products of electrolysis produce an opposing potential. In order to avoid these difficulties, frequently covered by the term *polarization*, the resistance

Figure 17-3. A cell for the measurement of conductance of electrolyte solutions, showing filling tubes (b), mercury contact tubes (a), and electrodes (e).

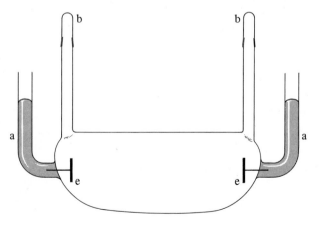

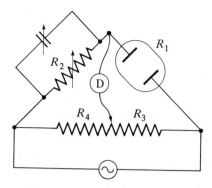

Figure 17-4. A Wheatstone bridge for the measurement of conductance.

of electrolytes is measured by applying an alternating potential difference, with a frequency of the order of kilohertz. The circuit used is a Wheatstone bridge; an example is shown schematically in Fig. 17-4. An oscillator is the source of the alternating potential difference, which is applied across the bridge, $R_3 + R_4$. The detector measures the voltage between the junction of R_2 and the cell R_1 and the variable contact of the bridge. The detector can be simply a set of earphones or an amplifier with an output that goes to a set of earphones or a cathode-ray oscilloscope. The variable resistance R_2 has a variable capacitor in parallel with it to balance the capacitance of the cell. At the balance point, when the detector indicates a minimum in the output signal, Eq. 17-9 is applicable and can be used to calculate R_1.

$$\frac{R_1}{R_2} = \frac{R_3}{R_4} \tag{17-9}$$

The precise measurement of resistance of electrolyte solutions is a complex process, and further details can be found in the references.[2]

Molar conductance. The conductivity κ of an electrolyte is a strongly concentration-dependent property. The molar conductance Λ, defined as

$$\Lambda = \frac{\kappa}{c} \tag{17-10}$$

2. T. Shedlovsky, in A. Weissberger, ed., *Technique of Organic Chemistry*, Vol. 1, *Physical Methods of Organic Chemistry*, 3rd ed., Part 4, Chapt. 45, Interscience Publishers, New York, 1960; D. A. MacInnes, *Principles of Electrochemistry*, Reinhold Publishing Corp., New York, 1939, Chapt. 3.

would be expected to vary less with concentration because κ is roughly proportional to the concentration. If we substitute for κ from Eq. 17-4, we obtain

$$\Lambda = \frac{j/E}{c} \qquad (17\text{-}11)$$

The current density j is the amount of charge passing through unit cross section per unit time and is given by the relation

$$j = c_+ z_+ F v_+ + c_- z_- F v_- \qquad (17\text{-}12)$$

where v_+ and v_- are the speeds of the positive and negative ions. In unit time, the number of moles of positive ions passing any plane of unit area perpendicular to the direction of current flow will be $c_+ v_+$, and the charge per mole is $z_+ F$, where F is the Faraday constant; a similar expression applies to the negative ions moving in the opposite direction. If v_+ and v_- are the number of moles of positive and negative ions that result from the complete ionization of a mole of electrolyte, and α is the fraction of the electrolyte ionized,

$$c_+ = \alpha v_+ c \qquad (17\text{-}13)$$

and

$$c_- = \alpha v_- c \qquad (17\text{-}14)$$

The *mobility* u_j of an ion is defined as

$$u_j = \frac{v_j}{E} \qquad (17\text{-}15)$$

Substituting from Eq. 17-12 through Eq. 17-15 in Eq. 17-11, we obtain

$$\Lambda = (v_+ z_+ u_+ + v_- z_- u_-)F\alpha \qquad (17\text{-}16)$$

$$= [v_+ \lambda_+ + v_- \lambda_-]\alpha \qquad (17\text{-}17)$$

where the molar conductance of an ion is

$$\lambda_j = z_j u_j F \qquad (17\text{-}18)$$

A measurement of transport numbers leads to values of the λ_j, since

$$t_j = \frac{v_j \lambda_j}{\Lambda} \qquad (17\text{-}19)$$

Exercise 17-4.   Verify Eq. 17-19.

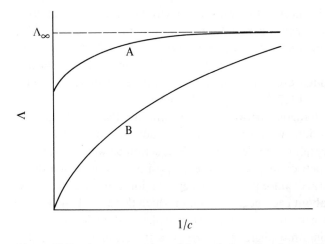

Figure 17-5. A representation of the variation of molar conductance Λ with dilution $1/c$ for curve A, strong electrolytes, B, weak electrolytes, where c is the concentration.

Thus, any variation of Λ with concentration must be due to a variation of α with concentration, a variation of u with concentration, or both.

Strong electrolytes. The molar conductance of strong electrolytes depends on concentration in the way shown schematically in curve A of Fig. 17-5. The variation with concentration is small and the value of Λ approaches a limiting value Λ_∞ asymptotically as the dilution increases. In contrast, the molar conductance of a weak electrolyte changes very greatly with concentration, and it is not possible to choose a limiting value at infinite dilution from the conductance curve, as shown in curve B of Fig. 17-5.

Until the early part of this century, it was thought that the variation of Λ with c for strong electrolytes reflected a variation in the fraction ionized. In 1909, Niels Bjerrum[3] proposed that all strong electrolytes consist entirely of ions in solution, a conclusion based primarily on his studies of the visible absorption spectra of hexaaquochromium salts. He found that the absorption per positive ion is independent of the concentration and independent of the nature of the negative ion, indicating that the electronic structure of the ions is independent of concentration and of the nature of the negative ion. He concluded that "the decrease in molecular conductivity . . . that accom-

3. N. Bjerrum, Proc. 7th International Congress of Pure and Applied Chemistry, London, 1909, Sec. X.

panies the increase in concentration must be due to the action of the electric charges of the ions on each other".[4] In other words, the mobility, not the degree of dissociation, varies with concentration. It is not surprising that charged ions interact in solution, but a theory to account explicitly for the concentration dependence of the mobility of strong electrolytes was not worked out until 1923–1927, by Debye and Hückel[5] and Onsager.[6]

If there were no interaction between ions in solution, the distribution of positive and negative ions would be perfectly random. Debye and Hückel suggested that, as a result of the electrostatic interaction between ions, there is a spherically symmetrical region about each positive ion, in which there is a higher than average concentration of negative ions, and a spherically symmetrical region about each negative ion, in which there is a higher than average concentration of positive ions. They called this region the "ion atmosphere". The ion atmosphere, of course, is a statistical average of the positions of the ions in their random thermal motion, and not a fixed array. Since any ion and its ion atmosphere are oppositely charged, there is an attractive force between them. In an external electric field, the ion atmosphere and its attendant solvent molecules would migrate in a direction opposite to that in which the central ion is moving, so that the motion of the central ion with respect to the solvent is slower than it would be in the absence of other ions. This effect was called the "electrophoretic" effect by Debye and Hückel.

Onsager pointed out that there is an additional mechanism by which the ion atmosphere decreases the mobility of an ion, which he called the "time of relaxation effect". In the absence of an external field, the ion atmosphere is spherically symmetrical. When an ion moves in an external electric field the ion atmosphere becomes distorted and a finite time interval, the time of relaxation, is required for the ion atmosphere to adjust to the motion of the ion. During the time of relaxation there is a greater concentration of opposite charge behind the ion than in front of it, with a net retardation of its motion.

In the 1870's, Kohlrausch found that the molar conductance of a number of strong electrolytes is a linear function of the square root of the concentration in dilute solution. The final result of the Debye-Hückel-Onsager theory is a limiting law for dilute solutions of the same form as Kohlrausch found empirically,

$$\Lambda = \Lambda_\infty - (A + B\Lambda_\infty)c^{\frac{1}{2}} \tag{17-20}$$

4. As quoted in E. A. Guggenheim and R. H. Stokes, *Equilibrium Properties of Aqueous Solutions of Single Strong Electrolytes*, Pergamon Press, Oxford, 1969, p. 3.
5. P. Debye and E. Hückel, *Physik. Z.*, *24*, 185, 305 (1923).
6. L. Onsager, *Physik Z.*, *27*, 388 (1926); *28*, 277 (1927).

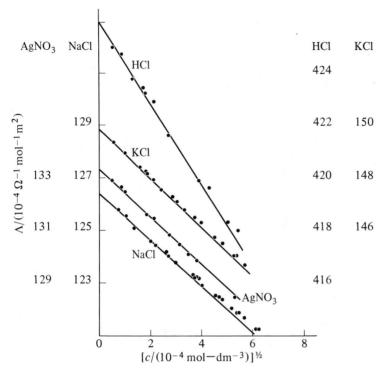

Figure 17-6. The molar conductance of several uni-univalent electrolytes as a function of the square root of the concentration (*c*). The lines represent the Onsager equation and the points are Shedlovsky's experimental results. (By permission, from D. A. MacInnes, *The Principles of Electrochemistry*, Reinhold Publishing Corp., New York, 1939.)

and a prediction of the numerical value of the constants *A* and *B*. The validity of the Onsager equation is strongly supported by the precise data obtained by Shedlovsky; it is shown graphically in Fig. 17-6. The lines are theoretical and the points represent experimental results.

Weak electrolytes. The conductance of weak electrolytes can be explained largely by the assumption that the dependence of Λ on concentration is a consequence of the variation of α with concentration (Eq. 17-16). The concentration dependence of the mobility must be taken into account for a precise description, however. In 1887, on the assumptions that the mobilities of the ions do not change with concentration and that dissociation is complete at infinite dilution, Arrhenius deduced from an equation of the form of Eq. 17-16 that

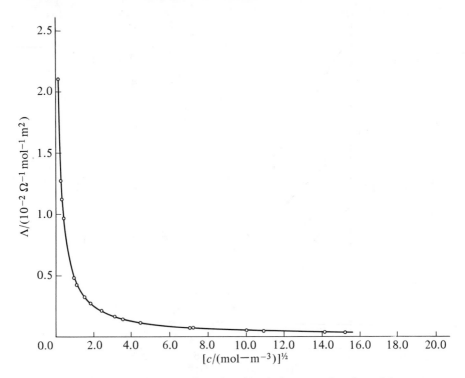

Figure 17-7. Molar conductance of acetic acid solutions as a function of the square root of the concentration (c). [From data of D. A. MacInnes and T. Shedlovsky, *J. Am. Chem. Soc.*, 54, 1429 (1932).]

$$\Lambda_\infty = (v_+z_+u_+ + v_-z_-u_-)F$$

and

$$\alpha = \frac{\Lambda}{\Lambda_\infty} \qquad (17\text{-}21)$$

When Λ for weak electrolytes is plotted against $c^{\frac{1}{2}}$, as in Fig. 17-7, it is not possible to choose a value of Λ_∞ by extrapolation because of the sharp change with concentration in dilute solution. The value of Λ_∞ is calculated from extrapolated values for strong electrolytes on the basis of Kohlrausch's *law of independent migration*, which states that λ_+ and λ_- (Eq. 17-17) are independent of the nature of the counterion at infinite dilution. Thus, for example, the value of Λ_∞ for acetic acid can be calculated from the values for HCl, NaCl, and NaOOCCH$_3$,

$$\Lambda_\infty(CH_3COOH) = \Lambda_\infty(HCl) + \Lambda_\infty(NaOOCCH_3) - \Lambda_\infty(NaCl) \qquad (17\text{-}22)$$

For the equilibrium

$$CH_3COOH(aq) + H_2O(l) = CH_3COO^-(aq) + H_3O^+(aq)$$

we can write an equilibrium constant on the concentration scale, as in Eq. 16-126. On the assumption that K_y can be set equal to one and that the solution is sufficiently dilute that $a\,(H_2O)$ is equal to one,

$$K_c = \frac{([H_3O^+]/c^\ominus)\,([CH_3COO]/c^\ominus)}{[CH_3COOH]/c^\ominus} \qquad (17\text{-}23)$$

If α is the degree of dissociation and c is the stoichiometric concentration of acetic acid,

$$K_c = \frac{(\alpha c)\,(\alpha c)/c^\ominus}{(1 - \alpha)c}$$

$$= \frac{\alpha^2\ c/c^\ominus}{1 - \alpha} \qquad (17\text{-}24)$$

which is the Ostwald *dilution law*, derived by Ostwald in 1888. Substituting from Eq. 17-21 in Eq. 17-24, we obtain

$$K_c = \frac{(\Lambda/\Lambda_\infty)^2\ c/c^\ominus}{(1 - \Lambda/\Lambda_\infty)} \qquad (17\text{-}25)$$

Precise measurements by MacInnes and Shedlovsky can be used to show that Eq. 17-25 yields fairly constant values of K_c at low concentrations, as shown in the fourth column of Table 17-1. There are two almost compensating errors in the use of Eq. 17-25, however.[7] The mobility of an ion is concentration dependent, so α should be calculated from Λ/Λ_e, where Λ_e is the sum of the molar conductances of the H_3O^+ and CH_3COO^- ions at the ionic concentrations at equilibrium. But, the ionic concentrations at equilibrium are not known until α is calculated. Therefore, an iterative procedure is used in which an approximate value α' is calculated from Λ/Λ_∞, an approximate value of Λ_e is calculated from the ionic concentration $\alpha'c$, and a new value of α' is calculated from Λ/Λ_e. This process is continued until two successive cycles show less than some predetermined change in the calculated value of α', which is then taken as the value of α.

The values of K_c'' in column six of Table 17-1 reflect a definite concentration dependence, the result of neglecting activity coefficients in Eq. 16-126. For electrolytes, the logarithm of the activity coefficient in dilute solution is found to be a linear function of the square root of the ion concentration. A plot of log K_c'' against the square root of the ion concentration is linear in

7. D. A. MacInnes and T. Shedlovsky, *J. Am. Chem. Soc.*, 54, 1429 (1932).

Table 17-1. Calculation of ionization constant of CH_3COOH^a

$c/(\text{mol-m}^{-3})$	$\Lambda/(\Omega^{-1}\text{mol}^{-1}\text{m}^2)$	$\alpha' = \Lambda/\Lambda_\infty$	$10^5 K_c' = \dfrac{(\alpha')^2\,(c/c^\ominus)}{1-\alpha'} \times 10^5$	$\alpha = \Lambda/\Lambda_e$	$10^5 K_c'' = \dfrac{\alpha^2\, c/c^\ominus}{1-\alpha} \times 10^5$
0.028014	0.021032	0.53847	1.760	0.53925	1.768
0.11135	0.012771	0.32697	1.769	0.32773	1.779
0.15321	0.011202	0.28680	1.767	0.28752	1.778
0.21844	0.0096466	0.24698	1.769	0.24767	1.781
1.02831	0.0048133	0.12323	1.781	0.12375	1.797
1.36340	0.0042215	0.10808	1.786	0.10857	1.803
2.41400	0.0032208	0.082460	1.789	0.082900	1.809
3.44065	0.0027191	0.069615	1.792	0.070022	1.814
5.91153	0.0020956	0.053652	1.798	0.054012	1.823
9.8421	0.0016367	0.041903	1.804	0.04225	1.832
12.829	0.0014371	0.036793	1.803	0.037095	1.834
20.000	0.0011563	0.029604	1.806	0.029875	1.840
50.000	0.0007356	0.018833	1.807	0.019048	1.849
52.303	0.0007200	0.018434	1.811	0.018649	1.854
100.000	0.0005200	0.013313	1.796	0.013496	1.846
119.447	0.0004759	0.012184	1.794	0.012359	1.847
200.000	0.0003650	0.009345	1.762	0.009495	1.821
230.785	0.0003391	0.008682	1.753	0.008827	1.814

$$\Lambda_\infty = 0.039059\ \Omega^{-1}\text{mol}^{-1}\text{m}^2;\ c^\ominus = 1000\ \text{mol-m}^{-3}$$

^{a}By permission, from conductance data of D. A. MacInnes and T. Shedlovsky, *J. Am. Chem. Soc.*, **54**, 1429 (1932).

dilute solution, as shown in Fig. 17-8, and can be extrapolated to zero ionic concentration to obtain the thermodynamic equilibrium constant. For acetic acid at 298 K, K_c is 1.753×10^{-5}.

The effects of high field and high frequency. The results of measurements at high field strength and at high frequency also support the Debye-Hückel-Onsager theory of conductance. Wien studied the effect of high field strength on the conductance of electrolytes, and the increase of conductance with increasing field strength at high fields is called the *Wien effect*. This effect is explained qualitatively by postulating that the ions move so rapidly under the influence of a very strong field that there is no time for the ion atmosphere to form. Thus, the electrophoretic effect and the time of relaxation effect have a decreasing influence with increasing field strength and the molar conductance increases with increasing field strength. The observation that the molar conductance of weak electrolytes does increase with increasing field strength even more than the molar conductance of strong electrolytes suggests that high field strengths increase the degree of dissociation of weak electrolytes in addition to their effect on the mobility of the ions.

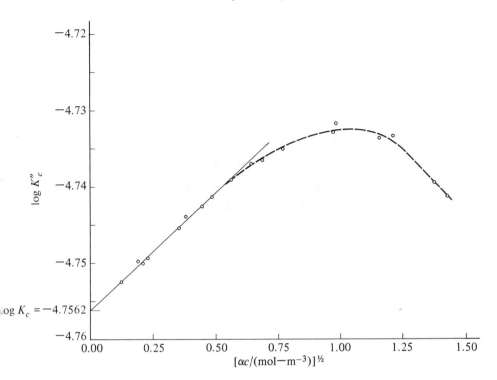

Figure 17-8. A plot of log K_c'' against the square root of the ionic concentration (αc), according to the data of D. A. MacInnes and T. Shedlovsky, *J. Am. Chem. Soc.*, *54*, 1429 (1932), as given in Table 17-1.

On the basis of the Debye-Hückel-Onsager theory, Debye and Falkenhagen predicted an increase in mobility of ions with increasing frequency of the applied field. If the direction of motion of the ions alternates in a period of time comparable to the relaxation time of the ion atmosphere, there is insufficient time for the dissymmetry that produces the time of relaxation effect to occur. Thus, increasing the frequency of an applied electric field eliminates the time of relaxation effect but not the electrophoretic effect. Their predictions have been confirmed experimentally.

17-3. Colligative properties of strong electrolytes

In 1887, Van't Hoff pointed out that any colligative property of a strong electrolyte in dilute solution is some multiple i of the same property for a nonelectrolyte, where i is called the *Van't Hoff factor*. For example, the vapor pressure lowering is

$$p_A{}^* - p_A = ip_A{}^* x_B \qquad (17\text{-}26)$$

Similarly, the freezing point depression is

$$\Delta T_f = iK_{f,A} m_B \qquad (17\text{-}27)$$

and the osmotic pressure is

$$\Pi = iRTc_B \qquad (17\text{-}28)$$

He also noted that the value of i is usually less than $v_+ + v_-$, the value expected if the strong electrolyte is completely dissociated. The difference between i and $v_+ + v_-$ was first attributed to incomplete dissociation, and a value of the degree of dissociation was calculated from the expression

$$\alpha = \frac{i - 1}{v_+ + v_- - 1} \qquad (17\text{-}29)$$

Exercise 17-5. Verify Eq. 17-29, assuming that the value of i is determined entirely by incomplete dissociation.

The values of α calculated from Eq. 17-29 were different from the values of α calculated from conductance data for the same solution by Eq. 17-21, and neither led to a constant value of K_c when substituted in Eq. 17-24 for varying values of c. The colligative properties of strong electrolytes, like the conductance behavior, were explained on the assumption of complete dissociation and interaction between ions.[8] Debye and Hückel were able to derive a limiting law for the activity coefficients of ions from the same model that had led to an understanding of the conductance of electrolytes.

The activity of strong electrolytes
Consider a uni-univalent electrolyte, such as NaCl, which ionizes completely in dilute aqueous solution (Na^+ and Cl^-). Equation 16-61 implies not only that $a(NaCl)$ is equal to $m(NaCl)/m^{\ominus}(NaCl)$ at infinite dilution, but also that the slope of a plot of $a(NaCl)$ against $m(NaCl)/m^{\ominus}(NaCl)$ is equal to one at infinite dilution. If the activity of NaCl in solution is calculated from the colligative properties as described in Sec. 16-5, a plot of $a(NaCl)$ against $m(NaCl)/m^{\ominus}(NaCl)$ has a limiting slope of zero instead of one. A nonzero

8. N. Bjerrum, *Z. Electrochem.*, *24*, 321–28 (1918).

finite slope is obtained only when the activity is plotted against the *square* of the molality, and then the slope is equal to one. This result suggests that the appropriate limiting law for the activity of NaCl is

$$\lim_{m(\text{NaCl}) \to 0} \frac{a(\text{NaCl})}{[m(\text{NaCl})/m^{\ominus}(\text{NaCl})]^2} = 1 \tag{17-30}$$

The empirical relation between the activity and the molality can be understood on the assumption that the chemical potential of the solute NaCl is the sum of the chemical potentials of the Na^+ and Cl^- ions. That is

$$\mu(\text{NaCl}) = \mu_+ + \mu_- \tag{17-31}$$

If we apply Eq. 16-51, the definition of the activity, the result is

$$\mu^{\ominus}(\text{NaCl}) + RT \ln a(\text{NaCl}) = \mu_+^{\ominus} + RT \ln a_+ + \mu_-^{\ominus} + RT \ln a_- \tag{17-32}$$

We can also assume that

$$\mu^{\ominus}(\text{NaCl}) = \mu_+^{\ominus} + \mu_-^{\ominus}$$

so that

$$a(\text{NaCl}) = (a_+)(a_-) \tag{17-33}$$

The individual ion activities should follow the limiting relations

$$\lim_{m(\text{NaCl}) \to 0} \frac{a_+}{(m_+/m_+^{\ominus})} = 1 \tag{17-34}$$

and

$$\lim_{m(\text{NaCl}) \to 0} \frac{a_-}{(m_-/m_-^{\ominus})} = 1 \tag{17-35}$$

and the product of the limits is

$$\lim_{m(\text{NaCl}) \to 0} \frac{(a_+)(a_-)}{(m_+/m_+^{\ominus})(m_-/m_-^{\ominus})} = \lim_{m(\text{NaCl}) \to 0} \frac{a(\text{NaCl})}{[m(\text{NaCl})/m^{\ominus}(\text{NaCl})]^2} = 1$$

which is the relation found empirically.

There is no method to determine the activity of a single ion experimentally, since there is no way to vary the amount of a single ion while keeping the amounts of the other ions constant. It is convenient to define a *mean ionic activity* $a_{\pm}$, which can be determined experimentally, by the relation

$$a_{\pm} = [(a_+)(a_-)]^{\frac{1}{2}} = (a_{\text{B}})^{\frac{1}{2}} \tag{17-36}$$

for a uni-univalent electrolyte. The single-ion activity coefficients, defined as

$$\gamma_+ = \frac{a_+}{(m_+/m_+^{\ominus})} \tag{17-37}$$

and

$$\gamma_- = \frac{a_-}{(m_-/m_-^{\ominus})} \tag{17-38}$$

cannot be measured experimentally, but the *mean ionic activity coefficient*, $\gamma_\pm$, defined as

$$\gamma_\pm = (\gamma_+\gamma_-)^{\frac{1}{2}} \tag{17-39}$$

for a uni-univalent electrolyte, can be measured. A mean ionic molality, $m_\pm$, is defined as

$$m_\pm = [(m_+)(m_-)]^{\frac{1}{2}} \tag{17-40}$$

so that

$$\gamma_\pm = \frac{a_\pm}{(m_\pm/m_\pm^{\ominus})} \tag{17-41}$$

The corresponding general definitions for an electrolyte that yields v_+ moles of positive ions and v_- moles of negative ions per mole of electrolyte on complete dissociation are

$$a_B = (a_+)^{v_+}(a_-)^{v_-} \tag{17-42}$$

$$a_\pm = (a_B)^{1/(v_+ + v_-)} = (a_B)^{1/v} \tag{17-43}$$

$$\gamma_\pm = [(\gamma_+)^{v_+}(\gamma_-)^{v_-}]^{1/v} \tag{17-44}$$

and

$$m_\pm = [(m_+)^{v_+}(m_-)^{v_-}]^{1/v}$$
$$= [(v_+)^{v_+}(m_B)^{v_+}(v_-)^{v_-}(m_B)^{v_-}]^{1/v} \tag{17-45}$$

and the appropriate limiting law consistent with experimental observation is

$$\lim_{m_B \to 0} \frac{a_B}{(m_B/m_B^{\ominus})^v} = (v_+)^{v_+}(v_-)^{v_-} \tag{17-46}$$

and, as before,

$$\gamma_\pm = \frac{a_\pm}{(m_\pm/m_\pm^{\ominus})}$$

Calculation of electrolyte activity from colligative properties

As we saw in Sec. 16-5, the values of colligative properties can be used to calculate the activity of the solvent. The Gibbs-Duhem relationship then provides corresponding values of the activity of the solute. For solutions of nonelectrolytes, we used the Gibbs-Duhem equation in the form (Eq. 16-92)

$$\delta \ln f_B = -\frac{x_A}{x_B} \delta \ln f_A$$

The activity coefficients of solute and solvent are of a comparable order of magnitude in dilute solutions of nonelectrolytes, so that Eq. 16-92 is a useful relationship. But, the activity coefficients of an electrolyte solute differ substantially from unity in very dilute solutions. The activity coefficient of the solvent in the same solutions differs from one by less than 1×10^{-3}. The data in the first three columns of Table 17-2 illustrate the situation. It can be seen that the calculation of the activity coefficient of solute from the activity coefficient of H_2O would be imprecise, at best.

Table 17-2. Activity coefficients and osmotic coefficients for aqueous potassium nitrate solutions[a]

$[KNO_3]/(\text{mol-dm}^{-3})$	$y(KNO_3)$	$f(H_2O)$	$g(H_2O)$
0.01	0.8993	1.00001	0.9652
0.05	0.7941	1.00005	0.9252
0.1	0.7259	1.0002	0.8965
1.0	0.3839	1.0056	0.6891

[a] By permission, from data of G. Scatchard, S. S. Prentiss, and P. T. Jones, *J. Am. Chem. Soc.*, 54, 2690–95 (1932). The activity coefficient y of the solute is on the concentration scale, whereas the activity coefficient f of the solvent is on the mole fraction scale.

To deal with this problem, Bjerrum[9] suggested that the deviation of solvent behavior from Raoult's law be described by the *osmotic coefficient g* rather than by the activity coefficient f_A. In the latter method, the chemical potential of the solvent is

$$\mu_A = \mu_A^\ominus + RT \ln x_A + RT \ln f_A \qquad (17\text{-}47)$$

and

$$\lim_{x_A \to 1} f_A = 1 \qquad (17\text{-}48)$$

In the former method, the chemical potential of the solvent is

$$\mu_A = \mu_A^\ominus + gRT \ln x_A \qquad (17\text{-}49)$$

9. N. Bjerrum, *Z. Electrochem.*, 24, 321–28 (1918); *Z. phys. Chem.*, 104, 406–32 (1932).

and

$$\lim_{x_A \to 1} g = 1$$

For a solution of electrolytes,[10]

$$x_A = \frac{n_A}{n_A + \sum_j n_{j,+} + \sum_j n_{j,-}}$$

$$= \frac{n_A}{n_A + \nu n_B} \tag{17-50}$$

The greater usefulness of g over f_A can be seen from the values in the third and fourth columns of Table 17-2.

Equating the expressions of μ_A from Eq. 16-51 and Eq. 17-49, we obtain

$$\ln a_A = g \ln x_A$$

or

$$g = \frac{\ln a_A}{\ln x_A} \tag{17-51}$$

When the activity of the solvent is determined from a colligative property, then g can be calculated. If we use the osmotic pressure (Eq. 16-107) as an example, the result is

$$g = \frac{-\mathscr{V}_A^* \Pi / RT}{\ln x_A} \tag{17-52}$$

From Eq. 17-50

$$\ln x_A = \ln \frac{n_A}{n_A + \nu n_B}$$

$$= \ln \frac{1}{1 + \nu n_B / n_A}$$

$$= -\ln\left(1 + \frac{\nu n_B}{n_A}\right) \tag{17-53}$$

For the dilute solutions for which the osmotic coefficient is most useful, the natural logarithm in Eq. 17-53 can be expanded in a Taylor's series, and terms of higher powers can be neglected. The result is

$$\ln x_A = -\frac{\nu n_B}{n_A} \tag{17-54}$$

10. D. P. Shoemaker and C. W. Garland, *Experiments in Physical Chemistry*, McGraw-Hill Book Co., New York, 1962, p. 142.

Substituting in Eq. 17-52, we obtain

$$\Pi = \left(\frac{gv}{RT}\right)\frac{n_B}{n_A\mathcal{V}_A{}^*} \qquad (17\text{-}55)$$

The product of $n_A\mathcal{V}_A{}^*$ is essentially equal to V, the volume of solution, for dilute solutions, so that Eq. 17-55 reduces to

$$\Pi = gv\frac{c_B}{RT} \qquad (17\text{-}56)$$

It can be seen that g is the ratio between the observed osmotic pressure and the osmotic pressure that would be observed for a completely dissociated electrolyte that follows Henry's law, hence the name osmotic coefficient. A similar result can be obtained for the freezing point depression and the vapor pressure lowering.

Exercise 17-6. Derive a relation between the osmotic coefficient and the freezing point depression, using Eq. 17-51, Eq. 17-54, Eq. 16-98, and Eq. 16-100.

Exercise 17-7. Derive a relation between the osmotic coefficient and the vapor pressure lowering using Eq. 17-51, Eq. 17-54, Eq. 16-78, and Eq. 16-87.

Once values of g as a function of solution composition have been obtained, the Gibbs-Duhem equation can be used to relate the osmotic coefficient of the solvent and the activity coefficient of the solute. For this purpose, the chemical potential of the solvent is expressed as in Eq. 17-49, with the approximation given in Eq. 17-54, so that

$$\mu_A = \mu_A^\ominus - RTvg\frac{n_B}{n_A} \qquad (17\text{-}57)$$

If we describe the composition of the solution in terms of molalities, Eq. 17-57 becomes

$$\mu_A = \mu_A^\ominus - RTvgm_B M_A$$

and

$$\delta\mu_A = -RTv M_A(g\,\delta m_B + m_B\,\delta g) \qquad (17\text{-}58)$$

The chemical potential of the solute is

$$\mu_B = \mu_B^\ominus + RT\ln a_B$$

$$= \mu_B^\ominus + RT\ln\left(\frac{m_\pm}{m_\pm^\ominus}\right)^v + RT\ln\,(\gamma_\pm)^v$$

$$= \mu^{\circ} + vRT \ln \left(\frac{m_{\pm}}{m_{\pm}^{\ominus}} \right) + vRT \ln \gamma_{\pm} \tag{17-59}$$

and

$$\delta \mu_B = vRT \, \delta \ln m_{\pm} + vRT \, \delta \ln \gamma_{\pm}$$

$$= vRT \, \delta \ln m_B + vRT \, \delta \ln \gamma_{\pm} \tag{17-60}$$

(See Eq. 17-45.)

Using the Gibbs-Duhem Equation in the form $\delta \mu_A = -(n_B/n_A)\delta \mu_B$, we obtain

$$\delta \ln \gamma_{\pm} = - \frac{1 - g}{m_B} \, \delta m_B + \delta g \tag{17-61}$$

which is *Bjerrum's equation*.

Exercise 17-8. Verify Eq. 17-61.

If Eq. 17-61 is integrated from the infinitely dilute solution to some finite but still dilute molality, the result is

$$\int_0^{\ln \gamma_{\pm}} d \ln \gamma_{\pm} = - \int_0^{m_B} \frac{1 - g}{m_B} \, dm_B + \int_1^g dg$$

or

$$\ln \gamma_{\pm} = - \int_0^{m_B} \frac{1 - g}{m_B} \, dm_B + (g - 1) \tag{17-62}$$

The integral in Eq. 17-62 is usually integrated graphically or numerically. Though both $1 - g$ and m_B go to zero as $m_B \to 0$, the ratio has a finite limit. Such a finite limit is not obvious from a plot of $(1 - g)/m_B$ against m_B for electrolytes, as can be seen from Fig. 17-9. But, if the integral in Eq. 17-62 is transformed to

$$2 \int_0^{(m_B)^{\frac{1}{2}}} \frac{(1 - g)}{m_B^{\frac{1}{2}}} \, d(m_B^{\frac{1}{2}}) \tag{17-63}$$

the finite limit can be seen clearly as in Fig. 17-10.

Exercise 17-9. Verify the transformation of the integral in Eq. 17-62 by substituting $m_B = (m_B^{\frac{1}{2}})^2$.

At the higher concentrations for which the osmotic coefficient expressions we have used are still valid, the activity coefficient of the solvent becomes significantly different from one. For still higher concentrations, Eq. 16-93

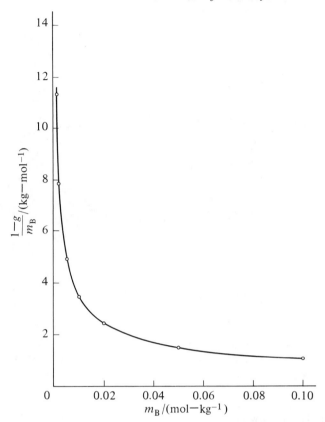

Figure 17-9. A plot of $(1 - g)/m_B$ against m_B from the freezing point data on KNO_3 solutions. [G. Scatchard, S. S. Prentice, and P. T. Jones, *J. Am. Chem. Soc.*, *54*, 2690–95 (1932).] See Eq. 17-62.

can be used to obtain the activity coefficient of the solute from the activity coefficient of the solvent.

Exercise 17-10. Derive the relationship between the osmotic coefficient and the activity coefficient of the solvent.

The Debye-Hückel theory

The chemical potential of a strong electrolyte as solute is given in Eq. 17-59. If the solute followed Henry's law of dilute solutions, the chemical potential would be

$$\mu_B^{dil} = \mu_B^{\ominus} + vRT \ln\left(\frac{m_{\pm}}{m_{\pm}^{\ominus}}\right)$$

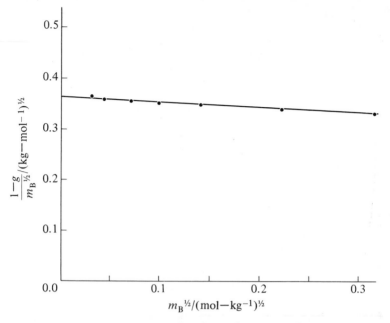

Figure 17-10. A plot of $(1 - g)/m_B^{\frac{1}{2}}$ against $m_B^{\frac{1}{2}}$ from the freezing point data on KNO_3 solutions. [G. Scatchard, S. S. Prentice, and P. T. Jones, *J. Am. Chem. Soc.*, *54*, 2690–95 (1932).] See Eq. 17-63.

The deviation from Henry's law is given by

$$\mu_B - \mu_{B,\,dil} = vRT \ln \gamma_\pm \qquad (17\text{-}64)$$

Equation 17-64 can also be interpreted as giving the Gibbs free energy change for the *transfer* of one mole of solute from a hypothetical solution in which it follows Henry's law to the real solution of the same molality.

Debye and Hückel assumed that the deviation from dilute solution behavior given in Eq. 17-64 is due entirely to electrical interactions among charges of the ions. From this assumption, the solution in which Henry's law is obeyed is one in which the ions are uncharged. Then we can say that $\mu_B - \mu_{B,\,dil}$ is equal to the Gibbs free energy of charging the ions, an electrical free energy, or $\mu_{B,\,el}$. Debye and Hückel proposed a model that led to an expression for $\mu_{B,\,el}$, and hence to $vRT \ln \gamma_\pm$.

The model is based on the same concept of the ion atmosphere that was used to explain the conductance data for strong electrolytes except that the average distribution of charges in the solution is an equilibrium distribution. An electrostatic potential corresponds to this average distribution, the value of which is a function of position in the solution. The electrostatic potential at a point is the potential energy per unit positive charge at that point for a charge small enough so that it does not perturb the potential.

Other workers had attempted to solve the problem of the ionic interactions in strong electrolytes, notably Milner in 1912 and Ghosh in 1918, but Debye and Hückel[11] overcame the mathematical difficulties by choosing a co-ordinate system with its origin at a single ion, called the central ion. They expressed the concentrations of the other ions and the electrical potential as functions only of distance from the central ion.

Derivation. Let n_+ be the average number of positive ions per unit volume in any volume element δV at a distance r from the central ion and let n_- be the average number of negative ions per unit volume in the same volume element. The charge density ρ is then

$$\rho = n_+ z_+ e + n_- z_- e \qquad (17\text{-}65)$$

where z_+ and z_- are the charge numbers of the positive and negative ions (with signs) and e is the protonic charge.

According to Gauss' law of electrostatics,[12]

$$\nabla \cdot \mathbf{E} = \frac{\rho}{D\varepsilon_0} \qquad (17\text{-}66)$$

where ∇ is the differential vector operator

$$\mathbf{i}\,\frac{\partial}{\partial x} + \mathbf{j}\,\frac{\partial}{\partial y} + \mathbf{k}\,\frac{\partial}{\partial z}$$

$\mathbf{E}$ is the electrical field vector, D is the dielectric constant of the medium, and ε_0 is the permittivity of vacuum (8.85×10^{-12} $C\text{-}V^{-1}\text{-}m^{-1}$). ($D$ is also called the relative permittivity, so that $D\varepsilon_0$ is equal to ε, the permittivity of the medium in which the charges exist.) The field E is the negative of the gradient of the electrostatic potential ψ; that is,

$$\mathbf{E} = -\,\mathbf{i}\,\frac{\partial \psi}{\partial x} - \mathbf{j}\,\frac{\partial \psi}{\partial y} - \mathbf{k}\,\frac{\partial \psi}{\partial z}$$

$$= -\nabla \psi = -\,\text{grad}\,\psi \qquad (17\text{-}67)$$

11. P. J. W. Debye and E. Hückel, *Physik, Z.*, *24*, 185 (1923); *The Collected Papers of Peter J. W. Debye*, Interscience Publishers, New York, 1954.
12. R. P. Feynman, R. B. Leighton, and M. Sands, *The Feynman Lectures on Physics*, Vol. II, Addison-Wesley, Reading, 1964, pp. 4–10.

If contour lines of constant potential are drawn, the field vector $\mathbf{E}$ is always in a direction normal to the equipotential lines. Substitution of Eq. 17-67 in Eq. 17-66 yields

$$-\nabla \cdot \nabla \psi = -\nabla^2 \psi$$

$$= -\frac{\partial^2 \psi}{\partial x^2} - \frac{\partial^2 \psi}{\partial y^2} - \frac{\partial^2 \psi}{\partial z^2}$$

$$= \frac{\rho}{D\varepsilon_0} \tag{17-68}$$

where the operator ∇^2 is the same one that appears in Schrödinger's equation (Eq. 4-35). This result is called *Poisson's equation*; it and its solutions form a complete mathematical description of electrostatic phenomena. The form of Poisson's equation in spherical coordinates when ψ is a function only of r is

$$\frac{d^2 \psi}{dr^2} + \frac{2}{r}\frac{d\psi}{dr} = -\frac{\rho}{D\varepsilon_0} \tag{17-69}$$

Exercise 17-11. Show that $\nabla^2 \psi$ reduces to the left side of Eq. 17-69 for a problem in which ψ depends only on r. Start with the expression for $\nabla^2 \psi$ in Eq. 4-39.

In order to solve Eq. 17-69, ρ must be known as a function of ψ. Debye and Hückel assumed that the distribution of ions about a central ion can be described by a Boltzmann function in which the exponential is the ratio of the potential energy in the electric field to kT; that is,

$$n_+ = n_+^\circ \exp\left(-\frac{z_+ e\psi}{kT}\right) \tag{17-70}$$

and

$$n_- = n_-^\circ \exp\left(-\frac{z_- e\psi}{kT}\right) \tag{17-71}$$

where n_+° and n_-° are the average concentrations of positive and negative ions in the solution. From this assumption, Eq. 17-65 becomes

$$\rho = n_+^\circ z_+ e \exp\left(-\frac{z_+ e\psi}{kT}\right) + n_-^\circ z_- e \exp\left(-\frac{z_- e\psi}{kT}\right) \tag{17-72}$$

Substitution of Eq. 17-72 in Eq. 17-69 gives

$$\frac{d^2 \psi}{dr^2} + \frac{2}{r}\frac{d\psi}{dr} = -\frac{1}{D\varepsilon_0}\left[n_+^\circ z_+ e \exp\left(-\frac{z_+ e\psi}{kT}\right) + n_-^\circ z_- e \exp\left(-\frac{z_- e\psi}{kT}\right)\right]$$

$$\tag{17-73}$$

which is the *Poisson-Boltzmann equation.*

Solving the Poisson-Boltzmann equation is a formidable task. To simplify the problem, Debye and Hückel considered the case of a solution sufficiently dilute that $ze\psi \ll kT$, so that the exponential terms can be expanded in a Taylor series and the higher powers can be neglected. The result is

$$\frac{d^2\psi}{dr^2} + \frac{2}{r}\frac{d\psi}{dr} = -\frac{1}{D\varepsilon_0}\left[n_+^\circ z_+ e\left(1 - \frac{z_+ e\psi}{kT}\right) + n_-^\circ z_- e\left(1 - \frac{z_- e\psi}{kT}\right)\right]$$

(17-74)

Exercise 17-12. Verify Eq. 17-74.

Since the solution is electrically neutral, $n_+^\circ z_+ = -n_-^\circ z_-$, and Eq. 17-74 reduces to

$$\frac{d^2\psi}{dr^2} + \frac{2}{r}\frac{d\psi}{dr} = \frac{e^2}{D\varepsilon_0 kT}[n_+^\circ z_+^2 + n_-^\circ z_-^2]\psi$$

(17-75)

The factor $(e^2/D\varepsilon_0 kT)[n_+^\circ z_+^2 + n_-^\circ z_-^2]$ is designated κ^2, a constant for any given solution, so that Eq. 17-75 is written

$$\frac{d^2\psi}{dr^2} + \frac{2}{r}\frac{d\psi}{dr} = \kappa^2\psi$$

(17-76)

The parameter κ has the dimensions of an inverse length, and $1/\kappa$ is called the *Debye length.* It is an approximate measure of the effective radius of the ionic atmosphere.

The left side of Eq. 17-76 can be shown to be equal to $(1/r)[d^2(r\psi)/dr^2]$, so that

$$\frac{d^2(r\psi)}{dr^2} = \kappa^2(r\psi)$$

(17-77)

Exercise 17-13. Verify Eq. 17-77.

The general solution of Eq. 17-77 is

$$r\psi = A\exp(-\kappa r) + B\exp(\kappa r)$$

Since ψ cannot become infinite at infinite r, B must be equal to zero, and

$$r\psi = A\exp(-\kappa r)$$

or

$$\psi = \frac{A}{r}\exp(-\kappa r)$$

(17-78)

Exercise 17-14. Show that Eq. 17-78 is a solution of Eq. 17-76.

The value of A can be obtained by observing that the sum of the charges other than the central ion must be equal in magnitude and opposite in sign to the charge of the central ion. Thus, if a positive ion is taken as the central ion, and if a is the distance of closest approach between ions,

$$
\begin{aligned}
-z_+e &= \int_a^\infty \rho \, dV \\
&= \int_a^\infty 4 \pi r^2 \rho \, dr
\end{aligned}
\tag{17-79}
$$

The integral in Eq. 17-79 can be evaluated by substituting for ρ from Eq. 17-69, Eq. 17-76, and Eq. 17-78. The result is

$$
z_+e = \int_a^\infty 4 \pi D\varepsilon_0 \kappa^2 \, Ar \exp(-\kappa r) \, dr
\tag{17-80}
$$

Exercise 17-15. Verify Eq. 17-80.

Integrating by parts, we obtain

$$
z_+e = 4 \pi D\varepsilon_0 A \exp(-\kappa a)(1 + \kappa a)
$$

or

$$
A = \frac{z_+e \exp(\kappa a)}{4 \pi D\varepsilon_0 (1 + \kappa a)}
\tag{17-81}
$$

Exercise 17-16. Verify the integration of Eq. 17-80.

Substitution from Eq. 17-81 for A in Eq. 17-78 yields

$$
\psi = \frac{z_+e \exp(\kappa a) \exp(-\kappa r)}{4 \pi D\varepsilon_0 r(1 + \kappa a)} \cdot
$$

At $r = a$, the distance of closest approach,

$$
\begin{aligned}
\psi(a) &= \frac{z_+e}{4 \pi D\varepsilon_0 a} \left(\frac{1}{1 + \kappa a} \right) \\
&= \frac{z_+e}{4 \pi D\varepsilon_0 a} \left(1 - \frac{\kappa a}{1 + \kappa a} \right) \\
&= \frac{z_+e}{4 \pi D\varepsilon_0 a} - \frac{z_+e}{4 \pi D\varepsilon_0} \frac{\kappa}{1 + \kappa a}
\end{aligned}
\tag{17-82}
$$

The first term on the right in Eq. 17-82 is the expression for the potential at the surface of a charged sphere due to its own charge, and this potential is independent of the concentration of the solution. The second term, which is concentration dependent, is the potential due to the other ions, the *ion atmosphere* of Debye and Hückel. It is this potential that can be used to calculate μ_{el}.

Guntelberg[13] and Müller[14] devised a model that leads to the calculation of μ_{el} from the Debye-Hückel expression for ψ. From Eq. 17-64 and the following text

$$\mu_{el} = \nu RT \ln \gamma_{\pm} \tag{17-83}$$

and μ_{el} is the Gibbs free energy change for the transfer of the ions from an infinitely dilute solution to the real solution. Though it is not possible to measure the activity of a single ion experimentally, the model calculation is carried out most simply in terms of a single ion activity coefficient. Let

$$\begin{aligned} \mu_{+, el} &= \mu_{+} - \mu_{+, dil} \\ &= RT \ln \gamma_{+} \end{aligned} \tag{17-84}$$

In this model $\mu_{+, el}$ is the free energy of transfer of a positive ion from a solution in which it is uncharged to one in which it is charged in the presence of the ion atmosphere. From Eq. 10-78

$$\delta\mu_{el} = \delta W_{net, rev, el} \tag{17-85}$$

Therefore, if there is a reversible process that is equivalent to the ion transfer, the work of that process will be equal to the required change in chemical potential.

If a solution were identical to an ionic solution in all respects except for the absence of a charge on the central ion, the central ion would have a chemical potential $\mu_{+, id}$, according to the assumptions of Debye and Hückel. It can be brought reversibly to its real chemical potential μ_{+} by bringing charge to the ion from infinity in infinitesimal increments $e\,\delta z_{+}$. Since the potential ψ at a point is the work required to transfer unit charge from infinity to that point,

$$\delta W_{el} = \psi\, e\, \delta z_{+} \tag{17-86}$$

In this case, ψ is the potential at $r = a$ due to the ion atmosphere, and the

13. E. Guntelberg, *Z. physik. Chem.*, *123*, 199 (1926).
14. H. Müller, *Physik. Z.*, *28*, 324 (1927).

value of ψ is a function of the charge on the central ion, as given by the second term on the right in Eq. 17-82. Thus,

$$RT \ln \gamma_+ = \mu_{el}$$

$$= \int_0^{\mu_{el}} d\mu$$

$$= \int_0^{z_+} \psi e \, dz_+$$

$$= \int_0^{z_+} -\frac{L z_+ e^2}{4\pi D \varepsilon_0} \left(\frac{\kappa}{1 + \kappa a}\right) dz_+$$

$$= -\frac{L e^2 z_+^2}{8\pi D \varepsilon_0} \left(\frac{\kappa}{1 + \kappa a}\right)$$

and

$$\ln \gamma_+ = -\frac{e^2 z_+^2}{8\pi D \varepsilon_0 kT} \frac{\kappa}{1 + \kappa a} \tag{17-87}$$

Similarly, if the negative ion is taken as the central ion,

$$\ln \gamma_- = -\frac{e^2 z_-^2}{8\pi D \varepsilon_0 kT} \left(\frac{\kappa}{1 + \kappa a}\right) \tag{17-88}$$

From Eq. 17-44,

$$\ln \gamma_\pm = \frac{1}{v}[v_+ \ln \gamma_+ + v_- \ln \gamma_-]$$

$$= -\frac{e^2}{8\pi D \varepsilon_0 kT} \left(\frac{\kappa}{1 + \kappa a}\right) \left(\frac{v_+ z_+^2 + v_- z_-^2}{v_+ + v_-}\right) \tag{17-89}$$

Since $v_+ z_+ = v_- z_-$, Eq. 17-89 can be written

$$\ln \gamma_\pm = -\frac{e^2}{8\pi D \varepsilon_0 kT} \left(\frac{\kappa}{1 + \kappa a}\right) \left[\frac{(-v_- z_-)z_+ + (-v_+ z_+)z_-}{v_+ + v_-}\right]$$

$$= \frac{e^2 z_+ z_-}{8\pi D \varepsilon_0 kT} \left(\frac{\kappa}{1 + \kappa a}\right) \tag{17-90}$$

Equation 17-90 is the Debye-Hückel expression for the activity coefficient of an electrolyte in dilute solution. A more approximate form that is applicable when $\kappa a \ll 1$ is the *Debye-Hückel limiting law*,

$$\ln \gamma_\pm = \frac{e^2 z_+ z_- \kappa}{8\pi D \varepsilon_0 kT} \tag{17-91}$$

From Eq. 17-75 and Eq. 17-76,

$$\kappa = \left[\frac{e^2}{D\varepsilon_0 kT} (n_+^\circ z_+^2 + n_-^\circ z_-^2) \right]^{\frac{1}{2}} \tag{17-92}$$

If we convert to concentrations and substitute for κ in Eq. 17-91, the result is

$$\ln \gamma_\pm = z_+ z_- \frac{e^2}{8\pi D\varepsilon_0 kT} \left[\frac{Le^2}{D\varepsilon_0 kT} (c_+^\circ z_+^2 + c_-^\circ z_-^2) \right]^{\frac{1}{2}} \tag{17-93}$$

From Eq. 16-9

$$c_j = m_j \rho_A$$

so that the activity coefficient can be expressed as a function of the molalities of the ions by

$$\ln \gamma_\pm = z_+ z_- \frac{e^2}{8\pi D\varepsilon_0 kT} \left[\frac{Le^2 \rho_A}{D\varepsilon_0 kT} (m_+^\circ z_+^2 + m_-^\circ z_-^2) \right]^{\frac{1}{2}} \tag{17-94}$$

The development of the Debye-Hückel theory provided a theoretical foundation for the empirical generalization proposed by Lewis and Randall that the activity coefficient of a given strong electrolyte is a function only of the *ionic strength I*, where

$$I = \frac{1}{2} \sum_j m_j z_j^2 \tag{17-95}$$

If we substitute numerical values for the fundamental constants in SI units and substitute the definition of the ionic strength from Eq. 17-95, Eq. 17-94 becomes

$$\ln \gamma_\pm = (1.33 \times 10^5) \frac{z_+ z_- (\rho_A)^{\frac{1}{2}}}{(DT)^{\frac{3}{2}}} (I)^{\frac{1}{2}} \tag{17-96}$$

and is applicable in this form to mixtures of electrolytes. For water at 298 K, with $D = 78.3^{15}$ and ρ_A equal to 997.1 kg-m^{-3},

$$\log \gamma_\pm = 0.510 \, z_+ z_- (I)^{\frac{1}{2}} \tag{17-97}$$

The generalization of Lewis and Randall is not strictly valid, though it holds at concentrations greater than those for which Eq. 17-97 is valid.

Exercise 17-17. Verify the numerical values in Eq. 17-96 and Eq. 17-97.

15. Malmberg and Maryott, *J. Res. Nat. Bur. Stand.*, 56, 1 (1956).

Exercise 17-18. Use Eq. 17-97 to derive an expression for the osmotic coefficient g of the solvent. Hint: Use Eq. 17-61 in the form

$$m_B \, \delta \ln \gamma_\pm = -\delta[m_B(1 - g)]$$

Experimental test. The solubility measurements of Bronsted and La Mer[16] provided strong support for the validity of the Debye-Hückel limiting law in dilute solutions. The measurements had been carried out to test the empirical generalizations of Lewis and Randall and of Bronsted about the ionic strength dependence of the activity coefficients of electrolytes. The results were published as one of the first tests of the new theory.

Bronsted and La Mer measured the solubility of a group of complex salts as a function of the ionic strength of the solution. They chose salts of very low solubility so that measurements could be made at very low ionic strength.

Let S_B represent the solubility of the salt, which is equal to the molality of the solution in equilibrium with excess solid. From Eq. 17-45,

$$m_B = \frac{m_\pm}{[(v_+)^{v_+} \, (v_-)^{v_-}\,]^{1/v}}$$

so that

$$S_B = \frac{m_\pm}{[(v_+)^{v_+} \, (v_-)^{v_-}\,]^{1/v}} \tag{17-98}$$

The chemical potential of the dissolved solute does not change with ionic strength, since the solute is in equilibrium with the same solid phase, independent of the ionic strength of the solution. Thus,

$$
\begin{aligned}
\mu_s &= \mu_B \\
&= \mu_B^\ominus + RT \ln a_B \\
&= \mu_B^\ominus + RT \ln a_\pm^v \\
&= \mu_B^\ominus + RT \ln (m_\pm/m_\pm^\ominus)^v + RT \ln \gamma_\pm^v \\
&= \mu_B^\ominus + RT \ln S_B^v \frac{[(v_+)^{v_+} \, (v_-)^{v_-}\,]}{(m_\pm^\ominus)^v} \\
&\quad + RT \ln \gamma_\pm^v
\end{aligned}
$$

and

$$\ln \frac{S_B}{m_\pm^\ominus} = \frac{\mu_s}{vRT} - \frac{\mu_B^\ominus}{vRT} - \ln [(v_+)^{v_+} \, (v_-)^{v_-}\,] + \ln \gamma_\pm \tag{17-99}$$

16. J. Bronsted and V. K. La Mer, *J. Am. Chem. Soc.*, 46, 555–73 (1924).

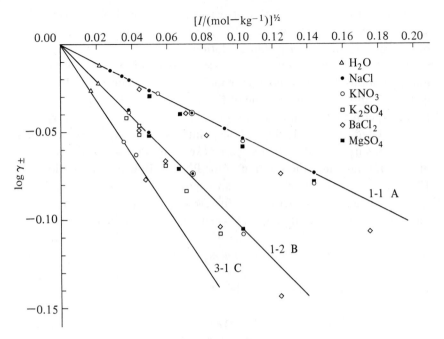

Figure 17-11. The activity coefficients of a 1-1 (A), a 1-2 (B), and a 3-1 (C) electrolyte as obtained by J. N. Bronsted and V. K. LaMer, *J. Am. Chem. Soc.*, **46**, 555–73 (1924), from solubility data for the electrolytes in solutions of the salts indicated. Curve A, $[Co(NH_3)_4(C_2O_4)]^+ [Co(NH_3)_2(NO_2)_2(C_2O_4)]^-$; B, $[Co(NH_3)_4(C_2O_4)]_2^+ [S_2O_6^{2-}]$; C, $[Co(NH_3)_6]^{+3}[Co(NH_3)_2(NO_2)_2(C_2O_4)]_3^-$. The points are experimental, and the lines are calculated from the Debye-Hückel limiting law, Eq. 17-97.

$$\ln \frac{S_B}{m_{\pm}^{\ominus}} = \ln \left(\frac{S_B^{\infty}}{m_{\pm}^{\ominus}} \right) + \ln \gamma_{\pm} \qquad (17\text{-}100)$$

The term $\ln(S_B^{\infty}/m_{\pm}^{\ominus})$ can be determined by plotting $\ln(S_B/m_{\pm}^{\ominus})$ against an appropriate function of the ionic strength and extrapolating to zero ionic strength, after which $\ln \gamma_{\pm}$ can be calculated. Both empirical generalization and the Debye-Hückel theory suggested $(I)^{\frac{1}{2}}$ as the appropriate function. Figure 17-11 shows the results of Bronsted and La Mer as a plot of $\log \gamma_{\pm}$ against $(I)^{\frac{1}{2}}$ for three complex salts of different charge types. The lines represent the Debye-Hückel limiting law and the points represent experimental results obtained when using the various salts indicated to control the ionic strength.

Limitations of the Debye-Hückel theory. A fundamental assumption of the Debye-Hückel theory is that the deviation from the behavior predicted by

Henry's law for solutions of electrolytes is the same for all electrolytes of the same charge. The results in Fig. 17-12 show that this assumption is correct only in the limit of extremely dilute solutions. From these results, we can also observe that the generalization of the ionic strength is strictly valid only in dilute solutions. Some of the discrepancies can be accounted for by the use of Eq. 17-90 by choosing a value of the distance of closest approach, a, to fit the experimental data, as shown in Table 17-3. The curves in Fig. 17-12 are calculated from Eq. 17-90, and the points represent experimental values.

It is not surprising that the Debye-Hückel limiting law holds only in very dilute solutions, since the solution of the Poisson-Boltzmann equation (Eq. 17-73) was achieved by expanding an exponential in $ze\psi/kT$ as a Taylor series and neglecting higher powers of $ze\psi/kT$. This approximation is valid only when $ze\psi \ll kT$; that is, only when ψ is small or when T is very large. It can be seen from the second term on the right in Eq. 17-82 that the concentration-dependent portion of ψ is proportional to κ, and hence to the concentration, when $\kappa a < 1$. Thus, the approximation used to simplify the Poisson-Boltzmann equation is valid only at very low concentration.[17]

The description of the solvent as a structureless dielectric continuum clearly is a weakness of the Debye-Hückel theory. Polar solvent molecules interact with ions through ion-dipole forces. We describe ions in solution as *solvated* to indicate that the interaction is sufficiently strong so that one or more solvent molecules is, at any given instant of time, effectively immobilized by attachment to an ion. Many attempts have been made to determine the number of solvent molecules immobilized by solute ions. For example, the mobilities[18] of the alkali metal ions increase in the order: $Li^+ < Na^+ < K^+ < Rb^+$, even though the sizes[19] of the ions increase in the order: $Li^+ < Na^+ < K^+ < Rb^+$. Thus, the smaller ions are more highly solvated, presumably because of their greater charge density. The Debye-Hückel theory predicts continually decreasing activity coefficients with increasing concentration of solute. The addition of solvation to the model provides a basis for explaining the increase in activity coefficient with increasing concentration in concentrated solutions, as illustrated in Fig. 17-13.

17. The approximate solution satisfies a fundamental requirement of electrostatics, that the potentials due to separate charges are additive, whereas the solutions of the original Poisson-Boltzmann equation do not. Thus, attempts to extend the Debye-Hückel theory to higher concentrations by including higher terms in the series expansion are questionable on theoretical grounds.
18. R. A. Robinson and R. H. Stokes, *Electrolyte Solutions*, 2nd ed., Butterworths, London, 1959, p. 463.
19. L. Pauling, *Nature of the Chemical Bond*, 3rd ed., Cornell University Press, New York, 1960, p. 514.

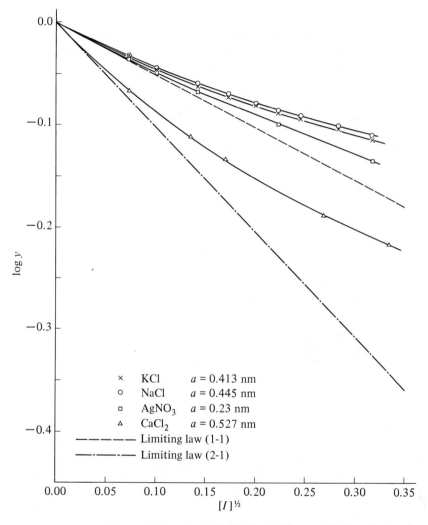

Figure 17-12. Activity coefficients for KCl, NaCl, AgNO$_3$, and CaCl$_2$ solutions from data on concentration cells with liquid junction, as listed in Table 17-3. The points are experimental, the dashed lines represent the Debye-Hückel limiting law, Eq. 17-97, and the curves represent Eq. 17-90 with the values of a given, chosen to produce a best fit to the data.

Table 17-3. A comparison of observed activity coefficients with values calculated from Eq. 17-90

c/(mol-dm⁻³)	KCl[a]		NaCl[b]		AgNO₃[c]	
	y(obs.)	y(calc.) $a = 0.413$ nm	y(obs.)	y(calc.) $a = 0.445$ nm	y(obs.)	y(calc.) $a = 0.23$ nm
0.005	0.9273	0.9276	0.9283	0.9281	0.922	0.922
0.01	0.9024	0.9024	0.9034	0.9036	0.892	0.894
0.02	0.8702	0.8701	0.8726	0.8726	0.858	0.857
0.03	0.8494	0.8490	0.8513	0.8515	—	—
0.04	0.8320	0.8321	0.8354	0.8354	—	—
0.05	0.8187	0.8183	0.8221	0.8221	0.795	0.794
0.06	0.8071	0.8067	0.8119	0.8110	—	—
0.08	0.7872	0.7874	0.7940	0.7925	—	—
0.10	0.7718	0.7719	0.7796	0.7779	0.733	0.735

c/(mol-dm⁻³)	CaCl₂[a]	
	y(obs.)	y(calc.) $a = 0.527$ nm
0.0018153	0.8588	0.8586
0.0060915	0.7745	0.7746
0.0095837	0.7361	0.7364
0.024167	0.6514	0.6510
0.037526	0.6097	0.6085
0.050000	0.5834	0.5810
0.096540	0.5275	0.5198

[a] T. Shedlovsky and D. A. MacInnes, J. Am. Chem. Soc., 59, 503–6 (1937).
[b] A. S. Brown and D. A. MacInnes, J. Am. Chem. Soc., 57, 1356–62 (1935).
[c] D. A. MacInnes and A. S. Brown, Chem. Rev., 18, 335–48 (1936).

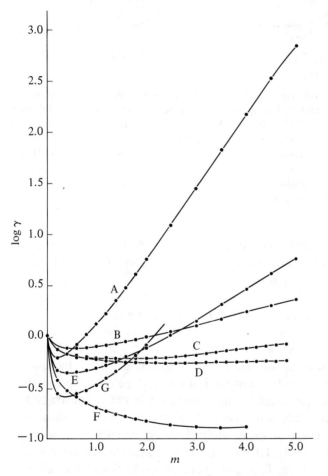

Figure 17-13. The activity coefficients of several electrolytes at 25 °C as function of molality. (From tabulations in R. A. Robinson and R. H. Stokes, *Electrolyte Solutions*, Butterworths, London, 1959, pp. 491–503.) Curve A, $UO_2(ClO_4)_2$; B, HCl; C, NaCl; D, KCl; E, $CaCl_2$; F, Na_2SO_4; G, $LaCl_3$.

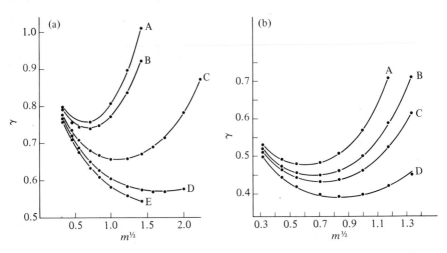

Figure 17-14. The activity coefficients of several electrolytes as function of the molality. The points are experimental, and the curves are according to the Stokes and Robinson extension of the Debye-Hückel theory that contains the parameter h, the number of moles of solvent immobilized per mole of solute. The value of h that gave the best fit to the data is as follows: (a) Curve A (HCl), 7.3; B (LiCl), 6.5; C (NaCl), 3.5; D (KCl), 1.9; E (RbCl), 1.25. (b) Curve A ($MgCl_2$), 13.3; B ($CaCl_2$), 11.9; C ($SrCl_2$), 10.3; D ($BaCl_2$), 8.4. [Reprinted with permission, from R. H. Stokes and R. A. Robinson, *J. Am. Chem. Soc.*, **70**, 1870–78 (1948). Copyright by the American Chemical Society.]

Stokes and Robinson[20] devised an extension of the Debye-Hückel theory that contains an adjustable parameter h, the number of moles of solvent immobilized per mole of solute. Figure 17-14 shows that their equation provides agreement with experiment for a number of 1–2 and 2–1 electrolytes to molalities greater than two, with reasonable values of h.

Bjerrum brought the story of strong electrolytes full circle when he suggested that strong electrolytes form ion pairs at sufficiently high concentrations, especially in solvents of low dielectric constant. In aqueous solutions, where the dielectric constant is too high to permit simple ion-pair formation, evidence for covalent bond formation in $PbCl_2$ and $Pb(NO_3)_2$ solution comes from the same kind of spectrophotometric studies that Bjerrum used to show that strong electrolytes are completely dissociated.

The strong acids HNO_3, $HClO_4$, and H_2SO_4 provide striking examples of incomplete dissociation. T. F. Young and his students used Raman spectroscopy to determine the concentrations of the species at equilibrium. Their

20. R. H. Stokes and R. A. Robinson, *J. Am. Chem. Soc.*, **70**, 1870–78 (1948).

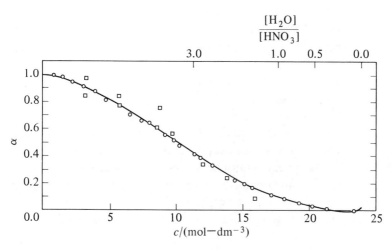

Figure 17-15. The degree of dissociation of HNO_3 at 298 K as a function of concentration. The Raman spectral data (○, curve) are from Young, Maranville, and Smith; the nuclear magnetic resonance data (□) are from G. C. Hood, O. Redlich, and C. A. Reilly, *J. Chem. Phys.*, *22*, 2067–71 (1954); degree of dissociation (α). (By permission, from T. F. Young, L. F. Maranville, and H. M. Smith, in W. J. Hamer, ed., *The Structure of Electrolytic Solutions*, John Wiley & Co., New York, 1959, p. 42.)

results for the degree of dissociation of HNO_3 are shown as circles and the curve in Fig. 17-15. Figure 17-16 shows the concentration of the various species in an aqueous solution of H_2SO_4 as a function of stoichiometric molarity. The squares in Fig. 17-15 are from the nuclear magnetic resonance experiments of Hood and associates.

Critical discussions of these aspects of solutions of strong electrolytes can be found in Robinson and Stokes[21] and Guggenheim and Stokes.[22]

17-4. Potentials of electrochemical cells

Electrochemical cells permit the calculation of the reversible electrical work that can be obtained from a chemical reaction, and, from Eq. 10-79, the free energy change of the reaction. When the reactants of a spontaneous oxidation-reduction reaction are mixed, the reaction proceeds irreversibly, and no net work is obtained. When the reactants are isolated from one

21. R. A. Robinson and R. H. Stokes, *Electrolyte Solutions*, Butterworths, London, 1959, Chapt. 14.
22. E. A. Guggenheim and R. H. Stokes, *Equilibrium Properties of Aqueous Solutions of Single Strong Electrolytes*, Pergamon Press, Oxford, 1969, Chapt. 9.

another but are allowed to react at electrodes in a galvanic cell, the electrons flowing in the external circuit can do work the magnitude of which depends on the rate of reaction. If the reaction is carried out reversibly, the magnitude of the work obtained is a maximum.

Measurement of cell potential

The potential of a cell must be measured under reversible conditions if thermodynamic results are to be calculated from the data. This requirement means that the chemical reactions occurring at the two electrodes must be reversible and that the potential must be measured by opposing the potential of the cell with an equal and opposite external potential. The second requirement is fulfilled by measuring the potential with a potentiometer, as illustrated schematically in Fig. 17-17, and the first requirement is checked by

Figure 17-16. Concentration of species (c_{sp}) in aqueous solutions of H_2SO_4 at 298 K, as determined by Raman spectroscopy; stoichiometric concentration c. Curve A, $HSO_4{}^-$; B, $SO_4{}^{2-}$; C, H_2SO_4. [By permission, from T. F. Young, *Rec. Chem. Prog.*, *12*, 81 (1951).]

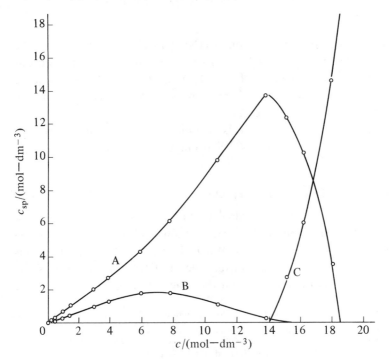

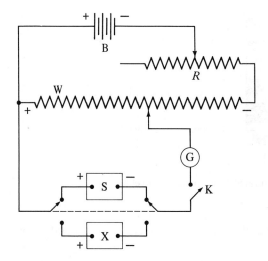

Figure 17-17. A diagram of a potentiometer; battery B, variable resistor R, slidewire W, galvanometer G, key K, standard cell S, and unknown cell X.

observing that the direction of current flow can be reversed by changing the value of the opposing potential about the balance point.

In the figure, the working battery B provides a variable potential across the slidewire W through the variable resistance R. The double-pole, double-throw switch permits either the standard cell S or the unknown cell X to be connected to the variable contact of the slidewire through the galvanometer G and the key K. With the standard cell in the circuit, the variable contact is placed at a position on the calibrated slidewire equal to the potential of the standard cell and the resistor R is adjusted until no current passes through the galvanometer when the key is depressed. The use of the key serves to minimize current flow through the cell when the potentiometer is not balanced. With the unknown cell in the circuit, the variable contact on the slidewire is moved until no current flows through the galvanometer when the key is depressed. The terminal connected to the negative end of the slidewire is defined as the negative terminal of the cell and vice versa.

The most commonly used standard cell is the Weston cell, shown schematically in Fig. 17-18. The cell reaction is

$$Cd(Hg) + Hg_2SO_4(c) + \tfrac{8}{3} H_2O(l) = CdSO_4 \cdot \tfrac{8}{3} H_2O(c) + 2 Hg(l)$$

The potential of the cell is remarkably stable, since all phases are either saturated or pure. For highest precision, the cell should be maintained at a

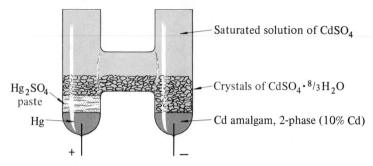

Figure 17-18. A diagram of a saturated Weston standard cell.

constant temperature. The potential is given as a function of temperature by the relation[23, 24]

$$\mathscr{E} = 1.01864 - 4.06 \times 10^5 \, (t - 20) - 9.5 \times 10^{-7} \, (t - 20)^2 + 1 \times 10^8 \\ (t - 20)^3$$

where t is the Celsius temperature. For routine measurements, an unsaturated Weston cell is used. It has a smaller temperature coefficient than the saturated cell, but its potential is not as stable.

Modern digital voltmeters have a sufficiently high input impedance that they can be used safely to measure potentials at essentially zero current flow if the reversibility of the electrodes has been established previously.

Chemical cells without liquid junction

From the standpoint of thermodynamic interpretation of the results, the simplest cell is one in which two different electrodes are immersed in the same solution. Such cells are called *chemical cells without liquid junction.* That is, there is no phase boundary between two liquid phases in the cell. The presence of such a boundary leads to the inherently irreversible process of diffusion between phases, and only in the absence of a liquid junction can rigorous thermodynamic results be obtained.

Consider the cell illustrated schematically in Fig. 17-19. This cell conventionally is represented by a diagram

$$\text{H}_2 \, (g, \, P) \, |\text{Pt}| \, \text{HCl}(m) | \text{AgCl}(c), \, \text{Ag}(c)$$

in which each substance present is indicated by a chemical symbol and the appropriate designation of its state and in which a single vertical bar rep-

23. G. W. Vinal *Primary Batteries*, Wiley & Sons, New York, 1950, Chapt. 6.
24. Script $\mathscr{E}$, denoting cell potential, should not be confused with the molar energy $\mathscr{E}$.

resents a boundary between gas and solid phases, or between liquid and solid phases. *By convention*, the chemical reaction at the left-hand electrode is written as an oxidation and the chemical reaction at the right-hand electrode is written as a reduction; that is

$$\frac{1}{2} H_2(g, P) + H_2O(l) = H_3O^+(m) + e^- \qquad (17\text{-}101)$$

and

$$AgCl(c) + e^- = Ag(c) + Cl^-(m) \qquad (17\text{-}102)$$

so that the net cell reaction is

$$\frac{1}{2} H_2(g, P) + H_2O(l) + AgCl(c) = H_3O^+(m) + Cl^-(m) + Ag(c) \qquad (17\text{-}103)$$

If, in measuring the potential of the cell, the left-hand electrode is negative and the right-hand electrode is positive, as implied by the reaction as written conventionally, the cell potential is given a positive sign. If the signs of the electrodes are reversed, the cell potential is given a negative sign.

The reaction in Eq. 17-103 is spontaneous. If the hydrogen gas that is bubbled over the platinum electrode came into contact with the AgCl coating on the Ag, AgCl electrode, they would react directly and no current would flow through the external circuit. Since it is necessary to prevent contact

Figure 17-19. A chemical cell without a liquid junction, with a Ag (c), AgCl(c)|Cl$^-$ electrode and a H_2 (g, P)|Pt|H_3O^+(aq.) electrode in a solution of HCl, molality m.

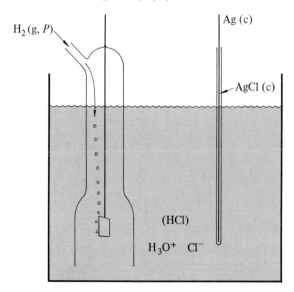

between reactants in the cell, most cells must be constructed with a liquid junction.

Exercise 17-19. Could the cell $H_2(g) |Pt|HCl, ZnCl_2| Zn$ be used to measure cell potentials and determine $\Delta\mathscr{G}$ for the reaction

$$Zn(c) + H_3O^+(aq) = Zn^{2+}(aq) + \tfrac{1}{2} H_2(g) + H_2O$$

Explain your answer.

Free energy and cell potentials. According to the conventional description of the cell in Fig. 17-19, electrons flow in the external circuit from the left-hand electrode, the *source* of electrons, to the right-hand electrode, the *sink* for electrons. The reversible external work done in a cell in which one mole of a chemical reaction occurs and n electrons appear in the chemical equation is

$$\mathscr{W}_{rev} = -nF(V_r - V_l) \tag{17-104}$$

where F is the Faraday constant, V_r is the potential at the right-hand electrode and V_l is the potential at the left-hand electrode. The quantity in parentheses is defined as the potential $\mathscr{E}$ of the cell, so that

$$\mathscr{W}_{rev} = -nF\mathscr{E} \tag{17-105}$$

From Eq. 10-79,

$$\Delta\mathscr{G} = \mathscr{W}_{rev}$$

so that

$$\Delta\mathscr{G} = -nF\mathscr{E} \tag{17-106}$$

Thus, when the reaction occurs in the cell as written, and is spontaneous, $\mathscr{E}$ is positive and $\Delta\mathscr{G}$ is negative. For the reactants and products in their standard states one can write

$$\Delta\mathscr{G}° = -nF\mathscr{E}° \tag{17-107}$$

If we apply Eq. 11-18, Eq. 11-73, and Eq. 16-51 to the reaction described in Eq. 17-103, the result is

$$\Delta\mathscr{G} = \mu°(H_3O^+) + \mu°(Cl^-) + \mu°[Ag(c)]$$
$$- \tfrac{1}{2}\mu°[H_2(g)] - \mu°(H_2O) - \mu°[AgCl(c)]$$
$$+ RT \ln \frac{a(H_3O^+)\, a(Cl^-)}{f[H_2(g)]/f°}$$

$$= \Delta \mathscr{G}^{\ominus} + RT \ln \frac{a(H_3O^+)\,a(Cl^-)}{f[H_2(g)]/f^{\ominus}} \tag{17-108}$$

since the pure solids are in their standard states and we can assume that the solution is sufficiently dilute so that the solvent is in its standard state. Substituting from Eq. 17-106 and Eq. 17-107 in Eq. 17-108, we obtain

$$\mathscr{E} = \mathscr{E}^{\ominus} - \frac{RT}{nF} \ln \frac{a(H_3O^+)\,a(Cl^-)}{f[H_2(g)]/f^{\ominus}} \tag{17-109}$$

the *Nernst equation* for the dependence of the potential of an electrochemical cell on the composition of the electrolyte.

Exercise 17-20. Convert the second term on the right in Eq. 17-109 to the common logarithm and calculate the numerical coefficient of the resulting logarithmic term for a one-electron reaction at 298 K.

The standard potential. When all the reactants and products are in their standard states, the activities in Eq. 17-109 are equal to one, $f/f^{\ominus}$ is equal to one, and $\mathscr{E}$ is equal to $\mathscr{E}^{\ominus}$. It is not possible, however, to determine $\mathscr{E}^{\ominus}$ by preparing a cell with all the reactants and products in their standard states and measuring the potential, since the standard states are hypothetical standard states (see Sec. 16-4).

If we substitute the molality and the activity coefficient in Eq. 17-109 according to Equations 17-37 and 17-38, and if we make the reasonable assumption that the fugacity of hydrogen gas is equal to the pressure at ordinary pressures, Eq. 17-109 becomes

$$\mathscr{E} = \mathscr{E}^{\ominus} - \frac{RT}{nF} \ln \frac{m/m^{\ominus}(H_3O^+)\,m/m^{\ominus}(Cl^-)}{p/p^{\ominus}[H_2(g)]}$$

$$- \frac{RT}{nF} \ln \gamma(H_3O^+)\,\gamma(Cl^-) \tag{17-110}$$

If we place all the measurable quantities on one side, the result is

$$\mathscr{E}' \equiv \mathscr{E} + \frac{RT}{nF} \ln \frac{m/m^{\ominus}(H_3O^+)\,m/m^{\ominus}(Cl^-)}{p/p^{\ominus}[H_2(g)]} = \mathscr{E}^{\ominus} - \frac{RT}{nF} \ln \gamma(H_3O^+)\,\gamma(Cl^-) \tag{17-111}$$

Only the product of the ionic activity coefficients appears in Eq. 17-111. Since there is no way to measure individual ion activity coefficients, it is convenient to express the cell potential as a function of the mean molality and the mean ionic activity coefficient. Substituting from Eq. 17-39 and Eq. 17-40, we have

$$\mathscr{E}' \equiv \mathscr{E} + \frac{2RT}{nF} \ln \frac{m_{\pm}/m_{\pm}^{\ominus}}{p/p^{\ominus}[H_2(g)]}$$

$$= \mathscr{E}^{\ominus} - \frac{2RT}{nF} \ln \gamma_{\pm} \qquad (17\text{-}112)$$

and $\mathscr{E}'$ can be calculated over a range of molalities. Since the activity coefficient has a limiting value of one at infinite dilution, the extrapolated value of $\mathscr{E}'$ at $m_{\pm} = 0$ is equal to $\mathscr{E}^{\ominus}$. When $\mathscr{E}'$ is plotted against $m_{\pm}$, the extrapolation is poor because the curve has a steep slope. The Debye-Hückel theory suggests that $\ln \gamma$ is a linear function of $m_{\pm}^{\frac{1}{2}}$ in a very dilute solution, and a plot of $\mathscr{E}'$ against $m_{\pm}^{\frac{1}{2}}$ provides a much more precise extrapolation. Figure 17-20 shows plots of $\mathscr{E}'$ against $m_{\pm}$ and $m_{\pm}^{\frac{1}{2}}$ based on data for the potential of the cell of Eq. 17-103 by T. F. Young and N. Anderson.

Once the value of $\mathscr{E}^{\ominus}$ has been determined, measurement of $\mathscr{E}$ with a cell containing electrolyte with a known value of $m_{\pm}$ permits a precise calculation of $\gamma_{\pm}$. The equilibrium constant of the reaction that occurs in the cell also can be calculated from the standard potential. Since (Eq. 17-107)

$$\Delta \mathscr{G}^{\ominus} = -nF\mathscr{E}^{\ominus}$$

and (Eq. 16-116)

$$\Delta \mathscr{G}^{\ominus} = -RT \ln K_a$$

$$\ln K_a = \frac{nF\mathscr{E}^{\ominus}}{RT} \qquad (17\text{-}113)$$

The same conclusion can be reached from Eq. 17-109, since the potential of a cell is zero when the reaction is at equilibrium. Thus,

$$0 = \mathscr{E}^{\ominus} - \frac{RT}{nF} \ln \left\{ \frac{a(H_3O^+)\, a(Cl^-)}{f/f^{\ominus}[H_2(g)]} \right\}_{equil}$$

$$= \mathscr{E}^{\ominus} - \frac{RT}{nF} \ln K_a$$

and

$$\ln K_a = \frac{nF\mathscr{E}^{\ominus}}{RT}$$

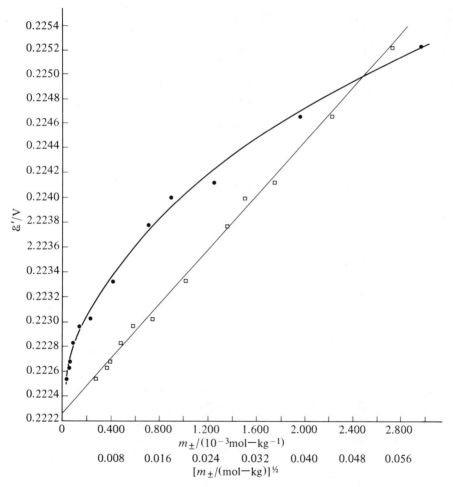

Figure 17-20. Values of $\mathscr{E}'$ of Eq. 17-112 for data on the cell in Eq. 17-103 from unpublished data of T. F. Young and N. Anderson, as given in I. M. Klotz and R. M. Rosenberg (*Chemical Thermodynamics*, 3rd ed., W. A. Benjamin, Menlo Park, 1972, p. 389). Values plotted against $m_\pm$ (●); against $(m_\pm)^{\frac{1}{2}}$ (□).

Cells without liquid junction can also be used to obtain precise values for the ionization constants of weak acids. In order to determine the equilibrium constant K_A of the weak acid HA one can use the cell

$$H_2\,(g, f/f^{\ominus} = 1)\,|\,Pt\,|\,HA\,(m_1),\ NaA\,(m_2),\ NaCl\,(m_3)\,|\,AgCl\,(c),\ Ag\,(c)$$

The cell reaction is the same as Eq. 17-103 and the cell potential is

$$\mathscr{E} = \mathscr{E}^{\ominus} - \frac{RT}{nF}\ln a(H_3O^+)\,a(Cl^-) \tag{17-114}$$

The equilibrium constant for the ionization of HA is

$$K_A = \frac{a(H_3O^+)\,a(A^-)}{a(HA)} \tag{17-115}$$

on the assumption that the solution is sufficiently dilute that $a(H_2O)$ is essentially equal to one. Substituting for $a(H_3O^+)$ from Eq. 17-115 in Eq. 17-114, we have

$$\mathscr{E} = \mathscr{E}^{\ominus} - \frac{RT}{nF}\ln\frac{K_A\,a(HA)\,a(Cl^-)}{a(A^-)}$$

$$= \mathscr{E}^{\ominus} - \frac{RT}{nF}\ln\frac{K_A\,m(HA)\,m(Cl^-)}{m(A^-)\,m^{\ominus}} - \frac{RT}{nF}\ln\frac{\gamma_{HA}\gamma_{Cl^-}}{\gamma_{A^-}} \tag{17-116}$$

From the stoichiometry of the ionization reaction

$$m(HA) = m_1 - m(H_3O^+)$$

$$m(A^-) = m_2 + m(H_3O^+)$$

and, since it does not participate in the ionization,

$$m(Cl^-) = m_3$$

The value of $m(H_3O^+)$ is small compared to m_1 and m_2, and its value can be estimated with sufficient accuracy from an approximate value of K_A, neglecting activity coefficients. If we place on the left all the terms of Eq. 17-116 the values of which are known, the result is

$$\mathscr{E} - \mathscr{E}^{\ominus} + \frac{RT}{nF}\ln\frac{m(HA)\,m(Cl^-)}{m(A^-)\,m^{\ominus}} = -\frac{RT}{nF}\ln K_A\frac{\gamma_{HA}\gamma_{Cl^-}}{\gamma_{A^-}}$$

$$= -\frac{RT}{nF}\ln K_A' \tag{17-117}$$

Since the quotient of activity coefficients approaches one as I approaches zero, a suitable extrapolation of K_A' to zero ionic strength should yield a

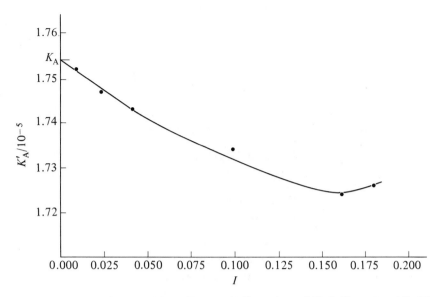

Figure 17-21. A plot of K_A' from Eq. 17-115 [from data of H. S. Harned and R. W. Ehlers, *J. Am. Chem. Soc.*, *54*, 1350–57 (1932)], for acetic acid at 25 °C. $K_A = 1.754 \times 10^{-5}$.

value of K_A. It has been found that a plot of K_A' against I yields a satisfactory value, as illustrated in Fig. 17-21 with Harned and Ehler's data for acetic acid at 298 K.

Concentration cells without liquid junction. The process of determining the standard potential of a cell to high precision is a difficult one. For some cells data cannot be obtained at a low enough concentration to carry out the extrapolation illustrated in Fig. 17-19. Nevertheless, if data are available for the activity coefficient in a solution of one concentration of an electrolyte, activity coefficients can be obtained for solutions at other concentrations from measuring the potential of a *concentration cell without liquid junction*. Consider the cell

$$H_2(g,P)|Pt|HCl(m_1)|AgCl(c), Ag(c) - Ag(c), AgCl(c)|HCl(m_2)|Pt|H_2(g,P)$$

which is just two chemical cells without liquid junction opposed in series. The reactions occurring at the electrodes are, by the usual convention,

$$\tfrac{1}{2} H_2(g,P) + H_2O(l) = H_3O^+(m_1) + e^-$$

$$AgCl(c) + e^- = Ag(c) + Cl^-(m_1)$$

$$\text{Ag(c)} + \text{Cl}^-(m_2) = \text{AgCl(c)} + e^-$$

$$\text{H}_3\text{O}^+(m_2) + e^- = \tfrac{1}{2}\,\text{H}_2(\text{g},P) + \text{H}_2\text{O(l)}$$

The net reaction is

$$\text{H}_3\text{O}^+(m_2) + \text{Cl}^-(m_2) = \text{H}_3\text{O}^+(m_1) + \text{Cl}^-(m_1) \qquad (17\text{-}118)$$

Since each of the two cells in series has the same pair of electrodes, but electrically opposed, $\mathscr{E}^\ominus$ is equal to zero for the entire cell, and

$$\mathscr{E} = -\frac{RT}{nF}\ln\frac{a_1(\text{H}_3\text{O}^+)\,a_1(\text{Cl}^-)}{a_2(\text{H}_3\text{O}^+)\,a_2(\text{Cl}^-)}$$

$$= -\frac{RT}{nF}\ln\frac{(m_1/m^\ominus)^2}{(m_2/m^\ominus)^2} - \frac{RT}{nF}\ln\frac{\gamma^2_{\pm,1}}{\gamma^2_{\pm,2}} \qquad (17\text{-}119)$$

It can be seen from Eq. 17-119 that the potential of the cell is positive and the change described in Eq. 17-118 is spontaneous, if $a_2 > a_1$. It can also be seen that the value of $\gamma_\pm$ in one solution can be calculated if the corresponding value in the other solution is known without having to know $\mathscr{E}^\ominus$ for an individual cell.

If two metal-amalgam electrodes of different composition are immersed in the same solution, one containing an ion of the metal in the electrodes, another kind of concentration cell without liquid junction is formed. Consider the cell

$$\text{Tl(amalgam, } x_1)\,|\,\text{TlCl}(m)\,|\,\text{Tl(amalgam, } x_2)$$

The equations of the reactions at the electrodes are

$$\text{Tl(amalgam, } x_1) = \text{Tl}^+(m) + e^-$$

and

$$e^- + \text{Tl}^+(m) = \text{Tl(amalgam, } x_2)$$

The net reaction is

$$\text{Tl(amalgam, } x_1) = \text{Tl(amalgam, } x_2)$$

The standard potential of the cell is zero if the same standard state is chosen for both electrodes. Thus the potential of the cell is

$$\mathscr{E} = -\frac{RT}{nF}\ln\frac{x_2 f_2}{x_1 f_1} \qquad (17\text{-}120)$$

and the measurement of the cell potential permits the calculation of the ratio of the activity coefficients in the two amalgams. When measurements of $\mathscr{E}$ are carried out over a range of values of x_1 at constant x_2, a suitable extrapolation of f_2/f_1 to $x_1 = 0$, as in Fig. 16-11, leads to values of f_2 and f_1.

Classification of reversible electrodes

1. Metal-Metal-Ion Electrodes. The most common type of electrode is that in which a metallic element is immersed in a solution containing a positive ion of the same metal. Common examples include Zn, Zn^{2+}; Cu, Cu^{2+}; Ag, Ag^+; and Cd, Cd^{2+}.

2. Metal Amalgam-Metal Ion Electrodes. These electrodes are similar to the metal-metal ion electrodes, but the metal is replaced by a solution of the metal in mercury, an *amalgam*. The potential of the cell in which the electrode is used depends on the composition of the amalgam as well as the composition of the solution. Some dilute amalgams of alkali metals can be used as electrodes in cells containing aqueous solutions when the pure metal cannot.

3. Nonmetallic Element-Platinum-Negative (or Positive) Ion Electrodes. In these electrodes, of which the hydrogen electrode is an example, an element is in contact with platinized platinum, which is immersed in a solution of the ion of the element. Common examples include $Cl_2(g)$ Pt, Cl^-; $Br_2(l)$, Pt, Br^-; and $I_2(c)$, Pt, I^-.

4. Metal-Insoluble Metal Salt-Negative Ion Electrodes. In these electrodes, a metal that is coated with an insoluble salt of the same metal is immersed in a solution of the negative ion of the salt. Common examples include Ag, AgCl, Cl^-; Pb, $PbSO_4$, SO_4^{2-}; Ag, AgI, I^-; and Hg, Hg_2Cl_2, Cl^-.

5. Oxidation-Reduction Electrodes. In these electrodes, platinum is immersed in a solution containing two ions of the same element but with different oxidation states. The most common is the Fe^{3+}, Fe^{2+} electrode.

Oxidation-reduction electrodes may also be prepared from more complex conjugate oxidant-reductant pairs, such as quinone and hydroquinone.

Standard electrode potentials. The standard potentials of chemical cells without a liquid junction provide an important source of precise thermodynamic data for oxidation-reduction reactions. As with other thermodynamic data (Sec. 11-5), it is more economical and more convenient to tabulate data for individual reactants and products or for conjugate oxidant-reductant pairs than to tabulate data for reactions. For example, the potentials for n individual electrodes would permit the calculation of the potentials of $n(n - 1)/2$ cells. It is not possible, however, to measure the potentials of individual electrodes, since a second lead from the potentiometer used to measure the potential would constitute a second electrode.

It is possible to assign a value to the standard potential of a single electrode by adopting an appropriate convention. The convention that is used has two parts:

1. Consistent with Eq. 17-104 and Eq. 17-105,

$$\mathscr{E}^{\ominus}_{cell} = \mathscr{E}^{\ominus}_r - \mathscr{E}^{\ominus}_l$$

in which $\mathscr{E}^{\ominus}_r$ and $\mathscr{E}^{\ominus}_l$ refer to the standard potentials of the right electrode and left electrode, respectively, when the cell diagram is written in the conventional way.

2. The standard hydrogen electrode, with the fugacity of hydrogen gas equal to 1.01×10^5 Pa, has $\mathscr{E}^{\ominus} \equiv 0$.

One can express the convention differently by stating that the standard potential of an electrode, the *standard electrode potential*, is equal to the standard potential of a cell in which the standard hydrogen electrode is the left electrode and the electrode in question is the right electrode.

Exercise 17-21. The standard potential of the cell the equation for which is Eq. 17-103 is 0.2224 V. The standard potential of the cell

$$Ag(c), AgCl(c) | ZnCl_2(aq) | Zn(c)$$

is -0.9855 V. Calculate $\mathscr{E}^{\ominus}$ for the Ag, AgCl electrode and $\mathscr{E}^{\ominus}$ for the Zn electrode.

Chemical cells with liquid junction

The practical applications of electrochemical measurements require a greater variety of cells than can be constructed without a liquid junction. Cells with liquid junctions are used, therefore, even if the measured potentials cannot be related in a rigorous way to thermodynamic quantities. A common example of a cell with a liquid junction is the glass electrode-reference electrode combination found in every commercial pH meter.

A cell with a liquid junction in which the two solutions contain the same solute at different concentrations and in which the same electrode is immersed in both solutions can be treated rigorously. We shall use an approximate treatment; a full discussion can be found in MacInnes[25] and Kortum.[26] Consider the cell

$$Ag(c), AgCl(c) | NaCl(m_1) | NaCl(m_2) | AgCl(c), | Ag(c)$$

25. D. A. MacInnes, *The Principles of Electrochemistry* Reinhold, New York, 1939, Chapts. 8, 13.
26. G. Kortum, *Treatise on Electrochemistry*, 2nd ed., Elsevier, Amsterdam, 1965, Chapt. 8.

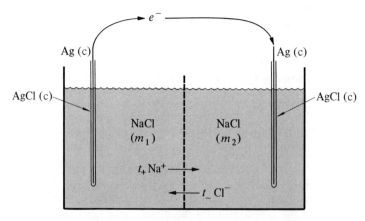

Figure 17-22. The flow of charge in a concentration cell with a liquid junction. The transport numbers of Na^+ and Cl^- are t_+ and t_-, respectively.

in which there is a liquid junction formed between two solutions of NaCl. The reactions that occur at the electrodes and at the liquid junction are

$$Ag(c) + Cl^-(m_1) = AgCl(c) + e^-$$
$$AgCl(c) + e^- = Ag(c) + Cl^-(m_2)$$
$$t_+ Na^+(m_1) = t_+ Na^+(m_2)$$
$$t_- Cl^-(m_2) = t_- Cl^-(m_1)$$

The flow of charge is shown schematically in Fig. 17-22. In describing the transport of ions across the liquid junction we have assumed that m_1 and m_2 differ so slightly that the transport numbers of Cl^- and Na^+ can be taken equal in the two solutions. The net result is

$$t_+ Na^+(m_1) + t_+ Cl^-(m_1) = t_+ Na^+(m_2) + t_+ Cl^-(m_2)$$

and the Nernst equation for the potential is

$$\mathscr{E} = -t_+ \frac{RT}{nF} \ln \frac{(m_2/m^\ominus)^2 \gamma_{\pm,2}^2}{(m_1/m^\ominus)^2 \gamma_{\pm,1}^2} \qquad (17\text{-}121)$$

If the transport numbers are known from another experiment, the measured potential can be used to calculate the ratio of the activity coefficients in the two solutions.

Exercise 17-22. If the liquid-junction potential of the cell the potential of which is given by Eq. 17-121 is defined as the difference between that potential and the potential of an equivalent concentration cell without a liquid junction, calculate $\mathscr{E}_{l,j}$ for the cell.

When different ions are present on two sides of a liquid junction, the liquid-junction potential depends on the structure of the liquid junction as well as on the concentrations of the ions, but there is no general theory that allows us to calculate liquid-junction potentials. To use cells with a liquid junction, it is necessary to design junctions that have as small a liquid junction potential as possible. The most common way to minimize the junction potential is through the use of a salt bridge. In such a device, a saturated solution of KCl or another salt in which the positive ion and negative ion have equal transport numbers forms a liquid junction with each of two half-cells. Since the concentrated salt solution carries the major part of the ion current across the phase boundaries, and since the positive ion and negative ion carry equal fractions of that current, it might be expected that the potentials set up at the boundaries by the diffusion of the positive ions and negative ions might be equal and opposite. Whether or not this is correct cannot be determined because liquid-junction potentials cannot be measured. Nevertheless, a large body of consistent data has been obtained using salt bridges to minimize the liquid junction potential.

The practical measurement of pH provides an excellent example of the use of cells with liquid junctions. The H_3O^+ ion is important in industrial processes and in the reactions that occur in living cells. Though the hydrogen electrode potential depends on the activity of H_3O^+ ions, it is not suitable for routine measurements because its platinized platinum surface is easily poisoned and because it requires the presence of highly purified hydrogen gas at a constant pressure.

In 1909, Haber and Klemensiewicz invented the glass electrode. A thin bulb of glass containing hydrochloric acid in which a platinum wire was immersed behaved as a hydrogen electrode. The modern version of this electrode is based on the work of MacInnes and Dole, who fused a thin membrane of soft glass on the end of a hard glass tube and used a Ag-AgCl electrode as the internal electrode. A second electrode, a *reference electrode*, is connected with the solution to be studied by means of a saturated KCl salt bridge, which may be a tiny fiber passing through the glass tube of the electrode, the space between the members of a ground joint, or the bore of a stopcock. The reference electrode is usually a Ag-AgCl electrode or a saturated calomel electrode. The latter has a platinum wire immersed in a mixture of Hg and Hg_2Cl_2 (calomel), in contact with a saturated solution of KCl. The cell can be diagrammed as

$$\text{Glass electrode} \mid H_3O^+(m_1), X^-(m_1) \mid KCl(\text{satd.}) \mid Hg_2Cl_2(c), Hg(l) \mid Pt$$

If the liquid junction potential is neglected (or assumed constant) and if the potential of the reference electrode is considered constant, the potential of the cell might be written as

$$\mathscr{E} = \mathscr{E}_x + \mathscr{E}_{ref} - \frac{RT}{F} \ln a(H_3O^+) \tag{17-122}$$

where $\mathscr{E}_x$ is a characteristic potential for each glass electrode-liquid junction combination. As we mentioned earlier, it is not possible to measure the activity of a single ion thermodynamically. Assuming that the activity coefficient of the H_3O^+ ion is equal to the mean activity coefficient of the electrolyte HX, we can write Eq. 17-122 as

$$\mathscr{E} = \mathscr{E}_x + \mathscr{E}_{ref} - \frac{RT}{F} \ln [H_3O^+] y_\pm \tag{17-123}$$

When Sorenson defined the symbol pH in 1909, he assumed that it represented the negative logarithm of the hydrogen ion concentration, calculated from the stoichiometric concentration of an acid and the degree of dissociation obtained from colligative properties or conductance. Since neither the activity nor the activity coefficient of a single ion can be measured, the concentration of the hydrogen ion cannot be related directly to electrochemical measurements with the glass electrode or the hydrogen electrode in cells with liquid junction. The term pH is now defined *operationally* in terms of the potential obtained in a glass-electrode or hydrogen-electrode cell with a standard buffer of *assigned* pH, compared to the potential in a solution the pH of which is to be measured. The mathematical form of the definition is

$$pH(X) = pH(S) + \frac{\mathscr{E}_X - \mathscr{E}_S}{(RT \ln 10)/F} \tag{17-124}$$

Modern pH meters are calibrated directly in pH units. The resulting values provide a practical scale of acidity even though they have no rigorous thermodynamic significance. Detailed discussion of the significance of pH can be found in Bates.[27]

Standard electrode potentials

Whether obtained from cells with liquid junction or cells without liquid junction, the basic form of equilibrium electrochemical data is *standard electrode potentials*, the standard potentials of cells in which the hydrogen electrode is written as the negative electrode and the other electrode as the

27. R. G. Bates, *Determination of pH, Theory and Practice*, 2nd ed., John Wiley & Sons, New York, 1973, Chapt. 2.

positive electrode. The sign of the potential indicates whether the reaction corresponding to this convention is spontaneous. Table 17-4 lists a number of standard electrode potentials and the corresponding half-reactions, taken from a recent compilation.

Table 17-4. Standard electrode potentials

Electrode	Half-reaction	$\mathscr{E}^{\ominus}/V$
$Li^+\|Li(c)$	$Li^+(aq) + e^- = Li(c)$	-3.045
$K^+\|K(c)$	$K^+(aq) + e^- = K(c)$	-2.925
$Ca^{2+}\|Ca(c)$	$Ca^{2+}(aq) + 2 e^- = Ca(c)$	-2.87
$Na^+\|Na(c)$	$Na^+(aq) + e^- = Na(c)$	-2.7141
$Mg^{2+}\|Mg(c)$	$Mg^{2+}(aq) + 2 e^- = Mg(c)$	-2.37
$OH^-\|H_2$, Pt	$2 H_2O(l) + 2 e^- = 2 OH^-(aq) + H_2(g)$	-0.8281
$Zn^{2+}\|Zn(c)$	$Zn^{2+}(aq) + 2 e^- = Zn(c)$	-0.7631
$Fe^{2+}\|Fe(c)$	$Fe^{2+}(aq) + 2 e^- = Fe(c)$	-0.440
$Cd^{2+}\|Cd(c)$	$Cd^{2+}(aq) + 2 e^- = Cd(c)$	-0.4019
$SO_4^{2-}\|PbSO_4(c), Pb(c)$	$PbSO_4(c) + 2 e^- = Pb(c) + SO_4^{2-}(aq)$	-0.3553
$I^-\|AgI(c), Ag(c)$	$AgI(c) + e^- = Ag(c) + I^-(aq)$	-0.1524
$H_3O^+\|H_2(g)$, Pt	$2 H_3O^+(aq) + 2 e^- = H_2(g) + 2 H_2O(l)$	0.0000
$Cl^-\|AgCl(c), Ag(c)$	$AgCl(c) + e^- = Ag(c) + Cl^-(aq)$	0.2224
$Cl^-\|Hg_2Cl_2(c), Hg(l)$	$Hg_2Cl_2(c) + 2 e^- = 2 Hg(l) + 2 Cl^-(aq)$	0.2681
$Cu^{2+}\|Cu(c)$	$Cu^{2+}(aq) + 2 e^- = Cu(c)$	0.337
$Fe^{3+}(aq), Fe^{2+}(aq)\|Pt$	$Fe^{3+}(aq) + e^- = Fe^{2+}(aq)$	0.771
$Ag^+\|Ag(c)$	$Ag^+(aq) + e^- = Ag(c)$	0.7991

By permission, from R. G. Bates, in *Techniques of Electrochemistry* (E. Yeager and A. J. Salkind, ed.), Wiley-Interscience, New York, 1972, pp. 27–29.

Problems

17-1. S. J. Bates and G. W. Vinal [*J. Am. Chem. Soc.*, **36**, 916 (1914)] carefully determined the relationship between the mass of silver deposited in electrolysis and the change Q passing through the circuit. Their data are given below. Calculate the value of the Faraday constant from each measurement, the mean value, and the standard deviation of the result.

$m'(Ag)/g$	Q/C
4.09903	3666.55
4.10523	3671.84
4.10475	3671.61
4.10027	3667.65
4.10516	3671.82

17-2. D. A. MacInnes and M. Dole [*J. Am. Chem. Soc.*, **53**, 1357 (1931)] carried out a precise determination of the transport numbers of K^+ and Cl^- by the Hittorf method in the apparatus shown schematically in Fig. 17-1. In one of their experiments, with a KCl solution of concentration 0.02 mol·dm^{-3}, a current of approximately 0.002 A was passed through the cell for 23 h at 298 K. The number of coulombs passed through the cell was determined by measuring the mass of silver deposited in two coulometers in series with the cell: 0.16024 g and 0.16043 g. After the electrolysis, the cell was divided into five compartments, as indicated in Fig. 17-1, and the following analyses of the contents of each were obtained.

Compartment	Mass % of KCl
1	0.19410, 0.19398
2	0.14939
3	0.14948
4	0.14932, 0.14932
5	0.10336, 0.10342

The contents of compartment 5 had a mass of 117.79 g and the contents of compartment 1 had a mass of 120.99 g. Calculate the transport numbers of K^+ and Cl^- in this solution by assuming that H_2O is a stationary solvent and that the central compartments represent the initial composition.

17-3. Longsworth [*J. Am. Chem. Soc.*, *54*, 2741 (1932)] calculated the transport numbers of Na^+ and Cl^- in NaCl solutions at 298 K by a moving boundary method, in which Cd was the positive electrode and Ag, AgCl the negative electrode. In an experiment with an NaCl solution of concentration 0.02 mol-dm^{-3}, he passed a constant current of 0.0016001 A through the cell. The cross-sectional area of the cell was 0.11122 cm². The distance moved by the boundary as a function of time is given below.

x/cm	t/s
0.500	172
1.000	344
1.500	516
2.000	689
3.000	1036
4.000	1380

Calculate the transport numbers of Na^+ and Cl^- in this solution. For comparison with the results of Problem 17-2, Longsworth obtained a value of $t_+ = 0.4901$ for 0.02 mol-dm^{-3} KCl.

17-4. Shedlovsky [*J. Am. Chem. Soc.*, *54*, 1411 (1932)] measured the molar conductance of several electrolytes at low concentrations in order to test the Onsager limiting law, Eq. 17-20. His data for KNO_3 and KCl at 298 K are listed below.

$c/(10^{-4}$ mol-dm$^{-3})$	$\Lambda/(10^{-4}\ \Omega^{-1}$-mol^{-1}-m²$)$
KNO_3	
0.69820	144.17
1.7613	143.62
3.8888	142.98
5.8651	142.61
8.6853	141.97

$c/(10^{-4}\ mol\text{-}dm^{-3})$	$\Lambda/(10^{-4}\ \Omega^{-1}\text{-}mol^{-1}\text{-}m^2)$
KCl	
0.32576	149.33
1.0445	148.91
2.6570	148.38
3.3277	148.19
3.5217	148.12

Plot the values of Λ against $c^{\frac{1}{2}}$ for both salts, obtain the slopes and intercepts, and calculate Λ_∞ for each salt and A and B, universal constants for uni-univalent salts.

17-5. Belcher [*J. Am. Chem. Soc.*, 60, 2744–47 (1938)] determined the ionization constant of butyric acid by measuring the conductance. (a) He obtained the following values for the conductance of sodium butyrate solutions

$c/(10^{-3}\ mol\text{-}dm^{-3})$	$\Lambda/(10^{-4}\ \Omega^{-1}\text{-}mol^{-1}\text{-}m^2)$
1.2132	80.14
2.5811	78.27
7.1049	76.72
11.963	75.21
16.858	74.04
29.336	71.86

Calculate Λ_∞ for sodium butyrate from these data. (b) Given the values of Λ_∞ for HCl, and NaCl as $426.04 \times 10^{-4}\Omega^{-1}\text{-}mol^{-1}\text{-}m^2$ and $126.42\ \Omega^{-1}\text{-}mol^{-1}\text{-}m^2$, respectively, calculate Λ_∞ for butyric acid. (c) Calculate $\alpha = \Lambda/\Lambda_\infty$ for butyric acid from the following conductance data for butyric acid, calculate K', and plot $\log K'$ against $c^{\frac{1}{2}}$ to obtain an extrapolated value of K at $c = 0$.

$c/(10^{-3}\ mol\text{-}dm^{-3})$	$\Lambda/(10^{-4}\ \Omega^{-1}\text{-}mol^{-1}\text{-}dm^{-3})$
0.29695	77.368
0.33908	72.904
0.43277	65.335
0.78715	49.671
0.83817	48.242
0.93748	45.846
1.0096	44.319

17-6. Harned and Ehlers [*J. Am. Chem. Soc.*, 54, 1350 (1932); 55, 2179 (1933)] measured the potential $\mathscr{E}$ of the cell $H_2 | Pt | HCl | AgCl$, Ag for various molalities of HCl at several temperatures at unit fugacity of hydrogen. Their data are shown below. (a) Plot $\mathscr{E}'$ of Eq. 17-111 against $m^{\frac{1}{2}}$ at each temperature to obtain $\mathscr{E}^{\ominus}$ and values of $\gamma_\pm$ at each concentration. (b) Plog $\log \gamma_\pm$ against $m^{\frac{1}{2}}$ and compare the limiting slope with the Debye-Hückel limiting law prediction. You will need to find values of D for H_2O at 288 K and 308 K. (c) Derive the relationships

$$\Delta\mathscr{S} = nF\left(\frac{\partial\mathscr{E}}{\partial T}\right)_P$$

and

$$\Delta \mathscr{H} = -nF\mathscr{E} + nF\left(\frac{\partial \mathscr{E}}{\partial T}\right)_P$$

starting with Eq. 17-106 and the properties of $\Delta \mathscr{G}$. (e) Estimate $\Delta \mathscr{G}^{\ominus}$, $\Delta \mathscr{S}^{\ominus}$, and $\Delta \mathscr{H}^{\ominus}$ for the cell reaction at 298 K and compare with the values obtained from tabulations of standard thermodynamic data.

$m/(10^{-4}$ mol-kg$^{-1})$	$\mathscr{E}/10^{-3}$ V 288 K	298 K	308 K
35.64	511.50	515.27	518.46
44.88	500.51	503.84	506.65
56.19	489.57	492.57	495.04
62.39	484.61	487.47	489.75
73.11	476.88	479.48	481.44
91.38	466.35	468.60	470.24
111.95	456.72	458.61	459.91

17-7. Calculate the values of $1/\kappa$ for aqueous solutions at 298 K for $I = 10^{-4}$, 10^{-3}, and 10^{-2}.

18 · Gas adsorption and heterogeneous catalysis

The study of the adsorption of gases by solid surfaces and of the catalytic effect of solid surfaces on reactions between gases provides an example of the application of the principles and techniques of physical chemistry. But these fields are also good examples of the stimulus provided to basic research by the need to solve such practical problems as prolonging the life of a tungsten filament in a lamp bulb, producing fertilizer cheaply, improving the effectiveness of gas masks, and devising a truly efficient catalytic converter for automobile exhaust pipes.

18-1. The solid-gas interface

The atoms and molecules in the surface of a solid, like the molecules in the surface of a liquid (Sec. 14-2), have valence forces that are not completely saturated because atoms and molecules in the surface are not surrounded completely by other like molecules or atoms. As a result, solid surfaces have a tendency to adsorb molecules from the phase with which they are in contact.

The *adsorption isotherm* describes the amount of gas adsorbed on a surface as a function of the equilibrium pressure of the gas at a fixed temperature. The conventional unit for adsorption is the volume of gas adsorbed (measured at 273.15 K and 1.01×10^5 Pa pressure) per unit mass of adsorbent. If a known amount of gas is admitted into an evacuated apparatus containing the adsorbent, measurement of the volume and pressure of the remaining gas at equilibrium determines the amount adsorbed. Alternatively, the mass of gas adsorbed on a sample suspended from a quartz spiral is measured by observing the extension of the spiral. Determining that adsorbent surfaces are "clean" initially is difficult. A common method for removing adsorbed gases is to "bake" the adsorbent at high temperature in a vacuum. Unless the surface is free of adsorbed material before adsorption is measured, the nature of the surface is ambiguous.

Adsorption of gases by solids is usually classified either as *physical adsorption* or *chemisorption*. The distinction is not always clear. Physical adsorption is due to forces of the kind that result in gas imperfection and the formation of liquids—London forces and dipole forces. It does not have an activation energy, and the rate of adsorption is limited only by the rate of diffusion to the surface and into the pores of the adsorbent. Chemisorption results from the formation of chemical bonds with surface atoms, and the energy of adsorption is much greater than that for physical adsorption. Chemisorption may involve an activation energy, so that the rate of chemisorption at low temperatures is very small. Thus, the measured amount of adsorption is small even though the equilibrium adsorption would be large if a sufficiently long time were allowed for the system to reach equilibrium.

The relation between the two forms of adsorption is clarified by the potential energy curves shown in Fig. 18-1. The zero of potential energy is the energy of the ground state of the gaseous molecules. Curves A (a) and (b), represent the potential energy of physical adsorption, with a shallow minimum at a relatively long distance from the surface. In this situation, the electronic state of the gas molecule is essentially unperturbed. Curves B (a) and (b), represent the potential energy of chemisorption. The energy at large distances from the surface is that of the rearranged electronic state of the chemisorbed molecule, which may be a dissociated state and is different from the ground state. At the point of intersection of the two curves, the two states have the same energy. As in the interaction of two atomic orbitals to form a bonding orbital (Chapter 5), the result is two different states, one with a higher and one with a lower energy, as shown by the dashed curves in Fig. 18-1. If the intersection occurs where the potential energy of physical adsorption is negative, the chemisorption does not have an activation energy; if the intersection occurs where the potential energy of physical adsorption is positive, the chemisorption has an activation energy, indicated as E_a in Fig. 18-1 (b).

The nature of the solid surface

A large surface area is the most important characteristic of solids that adsorb gases well. Most adsorbents are very finely divided, so that they would be expected to have a large surface area. For example, a cube 1 mm on an edge has a surface area of 6 mm², whereas the same material divided into cubes 10^{-3} mm on an edge has a surface area of 6×10^3 mm². The surface

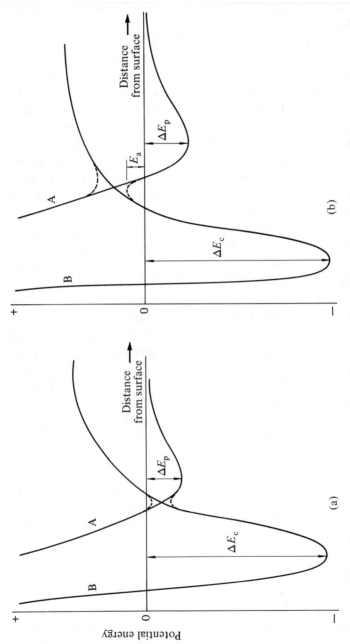

Figure 18-1. Potential energy curves for physical adsorption (A) and chemisorption (B). (a) The two potential energy curves cross at a negative potential energy, and there is no activation energy barrier for chemisorption. (b) The two potential energy curves cross at a positive potential energy, and chemisorption is activated. (By permission, from D. O. Hayward and B. M. W. Trapnell, *Chemisorption*, 2nd ed., Butterworths, Washington, 1964.)

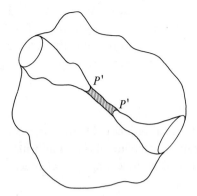

Figure 18-2. Diagram of a pore of variable size in an adsorbent particle. The vapor pressure in the partly filled pore is P', which is less than the vapor pressure P'', in equilibrium with the bulk liquid.

area of nonporous solids calculated from a microscopic determination of the particle size agrees with the surface area calculated from the amount of adsorbed N_2 gas that forms a monomolecular layer on the surface.[1] The gas adsorption method yields much higher values than the microscopic method, however, for solids that are known to be porous. This surface area is the pertinent one in any discussion of the chemical behavior of the surface.

Not only the total surface area but also the number of pores and the distribution of pore sizes are important characteristics of an adsorbent. The Laplace equation (Eq. 14-20) shows that there is a difference of pressure across a curved liquid surface. When enough gas has been adsorbed on parts of the wall of a pore, assumed to be essentially cylindrical, to close a portion of the pore, the resulting curved interface of the adsorbed material with the vapor phase has a radius approximately equal to the radius of the pore, as indicated in Fig. 18-2 (Eqs. 14-24 and 14-25). Lord Kelvin demonstrated that a variation in the radius of curvature of a liquid surface resulted in a variation of vapor pressure.[2] In the notation of Eq. 14-20, P'' corresponds to the vapor pressure in equilibrium with the bulk liquid, and P' corresponds to the vapor pressure in the partly filled pore; then

$$P'' - P' = \frac{2\gamma}{r}$$

1. S. Brunauer, *The Adsorption of Gases and Vapors*, Vol. I, Princeton University Press, Princeton, 1943, pp. 293–99.
2. W. Thompson, *Phil. Mag.* (4), *42*, 448 (1871); W. Moore, *Physical Chemistry*, 4th ed., Prentice-Hall, Englewood Cliffs, 1972, pp. 481–82.

where r is the radius of the pore and γ is the surface tension of the adsorbed vapor, assumed equal to that of the liquid at the same temperature. If we consider the pore size distribution to be essentially continuous, the change in pressure corresponding to an infinitesimal change in radius δr is

$$\delta P'' - \delta P' = -\frac{2\gamma}{r^2}\,\delta r \qquad (18\text{-}1)$$

At equilibrium, the chemical potential of the vapor is equal to the chemical potential of the adsorbed vapor. A change in the pressure of the equilibrium vapor leads to a new equilibrium state in which the changes of chemical potential are equal for the two phases. Thus,

$$\delta\mu'' = \delta\mu'$$

and, at constant temperature (see text below Eq. 15-10),

$$\mathcal{V}''\,\delta P'' = \mathcal{V}'\,\delta P'$$

Solving for $\delta P'' - \delta P'$ and substituting in Eq. 18-1, we obtain

$$\frac{\mathcal{V}' - \mathcal{V}''}{\mathcal{V}'}\,\delta P'' = -\frac{2\gamma}{r^2}\,\delta r \qquad (18\text{-}2)$$

Exercise 18-1.　Verify Eq. 18-2.

If we assume that $\mathcal{V}'$ is negligible with respect to $\mathcal{V}''$ and that $\mathcal{V}''$ can be represented by the ideal gas law expression RT/P'', Eq. 18-2 becomes

$$\frac{RT}{\mathcal{V}'}\frac{\delta P''}{P''} = \frac{2\gamma}{r^2}\,\delta r \qquad (18\text{-}3)$$

Exercise 18-2.　Verify Eq. 18-3.

Integration from vapor pressure P at radius r to the vapor pressure of the bulk liquid P_0 at $r = \infty$ yields

$$\frac{RT}{\mathcal{V}''}\ln\frac{P_0}{P} = \frac{2\gamma}{r} \qquad (18\text{-}4)$$

It can be seen from Eq. 18-4 that $P < P_0$ for pores of any finite size. Only pores of one size can be filling at equilibrium at any given vapor pressure. All pores less than this size are filled at a lower pressure, and larger pores are unfilled. According to this view of the adsorption process, the adsorption isotherm leads directly to a pore-size distribution curve. Though this treat-

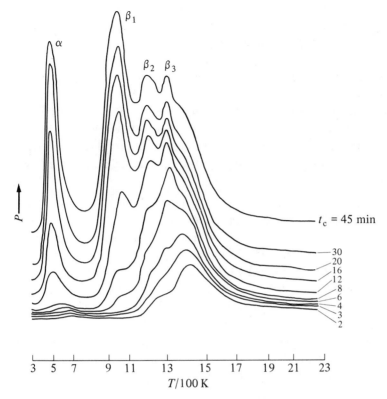

Figure 18-3. The pressure of carbon monoxide released from adsorption on tungsten as a function of temperature after various adsorption times at 300 K. The curves are displaced vertically to avoid overlap. [By permission, from P. A. Redhead, *Trans. Faraday Soc.*, *57*, 641 (1961).]

ment neglects adsorption on the pore walls other than through filling of the pores, the calculations of pore-size distribution are qualitatively correct.[3]

A third important characteristic of a solid surface is its heterogeneity. Real crystals are not the ideal structures that were discussed in Chapter 13; single crystals contain lattice defects, and polycrystalline particles contain grain boundaries. Taylor[4] pointed out in 1925 that edges, corners, grain boundaries, and lattice defects might differ in their adsorptive properties from other parts of the crystal surface. Since then it has been demonstrated

3. S. Brunauer, *The Adsorption of Gases and Vapors*, Vol. I, Princeton University Press, Princeton, 1943, pp. 385–93.
4. H. S. Taylor, *Proc. Roy. Soc., London*, *A108*, 105 (1925).

that, even in perfect single crystals, different crystal faces differ in their adsorptive properties[5], and this is also seen in oriented metal films, in which some crystal faces are preferentially exposed.[6]

Striking evidence for heterogeneity of adsorption sites was obtained by Redhead[7] in studies of the temperature dependence of desorption of carbon monoxide adsorbed on tungsten filaments. Figure 18-3 shows the pressure of CO released on desorption as a function of temperature, after adsorption for differing lengths of time at 300 K. For short adsorption times, only those sites are filled that require a high temperature for desorption. As more and more of the surface is covered, sites are filled that can be desorbed at lower temperatures.

Monolayer adsorption—The Langmuir isotherm

In some cases the adsorption isotherm indicates that the surface is saturated at sufficiently high pressure. The isotherms of NH_3 on charcoal at several temperatures are shown in Fig. 18-4. Irving Langmuir suggested a model to account for these results in 1918.[8] He assumed that the surface consists of a fixed number of adsorption sites M with equal energies of adsorption, of which a fraction θ are occupied at any pressure P of gas at equilibrium, and a fraction $(1 - \theta)$ are vacant. The rate of adsorption is proportional to $M(1 - \theta)$ and to the pressure P of the gas, whereas the rate of desorption is proportional to $M\theta$. At equilibrium the two rates are equal, and

$$k_1 M(1 - \theta) P = k_2 M\theta \qquad (18\text{-}5)$$

The proportionality constants k_1 and k_2 are assumed to be independent of θ. Solving for θ, we obtain

$$\theta = \frac{(k_1/k_2) P}{1 + (k_1/k_2) P}$$

$$= \frac{bP}{1 + bP} \qquad (18\text{-}6)$$

where $b = k_1/k_2$. At sufficiently high pressures, θ approaches one and the volume V of adsorbed gas approaches V_m, the maximum adsorbed volume.

Exercise 18-3. Verify Eq. 18-6.

5. H. Leidheiser and A. T. Gwathmey, *J. Am. Chem. Soc.*, 70, 1200–06, 1206 (1940); J. K. Roberts, *Proc. Roy. Soc.*, London, *A152*, 445 (1935).
6. O. Beeck and A. W. Ritchie, *Disc. Faraday Soc.*, 8, 159–66 (1950).
7. P. A. Redhead, *Trans. Faraday Soc.*, 57, 641 (1961).
8. I. Langmuir, *J. Am. Chem. Soc.*, 40, 1361 (1918).

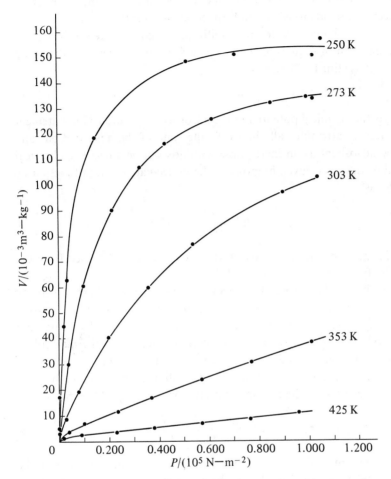

Figure 18-4. Adsorption isotherms of ammonia on charcoal at several temperatures. [From data of A. Titoff, *Z. physik. Chem.*, *74*, 641–78 (1910).]

Thus, V/V_m can be substituted for θ, and Eq. 18-6 becomes

$$V = \frac{V_m b P}{1 + bP} \qquad (18\text{-}7)$$

Though Langmuir derived Eq. 18-7 on the basis of a kinetic model, the final result is a thermodynamic relationship that is independent of kinetic assumptions. Fowler and Guggenheim[9] showed that the Langmuir isotherm

9. R. H. Fowler and E. A. Guggenheim, *Statistical Thermodynamics*, Cambridge University Press, Cambridge, 1952, Chapt. 10.

could be derived on a statistical thermodynamic basis for a model in which gas molecules are adsorbed on uniform fixed sites on the surface.

The thermodynamic criterion for equilibrium at constant temperature and pressure between molecules in the gas phase and molecules adsorbed on a solid surface is (Eq. 15-5)

$$\mu_g = \mu_s$$

where μ_s is the chemical potential of the adsorbed molecules. This statement of the criterion attributes all change to the state of the adsorbed gas and treats the adsorbent as an inert phase with adsorption sites. The chemical potential of the gas is given in terms of the molecular partition function by Eq. 11-65 as

$$\mu_g = - RT \ln \left(\frac{z_g'}{L} \right) + \mathscr{E}_{0,g}$$

The chemical potential of the adsorbed molecule can be obtained from the Helmholtz free energy, $A_s = E_s - TS_s$.

The entropy of the adsorbed molecule is the sum of the thermal entropy $S_{t,s}$ of the molecule on the adsorption site and the configurational entropy $S_{c,s}$ resulting from the possible arrangements of N adsorbed molecules and $M - N$ vacant sites on a surface containing M sites. Since the molecule is adsorbed on a localized and distinguishable site, and since its motion other than internal is vibration with respect to an equilibrium position on that site, we can use Eq. 8-38 for the entropy of oscillators in a crystal to write

$$S_{t,s} = kN \ln z_s + \frac{E_s - E_{0,s}}{T} \tag{18-8}$$

The partition function z_s is a summation of exponential terms over the electronic and vibrational modes of the adsorbed molecules and whatever rotational modes are possible within the restraints of the attachment to the surface. The configurational entropy is an entropy of mixing of N filled sites and $M - N$ vacant sites (Eq. 8-48), so that

$$S_{c,s} = - kN \ln \left(\frac{N}{M} \right) - k(M - N) \ln \left(\frac{M - N}{M} \right)$$

$$= - k \left[N \ln \left(\frac{N}{M - N} \right) + M \ln \left(\frac{M - N}{M} \right) \right] \tag{18-9}$$

Thus,

$$A_s = E_s - TS_s$$

$$= E_{0,s} - NkT \ln z_s + kT\left[N \ln\left(\frac{N}{M-N}\right) + M \ln\left(\frac{M-N}{M}\right)\right]$$

(18-10)

and

$$\mu_s = \left(\frac{\partial A_s}{\partial n}\right)_T$$

$$= L\left(\frac{\partial A_s}{\partial N}\right)_T$$

$$= \mathscr{E}_{0,s} - LkT \ln z_s + LkT \ln\left(\frac{N}{M-N}\right)$$

$$= \mathscr{E}_{0,s} - LkT \ln z_s + LkT \ln\left(\frac{\theta}{1-\theta}\right)$$

(18-11)

Exercise 18-4. Verify Eq. 18-11.

Equating μ_g and μ_s and solving for $\theta/(1-\theta)$, we obtain

$$\frac{\theta}{1-\theta} = \frac{z_s}{z_g'/L}\exp\left[\frac{-(\mathscr{E}_{0,s} - \mathscr{E}_{0,g})}{RT}\right]$$

(18-12)

Exercise 18-5. Verify Eq. 18-12.

If we assume that the gas is ideal, the partition function z_g' can be expressed as a product of the internal partition function $z_{i,g}$, a sum of exponentials over the electronic, vibrational, and rotational quantum levels of the molecule, and the translational partition function $z_{t,g}'$, given by Eq. 8-85 as

$$z_{t,g}' = \left(\frac{2\pi mkT}{h^2}\right)^{\frac{3}{2}}\mathscr{V}$$

Substituting in Eq. 18-12, we obtain

$$\frac{\theta}{1-\theta} = \frac{z_s h^3}{z_{i,g}(2\pi m)^{\frac{3}{2}}(kT)^{\frac{5}{2}}}\exp\left[\frac{-(\mathscr{E}_{0,s} - \mathscr{E}_{0,g})}{RT}\right]P$$

(18-13)

Exercise 18-6. Verify Eq. 18-13.

By comparison with Eq. 18-5 and Eq. 18-6, we see that

$$b = \frac{k_1}{k_2} = \frac{z_s h^3}{z_{i,g}(2\pi m)^{\frac{3}{2}}(kT)^{\frac{5}{2}}}\exp\left(\frac{-\Delta\mathscr{E}_{0,\text{ads}}}{RT}\right)$$

(18-14)

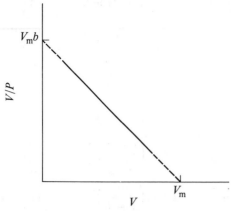

Figure 18-5. A schematic plot of V/P against V for a system that is represented by the Langmuir adsorption isotherm; slope $= -b$.

where

$$\Delta\mathscr{E}_{0,\,\text{ads}} = \mathscr{E}_{0,\,s} - \mathscr{E}_{0,\,g}$$

Thus, the form of the Langmuir isotherm does not depend on any assumptions about the kinetics of adsorption and desorption; it depends only on the assumptions that the gas is adsorbed to fixed sites and that the adsorbed molecules do not interact with one another.

It is hard to tell from a plot of V against P whether the experimental data fit Eq. 18-7 or some other equation that predicts saturation. Equation 18-7 is of the same mathematical form as the Michaelis-Menten equation (Eq. 12-91), and testing of the model and evaluation of the parameters b and V_m from experimental data are best done, as in that case, by converting Eq. 18-7 to a linear form. One that is particularly useful is[10]

$$\frac{V}{P} = V_m b - bV \tag{18-15}$$

A plot of V/P against V should be linear, as shown schematically in Fig. 18-5. The slope of the line is $-b$, the horizontal intercept is equal to V_m, and the vertical intercept is equal to $V_m b$. Figure 18-6 shows Titoff's data plotted this way. It can be seen that the data are linear at all temperatures at sufficiently high volumes adsorbed, but large deviations are found when only a small

10. J. M. Thomas and W. J. Thomas, *Introduction to the Principles of Heterogeneous Catalysis*, Academic Press, London, 1967, p. 38.

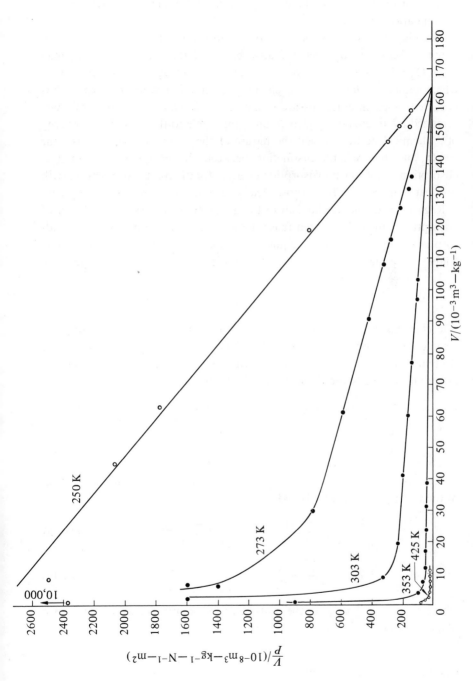

Figure 18-6. Titoff's data for ammonia on charcoal, plotted as V/P against V to test the applicability of the Langmuir adsorption isotherm.

673

fraction of the surface is occupied. All the curves extrapolate to the same horizontal intercept. This mode of plotting makes clear the extent of the extrapolation in each case.

Monolayer adsorption is most often observed when there is a highly specific interaction between the adsorbate and the surface. For example, when O_2 adsorbed on charcoal is removed by baking, it comes off as CO, which indicates that a strong chemical interaction with the surface has occurred. Such interactions are termed *chemisorption*, which is also characterized by an energy of adsorption comparable to the dissociation energy of a chemical bond. Though the nature of the adsorption process is clear for O_2 on charcoal, the distinction between chemisorption and physical adsorption is not sharp. Adsorption energies for different systems range with no clear demarcation from those clearly chemical to those clearly physical.

Regardless of the mechanism of the adsorption process, the enthalpy of adsorption can be calculated from the temperature dependence of the adsorption isotherm using an equation of the same form as the Clausius-Clapeyron equation (Eq. 15-11). At equilibrium at constant temperature and pressure (Eq. 15-5),

$$\mu_g = \mu_s$$

The chemical potential of the gas is a function of T and P, whereas the chemical potential of the adsorbed molecules is a function of T and θ. If the temperature is changed while θ is kept constant, the changes in chemical potential of the gas and the adsorbed molecules are

$$\delta\mu_g = -\mathscr{S}_g\delta T + \mathscr{V}_g\delta P \qquad (18\text{-}16)$$

and

$$\delta\mu_s = -\mathscr{S}_s\delta T \qquad (18\text{-}17)$$

When a new equilibrium state is reached, the changes in chemical potential of the two phases are equal, and

$$\frac{dP}{dT} = \frac{-(\mathscr{S}_s - \mathscr{S}_g)}{\mathscr{V}_g} = \frac{-\Delta\mathscr{S}_{ads}}{\mathscr{V}_g} \qquad (18\text{-}18)$$

$$= \frac{-\Delta\mathscr{H}_{ads}}{T\mathscr{V}_g} \qquad (18\text{-}19)$$

where $\Delta\mathscr{S}_{ads}$ and $\Delta\mathscr{H}_{ads}$ are the molar entropy change and enthalpy change of adsorption.

Exercise 18-7.   Verify Eq. 18-19.

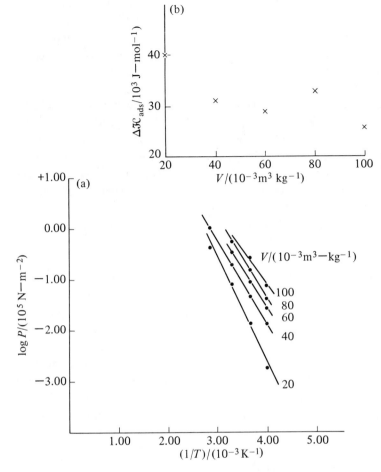

Figure 18-7. (a) Plots of log P against $1/T$ for several volumes from Titoff's data. (b) The resulting values of the isosteric $\Delta\mathcal{H}_{ads}$ as a function of volume, showing that the enthalpy of adsorption is a function of the fraction of surface covered.

If we assume that the gas phase behaves ideally, the result is

$$\frac{\delta P}{P} = -\frac{\Delta\mathcal{H}_{ads}}{RT^2}\,\delta T \qquad (18\text{-}20)$$

Integration of Eq. 18-20 yields

$$\ln P = \frac{\Delta\mathcal{H}_{ads}}{R}\,\frac{1}{T} + \text{constant} \qquad (18\text{-}21)$$

Thus, a plot of ln P against $1/T$ at a fixed value of θ should yield a straight line of slope $\Delta\mathcal{H}_{ads}/R$. The value of $\Delta\mathcal{H}_{ads}$ obtained this way is called the *isosteric* (constant volume) enthalpy of adsorption. The enthalpy of adsorption is always negative for systems that follow a Langmuir isotherm. The model in its simplest form requires that $\Delta \mathcal{H}_{ads}$ be independent of θ, indicating that the surface of the adsorbent is uniform and that no interaction occurs among the adsorbed molecules. Figure 18-7 shows a plot of log P against $1/T$ for four values of V from Fig. 18-4. The resultant values of $\Delta \mathcal{H}$ are plotted against V in the upper corner of the figure. It can be seen that $\Delta \mathcal{H}$ decreases with increasing V, though these data show considerable scatter. These results, together with those of Fig. 18-6, provide a rigorous test of the Langmuir model. They indicate either that the surface is heterogeneous, that adsorbed molecules interact, or both. Low values of $\Delta \mathcal{H}$ suggest that the process involved is not chemisorption. Though few systems fulfill all the requirements of the Langmuir model, the model is the basis for extensions to include consideration of heterogeneity of binding sites and interaction among adsorbed molecules.[11]

Multilayer adsorption—The Brunauer-Emmet-Teller isotherm
Physical adsorption has a lower enthalpy of adsorption than chemisorption, one comparable in magnitude to the enthalpy of liquefaction; it depends on an interaction between the adsorbent and the adsorbed species similar to those exhibited by non-ideal gases. Since the forces involved are similar to the forces involved in liquefaction, it is not surprising that the amount of gas adsorbed is sometimes not limited to a monomolecular layer. In some respects, the solid surface acts as a nucleus for liquefaction at temperatures and pressures at which the liquid would not form in the absence of the surface.

One of the characteristic types of multilayer adsorption isotherms is shown in Fig. 18-8, based on data for the adsorption of N_2 on rutile at 75 K. The initial rapid rise in adsorption represents the filling of the most active sites on the bare surface, followed by filling of the rest of the first layer. The rapid change in curvature followed by a more gradual slope represents the formation of second and higher layers. The slope increases very rapidly as the pressure of the gas approaches P_0, the equilibrium vapor pressure of the liquid.

The Brunauer-Emmett-Teller model for multilayer adsorption is a direct extension of the assumptions of the Langmuir isotherm to a more complex situation. The original derivation of the model[12] was based on equality of

11. A. W. Adamson, *Physical Chemistry of Surfaces*, 2nd ed., Interscience, New York, 1967, Chapt. XIII.
12. S. Brunauer, P. H. Emmett, and E. Teller, *J. Am. Chem. Soc.*, 60, 309 (1938); S. Brunauer, L. S. Deming, W. E. Deming, and E. Teller, *J. Am. Chem. Soc.*, 62, 1723 (1940).

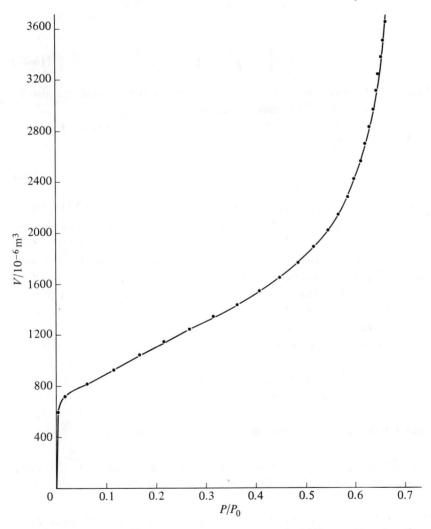

Figure 18-8. An isotherm characteristic of one type of multilayer adsorption. From data of L. E. Drain and J. A. Morrison [*Trans. Faraday Soc.*, *48*, 654–673 (1953)], on the adsorption of nitrogen (N_2) on rutile (TiO_2) at 75 K.

the rates of adsorption and desorption for each layer of adsorbed molecules. The simplifying assumption that the enthalpy of adsorption of the first layer is different from the enthalpy of liquefaction of the gas, whereas the enthalpy of adsorption of higher layers is equal to the enthalpy of liquefaction[13] enabled the authors of the model to derive the equation that bears their names,

$$V = \frac{V_m c P}{(P_0 - P)[1 + (c - 1)(P/P_0)]} \tag{18-22}$$

where

$$c = \exp\left[\frac{(\mathscr{E}_1 - \mathscr{E}_l)}{RT}\right] \tag{18-23}$$

Like the Langmuir model, the Brunauer-Emmett-Teller model can also be derived on a statistical thermodynamic basis.[14]

The parameters of the model are derived most easily from the following form of Eq. 18-22,

$$\frac{P}{V(P_0 - P)} = \frac{1}{V_m c} + \frac{c - 1}{V_m c}\frac{P}{P_0} \tag{18-24}$$

A straight line should be obtained when the left side of Eq. 18-24 is plotted against P/P_0. Figure 18-9 shows such a plot for the data used to plot the curve of Fig. 18-8. It can be seen that the plot is linear between P/P_0 equal to 0.05 and P/P_0 equal to 0.50, but not outside this range. These results suggest that the surface is not homogeneous. The adsorption at low surface coverage is stronger than that on the rest of the surface. The linear portion of the curve leads to a value of the heat of adsorption characteristic of the less active part of the surface.

The sum of the slope and the intercept of the linear part of the curve in Fig. 18-9 is equal to the inverse of V_m, the volume of adsorbed gas that fills a monomolecular layer on the surface. The parameter c is equal to one plus the ratio of slope to intercept.

Problem 18-8.   Verify the statements in the two preceding sentences.

The value of V_m for N_2, together with the data on the size of the N_2 molecule, is used routinely to calculate the surface area of adsorbents.

13. S. Brunauer, *The Adsorption of Gases and Vapors*, Vol. I, Princeton University Press, Princeton, 1943, p. 151.

14. T. L. Hill, *Statistical Thermodynamics*, Addison-Wesley, Reading, 1960, pp. 134–36.

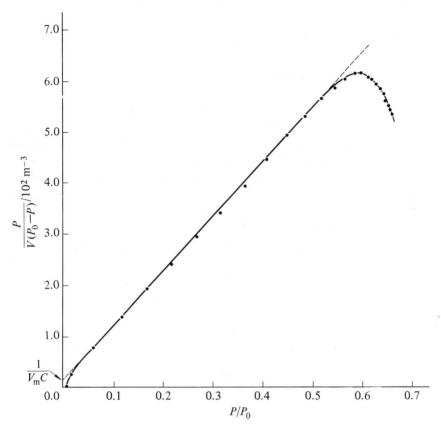

Figure 18-9. A plot of $P/V(P_0 - P)$ against P/P_0 for the data of Drain and Morrison (see Fig. 18-8), to test the applicability of the Brunauer-Emmet-Teller equation; slope $= (c - 1)/(V_m c)$, where V_m is the volume absorbed to form a monolayer and c is defined in Eq. 18-23.

Nature of the adsorbed species

Spectroscopic techniques are most useful in characterizing the nature of the adsorbed species, but difficult to apply to adsorbed layers. In 1956, Eischens and Pliskin reported a technique for measuring the infrared absorption spectra of gases adsorbed on finely divided metals.[15] They produced small metallic particles (5-10 nm in diameter) by a hydrogen reduction of metallic salts on the surface of finely divided alumina or silica. These particles were then placed on a CaF_2 plate in an infrared cell the temperature and gas phase composition of which could be controlled, as shown in Fig. 18-10. They

15. R. A. Eischens and W. A. Pliskin, *Adv. in Catal., 10*, 1–56 (1958).

showed that CO is chemisorbed on single sites with a carbon-metal bond similar to the bonds of simple metal-carbonyl coordination compounds, as well as on double sites, on which the carbon forms a bond to two adjacent metal atoms. Correlation of the infrared results with the desorption studies discussed above shows that the most easily desorbed molecules of CO are those on single sites, whereas the most strongly adsorbed molecules are on double (two-atom) sites. Physically adsorbed molecules, in contrast, exhibit an infrared spectrum that differs little from that of the same molecules in the gas phase, except for the appearance of new bands because the adsorbed molecule may have acquired a dipole moment (Sec. 6-1), a changing dipole moment in a particular mode of vibration (Sec. 6-2), or a changing polarizability (Sec. 6-2).

Photoelectron spectroscopy (Sec. 6-4) also provides structural information about the adsorbed species. X-ray photoelectron spectroscopy is an excellent means of identifying atoms on the surface through the presence of characteristic core electron energies and, thus, of determining whether a surface is clean. It can also be used to distinguish the different electronic states of adsorbed species, as in the results shown in Fig. 18-11 for CO adsorbed on tungsten.

Figure 18-10.   Scheme of an infrared cell used to measure the infrared spectrum of gas adsorbed on a finely divided solid (in this case, a catalyst). [By permission, from R. A. Eischens and W. A. Pliskin, *Adv. in Catalysis*, *10*, 1–56 (1958).]

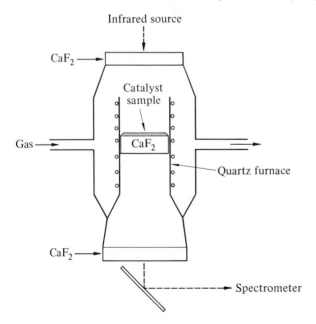

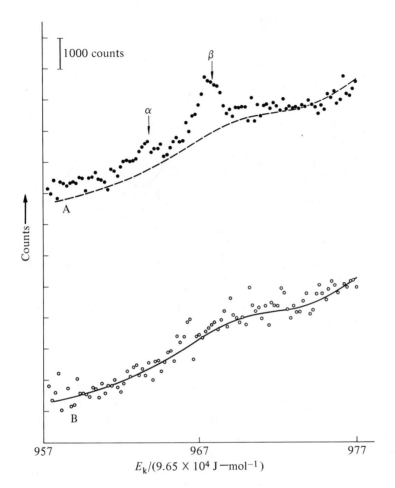

Figure 18-11. An X-ray photoelectron spectrum of carbon monoxide adsorbed on tungsten (at 300 K) showing different peaks for carbon monoxide adsorbed on α and β sites (curve A) and the absence of a spectrum when the surface is clean (B). [By permission, from J. T. Yates, Jr., T. E. Madey, and N. E. Erickson, *Surface Sci.*, *43*, 257 (1974).]

Ultraviolet photoelectron spectroscopy probes the valence electron levels in adsorbed species. Studies by Egelhoff and Linnett[16] of CH_3OH on tungsten 100 planes indicate that dissociation to CO and H occurs on the most strongly adsorbing sites, whereas species tentatively identified as CH_3OH and CH_3O are found when more weakly adsorbing sites are covered.

Low energy electron diffraction has been used to determine interatomic

16. As described by J. T. Yates, Jr., *Chem. Eng. News*, *52*, 19–29 (1974).

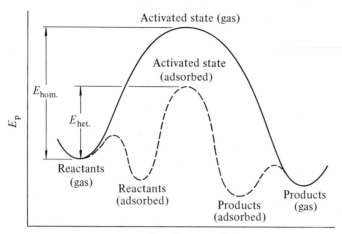

E_p

Activated state (gas)

Activated state (adsorbed)

$E_{hom.}$

$E_{het.}$

Reactants (gas)

Reactants (adsorbed)

Products (adsorbed)

Products (gas)

Reaction coordinate

Figure 18-12. Schematic curves of potential energy as a function of distance along reaction coordinate for a reaction in the gas phase and for the same reaction on a catalyst surface, showing a decrease in activation energy for the catalyzed reaction. (By permission, from S. Glasstone, K. J. Laidler, and H. Eyring, *The Theory of Rate Processes*, McGraw-Hill Book Co., New York, 1941.)

distances in adsorbed species. For example, it has been shown that,[17] for CO on palladium 100 planes, the C–O distance corresponds to a surface lattice spacing up to 50% of a mono-molecular layer. At θ equal to 0.5, there is a two-dimensional phase transition, and the C–O distance changes to a distance that no longer corresponds to the Pd lattice. The energy of adsorption decreases sharply at the same value of θ.

Additional information about the nature of adsorbed species has been obtained from the use of isotopically labeled adsorbates[18] by Madey and co-workers. When a mixture of $^{12}C^{18}O$ and $^{13}C^{16}O$ is adsorbed on tungsten, the molecules desorbed from the weaker binding sites show no isotopic mixing, whereas the molecules desorbed from the stronger binding sites show extensive isotopic mixing. These authors suggest that both C and O in CO are involved in the adsorption to the metal and that a bimolecular exchange on the surface and not dissociation is responsible for the isotopic mixing.

18-2. Adsorption and catalysis

Much of the impetus for study of the adsorption of gases by solids came

17. J. C. Tracy and P. W. Palmberg, *J. Chem. Phys.*, *51*, 4852 (1969).
18. T. E. Madey, J. T. Yates, Jr., and R. C. Stern, *J. Chem. Phys.*, *42*, 1372–78 (1965).

from the need to understand the action of heterogeneous catalysts, particularly the catalysts for the ammonia synthesis reaction and for the reactions and production of hydrocarbons from petroleum.

We would like to find out how a reaction can proceed more rapidly on the surface of the catalyst than in the gas phase. Part of the answer is that there is a higher probability of encounters between reactants when they are adsorbed on the catalyst surface than when they are in the gas phase; in effect, the adsorption process increases their concentration. In addition, since reactants are chemisorbed on the catalyst, the chemical state of the adsorbed reactants must be such that the activation energy of the reaction is smaller for the catalyzed reaction than for the gas phase reaction. A diagram of the difference in energy relations is shown in Fig. 18-12, and data for a number of heterogeneous reactions are given in Table 18-1.

Table 18-1. Activation energies for several catalyzed (heterogeneous) and gas phase (homogeneous) reactions

Reaction	Surface	$E_a(het)/(kJ \cdot mol^{-1})$	$E_a(hom)/(kJ \cdot mol^{-1})$	References[a]
$2\,HI = H_2 + I_2$	Au	104	184	1
$2\,N_2O = 2\,N_2 + O_2$	Au	121	244	2
$2\,NH_3 = N_2 + 3\,H_2$	W	163	>330	3
$CH_4 = C + 2\,H_2$	Pt	230–250	>330	4

S. Glasstone, K. J. Laidler, and H. Eyring, *The Theory of Rate Processes*, McGraw-Hill Book Co., New York, 1941, p. 390.
[a] Key to references:
 1. C. N. Hinshelwood and C. R. Prichard, *J. Chem. Soc.*, *127*, 1552 (1925).
 2. C. N. Hinshelwood and C. R. Prichard, *Proc. Roy. Soc., London, A108*, 211 (1925).
 3. C. N. Hinshelwood and R. E. Burk, *J. Chem. Soc.*, *127*, 1116 (1925).
 4. G.-M. Schwab and E. Pietsch, *Z. physik. Chem.*, *121*, 189 (1926).

The basic model for reactions that occur on the surfaces of solids was set forth by Langmuir and Hinshelwood[19] and involves the following sequence of steps:

1. Diffusion of reactants to the surface
2. Adsorption of reactants on the surface
3. Reaction on the surface
4. Desorption of the products
5. Diffusion of the products from the surface

Steps 1 and 5 are almost never rate-controlling in heterogeneous catalysis of reactions among gases. Though chemisorption frequently involves an activation energy, the activation energies of steps 3 and 4 are usually greater ($\Delta E_c + E_a$ in Fig. 18-1,B) and the latter are usually rate-controlling. Because they are difficult to separate in practice, they are considered as a combined

19. C. N. Hinshelwood, *Kinetics of Chemical Change*, 2nd ed., Oxford University Press, Oxford, 1940, p. 187.

step. From the Langmuir-Hinshelwood model, we assume that the rate of a catalyzed reaction depends on the surface concentrations of the reactants, expressed in moles per unit area. The surface concentrations, in turn, are determined by the pressures of the reactants in the gas phase. The Langmuir isotherm is used to relate surface concentration to gas pressure, since chemisorption is usually involved in catalytic reactions.

Consider the example of a simple bimolecular reaction on the surface. The probability that the two reactants are adsorbed on adjacent sites is proportional to the product of the surface concentrations of the reactants. If the two gaseous reactants are at such pressures so that very little of the surface is occupied, the surface concentration of each is proportional to the pressure, and the rate of the surface reaction is proportional to the product of the gas pressures, a second-order reaction. If the surface is almost saturated with both reactants, the surface concentrations are independent of the gas pressures, and the reaction is zero order. When one reactant is more strongly adsorbed than the other, and both reactants are competing for the same set of sites on the surface of the catalyst, the number of sites available to the second reactant decreases with increasing pressure of the first, and the reaction rate may be proportional to the ratio of the gas pressures rather than their product. A detailed discussion of a number of possible forms for the rate equations for surface catalyzed reactions has been presented by Laidler.[20]

Kinetic data for catalytic reactions must be supplemented with other data so that a mechanism can be chosen, since the rate equations are quite complex and the rate equations for different mechanisms are sometimes difficult to distinguish on the basis of kinetic data alone.

Infrared spectroscopy is one of the most important of these techniques. Eischens and Pliskin[21] have shown that when ethylene is adsorbed on metals that catalyze hydrogenation, two kinds of adsorption occur. When hydrogen is preadsorbed on the surface, ethylene is adsorbed associatively, with its two carbon atoms bound to the metal surface. When hydrogen is not present, ethylene is adsorbed dissociatively, as acetylene plus H atoms, and self-hydrogenation occurs to form ethane.

Leftin[22] used visible and ultraviolet spectroscopy to demonstrate that carbonium ions are present when triphenylmethane is adsorbed on silica-alumina catalysts that are used in catalytic cracking of hydrocarbons. This work provided strong support for the importance of these species as intermediates in cracking reactions.

20. K. Laidler, *Chemical Kinetics*, 2nd ed., McGraw-Hill Book Co., New York, 1965, Chapt. 6.
21. R. P. Eishens and W. A. Pliskin, *Adv. Cataly.*, *10*, 1–56 (1958).
22. H. P. Leftin, *J. Phys. Chem.*, *64*, 1714 (1960).

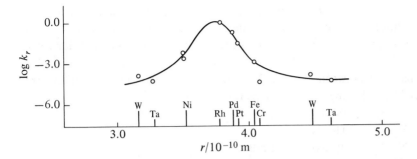

Figure 18-13. The logarithm of the rate constant for the hydrogenation of ethylene as a function of lattice spacing of the metallic catalyst. [By permission, from O. Beeck, *Disc. Faraday Soc.*, *8*, 118–28 (1950); O. Beeck and A. W. Ritchie, *Disc. Faraday Soc.*, *8*, 159–66 (1950).]

Direct observation of the crystal faces of a catalyst surface and of adsorbed molecules is possible through the use of the field emission microscope[2][3] and low-energy electron diffraction.[24] Beeck and co-workers used oriented metallic films that exposed certain crystal faces preferentially to demonstrate the importance of lattice spacing of the surface atoms in catalysis.[2][5] They found a clear dependence of the activity of transition metal films on the lattice spacing, with rhodium 10^3 times as active as nickel, as shown in Fig. 18-13. They also found that oriented films of platinum, in which the 110 planes predominated, are only one-tenth as effective in the dehydrogenation of cyclohexane as unoriented films, indicating the importance of the 111 planes present in the unoriented films. Leidheiser and Gwathmey used highly polished spheres cut from single crystals of copper to study the efficiency of different crystal faces in catalysis.[26] They found measurable differences in the rate of the catalytic oxidation of hydrogen on different faces, and they were able to observe distinct physical changes on those faces that were most active.

The catalytic synthesis of ammonia is a well-known and economically important catalytic reaction. High-yield modern agriculture depends heavily on ammonia production. The reaction is complex but reasonably well under-

23. R. Gomer, *Field Emission and Field Ionization*, Harvard University Press, Cambridge, 1961, Chapt. II.
24. L. H. Germer, *Adv. Cataly.*, *13*, 191 (1962).
25. O. Beeck, *Disc. Faraday Soc.*, *8*, 118–28 (1950); O. Beeck and A. W. Ritchie, *Disc. Faraday Soc.*, *8*, 159–66 (1950).
26. H. Leidheiser and A. T. Gwathmey *J. Am. Chem. Soc.*, *70*, 1200–6 (1948).

stood.[27] The catalyst mixture most commonly used is iron and iron oxide with alumina (aluminum oxide) and potassium oxide as promoters. The alumina seems to provide a large surface area; the potassium oxide seems to regulate the basicity. On the basis of extensive kinetic studies and of comparisons between H_2 and D_2 as reactants, Taylor and co-workers proposed the following series of steps for the reaction:

1. $N_2(g) = 2 N(ads)$
2. $H_2(g) = 2 H(ads)$
3. $N(ads) + H(ads) = NH(ads)$
4. $NH(ads) + H(ads) = NH_2(ads)$
5. $NH_2(ads) + H(ads) = NH_3(ads)$
6. $NH_3(ads) = NH_3(g)$

It was learned early that catalysts for ammonia synthesis also adsorb nitrogen as atoms but do not form nitrides. The exchange reaction between $^{14}N_2$ and $^{15}N_2$ on the catalyst surface occurs at a rate comparable to that of the synthesis reaction, whereas the $H_2 - D_2$ exchange is much faster. These observations support the hypothesis that the dissociative chemisorption of N_2 is the rate-determining step in the reaction.

Problems

18-1. C. E. H. Bawn [*J. Am. Chem. Soc.*, **54**, 72 (1932)] reported the following data for the adsorption of CO on a mica sample with a surface area of 0.6239 m². The first column at each temperature represents a blank run to determine the amount of CO adsorbed on the glass vessel.

Glass vessel		Vessel and mica	
$P/133$ Pa	$V/10^{-8}$ m³ (STP)	$P/133$ Pa	$V/10^{-8}$ m³ (STP)
	90 K		
0.0132	0.498	0.0056	10.82
0.0344	0.797	0.0105	13.39
0.0852	1.114	0.0453	17.17
		0.0545	17.69
		0.0791	18.89
		0.1059	19.60
	193 K		
0.0252	0.080	0.0121	0.075
0.0534	0.283	0.0597	0.624
0.0829	0.434	0.138	1.50
0.120	0.553	0.225	2.29

27. J. M. Thomas and W. J. Thomas, *Introduction to the Principles of Hetergeneous Catalysis*, Academic Press, London, 1967, pp. 408–22; A. Adamson, *Physical Chemistry of Surfaces*, 2nd ed., Interscience, New York, 1967, pp. 686–88.

(a) Plot the raw data at each temperature for sample and blank, draw smooth curves, and calculate the smoothed values of the gas adsorbed at a number of points at each temperature. (b) Test the data for fit to the Langmuir adsorption isotherm by plotting V/P against V and calculate V_m and b. (c) Calculate isosteric $\Delta \mathcal{H}_{ads}$ at several values of V. (d) Assuming a molecular diameter of 0.35 nm for CO, calculate the surface area from V_m.

18-2. Wooten and Brown [*J. Am. Chem., Soc.*, **65**, 113 (1943)] obtained the following data for the volume of ethylene gas adsorbed on mixed $BaCO_3$ and $SrCO_3$ at 90 K. The vapor pressure of ethylene at 90 K is 4.07 Pa. The volume of gas adsorbed is corrected to 298 K and 133 Pa pressure.

$P/133 \times 10^{-3}$ Pa	$V/10^{-6}$ m³
22.30	10.34
12.70	7.85
7.30	6.42
4.48	5.52
2.74	4.98
1.85	4.60
1.32	4.33

Compare the values for the volume of gas adsorbed for monolayer coverage calculated from the Langmuir isotherm and the Brunauer-Emmett-Teller equation. Reprinted by permission. Copyright by the American Chemical Society.

18-3. Hinshelwood and Burk [*J. Chem. Soc.*, **127**, 1116 (1925)] measured the rate of decomposition of NH_3 to H_2 and N_2 at 1129 K on the surface of a tungsten wire. In one experiment they found the following pressures as a function of time.

t/s	$P/133$ Pa	t/s	$P/133$ Pa
0	200	800	292
100	214	1000	312
200	227	1200	332
300	238	1400	349
400	248.5	1800	378
500	259	2000	387
600	270		

What can you conclude about the surface coverage during the reaction from these data? Added N_2 or H_2 had a negligible effect on the rate.

18-4. Hinshelwood and Burk also studied the rate of decomposition of NH_3 on Pt at 1411 K and in particular the effect of added H_2 on the rate. The change in pressure in time intervals of 120 s as a function of H_2 pressure at a fixed initial pressure of NH_3 of 1.33×10^4 Pa were as follows.

$\Delta P/133$ Pa	$P(H_2)/133$ Pa
33	50
27	75
16	100
10	150

What can you conclude about the surface coverage by NH_3 and H_2 from these data?

19 · Colloids
and macromolecules

Certain properties of systems, such as light scattering and sedimentation in a gravitational or centrifugal field, become particularly prominent when one or more of the components occur as particles in the size range from about 1 nm to about 1 m. Systems of this kind have been denoted *colloidal* historically. There is no general agreement with respect to the definition of colloidal systems or to classification of colloids into subgroups. We shall limit our consideration to two arbitrarily chosen classes, which we shall denote *sols* and *macromolecular solutions*.[1] Extensive discussions of a variety of colloidal systems can be found in the references.[2]

19-1. Sols

At a time when other chemists still considered the *colloidal state* fundamentally different from the better understood *crystalloidal state*, von Weimarn suggested that the colloidal state is characterized only by the size of the suspended particles and that any substance can in principle be prepared in the colloidal state.[3] He pointed out that colloids can be prepared either by dispersing macroscopic samples in a solution until they are fine enough to be

1. "When I use a word, it means just what I choose it to mean, neither more nor less." Lewis Carroll, *Through the Looking Glass.*
2. A. Adamson, *The Physical Chemistry of Surfaces*, 2nd ed., Interscience, New York, 1967; K. J. Mysels, *Introduction to Colloid Chemistry*, Interscience Publishers, New York, 1959; A. E. Alexander and P. Johnson, *Colloid Science*, Clarendon Press, Oxford, 1949.
3. P. P. von Weimarn, in *Colloid Chemistry*, (J. Alexander, ed.), Vol. I, Chemical Catalogue Co., New York, 1925, pp. 27–101.

considered colloids or by precipitating a substance in a finely divided state. The resulting colloids are *sols*, in contrast to *macromolecular solutions*, which are simply solutions of molecules of colloidal size (Sec. 19-2). Macromolecules are distinguished from the dispersed phase of sols by their relative ease of reversible solution and precipitation and by a particle mass that is defined primarily by the conditions of chemical synthesis or biosynthesis rather than by the conditions of solution preparation. Sols may be either crystalline particles, e.g., metallic sols or insoluble salts, or amorphous particles, e.g., silica and other hydrous oxides.

Preparation of sols

Sols can be prepared by grinding up macroscopic particles in the presence of a *dispersion medium*, a solvent in which the colloid is ordinarily insoluble. Since mechanical grinding in a colloid mill and ultrasonic dispersion also promote reaggregation of small particles into larger ones, a solid inert diluent is sometimes added during the dispersion process to facilitate colloid preparation.

Sols are most often prepared by condensation. Sol formation depends on the precipitation of particles the sizes of which are less than or equal to the size of colloids. A chemical reaction, addition of a poor solvent to a solution, and electrochemical reactions are all used. The Bredig arc method for the preparation of metallic sols combines dispersion with condensation; metallic electrodes are made to arc under the surface of the solvent. Metal atoms evaporate from the arc and condense in the solvent to form a sol.

Conditions governing the formation of a colloidal precipitate were set forth by von Weimarn.[4] The rate of precipitation must be very rapid in order to obtain a large number of nuclei that exhaust the remaining solute before too much particulate growth occurs. Rapid precipitation can be achieved by working at a high degree of supersaturation, defined as $(C - S)/S$, where C is the concentration of solute to be precipitated and S is the solubility. If supersaturation is greatly increased by increasing C, the nuclei formed may be close enough together to form a *gel*, in which the solvent is held by a continuous matrix of colloidal particles, instead of a sol. It is preferable, therefore, to work under conditions in which S is as small as possible so that the nuclei will remain highly dispersed.

Properties of sols

Sedimentation Properties. Under the influence of the earth's gravitational

4. P. P. von Weimarn, in *Colloid Chemistry*, (J. Alexander, ed.), Vol. I, Chemical Catalogue Co., New York 1926, pp. 27–101.

field a sol tends to settle to the bottom of its container, but the Brownian motion of the solvent particles tends to keep them uniformly distributed. Very large particles settle out and very small particles remain uniformly distributed. For an intermediate range of particle size, a Boltzmann distribution of particle concentration with height is set up at equilibrium. The potential energy of a colloidal particle of mass m and density ρ at a height h above some arbitrary zero of potential energy in a medium of density ρ_0 is $mgh(1 - \rho_0/\rho)$, where the term in parentheses reflects the buoyant effect of the medium. The number of particles per unit volume $N(h)$ at any height is given by the Boltzmann expression,

$$N(h) = N(0) \exp\left[\frac{-mgh(1 - \rho_0/\rho)}{kT}\right] \qquad (19\text{-}1)$$

Jean Perrin used this equation to determine Boltzmann's constant and thereby Avogadro's constant.[5] He prepared a suspension of gum gamboge, obtained from the dried latex of *Garcinia morella*, in which the particles were relatively uniform in size. He was able to count the number of particles at different heights in a cell 0.1 mm deep on the stage of a microscope by bringing successive layers in the suspension into focus. Though the particles could not be seen, even under a microscope, he could observe the light they scattered. He calculated the mass of a particle by weighing a sample in which he had already counted the particles. Despite the difficulties of the experiment, he obtained a value of 7×10^{23} for Avogadro's constant, which is very close to the present value of 6.02×10^{23}. Perrin received the Nobel prize in physics in 1926 for this work, work in which he not only determined the value of Avogadro's constant but showed that colloidal particles are described by the same kinetic theory that describes gas molecules and molecules in solution. Perrin's particles had a mass approximately 10^{10} times as great as a hydrogen atom. But to observe sedimentation equilibrium with smaller colloidal particles, fields much larger than those provided by the earth's gravitation are required.

Exercise 19-1. For a particle of molar mass equal to 10^{10} g-mol^{-1}, suspended in H_2O at 298 K, with a particle density of $\rho = 1.33$ kg-dm^3, calculate $N(h)/N(0)$ for $h = 1 \times 10^{-5}$ m.

Optical properties. Sols are slightly opalescent due to the Rayleigh scattering of light by the suspended particles (Sec. 6-2). Since scattering is inversely

5. J. Perrin, *Brownian Motion and Molecular Reality* (trans. F. Soddy), Taylor and Francis, London, 1908; H. A. Boorse and L. Motz, *The World of the Atom*, Vol. I, Basic Books, New York, 1966, pp. 625–40.

proportional to the fourth power of the wavelength, sols may have a bluish tinge when viewed at right angles to an incident light beam and a yellowish-red tinge in transmitted light. The blue color of the sky at midday and the red color of sunrise and sunset arise from this phenomenon, though the scattering is from regions of fluctuating density in the atmosphere rather than from colloidal particles. Metallic sols, such as colloidal gold, show a range of brilliant colors that vary with particle size. These colors cannot be explained by Rayleigh scattering; they have been explained by Mie in terms of the optical properties of the bulk metal.[6] Very large nonmetallic colloidal particles of uniform size exhibit colors that vary with the angle of observation relative to the incident beam. Such scattering can also be explained by the Mie theory.[7]

Electrical properties. All colloidal particles in sols carry an electric charge, even those particles that are not electrolytes. The most direct evidence of this comes from the observation of the migration of colloidal particles in an electrical field. Electrophoresis in sols can be followed by direct microscopic observation of the light scattered by the particles.

Most colloids in sols are not ionic, and the charge on the colloidal particle must arise from the adsorption of small ions from the solution. Ions may interact with dipoles in the surface of the colloidal particle or ions may induce dipoles in the surface, which in turn polarize the ions. Anion adsorption is usually stronger than cation adsorption, probably because anions can be polarized to a greater extent. Metallic sols, such as gold, may form covalent bonds with anions much as individual gold ions form coordination complexes with anions. In the case of such an ionic colloid as $AgCl$, the colloid can be positively or negatively charged, depending on the relative concentrations of Ag^+ and Cl^- ions in the solution.

The charge on a colloidal particle, like the charge on the central ion in the Debye-Hückel theory (Sec. 17-3), influences the distribution of ions in the neighboring solution. The commonly used models[8] treat the surface of a colloidal particle as an infinite uniformly charged plane. A Boltzmann distribution of ions as a function of potential in the solution and the Poisson

6. G. Mie, *Ann. Phys. Leipzig*, *25*, 377 (1908); A. E. Alexander and P. Johnson, *Colloid Science*, Vol. I, Clarendon Press, Oxford, 1949, pp. 428–30.

7. E. M. Zaiser and V. K. LaMer, *J. Colloid Sci.*, *3*, 571 (1948); K. J. Mysels, *Introduction to Colloid Chemistry*, Interscience Publishers, New York, 1959, pp. 429–32.

8. G. Gouy, *J. Phys.*, *9*, 457 (1910); *Ann. Phys.*, *7*, 129 (1917); D. L. Chapman, *Phil. Mag.*, *25*, 475 (1913); P. Debye and E. Hückel, *Phys. Z.*, *24*, 185 (1923); P. Debye, *Phys. Z.*, *25*, 93 (1924); A. Adamson, *The Physical Chemistry of Surfaces*, 2nd ed., Interscience, New York, 1967, Chapt. IV.

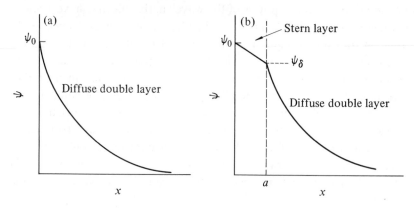

Figure 19-1. (a) The distribution of electrostatic potential in the Gouy model of the diffuse double layer. (b) The distribution of electrostatic potential in the Stern model of the double layer.

equation for the spatial dependence of the potential are assumed in these models. For low concentrations of ions and a low surface density of charge, the result is an exponential variation of potential with distance from the surface, as indicated mathematically in Eq. 19-2 and schematically in Fig. 19-1, a. This curve describes the *diffuse double layer*, whose effective thickness is taken as $1/\kappa$.

$$\psi = \psi_0 \exp\left(-\kappa x\right) \tag{19-2}$$

The Debye length, $1/\kappa$, corresponds to the distance at which ψ is equal to $(1/e)\psi_0$ and the center of gravity of electrical charge in the double layer. Stern[9] suggested that a distinction be made between an inner layer of adsorbed ions, the centers of which are at a distance a of closest approach to the surface, and the outer diffuse layer. This modification takes into account the finite size of ions and their inability to approach the surface more closely than the distance a. The potential variation corresponding to this model is shown in Fig. 19-1, b. At high ionic strength, $1/\kappa$ is very small and only the Stern layer is important; at very low ionic strength, the Stern layer is insignificant and only the diffuse double layer is important.

The stability of sols depends on the maintenance of repulsive forces between particles. In the absence of repulsive forces, attractive London

9. O. Stern, *Z. Electrochem.*, **30**, 508 (1924); J. O'M. Bockris and A. K. N. Reddy, *Modern Electrochemistry*, Vol. 2, Plenum Press, New York, pp. 733–39.

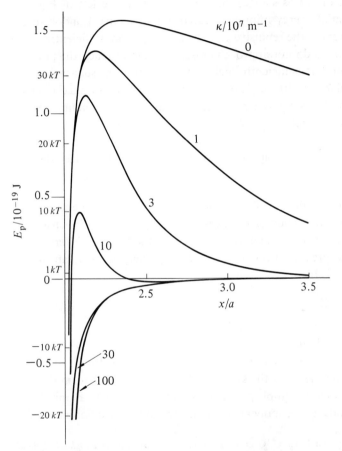

Figure 19-2. Results of the calculation of Verwey and Overbeeck for the long range electrostatic potential energy of two spherical colloidal particles as a function of distance. In this example, $a = 100$ nm, $T = 298$ K, $\psi_0 = kT/e$, and each curve corresponds to a different value of the Debye length $1/\kappa$. (By permission, from E. J. W. Verwey and J. Th. G. Overbeeck, *Theory of the Stability of Lyophobic Colloids*, Elsevier, Amsterdam, 1948.)

forces between particles would result in a flocculation of the colloid. The double layer of adsorbed ions and the associated ion atmosphere provides this repulsive force. Verwey and Overbeeck[10] developed a theory of the stability of colloids that describes the balance between attractive London forces and repulsive charge interactions.[11] Figure 19-2 shows an example of the results of their calculations of the potential energy of two spherical colloidal particles as a function of interparticle distance at 298 K and several values of κ. At distances shorter than the 200 nm at the left of Fig. 19-2 there is a potential energy minimum due to the attractive London forces. At low ionic strength the repulsive electrostatic forces are sufficiently strong so that the particles do not approach closely enough to reach the potential energy minimum. At sufficiently high salt concentration, such as one for which κ is equal to 3×10^8 m^{-1}, there is no potential energy maximum and the colloidal particles approach closely, remain at the potential energy minimum, and flocculate.

Exercise 19-2. Calculate the ionic strength in H_2O corresponding to the values of κ given in Fig. 19-2.

The long-known effect of flocculation of colloids by concentrated salt solutions and the dispersion of some precipitates to colloidal dimensions by low ionic strength solvents can be explained by the theory of Verwey and Overbeeck. The greater effectiveness of highly charged ions can also be understood by this theory.

19-2. Macromolecules

Synthetic macromolecules
Our civilization is distinctive in its use of materials not found in nature; prominent among these are the synthetic *high polymers* that form the basis of the plastics industry. Simple polymers contain many identical units linked by covalent bonds, and copolymers contain two or more different kinds of

10. E. J. W. Verwey and J. Th. G. Overbeeck, *Theory of the Stability of Lyophobic Colloids*, Elsevier, Amsterdam, 1948.
11. An impressive aspect of this work is the set of experiments in which Overbeeck and his colleagues and other workers have measured the London forces between macroscopic flat plates of glass and quartz as a function of distance. See W. Black, J. G. V. de Jongh, J. Th. G. Overbeeck, and M. J. Sparnaay, *Trans. Faraday Soc.*, *56*, 1597 (1960); B. V. Derjaguin, I. I. Abrikossova, and E. M. Lifshitz, *Quart. Rev.* (*London*), *10*, 292 (1956); J. A. Kitchener and A. P. Prosser, *Proc. Roy. Soc.*, London, *A242*, 403 (1957).

units linked by covalent bonds. Polymers can be formed by *addition* reactions, as in the synthesis of polyethylene,

$$H_2C=CH_2 + H_2C=CH_2 \rightarrow H_3C-CH_2-CH=CH_2$$

$$H_3C-CH_2-CH=CH_2 + H_2C=CH_2$$
$$\rightarrow H_3C-CH_2-CH_2-CH_2-CH=CH_2 \text{, etc.}$$

or by *condensation*, as in the synthesis of polyesters,

$$HO-CH_2-CH_2-OH + HOOC-\langle \rangle \quad -COOH$$
$$\rightarrow HO-CH_2-CH_2-O-\underset{O}{\overset{}{C}}-\langle \rangle \quad -COOH + H_2O$$

$$H-O-CH_2-CH_2-O-\underset{O}{\overset{}{C}}-\langle \rangle \quad -COOH + HO-CH_2-CH_2-OH$$
$$\rightarrow HO-CH_2-CH_2-O-\underset{O}{\overset{}{C}}-\langle \rangle \quad -\underset{O}{\overset{}{C}}-O-CH_2-CH_2-OH + H_2O, \text{etc.}$$

Addition polymerization can also be classified as *free-radical chain polymerization*, in which the reaction of one radical results in a product radical that continues the reaction. In condensation polymerization, the successive steps are independent of one another and lead to a polymer because the reactants are bifunctional.

The reactions shown above lead to *linear* polymers. Two-dimensional and three-dimensional polymer networks can be obtained by the use of such cross-linking reagents as divinylbenzene or other dienes for addition reactions and by the use of such trifunctional reagents as glycerol for condensation reactions.

The properties of a polymer are determined by the molecular mass, which determines the melting point for a given type of polymer, by the degree of cross-linking, which determines rigidity, by the nature of the functional groups, which determines intermolecular forces, and by the degree of crystallinity, which affects the melting point and rigidity.

Like other chemical reactions, polymerization occurs by random collisions of molecules, and the degree of polymerization varies from one molecule to another in any sample. Thus there is no unique molecular mass for the sample, but the molecular mass distribution can be characterized by several average molecular masses. The *number-average molecular mass M_n* is defined as

$$M_n = \frac{\sum_j M_j c_j}{\sum_j c_j} = \frac{\sum_j c'_j}{\sum_j (c'_j/M_j)} \tag{19-3}$$

where M_j is the molecular mass of the jth kind of molecule, c_j is the concentration in moles per unit volume, and c'_j is the concentration in mass per unit volume. The *mass-average molecular mass* M_m is defined as

$$M_m = \frac{\sum\limits_j (c_j M_j) M_j}{\sum\limits_j c_j M_j} = \frac{\sum\limits_j c'_j M_j}{\sum\limits_j c'_j} \tag{19-4}$$

The *z-average molecular mass* M_z is defined as

$$M_z = \frac{\sum\limits_j (c_j M_j^2) M_j}{\sum\limits_j c_j M_j^2} \tag{19-5}$$

We shall see that the different average molecular masses are obtained from different physical measurements on polymer solutions. The differences among the different molecular mass averages provide a measure of the breadth of the molecular mass distribution.

Exercise 19-3.   Show that $M_n = M_m = M_z$ for a sample in which all the molecules have the same molecular mass.

The *configuration* of polymer molecules in solution is fixed by the covalent bonds that constitute the molecular structure. The *conformation* of polymer molecules in solution is determined by the degree of free rotation about single bonds and the interactions of the units of the polymer with each other and with the solvent. Paul Flory and Maurice Huggins developed statistical theories for the molecular mass distribution that results from a polymerization process and for the distribution of conformations that result when a polymer molecule is in solution.[12]

Consider a sample of n_0 moles of monomer of molecular mass M_0, with reactive groups A and B on each monomer (for example, a hydroxy acid or an amino acid). If, at equilibrium, a fraction p of the A groups and an equal fraction of the B groups have reacted, p is called the *extent of reaction* or the *probability* of forming an AB bond. Then the fraction of unreacted A or B groups is $1 - p$, the probability of finding an unreacted A or B. The probability of finding a linear dimer with one bond A–B and one unreacted group of each kind is $p(1 - p)$, since the probability of two independent events occurring together is the product of the probabilities of the individual events. Similarly, the probability of a trimer is $p^2(1 - p)$, and the probability of a j-mer is $p^{j-1}(1 - p)$. Since p is less than one, the probability p_j of finding

12. P. J. Flory, *Principles of Polymer Chemistry*, Cornell University Press, Ithaca, 1953, pp. 318–23.

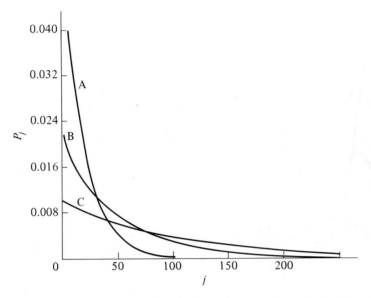

Figure 19-3. The probability of finding a j-mer, or the mole fraction of j-mer, as a function of j in a linear condensation polymer for several values of the extent of reaction. Extent of reaction for curve A, 0.95; B, 0.98; C, 0.99. [By permission, from P. J. Flory, *J. Am. Chem. Soc.*, *58*, 1877–85 (1936).]

a j-mer is a monotone decreasing function of j, as shown in Fig. 19-3. This quantity is also the mole fraction x_j of j-mer. Since each molecule at equilibrium has one unreacted A group and one unreacted B group, the total number of molecules at equilibrium is $n_0(1 - p)$ and the number of molecules of j-mer is

$$x_j (n_0) (1 - p) = n_0 p^{j-1} (1 - p)^2 \tag{19-6}$$

The mass fraction of j-mer is

$$m_j = \frac{n_0 \, p^{j-1} \, (1 - p)^2 \, (j \, M_0)}{n_0 M_0} = j \, p^{j-1} \, (1 - p)^2 \tag{19-7}$$

This function goes through a maximum, as shown in Fig. 19-4, but the distribution is very broad. If a narrow range of molecular masses is desired, it must be obtained either by a fractionation procedure or by polymerization in the presence of a controlled number of initiating centers.

When a polymer is dissolved in a good solvent, one in which the interactions between the monomer units and solvent molecules are as strong as

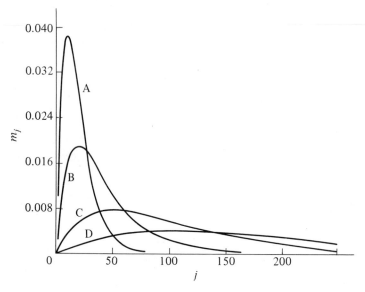

Figure 19-4. The mass fraction (m_j) of j-mer as a function of j in a linear condensation polymer for several values of the extent of reaction. Extent of reaction for curve A, 0.90; B, 0.95; C, 0.98; D, 0.99. [By permission, from P. J. Flory, *J. Am. Chem. Soc.*, *58*, 1877–85 (1936).]

those between the monomer units, the polymer molecules exhibit an extended random coil conformation. When a polymer is dissolved in a poor solvent, one in which the interactions between the monomer units are much stronger than those between monomer units and solvent molecules, the polymer molecules exhibit a compact conformation.

A random coil polymer in solution exists in many possible conformations. As a zeroth approximation, one can assume that the segments of the polymer chain rotate freely with respect to one another, and that there is no excluded volume effect; that is, the segments act essentially as mass points joined by rods of zero thickness and length l. It can be shown[13] that the probability of finding one end of the chain at a distance between r and $r + \delta r$ from the other end is

$$W(r)\delta r = \left(\frac{\beta}{\pi^{\frac{1}{2}}}\right)^3 \exp\left(-\beta^2 r^2\right) \cdot 4\,\pi r^2\,\delta r \qquad (19\text{-}8)$$

13. P. J. Flory, *Principles of Polymer Chemistry*, Cornell University Press, Ithaca, 1953, Chapt. X.

where

$$\beta = \left(\frac{3}{2}\right)^{\frac{1}{2}} (n^{\frac{1}{2}} l)^{-1} \tag{19-9}$$

n is the number of segments in the chain, and l is the length of a segment. The most probable end-to-end distance for this model is

$$\frac{1}{\beta} = \left(\frac{2}{3}\right)^{\frac{1}{2}} n^{\frac{1}{2}} l \tag{19-10}$$

A more realistic model can be constructed by considering barriers to rotation about bonds between segments and the excluded volume due to the finite size of the segments.[14]

Biological macromolecules

Many of the molecules synthesized by living cells are macromolecules. Some, like starch, glycogen, cellulose, and rubber, are composed of repeating units of a single monomer. Others, like proteins and nucleic acids, are composed of a relatively small number of similar units. The first group has a distribution of molecular masses similar to that of synthetic high polymers. The second group consists of identifiable discrete molecules of fixed molecular masses, with a composition determined by the genetic information carried in the DNA of each cell. Many proteins are now readily available from commercial sources as crystalline material of a purity and identity comparable to that of small organic molecules.

Physical methods for the study of macromolecules

Colligative properties. Osmotic pressure is the only colligative property that is useful for the study of macromolecules, for the reasons given in Sec. 16-4. If we assume that the osmotic pressure due to molecules of different molecular mass are additive in dilute solution, we can show that the osmotic pressure of a mixture of macromolecules leads to the number average molecular mass. For a given kind of molecule, from Eq. 16-110,

$$\pi_j = \frac{c'_j RT}{M_j} \tag{19-11}$$

in sufficiently dilute solution. Then

$$\pi = \sum_j \pi_j$$

14. P. J. Flory, *Principles of Polymer Chemistry*, Cornell University Press, Ithaca, 1953; C. Tanford, *Physical Chemistry of Macromolecules*, John Wiley & Sons, New York, 1961.

$$= RT \sum_j \frac{c_j}{M_j} \tag{19-12}$$

If we divide both sides of Eq. 19-12 by $\sum_j c'_j = c'$, the total mass concentration, we obtain

$$\frac{\pi}{c'} = \frac{RT \sum_j (c'_j/M_j)}{\sum_j c'_j}$$

$$= \frac{RT}{\sum_j c'_j / \sum_j (c'_j/M_j)} \tag{19-13}$$

We see from Eq. 19-3 that the denominator on the right in Eq. 19-13 is just M_n, so that

$$\frac{\pi}{c'} = \frac{RT}{M_n} \tag{19-14}$$

Sedimentation properties.   Unlike the large colloidal particles of sols, individual macromolecules in solution cannot be observed by scattered light. A number of ingenious optical systems have been devised, however, so that both the sedimentation equilibrium and sedimentation velocity behavior of macromolecules can be observed. Macromolecules in solution are not heavy enough to sediment in the earth's gravitational field. In order to observe their sedimentation, Svedberg invented the analytical ultracentrifuge. In a modern analytical ultracentrifuge, a rotor of aluminum or titanium is rotated at a high, controlled speed in an evacuated chamber. Mounted in the rotor is a sector-shaped cell of aluminum or a rigid plastic with quartz or sapphire windows. The cell is almost filled with a solution of the macromolecule to be studied and sealed to withstand the pressure developed in the intense centrifugal field.

At equilibrium in the centrifugal field, the concentration of macromolecules in the cell is described by a Boltzmann expression analogous to Perrin's equation for the gravitational field (Eq. 19-1). The centrifugal acceleration $\omega^2 x$ is substituted for the gravitational acceleration g, where ω is the rotational speed in radians per second, and x is substituted for the height h. The difference in potential energy per mole of macromolecule between two points x_2 and x_1 in the centrifugal field is then

$$- M\omega^2 \left(\frac{1 - \rho_0}{\rho} \right) (x_2^2 - x_1^2)$$

and the Boltzmann expression is

$$c'_2 = c'_1 \exp \left[\frac{M\omega^2 (1 - \rho_0/\rho) (x_2^2 - x_1^2)}{RT} \right] \tag{19-15}$$

Taking logarithms of both sides, we obtain

$$\ln c_2' = \ln c_1' + \frac{M\omega^2 (1 - \rho_0/\rho)}{RT} (x_2^2 - x_1^2) \qquad (19\text{-}16)$$

For macromolecules in solution, it is conventional to use the partial specific volume $v = 1/\rho$, so that Eq. 19-16 becomes

$$\ln c_2' = \ln c_1' + \frac{M\omega^2 (1 - v\rho_0)}{RT} (x_2^2 - x_1^2) \qquad (19\text{-}17)$$

A plot of $\ln c'$ against x^2 should be linear, with a slope equal to $M\omega^2 (1 - v\rho_0)/RT$ in which everything but the value of M is known. Equilibrium ultracentrifugation is a very versatile technique, though skill in it is essential for good results; the reader is referred to Chervenka's review for discussions of the various methods that have been developed and for references to the literature.[15]

The methodology of equilibrium ultracentrifugation is quite rigorous and precise; it is the method of choice for the determination of the molecular weight of a macromolecule available in a high state of purity that can be studied in sufficiently dilute solution so that deviations from ideality need not be considered. Because these conditions are not always met and because equilibrium measurements in the past were quite time-consuming, sedimentation velocity measurements are used more than sedimentation equilibrium measurements. The speed of rotation is chosen to be high enough so that there is a net migration of the molecule in the direction of the centrifugal field. The centrifugal force on the macromolecule is $(M/L) (1 - \rho_0 v)\omega^2 x$. When a steady velocity has been reached, the centrifugal force is equal in magnitude to the frictional force, $f(dx/dt)$, where f is the frictional coefficient of the molecule. Equating the two expressions, we have

$$\left(\frac{M}{L}\right)(1 - \rho_0 v) \omega^2 x = f\left(\frac{dx}{dt}\right) \qquad (19\text{-}18)$$

or

$$M = \frac{Lf(dx/dt)}{(1 - \rho_0 v) \omega^2 x} \qquad (19\text{-}18)$$

The ratio of (dx/dt) to $\omega^2 x$ is defined as the *sedimentation coefficient s*, so that Eq. 19-18 becomes

$$M = \frac{Lfs}{(1 - \rho_0 v)} \qquad (19\text{-}19)$$

15. C. Chervenka, *A Manual of Methods for the Analytical Utracentrifuge*, Beckman Instrument Co., Palo Alto, 1970.

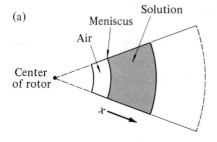

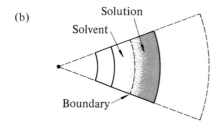

Figure 19-5. (a) Solution of a macromolecule in the sector-shaped cell of an ultra-centrifuge before sedimentation. (b) Boundary between solution of sedimenting macro-molecule and solvent after sedimentation for some time.

The calculation of s from experimental data depends on integrating the defining equation. Thus,

$$\frac{dx}{x} = \omega^2 s\, dt$$

and

$$\ln x = \omega^2 s t + \text{constant} \qquad (19\text{-}20)$$

The position of the macromolecule in a centrifugal field must be measured as a function of time. If $\ln x$ is plotted as a function of t, the curve should be linear and the slope is equal to $\omega^2 s$.

Figure 19-5, a is a schematic representation of the sector-shaped ultra-centrifugal cell in the rotor as viewed from above. Initially, the solution is uniform. As sedimentation proceeds, the macromolecule moves toward the edge of the rotor in the direction of the centrifugal field. After some time, the fraction of the solution closest to the center of the rotor becomes free of macromolecule and the boundary between pure solvent and solution has moved down the cell as indicated in Fig. 19-5, b. The rate of movement of this boundary is taken as the speed of the macromolecule in the centrifugal field. Figure 19-6, a shows the concentration of macromolecule as a function of position in the cell, initially and after some time t. Figure 19-6, b shows the concentration gradient dc/dx at time t. Absorption optics in the ultracentrifuge allows a measurement of the concentration as a function of position, as do interference optics; Schlieren optics measure the refractive-index gradient and hence the concentration gradient. Either the position of the peak in Fig. 19-6, b or the inflection point in Fig. 19-6, a is usually taken as the position of the boundary. Figure 19-7 shows examples of photographs obtained with absorption, interference, and Schlieren optics.

According to Eq. 19-19, one must know the frictional coefficient f as well as the sedimentation coefficient s in order to calculate the molecular mass.

Figure 19-6. (a) Plot of concentration of macromolecule as a function of position in the ultracentrifuge cell, initially (t_0) and after sedimentation for some time (t). (b) Plot of the concentration gradient dc/dx as a function of position in the cell after sedimentation for time t.

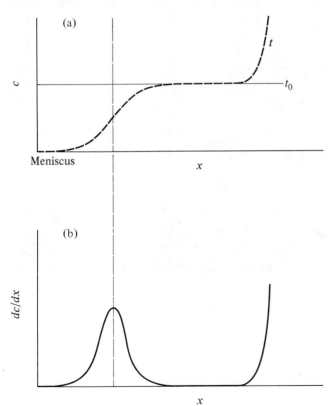

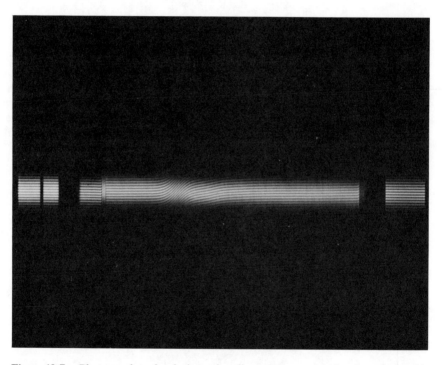

Figure 19-7. Photographs of solution of sedimenting macromolecules taken with (a) absorption optical system, (b) interference optical system, and (c) Schlieren optical system. All photographs are of a sample of immunoglobulin that contains some dimeric species and perhaps higher polymers in addition to the major component. The direction of sedimentation is from left to right. Runs were made at 60,000 rpm at 293 K. Each photograph was taken after 35 min. The protein concentrations were (a) 0.5 g-dm^{-3}, (b) 3.0 g-dm^{-3}, and (c) 7.5 g-dm^{-3}. (Courtesy of Dr. C. H. Chervenka, Spinco Division, Beckman Instruments, Inc.)

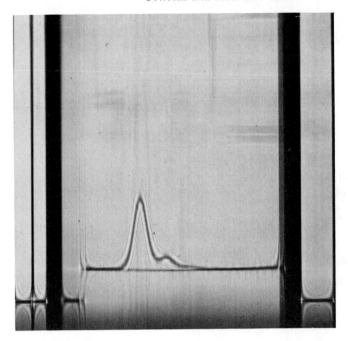

The measurement of the diffusion of the macromolecule leads to a value of f. In a diffusion experiment, a sharp boundary is formed between a solution of macromolecules and the solvent [by techniques such as those used in moving boundary experiments (Sec. 17-2)]. As diffusion proceeds, the broadening of the initially sharp boundary is observed by one of the optical methods used to follow sedimentation velocity. According to *Fick's first law of diffusion*, the rate of flow of solute across the boundary is proportional to the cross-sectional area and the concentration gradient. The coefficient of proportionality is the *diffusion coefficient D*.

Thus,

$$\frac{\partial m}{\partial t} = DA\frac{\partial c'}{\partial x} \tag{19-21}$$

where m is the mass transported and c' is the mass concentration.[16] Partial derivatives are used because m and c' are functions of both x and t. Because it is easier experimentally to measure changes of concentration with time

16. D is called the diffusion coefficient rather than the diffusion constant because it frequently depends on the concentration.

than to measure mass flow, we differentiate Eq. 19-21 with respect to x to obtain (assuming that D is a constant),

$$\frac{\partial}{\partial x}\left(\frac{\partial m}{\partial t}\right) = DA\frac{\partial^2 c'}{\partial x^2}$$

or

$$\frac{\partial}{\partial t}\left(\frac{\partial m}{\partial x}\right) = DA\frac{\partial^2 c'}{\partial x^2}$$

since the order of differentiation is immaterial for a function of two variables. Rearranging slightly, we have

$$\frac{\partial}{\partial t}\left(\frac{1}{A}\frac{\partial m}{\partial x}\right) = D\frac{\partial^2 c'}{\partial x^2} \tag{19-22}$$

Since $A\partial x$ is a volume, the term in parentheses is a concentration, and Eq. 19-22 can be written as

$$\frac{\partial c'}{\partial t} = D\frac{\partial^2 c'}{\partial x^2} \tag{19-23}$$

which is *Fick's second law of diffusion.* The boundary conditions for the experiment described above are (a) $c' = c'_0$ at $x = 0$ and $t = 0$, (b) $c' = 0$ at $-\infty$, and $c' = c'_0$ at ∞ at all t. It can be shown[17] that the solution to Eq. 19-23 with these boundary conditions is

$$\left(\frac{\partial c}{\partial x}\right) = \frac{c'_0}{2(\pi Dt)^{\frac{1}{2}}}\exp\left(-\frac{x^2}{4Dt}\right) \tag{19-24}$$

This function is of the same form as the Gaussian error curve. The maximum height of the curve is given by

$$H = \frac{c'_0}{2(\pi Dt)^{\frac{1}{2}}} \tag{19-25}$$

and the area under the curve is

$$A = \int_{-\infty}^{\infty}\frac{\partial c}{\partial x}\,dx = c'_0 - 0 = c'_0 \tag{19-26}$$

Thus,

$$\frac{A}{H} = 2(\pi Dt)^{\frac{1}{2}} \tag{19-27}$$

17. J. Crank, *The Mathematics of Diffusion*, Clarendon Press, Oxford, 1956.

A plot of $(A/H)^2$, taken from Schlieren photographs of the refractive-index gradient curve, against t should be a straight line with slope $= 4\pi D$.

Einstein[18] derived the relation between the diffusion coefficient and the frictional coefficient,

$$D = \frac{RT}{Lf} \tag{19-28}$$

Substituting for f from Eq. 19-28 in Eq. 19-19, we obtain

$$M = \frac{RT}{(1 - \rho_0 v)} \frac{s}{D} \tag{19-29}$$

Thus, the molecular mass of a macromolecule can be calculated from combined sedimentation and diffusion measurements.

Viscosity. The difference between the viscosity of a macromolecular solution and the viscosity of the solvent provides considerable information about the conformation of the macromolecule and, in the case of a random coil, about the molecular mass. If η_0 is the viscosity of the solvent and η is the viscosity of the solution, the *relative viscosity* η_r is equal to η/η_0, the *specific viscosity* η_{sp} is equal to $\eta_r - 1$, and the *intrinsic* viscosity is

$$[\eta] = \lim_{c' \to 0} \frac{\eta_{sp}}{c'} \tag{19-30}$$

Einstein developed a theory of the effect of spherical solute particles on the viscosity of the solvent in 1906.[19] His results were derived for solutions so dilute that there is no overlapping between the regions of flow disturbance about different solute molecules. His result is

$$\eta_{sp} = 2.5 \, \phi \tag{19-31}$$

where ϕ is the volume fraction of the macromolecule. If η_{sp} is found to follow Eq. 19-31, there is a strong basis for concluding that the macromolecule is a rigid sphere, but no information about the size of the particle can be obtained from viscosity measurements. If the particle is asymmetrical and/or solvated, the coefficient in Eq. 19-31 is greater than 2.5. Oncley[20] has calculated values of η_{sp}/ϕ for various values of axial ratio of prolate and oblate ellipsoids and for various degrees of hydration for protein molecules.

18. A. Einstein, *Z. fur Electrochem.*, *14*, 235–39 (1908); H. A. Boorse and L. Motz, *The World of the Atom*, Vol. I, Basic Books, New York, pp. 587–96.
19. A. Einstein, *Ann. Phys. Leipzig*, *19*, 289 (1906); *34*, 591 (1911).
20. J. L. Oncley, *Ann. N. Y. Acad. Sci.*, *41*, 121 (1941).

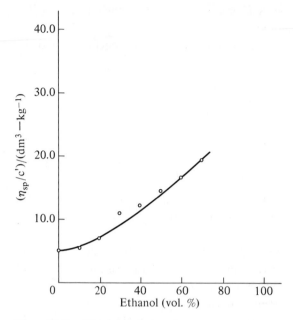

Figure 19-8. The reduced viscosity, η_{sp}/c', as a function of volume percent ethanol for a 0.01 kg-dm^{-3} solution of bovine serum albumin at pH 4.0 and 303 K. [By permission, from R. M. Rosenberg, D. W. Rogers, J. E. Haebig, and T. L. Steck, *Arch. Biochem. Biophys.*, *97*, 433 (1962).]

If data are available on hydration, viscosity measurements permit the calculation of two possible values of axial ratio, one for a prolate and one for an oblate ellipsoid.

Viscosity data are most useful quantitatively in the case of random coil polymers. Flory found empirically that the relation

$$[\eta] = KM^{\alpha} \tag{19-32}$$

holds for a given polymer-solvent pair over the molecular mass range of 10–1000 kg-mol^{-1}. The constants K and α must be determined in each case with carefully fractionated samples, and viscosity measurements can then be used to determine molecular masses. For $\alpha = 1$, the value of M obtained is M_m, whereas for $0 < \alpha < 1$,

$$M_n < M_{\eta} < M_m$$

Viscosity measurements provide a precise method for monitoring conformational changes in such macromolecules as proteins and nucleic acids, though it is difficult to interpret the nature of the changes rigorously. When the rigid globular conformation of a protein is changed by such denaturing agents as urea, alcohol, or detergents, to one that approaches a random coil, the increase in the effective volume for interference with the flow of solvent is reflected in a substantial increase in viscosity. Figure 19-8 shows data of this kind for the denaturation of serum albumin by ethanol in acid solution.

Problems

19-1. In one of Perrin's experiments [*Ann. Chim. Phys.*, *18*, 55–63 (1909)] he counted a total of 13,000 particles of gamboge and found average concentrations proportional to the following numbers at the given heights above the bottom of the vessel.

$h/10^{-6}$ m	Relative concentration
5	100
35	47
65	22.6
95	12

The experiment was carried out at 293 K and the density of the particles was 0.2067 kg-dm^{-3} greater than that of the water in which they were suspended. The mean radius of the particles was 2.12×10^{-7} m. Calculate Avogadro's constant from these data.

19-2. K. O. Pederson, a colleague of Svedberg's, studied the sedimentation equilibrium of lactoglobulin, a protein from milk, in the analytical ultracentrifuge [*Biochem. J.*; *30*, 967 (1936)]. By graphical integration of the concentration gradient curve obtained with a Schlieren optical system, he calculated the concentration of protein as a function of position in the cell

$x/10^{-3}$ m	$c'/(g\text{-dm}^{-3})$
49.0	1.30
49.5	1.46
50.0	1.64
50.5	1.84
51.0	2.06
51.5	2.31

The partial specific volume of the protein was found to be 0.7514 dm^3-kg^{-1}, and the density of the solution was 1.034 kg-dm^{-3}. The rotor was turning at a rate of 182.8 rps. Calculate the molecular mass.

19-3. Williams, Baldwin, Saunders, and Squire [*J. Am. Chem. Soc.*, *74*, 1542–48 (1952)] determined both the sedimentation coefficient and the diffusion coefficient from sedimentation velocity experiments on globulins that had been partially digested by the proteolytic enzyme pepsin. They obtained values of $s = 5.6 \times 10^{-13}$ s and

$D = 5.5 \times 10^{-11}$ m²-s⁻¹ at 293 K. Calculate the molar mass of the protein; assume $v = 0.75$ dm³-kg⁻¹.

19-4. Tanford, Kawahara, and Lapanje [*J. Am. Chem. Soc.*, **89**, 729–36 (1967)] have determined the intrinsic viscosities of a number of proteins denatured by guanidinium hydrochloride and with disulfide bridges reduced, so that they behave as random coils. Use their data to calculate the constants K and α of Eq. 19-32 for random coil polypeptide chains. In this case, the number of amino acid residues n is used in place of the molar mass M, since there is not a single monomer unit.

Protein	$M/(\text{g-mol}^{-1})$	Residues per chain	$[\eta]/(\text{dm}^3\text{-kg}^{-1})$
Insulin	2,970	26	6.1
Ribonuclease	13,680	124	16.6
Hemoglobin	15,500	144	18.9
Myoglobin	17,200	153	20.9
β-Lactoglobulin	18,400	162	22.8
Chymotrypsinogen	25,700	245	26.8
Pepsinogen	40,000	365	31.5
Aldolase	40,000	365	35.3
Serum albumin	69,000	627	52.2

Reprinted by permission. Copyright by the American Chemical Society.

19-5. The following results were obtained in a sedimentation velocity experiment with a 0.385 g-dm⁻³ solution of the enzyme aspartate aminotransferase in tris(hydroxymethyl)aminomethane buffer, pH 8.3, 1 M NaCl, at a rotor speed of 48,500 rpm.

t/min	Distance from center/cm
0	6.57
12	6.62
24	6.67
36	6.72
44	6.76
52	6.80
60	6.84
68	6.86

Calculate the sedimentation coefficient of the enzyme at the experimental conditions.

19-6. Martinez-Carrion, Sator, and Raftery [*Biochem. Biophys. Res. Comm.*, **65**, 129–37(1975)] obtained the following values for the osmotic pressure of acetylcholine receptor protein at 300 K.

$c'/(\text{g-dm}^{-3})$	14.9	9.93	4.97	3.31
$\pi/(\text{cm H}_2\text{O})$	1.32	0.87	0.44	0.28

The solvent density is 1.027×10^3 g-dm⁻³. Calculate the molar mass of the protein.

Appendix A
Physical quantities and units

This appendix is adapted from *Quantities, Units and Symbols*, 2nd Ed., a report of the Symbols Committee of the Royal Society, London, 1975.

The value of a *physical quantity* is equal to the product of a *numerical value* and a *unit*.

Neither any physical quantity, nor the symbol used to denote it, implies a particular choice of unit.

Operations on equations involving physical quantities, units, and numerical values follow the ordinary rules of algebra. For example, the physical quantity called the wavelength λ of one of the yellow lines of sodium has the value

$$\lambda = 5.896 \times 10^{-7} \text{ m}$$

where m is the symbol for the unit of length called the meter. Alternative forms are

$$\frac{\lambda}{\text{m}} = 5.896 \times 10^{-7}$$

and

$$\frac{\lambda}{10^{-7} \text{ m}} = 5.896$$

The latter forms are appropriate for the headings in tables of numerical values and for the labels of graphs, since pure numbers are listed and plotted.

A physical quantity is defined by a complete specification of the operations

711

used to measure the ratio (a pure number) of two particular values of that physical quantity, of which one is arbitrarily chosen as the *unit value*.

By international agreement, seven physical quantities are chosen for use as dimensionally independent *base quantities*:

Physical quantity	Symbol
length	l
mass	m
time	t
electric current	I
thermodynamic temperature	T
luminous intensity	I_V
amount of substance	n

All other physical quantities are regarded as being *derived* from the base quantities.

The SI units of the base quantities are defined as follows:

meter: The meter (m) is the length equal to 1,650,763.73 wavelengths in vacuum of the radiation corresponding to the transition between the levels $2p_{10}$ and $5d_5$ of the krypton-86 atom.

kilogram: The kilogram (kg) is the unit of mass; it is equal to the mass of the international prototype of the kilogram.

second: The second (s) is the duration of 9,192,631,770 periods of the radiation corresponding to the transition between the two hyperfine levels of the ground state of the cesium-133 atom.

ampere: The ampere (A) is that constant current, which, if maintained in two straight parallel conductors of infinite length, of negligible cross section, and placed 1 m apart in vacuum, would produce between these conductors a force equal to 2×10^{-7} newton/m of length.

kelvin: The Kelvin (K), unit of thermodynamic temperature, is the fraction 1/273.16 of the thermodynamic temperature of the triple point of water.

candela: The candela (cd) is the luminous intensity, in the perpendicular direction, of a surface of 1/600,000 m² of a black body at the freezing temperature of platinum under a pressure of 101,325 newtons/m².

mole: The mole (mol) is the amount of substance of a system that contains as many elementary entities as there are atoms in 0.012 kg of carbon 12.

When the mole is used, the elementary entities must be specified and may be atoms, ions, molecules, electrons, other particles, or specified groups of such particles. For example:

1 mole of HgCl has a mass equal to 0.23604 kg

1 mole of Hg_2Cl_2 has a mass equal to 0.47208 kg

1 mole of e^- has a mass equal to 5.4860×10^{-7} kg

1 mole of a mixture containing $2/3$ mole of H_2 and $1/3$ mole of O_2 has a mass equal to 0.0120102 kg

The derived SI units include:

Physical quantity	Symbol of physical quantity	Name of SI unit	Symbol of SI unit	Equivalent form of SI unit
force	**F**	newton	N	m-kg-s^{-2}
energy	E, ε	joule	J	N-m
pressure	P, p	pascal	Pa	N-m^{-2}, J-m^{-3}
power	P	watt	W	J-s^{-1}
electric charge	Q, q	coulomb	C	A-s
electric potential difference	$\mathscr{E}$	volt	V	J-C^{-1}
electric resistance	R	ohm	Ω	V-A^{-1}
electric conductance	L	siemens	S	Ω^{-1}
electric capacitance	C	farad	F	C-V^{-1}
magnetic flux	Φ	weber	Wb	V-s
magnetic flux density	**B**	tesla	T	Wb-m^{-2}
frequency	ν	Herz	Hz	s^{-1}

The units of other derived quantities can be obtained from their definitions in terms of those we have listed.

The following prefixes may be used to construct decimal multiples of units:

Multiple	Prefix	Symbol	Multiple	Prefix	Symbol
10^{-1}	deci	d	10	deca	da
10^{-2}	centi	c	10^2	hecto	h
10^{-3}	milli	m	10^3	kilo	k
10^{-6}	micro	μ	10^6	mega	M
10^{-9}	nano	n	10^9	giga	G
10^{-12}	pico	p	10^{12}	tera	T
10^{-15}	femto	f			
10^{-18}	atto	a			

In the case of mass units, decimal multiples should be formed by attaching a SI prefix to g rather than kg, even though kg is the base unit.

Certain non-SI units that have been used in the past and may continue to be used in the future are defined as follows in terms of the corresponding SI units:

Physical quantity	Name of unit	Symbol for unit	Definition of unit
length	Angstrom	Å	10^{-10} m $= 10^{-1}$ nm
volume	liter	l	10^{-3} m^3 $=$ dm^3
force	dyne	dyn	10^{-5} N
pressure	bar	bar	10^5 Pa
	atmosphere	atm	101325 Pa
	torr	Torr	≈ 133.32 Pa
	mm Hg	mm Hg	≈ 133.32 Pa
energy	erg	erg	10^{-7} J
	calorie	cal	4.184 J
	electronvolt	eV	$\approx 1.602 \times 10^{-19}$ J
viscosity	poise	P	10^{-1} Pa-s

Partial molar quantities, $(\partial X/\partial n_j)_{T, P, n_{\omega}}$, are denoted by the symbol $\mathscr{X}_j$. The corresponding value for a pure substance is the *molar quantity* $\mathscr{X}_j^*$, or $\mathscr{X}$ if there is no ambiguity. The superscript $\ominus$ is used to represent a standard value.

The following are the recommended values of physical constants.[1]

Quantity	Symbol	Value
speed of light in vacuum	c	2.997924580×10^8 m-s^{-1}
permittivity of a vacuum	ε_0	$8.85418782 \times 10^{-12}$ C-V^{-1}-m^{-1}
proton charge	e	$1.6021892 \times 10^{-19}$ C
Planck constant	h	6.626176×10^{-34} J-s
Avogadro constant	L	6.022045×10^{23} mol^{-1}
rest mass of electron	m_e	9.109534×10^{-31} kg
rest mass of proton	m_p	$1.6726485 \times 10^{-27}$ kg
rest mass of neutron	m_n	$1.6749543 \times 10^{-27}$ kg
Faraday constant	F	9.648456×10^4 C-mol^{-1}
Bohr radius	a_0	$5.2917706 \times 10^{-11}$ m
Bohr magneton	μ_B	9.274078×10^{-24} J T^{-1}
gas constant	R	8.31441 J-K^{-1}-mol^{-1}
Boltzmann constant	k	1.380662×10^{-23} J-K^{-1}

Additional information on physical quantities and units can be found in:

M. L. McGlashan, *Ann. Rev: Phys. Chem.*, *24*, 51–76 (1973).

International Union of Pure and Applied Chemistry, *Manual of Symbols and Terminology for Physicochemical Quantities and Units*, Butterworths, London, 1970.

Policy for National Bureau of Standards Usage of SI units, *J. Chem. Educ.*, *48*, 569 (1971).

A. C. Norris, *J. Chem. Educ.*, *48*, 797 (1971).

1. E. R. Cohenand B. N. Taylor, *J. Phys. Chem. Ref. Data*, *2*, 663 (1973).

Appendix B
Atomic masses of the elements

The numbers[1] in the table are dimensionless quantities, representing the mass of an atom or the average mass of an average isotopic distribution of atoms relative to the mass of an atom of carbon $12 = 12$.

Element	Symbol	Atomic number	Atomic mass
Actinium	Ac	89	(227)
Aluminum	Al	13	26.9815
Americium	Am	95	(243)
Antimony	Sb	51	121.7_5
Argon	Ar	18	39.94_8
Arsenic	As	33	74.9216
Astatine	At	85	~210
Barium	Ba	56	137.3_4
Berkelium	Bk	97	(247)
Beryllium	Be	4	9.01218
Bismuth	Bi	83	208.9806
Boron	B	5	10.81
Bromine	Br	35	79.904
Cadmium	Cd	48	112.40
Calcium	Ca	20	40.08
Californium	Cf	98	(251)
Carbon	C	6	12.011
Cerium	Ce	58	140.12
Cesium	Cs	55	132.9055
Chlorine	Cl	17	35.453
Chromium	Cr	24	51.996

1. *Handbook of Chemistry and Physics*, 54th Ed., R. C. Weast, Ed., © CRC Press, Inc., 1973. Used by permission of CRC Press, Inc.

Element	Symbol	Atomic number	Atomic mass
Cobalt	Co	27	58.9332
Copper	Cu	29	63.546
Curium	Cm	96	(247)
Dysprosium	Dy	66	162.50
Einsteinium	Es	99	(254)
Erbium	Er	68	167.26
Europium	Eu	63	151.96
Fermium	Fm	100	(257)
Fluorine	F	9	18.9984
Francium	Fr	87	(223)
Gadolinium	Gd	64	157.2_5
Gallium	Ga	31	69.72
Germanium	Ge	32	72.5_9
Gold	Au	79	196.9665
Hafnium	Hf	72	178.4_9
Holmium	Ho	67	164.9303
Hydrogen	H	1	1.0080
Indium	In	49	114.82
Iodine	I	53	126.9045
Iridium	Ir	77	192.2_2
Iron	Fe	26	55.84_7
Krypton	Kr	36	83.80
Lanthanum	La	57	138.905_5
Lawrencium	Lr	103	(257)
Lead	Pb	82	207.2
Lithium	Li	3	6.941
Lutetium	Lu	71	174.97
Magnesium	Mg	12	24.305
Manganese	Mn	25	54.9380
Mendelevium	Md	101	(256)
Mercury	Hg	80	200.5_9
Molybdenum	Mo	42	95.9_4
Neodymium	Nd	60	144.2_4
Neon	Ne	10	20.17_9
Neptunium	Np	93	237.0482
Nickel	Ni	28	58.7_1
Niobium	Nb	41	92.9064
Nitrogen	N	7	14.0067
Nobelium	No	102	(254)
Osmium	Os	76	190.2
Oxygen	O	8	15.999_4
Palladium	Pd	46	106.4
Phosphorus	P	15	30.9738
Platinum	Pt	78	195.0_9
Plutonium	Pu	94	(244)
Polonium	Po	84	(≈ 210)
Potassium	K	19	39.10_2
Praseodymium	Pr	59	140.907_7
Promethium	Pm	61	(145)
Protoactinium	Pa	91	231.0359

Element	Symbol	Atomic number	Atomic mass
Radium	Ra	88	226.0254
Radon	Rn	86	(≈ 222)
Rhenium	Re	75	186.2
Rhodium	Rh	45	102.9055
Rubidium	Rb	37	85.467_8
Ruthenium	Ru	44	101.07
Samarium	Sm	62	150.4
Scandium	Sc	21	44.9559
Selenium	Se	34	78.9_6
Silicon	Si	14	28.08_6
Silver	Ag	47	107.868
Sodium	Na	11	22.9898
Strontium	Sr	38	87.62
Sulfur	S	16	32.06
Tantalum	Ta	73	180.947_9
Technetium	Tc	43	98.9062
Tellurium	Te	52	127.6_0
Terbium	Tb	65	158.9254
Thallium	Tl	81	204.3_7
Thorium	Th	90	232.0381
Thulium	Tm	69	168.9342
Tin	Sn	50	118.6_9
Titanium	Ti	22	47.9_0
Tungsten	W	74	183.8_5
Uranium	U	92	238.029
Vanadium	V	23	50.941_4
Xenon	Xe	54	131.30
Ytterbium	Yb	70	173.0_4
Yttrium	Y	39	88.9059
Zinc	Zn	30	65.38
Zirconium	Zr	40	91.22

Appendix C
Some mathematical
information

C-1. Taylor's series

If a function $f(x)$ can be represented by a polynomial in x, as

$$f(x) = a_0 + a_1 x + a_2 x^2 + \ldots$$

then the derivatives of $f(x)$ are

$$f'(x) = a_1 + 2a_2 x + \ldots$$

$$f''(x) = 2a_2 + (3)(2)a_3 x + \ldots$$

$$f'''(x) = (3)(2)a_3 + (4)(3)(2)a_4 x + \ldots$$

and so forth.

The value of the function and the derivatives at $x = 0$ are

$$f(0) = a_0, \qquad f'(0) = a_1, \qquad f''(0) = 2a_2,$$

$$f'''(0) = (3)(2)a_3$$

and so forth.

Then the polynomial in x can be expressed as

$$f(x) = f(0) + f'(0)x + \frac{f''(0)}{2!} x^2 + \frac{f'''(0)}{3!} x^3 + \ldots$$

which is the form of a Taylor's series about $x = 0$. By a change of coordinates, the series can be converted to one about an arbitrary point, $x = a$, as

$$f(x) = f(a) + f'(a)(x - a) + \frac{f''(a)}{2!} (x - a)^2 + \frac{f'''(a)}{3!} (x - a)^3 + \ldots$$

C-2. Graphical and numerical integration

Frequently, it is necessary to evaluate an integral when the analytical form of the integrand is not known, as in the application of the Gibbs-Duhem equation in Chapter 16 or in the calculation of standard entropies from heat capacity data in Chapter 10. In such a circumstance, graphical and numerical integration are useful techniques.

In the calculation of standard entropies, it is necessary to calculate the value of the integral

$$\int_{T_1}^{T_2} \left(\frac{\mathscr{C}_p}{T} \right) dT$$

which is equal to the area under the curve of $\mathscr{C}_p/T$ against T between T_1 and T_2, or the area under the curve $\mathscr{C}_p$ against $\ln T$ between $\ln T_1$ and $\ln T_2$. If the experimental heat capacity data are plotted in either of these ways, the area can be determined by means of a planimeter or by cutting out the area and comparing the mass of the paper to that of a comparable sheet of known area.

Given a positive function $f(x)$ between the limits $x = a$ and $x = b$, the integral of $f(x)$ between these limits is defined as

$$\int_a^b f(x)\,dx = \lim_{\Delta x_v \to 0} \sum_{v=1}^{n} f(x_v)\,\Delta x \qquad \text{(C-1)}$$

where Δx_v is one of the v subintervals into which the interval from a to b is divided, and x_v is an arbitrary value of x in each interval. The sum on the right in Eq. C-1 is an approximation to the integral for any finite values of Δx_v, and the approximation can be made as close as desired by choosing sufficiently small values of Δx_v.

When the integral of a function obtained from experimental results is to be calculated, the experimental points are plotted and a smooth curve is fitted to the points. Values of the function are read from the curve at desired intervals, and the sum in Eq. C-1 is calculated. Alternatively, the experimental results are fitted to an empirical function, such as that the coefficients of which are given in Table 9-5. Then the value of the function is calculated at desired intervals and the sum obtained. The latter method is simply executed by a digital computer.

C-3. Vectors

Scalars are quantities that have magnitude but not direction, whereas *vectors* are quantities that have magnitude and direction. The distance between two points is a scalar, whereas the displacement from one point to another is a vector.

Though the magnitude and direction of a vector is independent of the coordinate system, it is convenient to describe vectors in terms of the coordinate system. Consider a vector **A** from the origin to the point (A_x, A_y, A_z) of a Cartesian coordinate system. Then the length of the vector is given as

$$|\mathbf{A}| = A = (A_x^2 + A_y^2 + A_z^2)^{\frac{1}{2}}$$

The quantities A_x, A_y, and A_z also represent the projections of the vector **A** on the x, y, and z axes, respectively. The vectors corresponding to these projections are $A_x\mathbf{i}$, $A_y\mathbf{j}$, and $A_z\mathbf{k}$, where **i**, **j**, and **k** are the unit vectors in the x, y, and z directions, respectively. By the rules of vector addition

$$\mathbf{A} = A_x\mathbf{i} + A_y\mathbf{j} + A_z\mathbf{k}$$

A vector **A** is thus determined completely by its components A_x, A_y, and A_z.

Two vectors can be multiplied in two different ways. In the first, the result of the multiplication is the *scalar product*, symbolized as **A** · **B**, and defined as

$$\mathbf{A} \cdot \mathbf{B} = (A)(B) \cos \delta$$

where δ is the angle between the vectors. Examples of the scalar product include the product of a force by a displacement to yield the work done (Chapter 9), and the product of a magnetic moment by the magnetic field strength to yield the potential energy of the moment in the magnetic field (Chapter 4).

The scalar product also can be expressed in terms of the components of **A** and **B**. Thus,

$$\mathbf{A} \cdot \mathbf{B} = A_xB_x(\mathbf{i}\cdot\mathbf{i}) + A_yB_y(\mathbf{j}\cdot\mathbf{j}) + A_zB_z(\mathbf{k}\cdot\mathbf{k}) + A_xB_y(\mathbf{i}\cdot\mathbf{j}) + A_xB_z(\mathbf{i}\cdot\mathbf{k})$$
$$+ A_yB_x(\mathbf{j}\cdot\mathbf{i}) + A_yB_z(\mathbf{j}\cdot\mathbf{k}) + A_zB_x(\mathbf{k}\cdot\mathbf{i}) + A_zB_y(\mathbf{k}\cdot\mathbf{j})$$

Since the scalar product depends on cos δ, only the product of identical unit vectors has a nonzero value and they have the value one. Then

$$\mathbf{A} \cdot \mathbf{B} = A_xB_x + A_yB_y + A_zB_z$$

The *vector product* of two vectors **A** × **B** is a vector with a magnitude

$$|\mathbf{A} \times \mathbf{B}| = (A)(B) \sin \delta$$

and a direction normal to the plane defined by the vectors **A** and **B**. The orientation of the product vector is determined by a right-hand rule. If one rotates vector **A** into vector **B**, the product points in the direction a right-hand screw would advance if so turned. It is clear from this definition that

$$\mathbf{B} \times \mathbf{A} = -\mathbf{A} \times \mathbf{B}$$

An area can be thought of as a vector product of two displacements at right angles to one another. Thus, the area vector is normal to the geometric area. In this way, the force vector **F** can be expressed as the product of the scalar quantity pressure (P) and the vector quantity area (**A**), and $P = F/A$, as in Eq. 1-1.

C-4. The method of least squares

Whenever we carry out an experiment, we obtain a small sample of values that we hope is representative of a larger universe of values, and we try to obtain from the experimental sample an estimate of the characteristics of the larger universe. The problem is to obtain the most probable values of the characteristic parameters of that universe in the light of our *experiments* and of our *assumptions* about the nature of the universe of measurements.

A common method of estimating the most probable values is the *method of least squares*, which can be stated as follows:

The most probable value of a quantity is that value for which the sum of the squares of the deviations of the observed values from the most probable value is a minimum.

If we assume that a series of observations represents random variations in the experimental value of a single quantity, then the quantity to be minimized is

$$\sum_{j=1}^{n} (x_j - X)^2$$

where x_j is an observed value, X is the most probable value sought, and n is the number of observations. The x_j are constants, and the summation is to be minimized with respect to variation of the parameter X. Thus,

$$\frac{d}{dX}\left[\sum_{j=1}^{n}(x_j - X)^2 \right] = \frac{d}{dX}\left[\sum_{j=1}^{n}(x_j^2) - 2\sum_{j=1}^{n} x_j X + \sum_{j=1}^{n} X^2 \right]$$

$$= -2\sum_{j=1}^{n} x_j + 2nX = 0$$

and

$$X = \frac{\sum_{j=1}^{n} x_j}{n}$$

By the method of least squares, then, the most probable value for a series of observations *assumed* to represent a single quantity is the *arithmetic mean*.

If we *assume* that two variables y_j and x_j are connected by the linear relationship

$$y = mx + b$$

then the quantity to be minimized is

$$R = \sum_{j=1}^{n} [y_j - (mx_j + b)]^2$$

The y_j and x_j are known constants obtained from experiment, and the quantity is minimized with respect to variation in m and b. The conditions for the minimum are

$$\frac{\partial R}{\partial m} = 0 \quad \text{and} \quad \frac{\partial R}{\partial b} = 0$$

The results of such a calculation are

$$m = \frac{n \sum_{j=1}^{n} (x_j y_j) - \left(\sum_{j=1}^{n} x_j \right) \left(\sum_{j=1}^{n} y_j \right)}{n \sum_{j=1}^{n} (x_j^2) - \left(\sum_{j=1}^{n} x_j \right)^2}$$

$$b = \frac{\left(\sum_{j=1}^{n} x_j^2 \right) \left(\sum_{j=1}^{n} y_j \right) - \left(\sum_{j=1}^{n} x_j \right) \left(\sum_{j=1}^{n} x_j y_j \right)}{n \sum_{j=1}^{n} (x_j^2) - \left(\sum_{j=1}^{n} x_j \right)^2}$$

The result of a least squares treatment gives the most probable values of the parameters of an *assumed* relationship; it does *not* show that the assumed relationship is correct.

A more detailed discussion can be found in the references.[1]

1. E. B. Wilson, *An Introduction to Scientific Research*, McGraw-Hill, New York, 1952; D. P. Shoemaker and C. W. Garland, *Experiments in Physical Chemistry*, McGraw-Hill, New York, 1962, Chapt. II; A. G. Worthing and J. Geffner, *Treatment of Experimental Data*, John Wiley & Sons, New York, 1943; W. R. Steinbach and D. M. Cook, *Am. J. Phys.*, *38*, 751–54 (1970).

Appendix D
Bibliography

Chapter 1 and Chapter 2
L. P. Hammett, *Introduction to the Study of Physical Chemistry*, McGraw-Hill, New York, 1952.
W. Kauzmann, *Kinetic Theory of Gases*, W. A. Benjamin, New York, 1966.

Chapter 3
H. A. Boorse and L. Motz, *World of the Atom*, Basic Books, New York, 1966; chapters on Planck, Einstein, Bohr, and De Broglie.
W. H. Cropper, *The Quantum Physicists*, Oxford University Press, New York, 1970.

Chapter 4 and Chapter 5
W. H. Cropper, *The Quantum Physicists*, Oxford University Press, New York, 1970.
H. A. Boorse and L. Motz, *World of the Atom*, Basic Books, New York, 1966; chapters on Schrödinger, Born, and Heisenberg.
J. W. Linnett, *Wave Mechanics and Valency*, Methuen, London, 1960.
P. Atkins, *Molecular Quantum Mechanics*, Clarendon Press, Oxford, 1970.

Chapter 6
P. B. Ayscough, *Electron Spin Resonance in Chemistry*, Methuen, London, 1967.
A. Carrington and A. D. McLachlan, *Introduction to Magnetic Resonance*, Harper & Row, New York, 1967.
G. M. Barrow, *Molecular Spectroscopy*, McGraw-Hill, 1962.
J. A. Pople, W. G. Schneider, and H. J. Bernstein, *High Resolution Nuclear Magnetic Resonance*, McGraw-Hill, New York, 1959.

Chapter 7
S. H. Bauer, "Diffraction of Electrons by Gases", in *Physical Chemistry, an Advanced Treatise*, Vol. IV (D. Henderson, ed.), Academic Press, New York, 1970.

723

Chapter 8

L. H. Hammett, *Introduction to the Study of Physical Chemistry*, McGraw-Hill, New York, 1952.

E. A. Guggenheim, *The Boltzmann Distribution Law*, Interscience Publishers, New York, 1955.

R. P. H. Gasser and W. G. Richards, *Entropy and Energy Levels*, Clarendon Press, Oxford, 1974.

Chapter 9, Chapter 10, and *Chapter 11*

K. Denbigh, *The Principles of Chemical Equilibrium*, Cambridge University Press, Cambridge, 1961.

M. Born, *Natural Philosophy of Cause and Chance*, Clarendon Press, Oxford, 1949.

G. N. Lewis and M. Randall, *Thermodynamics*, McGraw-Hill, New York, 1923.

H. A. Bent, *The Second Law*, Oxford University Press, New York, 1965.

Chapter 13 and *Chapter 14*

T. L. Hill, *Matter and Equilibrium*, W. A. Benjamin, New York, 1966.

D. Dreisbach, *Liquids and Solutions*, Houghton Mifflin, Boston, 1966.

Chapter 19

K. E. Van Holde, *Physical Biochemistry*, Prentice-Hall, Englewood Cliffs, N. J., 1971.

General References

C. N. Hinshelwood, *The Structure of Physical Chemistry*, Clarendon Press, Oxford, 1951.

J. H. Wolfenden, R. E. Richards, and E. E. Richards, *Numerical Problems in Advanced Physical Chemistry*, Clarendon Press, Oxford, 1964.

E. A. Guggenheim and J. E. Prue, *Physicochemical Calculations*, North Holland, Amsterdam, 1955.

L. G. Sillen, P. W. Lange, and C. O. Gabrielson, *Problems in Physical Chemistry*, Prentice-Hall, New York, 1952.

J. R. Partington, *An Advanced Treatise on Physical Chemistry*, Longmans, Green & Co., London, 1949–1954.

Index